STUDY GUIDE TO ACCOMPANY

SECOND EDITION

CHEMISTRY

STANLEY R. RADEL
City College of the City University of New York

MARJORIE H. NAVIDI
Queens College of the City University of New York

Written by
Marion A. Brisk, Ph. D.
The City College of New York

WEST PUBLISHING COMPANY
Minneapolis/St. Paul New York Los Angeles San Francisco

WEST'S COMMITMENT TO THE ENVIRONMENT

In 1906, West Publishing Company began recycling materials left over from the production of books. This began a tradition of efficient and responsible use of resources. Today, up to 95% of our legal books and 70% of our college texts and school texts are printed on recycled, acid-free stock. West also recycles nearly 22 million pounds of scrap paper annually—the equivalent of 181,717 trees. Since the 1960s, West has devised ways to capture and recycle waste inks, solvents, oils, and vapors created in the printing process. We also recycle plastics of all kinds, wood, glass, corrugated cardboard, and batteries, and have eliminated the use of Styrofoam book packaging. We at West are proud of the longevity and the scope of our commitment to the environment.

Production, Prepress, Printing and Binding by West Publishing Company.

TEXT IS PRINTED ON 10% POST CONSUMER RECYCLED PAPER

PRINTED WITH SOY INK™

Printed in the United States of America
01 00 99 98 97 96 95 94 8 7 6 5 4 3 2 1 0

ISBN 0–314–03526–5

Table of Contents

Preface

The purpose of your study guide is to help you learn chemistry more effectively. Important principles and concepts are emphasized and summarized so that they can be easily understood. Your study guide will also help you to develop *study techniques* that will improve your mastery of any scientific discipline. The "How to Study Chemistry Effectively" chapter at the beginning of your study guide describes these study techniques.

Within each chapter of your study guide many problems are worked-out for you in a simple step by step fashion. Many chapters contain **Problem-solving Tips** which summarize the different kinds of problems that you must be able to solve and the strategies you can use to solve them. Only by working on problems and thinking about them can you develop an understanding of chemistry.

Using Your Study Guide

Each chapter is keyed to a chapter in your textbook and contains several features which you will find helpful in learning the material. At the beginning of each chapter you will find a **Self Assessment** which contains questions and problems along with answers and selected solutions. These exercises will help you identify those topics which require additional review.

Each chapter contains a **Review** which presents the important concepts and many worked-out Example problems. Often these problems are followed by **Practice Drills** that test your understanding of these important concepts. Each drill concentrates on a particular skill that you will need in order to solve problems in that chapter.

Problem-Solving Tips are given throughout your study guide. Some summarize the important formulas and their meanings; others list the steps needed to solve specific problems and the reasons for each step.

A **Self-Test** along with answers and selected solutions is provided at the end of each chapter. This section will allow you to test your knowledge after working through the chapter in the study guide and textbook.

The author wishes to express thanks to Dr. Marjorie Navidi for her invaluable suggestions offered with unfaltering humor, to Gail Farley, my friend and assistant, who typed this manuscript with tireless effort, to Dr. Honi Gluck for much needed constructive criticism and encouragement, and especially to my mother, Candida Valentino Brisk, who told me that "girls are good in math." I would also like to thank the many special students of the Sophie Davis Center of Biomedical Education at CCNY who stirred me to think about how students learn. Appreciation is expressed to Dr. David Adams, Nancy Hill-Whilton of West Publishing Company, and Dr. Michael Stoloff for their efforts in developing and producing this study guide.

HOW TO STUDY CHEMISTRY EFFECTIVELY

Marion A. Brisk, Ph.D.
The City College of New York

Stefan Bosworth, Ph.D.
University of California, San Diego

1. How to Read and Understand Your Textbook
2. Getting the Most Out of the Lecture
3. Solving Problems
4. Managing Your Time
5. Using Study Groups
6. Preparing For and Taking Exams
7. Tips on Creating a Study Area
8. The Laboratory Period

Introduction

Every student would like an A in chemistry. Many students, however, need to learn the techniques that will help them to earn high grades. If you were successful in high school without ever thinking about how to study, you might not be prepared for the fact that college courses require a great deal of independent learning, and that you have to integrate material from lecture, textbooks, handouts, and problems with what seems like little guidance compared to what you may have been used to. *The techniques described here will help you to study chemistry and other sciences more effectively.* They will not only improve your grades but will give you more confidence as well, so that learning chemistry can be a pleasure rather than a chore.

How to Read and Understand Your Textbook

Many students make the mistake of reading a textbook like a novel; they read an entire chapter once and then attempt to do the problems. It's not surprising that the problems appear too difficult and seem to belong to some other chapter. To learn the most from a textbook you must *actively* read, that is, you must constantly be thinking about what you are reading, pausing to relate it to what you have just read before, and making sure that you understand its applications. Ask yourself questions to make sure you are understanding the main ideas of the paragraphs; turn section headings into questions and then relate sections to each other and to the main topic of the reading assignment. Asking and answering questions helps you to not only concentrate on the main ideas but also increases your retention of the material. Your textbook provides you with learning objectives for each section; use those objectives to focus your attention on what you need to know. The process of self-testing is a skill that scientists at all levels use to learn new concepts. For the student it is particularly helpful since it is also a form of exam preparation; answering questions on exams will be less anxiety producing when you have been answering questions all along.

To help yourself read actively you can also underline or even better take notes in outline form on the text material. Your outline should include main ideas, important formulas, and their applications.

Survey chapters before you begin to actively read; that is note the main headings and sub-headings, read the introduction (previews in your text) to see how this chapter relates to previous chapters, read the lists of learning objectives, turn to the key terms and important formulas at the end of the chapter; these will all direct your attention to what you must learn from the text. Read each paragraph first and then go back and outline or underline only the important material. Review your notes after you finish briefly going over main ideas and examples. Make an effort to understand and retain the material by engaging as many senses as possible as you actively read. Try to visualize many of the principles and examples described in the text. Remember chemistry describes the world in which we all live so that much of what you learn you can apply to familiar objects and situations.

Make sure you spend a *significant portion* of your study time *doing problems.* Your textbook provides you with many clearly worked-out example problems followed by practice problems. The only way you can be sure you really understand the problems is by doing them yourself. Even if you follow the solutions in your text this is no guarantee that you understand the problem well enough to do one like it on your own. By doing the practice problems you not only ensure that you really understand the example problems but you increase your chances of remembering how to solve similar problems on an exam. After completing a section, try to do the exercises assigned to you for that section. Keep in mind that chemistry is a problem-solving discipline so that the more problems you solve the better you will understand the material. As you read, make a note of the parts of the text that are not clear to you and make a note of the problems you are not sure of or can't do at all. Consult 1) lecture notes, 2) study guide, 3) instructors and 4) classmates to clear up what you don't understand. Never let your questions go unanswered; if you do you not only decrease your chances of doing well on exams but you jeopardize your future understanding of chemistry since new topics very often depend on your understanding of past topics.

1. Survey the chapter.
2. Outline or underline as you read actively.
3. Do practices and other problems.
4. Keep a record of questions, and any problems you don't fully understand so that you can consult instructors, classmates, textbooks, or lecture notes for answers.

Getting the Most Out of the Lecture

In listening as in reading, you must be actively involved to get the most out of it. If you actively listen to the lecture, your notes will be accurate and complete and your time in the lecture room will be well-spent.

Read the assigned material before you come to lecture. If you don't much of your lecture time will be wasted. Without some idea of the topic being discussed you will find it difficult to maintain your attention and focus on the important ideas of the lecture. Your notes will be incomplete and you won't be able to ask questions; the lecture will not help you to learn the material. On the other hand, if you arrive prepared you will be able to determine and record the important information so that your notes will be useful for studying and for exam preparation.

Make sure you record all information written on the blackboard or overhead projector (except for Tables and figures taken from your text). Try to take notes in an outline form that shows major topics, sub-topics, and relationships between them. Pay close attention to the examples that were worked out in class and try to re-do them after class, perhaps changing values for practice.

Read over your notes as soon as possible; preferably the same day. Rewrite what is unclear. If you are uncertain about parts of what was said compare your notes with those of other students or ask your instructor. Then go back and include that material in your notes. Think of your notes as a handwritten book and strive to make them accurate and complete.

1. Read the assignment before the lecture.
2. Try to take notes in outline form showing major topics, sub-topics and their relationships. Include examples.
3. Read over notes the same day. Re-write, change and add to notes where necessary.
4. Make sure your notes are complete. If you missed part of the lecture find out what it was you omitted and fill it in.

Solving Problems

Most of your time in chemistry should be spent solving problems that are applications of concepts and formulas learned in lecture and the text. You can improve your ability to solve problems by learning how to think about the problems that are solved for you in the text, study guide, and in the lecture, and also by learning how to think about the many relationships (formulas) used in the problem.

Understanding relationships and formulas is crucial to learning chemistry. Many students memorize relationships and formulas without taking the time and energy to think about them. This often leads to inappropriate applications and incorrectly solved problems. Ask yourself the following questions whenever you learn a new formula.

1. What system or change does this formula describe?
   What do the variables mean and what are their units?
2. When does it apply?

3. What are some examples of its application? What is its significance?

Formulas are listed for you at the end of each chapter. Ask yourself these questions for all of these relationships. In the study guide these questions are answered for you for some of the formulas you need to understand. Read these "Important Formula" charts carefully and add to them. If you actually think about relationships as you learn them, it will be easier to see how to apply them.

Note carefully which concepts or relationships are used in the worked out problems in the text, study guide, and solutions manual. Why was this formula used and not one of the others in the chapter? What information given in the problem indicates that the problem should be solved in this way? You will find answers to these kinds of questions in the "Problem Solving Tips" given in the study guide.

When you start to work on a problem it is critical to first write down the information that is given along with units, and to identify the unknown. Use a diagram whenever possible to show what is being described in the problem indicating the given and unknown values. Then try to plan out your solution before you start doing any calculations. If you aimlessly calculate values, you may generate superfluous data which may be confusing. To solve a problem you must determine which relationships are relevant out of all those you have learned. Think of relationships that involve your unknown. Which ones include all or some of the given data? Be sure the relationships you use apply to the system as described in the problem; for example, don't use formulas for gases in a problem about liquids. If you still can't see a method, think about relationships that involve the other values given in the problem. For example, if the volume and density of a solution are provided, then the mass is also known ($d = m/V$). If you now include the mass with known information the solution may become apparent.

Think about the sample problems you have studied. Solutions to previous problems may provide hints to solving new problems.

After you have planned the solution, then do the calculations. Use a calculator to save time and eliminate arithmetic errors. Be sure that the values you use have the appropriate units for the formula you are applying. Check your answers for the following:

1. Make sure your answer is what the problem asked for.
2. Make sure your answer is reasonable. When you study sample problems in your text and study guide think about the magnitudes of the answers so that you will have some concept of reasonable answers. If you calculate that the mass of a molecule is 10 kg it is clear that you have made an error. However, if you calculate that 10 kJ/mol of heat are released by a reaction you will not realize that you have made an error unless you have previously noted reasonable values for heat being released from chemical reactions.
3. Make sure your answer has the correct number of significant figures.

Problem Solving Tips:

1. Identify the known quantities and the unknown quantities asked for.
2. Plan the solution: What do you know about the unknown that might link it to given information?
3. Perform calculations.
4. Check your answers.

Managing Your Time

Learn to manage your time. This skill will be invaluable, especially if you plan a career in science. Learning chemistry takes time and energy, and you should try to study some chemistry nearly everyday. Devise a study schedule at the beginning of each week so that you will be studying chemistry throughout the week. Include periods for textbook reading before lecture, review of lecture notes, problem solving, and review for quizzes and exams. Make sure that your study schedule not only includes enough time to study chemistry, but also allocates sufficient time for other activities that you must complete during the course of the week.

Be specific in constructing your schedule; indicate which chapters or sections you plan to study, which set of problems to work out, and which topics to review for an approaching exam. Try not to schedule very long blocks of time for studying chemistry: one or two hour blocks of time interspersed with other work for different courses is best.

Devise your schedule at the beginning of each week by first looking at all of your assignments and then allowing enough time to complete them. Students tend to complete their assignments more often when they schedule chapter reading and problem solving at the beginning of the week and reviewing towards the end. Always allow yourself more time than you think you will actually need to complete the assignment. It is always better to overestimate the time you will need than to find out that you are short of time later.

It is critical that you follow your schedule and don't permit yourself to be distracted. If you carefully construct and follow your schedule and make necessary adjustments to accommodate each week's requirements you should find that the actual amount of free time will increase.

1. Construct a study schedule at the beginning of each week.
2. Be specific as to what you plan to do during your study sessions.
3. Overestimate the time you will need to complete each assignment.
4. Be sure to include in your schedule enough time to complete all other necessary tasks as well as time for leisure activities.

Tips on Creating a Study Area

To help your concentration create a study space. Study in an area where lighting is adequate and distractions are few. Try to create an environment that is pleasant but without items that might divert your attention such as **a** radio, stereo, television or telephone. Make sure that everything that you might need for studying is in your study area. This should include: paper, pens and pencils, calculator, computer, notes, outlines, and completed assignments as well as all textbooks and reference books that might be needed. Try to use your study area solely for the purpose of studying.

By creating a study space you make your studying more efficient. You will reduce wasted time searching for needed material and minimize distractions, thus improving your concentration.

1. Study in a comfortable but efficient area minimizing distractions.
2. Make sure that your study area is equipped with all the items you will need for studying.

Using Study Groups

The effective use of study groups can be an important part of an overall study program that will lead to success in General Chemistry. However, many students have tried to use study groups, only to find that they were not helpful. Generally study groups fail when students either do not know how to form an effective group or do not know what tasks a study group is best suited to perform.

Study groups should consist of three to six members who are serious chemistry students and committed to making the group meetings effective. It's generally a good idea to establish a consistent study time and place: two hours at the end of the week with more frequent meetings as an examination approaches at your college or a study area of a dormitory usually suffices. Study groups are most effective for reviewing material that each student has previously studied on their own such as assigned chapters, problems, and old exams. If members do not individually prepare the group meetings will not be helpful.

Each participant should be assigned a specific task to complete and present to the group. These tasks include solving specific problems, explaining sections of a chapter or of a lecture, preparing a chapter outline etc. By each member being responsible for covering different aspects of the material the group is assured of covering all of the assignment. In addition, each member is accountable to the group as a whole which has the effect of encouraging students to keep up with the course work.

Study groups acquire particular significance when preparing for a chemistry exam. While they are not a substitute for the review that would normally take place before an exam, they can be an important addition to that review process. Usually it is a good idea to increase the amount of time a study group meets about two weeks prior to a major chemistry exam (usually from two to four hours a week). The additional time should be used for a general review of the material and for problem solving. Remember chemistry exams usually emphasize problem solving so that the group should spend most of its preparation seeking and understanding solutions to problems. Each member of the study group should prepare and explain to the study group a section of the material that is to be covered on the exam and also to develop a set of problems and answers that encompasses the same material. Problems should be distributed to the rest of the group without the answers and groups members should complete the problems during the study group meeting. Group members who have difficulty with certain problems can rely on the member of the group who developed the problem to explain the answer. This kind of studying for exams is extremely effective because it puts you, the student, in the role of the professor, deciding what information is important and likely to be covered on an exam and what is not. The process is also effective because it requires you to be actively involved in the learning of formulas and problem solving. Further, this process lets you know what chemistry material you need to spend more time reviewing and what material you already know sufficiently.

1. Form a study group consisting of three to six serious chemistry students.
2. Try to convene at the same place and same times weekly.
3. Assign specific tasks to members.
4. Spend most of your time discussing and solving problems.
5. Increase the time for your sessions to prepare for exams. Each member should be responsible for preparing and presenting material that will appear on the upcoming test.

Preparing For and Taking Exams

To adequately prepare for an exam you must first organize your time in advance. Start studying a few weeks before an exam by adding time to your usual study schedule and use this additional time for exam preparation only. Survey the material that will be covered on the exam and divide the relevant topics or chapters into categories based on your present level of understanding. For example, identify chapters that you know well, those that you have a limited mastery of, and finally topics that you do not understand at all. Write out an approximate but detailed schedule including not only main chapters or topics you plan to cover during each week but also specifically what you plan to study each day. Leave more time for the material you are unsure of as well as for the more lengthy and complex topics and *spend most of your time solving problems.*

Because chemistry exams generally consist of mostly problem solving, your preparation can only be effective if you actually solve many relevant problems. Review the assigned problems and solve additional problems in the text and study guide as well as from previous exams if available.

Be careful not to just review the solutions of problems. Students who just review the solutions following each step often find that they cannot solve problems on exams. The only way that you can be sure that you adequately understand any problem is by solving it yourself writing out each step. For problems already solved, simply change given values and re-work the problem finding a different answer. *Remember, to be successful on a problem solving exam you must have the experience of solving many problems yourself.*

Review your old exams to re-familiarize yourself with the kinds of questions your professor asks. Identify the questions you were most successful answering a well as those you could not correctly complete. Try to emphasize problems that resemble those that were particularly difficult for you in the past.

Try to work in a study group where you can solve problems, review lecture and textbook notes together. You will find it helpful to construct outlines of the work being covered; each member of your group can contribute outlines of specific topics a well as presenting solutions of the relevant problems. Remember, the more you write and think about the topics, the more you will retain and understand. Study group members can also develop new problems that the entire group can work on under simulated exam conditions. Such exercises will help to reveal your weak areas and develop test taking skills.

1. Devise a schedule for preparation a few weeks before the exam.
2. Determine which topics you know well, those you do not understand completely, and material you do not know at all.
3. Spend most of your time actually solving assigned problems, as well as new problems from your text, study guide, and previous exams.
4. Review previous exams to remind yourself of how your knowledge will be tested.
5. Work with your study group as much as possible developing outlines, solving problems, and creating practice exams.

The Night Before the Exam

The night before the exam should be used for a quick review of the more important topics or a review of the material you are still uncertain about. At this point you should have studied all of the relevant topics and associated problems. Chemistry cannot be learned overnight and any attempts at cramming the work will only result in more anxiety during the exam. Learning any science is a gradual process requiring much time and energy. Remember your exam grades will reflect your study techniques. Adequate preparation not only increases your knowledge and improves the skills required for the exam, but also reduces your anxiety level so that you can think more clearly.

Taking the Exam

Read the directions carefully and work first on the problems or questions you think you can answer correctly. Leave the problems you are very uncertain about for last. In this way you will ensure that you receive credit for what you know and also you will elevate your confidence level to help you tackle those problems that you find challenging.

Try to show clearly each step you take in solving a problem so that you can check your work more efficiently and also so that your instructor can assign partial credit if that is his or her policy. After you solve a problem make sure that you calculated the required quantity and that your value is reasonable. Make sure your calculator can perform all required operations and replace your batteries before the exam.

1. Read instructions carefully.
2. Answer the questions or solve the problems you feel sure about first.
3. Show all work clearly.
4. Use a calculator with all required functions.
5. Check your answers to see if they match the questions and if they are reasonable.

The Laboratory Period

Many chemistry courses have laboratory components. Students often do not consider the laboratory exercises important and consequently do not benefit from the experiments. However, laboratory exercises usually emphasize important concepts and also introduce laboratory skills that will be needed for more advanced courses. If you appropriately prepare, perform, and write up the laboratory experiments, you will benefit in the following ways:

1. You will develop a solid understanding of the concepts emphasized in the laboratory.
2. You will earn high grades for the lab component of your course.
3. You will learn lab techniques that you will need in future science courses and in scientific research.
4. You will acquire an understanding of the scientific method which is necessary in order to understand scientific journal articles and to conduct research.

Preparing for the Laboratory Exercise

Make sure you understand the objective and procedures of the experiment before the lab. It is **a** good idea to outline the procedure so that you can spend your time in lab actually learning and not blindly following the lab manual or trying to figure out what to do next. Your outline will also be helpful in writing up the lab later if that is required. Also by taking the time to understand the lab in advance you are less likely to make mistakes and your data will be more accurate. Try to plan your time efficiently. If certain instruments require use by the entire class, try to use them early before the lines form.

Prepare tables for data before the lab so that they can be neatly filled in. Students often jot down measurements during the experiment only to find later that they cannot identify these values. If you are uncertain about any part of the procedure or analysis of results ask the *lab instructor* before you begin. Do not take the chance of making **a** mistake and obtaining inaccurate data as a result, or even worse, injuring yourself or your classmates. Also make sure you know how to safely handle all chemicals and equipment.

1. Read and understand the lab in advance.
2. Write up objectives and procedures of the lab as well as tables needed for data.
3. Ask your lab instructor questions before you begin.
4. Know how to safely handle all chemicals and equipment.

Performing the Laboratory Exercise

Familiarize yourself with the lab room. Know the locations of first aid equipment, fire extinguishers, hoods, gas and water shut off valves, reagents, equipment, and instruments. Know required general procedures such as the wearing of safety glasses and storage of coats and books. Try to concentrate on your own work and if a problem develops ask the lab instructor not one of your classmates. Make sure you properly dispose of reagents avoiding spilling harmful chemicals into sinks.

In some labs you may work with a partner. It is a good idea to allocate tasks beforehand to avoid repetition and maximize your efficiency. Be sure that you both learn all lab skills as they may be used in future lab sessions and may be tested in practical exams which often are given at the end of the semester.

1. Become familiar with the lab room and general procedures.
2. If a problem or question arises during the experiment do not hesitate to ask your lab instructor.
3. Clearly record all lab data. Do not use loose scraps of paper.
4. Appropriately handle all reagents, equipment, and instruments.
5. Follow all safety precautions and appropriately dispose of chemicals.
6. Share the lab work equally with your partner and make sure you learn all of the laboratory skills.

Writing the Lab Report

Try to write your lab report as soon as possible after you have completed the experiment. The lab will be fresh in your mind and you will not need to waste time reviewing what the lab is about and what you actually did before you can write the report. When writing the report carefully follow the requirements given to you by your instructor. If you are unsure about content, talk with your instructor and perhaps request a sample lab report. If you work with a partner most likely your instructor still requires individual reports. Make sure that your lab reports share only the lab data.

Your report should be grammatically correct, be well organized, and contain all of the information required. Lab instructors generally emphasize your data and analysis but also take into account the effectiveness of your writing technique.

1. Include only what is asked for by your instructor.
2. Write your own report.
3. Use tables and graphs to present data.
4. Review your returned lab reports and determine how you may improve future reports.

CHAPTER 1
CHEMISTRY AND MEASUREMENT

SELF-ASSESSMENT

For questions 1-3 select the best answer.

Matter and Energy (Section 1.1)

1. Which of the following statements applies to either matter or energy?
   (a) It consists of atoms.
   (b) It can change from one form into another.
   (c) Its amount in the universe is constant.
   (d) It is measured in amperes.
   (e) None of the above apply.

Measurement and the Scientific Method (Section 1.2)

2. All of the following are true about the scientific method except:
   (a) Data or facts are collected.
   (b) Laws are formulated from the data.
   (c) Theories explain these laws.
   (d) Laws are based on theories.
   (e) The scientific method is a systematic procedure for studying nature.

3. Which of the following is not a fundamental quantity in the SI system?
   (a) length
   (b) time
   (c) mass
   (d) temperature
   (e) surface area

Length and Volume (Section 1.3)

4. The SI unit for volume is ____________.

5. Approximating the earth as a sphere, calculate its volume in SI units given an average radius of 6.38×10^6 m. ($V = 4/3\pi r^3$)

Reliability of Measurements (Section 1.4)

6. A reliable measurement is both ____________ and ____________.

7. The Environmental Protection Agency (EPA) tests competence of chemical laboratories by requiring them to analyze a known sample. The sample used for Lab A and Lab B contains 10.02 g of iron. Lab A reported 12.21 g while Lab B reported 10.0 g.
(a) Which laboratory gave the more precise result?
(b) Which laboratory gave the more accurate result?

8. How many significant figures are in the following length measurements?
(a) 3.2 m
(b) 3.20 m
(c) 0.032 m
(d) 320 m

9. Complete the following chart, using the rules for significant figures in calculations.

Problem	Answer	Number of significant figures in answer
10.0 + 0.55 + 2	____________	____________
100.0 – 0.55	____________	____________
40.0 ÷ 4.0	____________	____________
Determine the length in centimeters of a 3.0-inch common nail. One inch is exactly 2.54 cm.	____________	____________

Dimensional Analysis (Section 1.5)

10. Your living room is 21 feet long and 13 feet wide. What are the dimensions of the room in meters? (1 m = 39.37 in)

11. The speed of a runner preparing for the New York City marathon is 230 m/min. Convert this speed into miles per hour. (1.609 km = 1 mile)

Mass, Force, and Weight (Section 1.6)

12. One gram of a particular brand of chewing tobacco contains 63 ng of a cancer-causing chemical. How many milligrams of chewing tobacco contain 30 ng of this chemical?

13. Underline the correct word: As you travel from New York City to the equator your (mass, weight) will change slightly while your (mass, weight) remains the same.

14. Express your weight in newtons. ($g = 9.80\ m/s^2$; 1 kg = 2.205 lb)

15. A bowling ball weighs 15.8 pounds and has a circumference of 28 inches. Calculate the average density in grams per milliliter of the bowling ball. (The circumference of a circle is $2\pi r$; volume = $4/3\pi r^3$; 1 in = 2.54 cm; 1 lb = 454 g)

16. Calcium has recently been shown to lower blood pressure in hypertensive patients. How many grams of calcium are in one quart of cows' milk if there are 1.14 mg of calcium in one gram of milk? Assume milk has the same density as water. (1 L = 1.057 quarts)

17. The amount of dioxin, a potentially harmful pollutant, in one liter of a certain lake is 1.2 ng. How many kilograms of dioxin are in the entire lake if the lake contains 5×10^{12} kg of water? (Specific gravity of lake water is 1.0.)

Heat, Work, and Other Forms of Energy (Section 1.7)

18. How many kilojoules of work does an athlete do when lifting a 300-lb weight one meter against gravity? (1 lb = 4.45 N)

19. What is the kinetic energy in joules of
    (a) a 500-lb race horse running at a speed of 50 mph?
    (b) a 1000-lb race horse running at a speed of 40 mph?
    (c) Which horse will win the race?
    (Hint: 1 mph = 0.447 m/s and 1 lb = 2.2 kg)

20. As the distance between a proton and an electron increases, the potential energy (decreases, increases). As the distance between two electrons increases, the potential energy (decreases, increases).

21. Oxygen's boiling point is -297°F or ____________ °C.

22. If a lake and a glass of water both undergo the same increase in temperature, the lake absorbs (more, less) heat than the glass of water.

ANSWERS TO SELF-ASSESSMENT PROBLEMS

If you missed an answer be sure to study the relevant section in the textbook and study guide.

1. (b); 2. (d); 3. (e)

4. m^3 (not the liter)

5. 1.09×10^{21} m^3

6. accurate, precise

7. (a) Lab A (b) Lab B

8. (a) 2 (b) 3 (c) 2 (d) 2 or 3

9. 13 (2); 99.4 (3); 10 (2); 7.6 cm (2)

10. 6.4 m long; 4.0 m wide

11. 8.58 mph

12. 4.8×10^2 mg chewing tobacco

13. weight, mass

14. A 140-lb person weighs 622 N.

15. 1.2 g/mL

16. $1 \text{ quart} \times \frac{1 \text{ L}}{1.057 \text{ quarts}} \times \frac{1000 \text{ g milk}}{1 \text{ L}} \times \frac{1.14 \text{ mg calcium}}{1 \text{ g milk}} \times \frac{1 \text{ g}}{1000 \text{ mg}} = \underline{1.08 \text{ g calcium}}$

17. 6 kg dioxin

18. 1.34 kJ

19. (a) 5.7×10^4 J (b) 7.3×10^4 J (c) 500-lb horse

20. increases, decreases

21. -183°C

22. more

REVIEW

Matter and Energy (Section 1.1)

Learning Objectives

1. Give examples of matter and energy.

Review

Matter is anything that has mass and occupies space. Matter is constantly changing forms: snow melts in the spring into water, filling rivers and streams; neutrons and protons form carbon, oxygen, and nitrogen nuclei in the centers of stars; stars die ejecting nuclei into space where they may eventually be incorporated into living material.

Energy is the ability to do work. Matter changes form because energy can move objects and do work. Some forms of energy are light, wind, and electricity.

Measurement and the Scientific Method (Section 1.2)

Learning Objectives

1. Describe the steps of the scientific method.
2. State the two ways in which numerical data are obtained.
3. State the SI units of length, mass, and time.

Review

The scientific method is a systematic, logical procedure that scientists follow in order to understand nature. The steps of the scientific method are:

1. Collect data or facts.

2. Formulate laws. Laws summarize facts.
3. Devise models that will explain the laws. These models of nature are called theories.

Data are collected either by counting individual items or by measurement. A measurement is the magnitude of a quantity based on a defined unit that serves as a standard. Measuring devices are often marked off in multiples or fractions of the basic unit; a ruler measures length because it is marked off in fractions of a meter or a foot. Two principal unit systems are in use in the U.S.: metric and English. The SI system of metric units is the official system of the international scientific community and is based on seven fundamental quantities: length, mass, time, temperature, electric current, intensity of light, and number of particles. (See Table 1A) The SI units of the first four quantities are the meter (m), kilogram (kg), second (s), and kelvin (K).

All other units are combinations of the fundamental units: for example, the combined unit of area in the SI system is m^2, which is derived from the meter, the unit of length;

$$\underset{\text{(fundamental unit)}}{m \times m} = \underset{\text{(combined unit)}}{m^2}$$

Table 1A: Metric Units of Some Physical Quantities and Their English Equivalents

Quantity	Metric Unit	English Equivalent
length	*meter (m)	1 m = 39.37 inches
	centimeter (cm)	(exact) 2.54 cm = 1 inch
	kilometer (km)	1.609 km = 1 mile
mass	*kilogram (kg)	1 kg = 2.205 pounds
	gram (g)	453.6 g = 1 pound
time	*second (s)	
temperature	*kelvin (K)	
	Celsius (°C)	
number of particles	*mole (mol)	
electric current	*ampere (A)	
luminous intensity	*candela (cd)	
volume	cubic meters (m^3); (1 m^3 = 1000 L)	
	liter (L)	1 L = 1.057 quarts 3.785 L = 1 gallon
	milliliter (mL); (1 mL = 1 cm^3)	

Your Additions:

*SI units of fundamental quantities

Length and Volume (Section 1.3)

Learning Objectives

1. Visualize the magnitude of a meter, a centimeter, and a millimeter.
2. Visualize the magnitude of a liter and a cubic centimeter.
3. Convert multiples and fractions of the meter (kilometers, centimeters, etc.) to meters.

Review

The meter is the SI unit for length. Smaller units are fractions of the meter, while larger units are multiples of the meter. All of these length units are related by powers of ten. Table 1B shows the prefixes associated with each fraction and multiple of the meter.

Table 1B: Common Prefixes

Prefix	Symbol	Relation to a Fundamental SI Unit	Example
mega	M	10^6	1 M = 10^6 m
kilo	k	10^3	1 kg = 10^3 g
deci	d	10^{-1}	1 dL = 10^{-1} L
centi	c	10^{-2}	1 cm = 10^{-2} m
milli	m	10^{-3}	1 mL = 10^{-3} m
micro	μ	10^{-6}	1 μm = 10^{-6} m
nano	n	10^{-9}	1 ng = 10^{-9} g
pico	p	10^{-12}	1 ps = 10^{-12} s

Example 1.1: The diameter of a Herpes virus is 200 nm. Express this measurement in meters.

Answer: From Table 1B: 1 nm = 10^{-9} m. Therefore, 200 nm = 200 × 10^{-9} m or 2.00×10^{-7} m in scientific notation.

Practice Drill on Interconverting Metric Length Units: Fill in the correct exponents:

(a) 10^1 cm = $10^?$ nm = $10^?$ km = $10^?$ m
(b) 10^{-3} m = $10^?$ cm = $10^?$ μm = $10^?$ pm
(c) 10^2 mm = $10^?$ m = $10^?$ km = $10^?$ nm

Answers: (a) 8, -4, -1 (b) -1, 3, 9 (c) -1, -4, 8

The volume occupied by some objects can be calculated from their linear dimensions. Table 1.5 (p. 5) in your text gives the formulas for some common shapes. Note that these formulas all yield units that are the cubes of length units; the SI unit for volume is therefore m^3.

Note: The liter is a non-SI unit:

1 L = 1000 cm^3 = 1000 mL

Dividing both sides by 1000 gives

1 cm^3 = 1 mL

Study the following list to help you visualize some of the metric units. To check the "reasonableness" of an answer to a problem, you need a general idea of the magnitude of your units.

mass of a MacIntosh apple = 150 g
mass of a 16-lb bowling ball = 7.3 kg
height of a 5-ft person = 1.5 m
diameter of a quarter = 2.4 cm
one quart of milk = 0.95 L
a teaspoon of liquid = 5 mL
mass of a lifesaver candy = 0.5 g

Practice: Which of the following statements must be incorrect?

(a) height of a man = 11 m
(b) mass of an atom = 5.0 g
(c) total blood volume of an adult human = 0.5 mL
(d) volume of an atom = 1.2 mL
(e) a teaspoon of liquid = 5 cm^3

Answers: a-d

In solving the following problem the steps described in the introductory chapter of the study guide are clearly shown. We will continue to use this strategy to solve problems throughout the study guide.

Example 1.2: A can of soda is 13 cm long with a radius of 3.0 cm. How many liters of soda are in two cans?

(What information is supplied?)	Given:	h = 13 cm r = 3.0 cm
(What am I being asked to calculate? Include units.)	Unknown:	volume of two cans of soda in liters.
(How is the unknown related to given information?)	Plan:	From Table 1.5 in your text V (cylinder) = $\pi r^2 \times h$. Therefore, volume of two cans = $2 \times \pi r^2 \times h$
(Substitute given values and solve for unknown.)	Calculations:	volume of two cans $= 2 \times 3.14\,(3.0 \text{ cm})^2 \times 13 \text{ cm} = 7.3 \times 10^2 \text{ cm}^3$
(Check units.)	Answer Check:	Our units are now cubic centimeters. The question asks for liters. Since 1000 cm^3 = 1 L, $7.3 \times 10^2 \text{ cm}^3 = \frac{7.3 \times 10^2}{1000} \text{ L} = 0.73 \text{ L}$
(Is the answer reasonable?)		Since 1 L is approximately 1 quart, the calculated value is reasonable for two beverage cans.

Practice: What is the volume in milliliters of a sphere with a 6.28-cm circumference?

Answer: 4.17 mL. Try to follow the indicated steps as you solve the problem.

Given:

Plan:

Calculations:

Answer Check:

Reliability of Measurements (Section 1.4)

Learning Objectives

1. Distinguish between accuracy and precision.
2. State the number of significant figures in a measurement.
3. Perform calculations to the correct number of significant figures.
4. Round off a number to any specified number of digits.
5. Distinguish between systematic and random errors.

Review

A reliable measurement is both accurate and precise. The accuracy of a measurement refers to how close the measured value is to the true value. Precision refers to both the number of digits in the value and the reproducibility of the value. When several measured values of the same quantity agree closely, the measurements are said to be reproducible and therefore precise. Precise values are not necessarily accurate.

Example 1.3: Two different diets are presented to a patient for weight reduction. Diet A lists amounts of suggested food to the nearest tenth of a gram while Diet B lists foods to the nearest gram. Which diet requires more precise measurements?

Answer: Diet A requires more precision because more digits are required in the measurement to comply with the diet.

The number of significant figures in a measurement is a rough indication of precision. Table 1C summarizes the rules for determining the number of significant figures in a measurement.

Table 1C: Rules for Determining Significant Digits

	Rule	Example	Number of Significant Digits
1.	All nonzero digits are significant.	3.2	2
2.	A zero between two nonzero digits is significant.	320.2	4
3.	Final zeros to the right of the decimal point are significant.	3.20	3
4.	Zeros appearing to the left of the first nonzero digit are not significant.	0.032	2
*5.	Final zeros in a number with no decimal point may or may not be significant.	320	2 or 3

*In this case you need to know something about the measurement. For example, if a ruler measures millimeters to the nearest integer, then 320 mm contains three significant figures. If however, the ruler measures only to the nearest centimeter then 320 mm = 32 cm and only two significant figures are in the measurement.

Practice Drill on Significant Figures: Give the number of significant figures for each of the following measurements:

(a) 1.01 m ____________

(b) 3.20×10^5 kg ____________

(c) 0.0128 g ____________

(d) radius of sun = 7×10^8 m ____________

(e) speed of light 1.86×10^5 mile/s in meters per second ____________

Answers: (a) 3 (b) 3 (c) 3 (d) 1 (e) 3

The number of significant figures in a calculated value is determined by the rules in Table 1D:

Table 1D: How to Use Significant Figures in Calculations

Operation	Rule	Examples	Number of Significant Figures
Addition and Subtraction	1. First add or subtract values as they appear.	10.0 + .55 + 2 = 12.55	
	*2. Round off the answer so that it has the same number of places as the value with the least number of decimal places.	12.55 = 13	2
Multiplication and Division	1. Multiply or divide values.	40.0 ÷ 4.0 = 10.0	
	*2. Round off the answer so that it has the same number of significant figures as the measurement with the least number of significant figures.	10.0 = 10	2

*<u>Rules for Rounding Off</u>

1. When the digits to be dropped are greater than five or five followed by zeros, the last remaining digit is increased by 1.
2. When the digits to be dropped are less than five or five followed by zeros, the last remaining digit is retained.
3. When the digits to be dropped are a five or a five followed by zeros, the last remaining digit is rounded to the nearest even number.

Practice: Round off the following values to two significant figures:

(a) 0.0445 _____ (Apply Rule 3) (c) 0.0446 _____ (Apply Rule 1)

(b) 0.0444 _____ (Apply Rule 2) (d) 0.0455 _____ (Apply Rule 3)

Answers: (a) 0.044 (b) 0.044 (c) 0.045 (d) 0.046

Practice: Perform the following operations to the correct number of significant figures.

(a) $(1.0 \times 10^5) \times (5.15 \times 10^{-8}) =$ __________

(b) $\dfrac{5.650 \times 10^{-5}}{7.01 \times 10^9} =$ __________

(c) $\dfrac{2.3201 \times 5.2}{0.081} =$ __________

(d) $\dfrac{(1.03 + 0.001 + 0.72)}{0.0056} =$ __________

Answers: (a) 5.2×10^{-3} (b) 8.06×10^{-15} (c) 1.5×10^2 (d) 3.1×10^2

Errors in measurement are either systematic or random.

Systematic errors arise from a constant imperfection in either the measuring device or in the experimental method; hence, systematic errors reduce accuracy. Systematic errors can be eliminated by calibrating the instrument or improving the experimental method.

Random errors are the small variations in measured values that are caused by chance. Random errors reduce precision. They can be minimized by averaging a large number of measurements.

Dimensional Analysis (Section 1.5)

Learning Objectives

1. Use units and conversion factors in calculations.

Review

A conversion factor is a fraction whose numerator and denominator are equal in value but have different units. For example, the equality 1 meter = 39.37 inches will give rise to two conversion factors:

$$\frac{1 \text{ m}}{39.37 \text{ in}} \text{ and } \frac{39.37 \text{ in}}{1 \text{ m}}$$

Conversion factors are used to convert from one unit to another. A conversion factor is chosen so that the given unit or old unit is cancelled and is replaced by the new unit. For example, if you are changing a length in meters (old) into inches (new), your conversion factor must have inches in the numerator and meters in the denominator.

Example 1.4: Convert 1.0 m to inches.

The conversion factor on the right above is used to convert meters into inches. Meters will cancel, and the answer will be in inches:

$$1.0\text{ m} \times \frac{39.37\text{ in}}{1\text{ m}} = \underline{39\text{ in}}$$

Practice Drill on Conversion Factors: What conversion factor would you use in each of the following given the relationships:

1.057 quart = 1 L	1.609 km = 1 mile
2.2 lb = 1 kg	6 ft = 1 fathom
1 mL = 1 cm^3	

Given (old)	Change to (new)	Conversion factor
(a) 2.00 mile	km	____________
(b) 5.21 L	quarts	____________
(c) 100 lb	kg	____________
(d) 1005 mL	cm^3	____________
(e) 5.00 fathoms	ft	____________

Answers: (a) $\frac{1.609\text{ km}}{1\text{ mi}}$ (b) $\frac{1.057\text{ quart}}{1\text{ L}}$ (c) $\frac{1\text{ kg}}{2.2\text{ lb}}$ (d) $\frac{1\text{ cm}^3}{1\text{ mL}}$ (e) $\frac{6\text{ ft}}{1\text{ fathom}}$

PROBLEM-SOLVING TIP

Using Conversion Factors

Some problems require several conversion factors. For these problems it is a good idea to plan the necessary conversions before carrying out the calculations. Be sure to indicate where you started (given) and where you want to end up (unknown). Show your steps clearly so that you can check to see that you have included all necessary conversions.

Example 1.5: Convert a length measurement of 10 m into ft using 2.54 cm = 1 in. Using arrows to represent conversion factors, you could write your plan as:

10 m → cm → in → ? ft

The first conversion will have cm in the numerator and m in the denominator; the second conversion will have in in the numerator and cm in the denominator and the third will have ______________ in the numerator and ______________ in the denominator. Answer: ft, in

$$10 \text{ m} \times \frac{100 \text{ cm}}{1 \text{ m}} \times \frac{1 \text{ in}}{2.54 \text{ cm}} \times \frac{1 \text{ ft}}{12 \text{ in}} = \underline{33 \text{ ft}}$$

(given) (unknown)

Notice that your plan was determined by the information available to you. For example, if you were also given that 39.37 in = 1 m, then your plan for the above problem could be

10 m → in → ? ft

$$10 \text{ m} \times \frac{39.37 \text{ in}}{1 \text{ m}} \times \frac{1 \text{ ft}}{12 \text{ in}} = \underline{33 \text{ ft}}$$

(given) (unknown)

If you are having trouble with a conversion , try to think of conversion factors that can be derived from information given in the problem. Identify conversion factors whose denominator has the same unit as the given information and factors whose numerator contains the same unit as the unknown quantity. Once you know these factors, the required conversions will be more apparent. Remember it is generally assumed that you know the metric and English systems. Therefore, you should be able to do any conversions necessary within the metric or English system without being given specific information.

Practice Drill on Conversions: Do the following conversions. Observe significant figures.

(a) 2640 ft → ____________ km (1 mile = 5280 ft; 1 mile = 1.609 km)

(b) 4.0 gal → ____________ L (0.946 L = 1 quart)

(c) 11 fluid ounces → ____________ L (29.51 mL = 1 fluid ounce)

(d) 908 g of paper → ____________ number of reams of paper
(1 ream = 500 sheets; 1 sheet = 0.500 oz; 16 oz = 1 lb; 454 g = 1 lb)

Answers: (a) 0.8035 km (b) 15 L (c) 0.32 L (d) 0.128 reams

Mass, Force, and Weight (Section 1.6)

Learning Objectives

1. Perform calculations involving units of mass.
2. Distinguish between mass and weight.
3. Calculate the weight of an object from its mass and vice versa.
4. Calculate the mass, volume, or density of an object, given two of the three quantities.
5. Distinguish between density and specific gravity.

Review

Mass

Mass is the amount of matter in an object. Common metric units are the kilogram (SI unit) and the gram (1/1000 kilogram).

Example 1.6: What is the mass of a 150-lb woman in kilograms? (1 kg = 2.205 lb)

Answer: 150 lb → ____________ kg

$$150\text{ lb} \times \frac{1\text{ kg}}{2.205\text{ lb}} = \underline{68.0\text{ kg}}$$

The conversion factor, 1 kg/2.205 lb, is always valid on earth. However, some factors apply only to specific problems and are not always valid:

Example 1.7: There are 190 mg of sodium in one ounce of a popular brand of potato chips. How many milligrams of sodium would you ingest if you consumed a 3-ounce bag of potato chips?

Answer: To calculate the milligrams of sodium in a 3 ounce bag you need a factor whose numerator has milligrams of sodium (new unit) and denominator has ounces of potato chips (given or old unit).

3 ounces of potato chips → ____________ mg of sodium

$$3\text{ ounces of potato chips} \times \frac{190\text{ mg sodium}}{1\text{ ounce potato chips}} = \underline{570\text{ mg sodium}}$$

In this problem, 190 mg of sodium is equated with 1 ounce of potato chips to get the conversion factor.

Example 1.8: A 384-g box of high-fiber cereal contains 3.8 g of sodium. How many servings of cereal contain 1.1 g of sodium? (1 serving = 28 g cereal)

Answer: 1.1 g sodium → ____________ g cereal → ____________ servings

$$1.1\text{ g sodium} \times \frac{384\text{ g cereal}}{3.8\text{ g sodium}} \times \frac{1\text{ serving}}{28\text{ g cereal}} = \underline{4.0\text{ servings}}$$

In this example 3.8 g of sodium is equated with 384 g of cereal to get the conversion factor.

Caution: Be careful to label units completely; that is, to write g cereal and g sodium and not just g. Complete labeling is especially important when different quantities have the same units. Remember grams of cereal will cancel only grams of cereal not grams of some other quantity.

Force and Weight

A force is a push or pull. The force that pulls objects downward is the gravitational force. This force is proportional to the mass of the object:

$f = m \times g$

- m = mass of object on which force acts (kg)
- g = gravitational acceleration (m/s^2)
- f = gravitational force (N)

The gravitational acceleration, g, is not a constant; the value of g depends on the location of the object on the surface of the earth. However, a standard value has been defined that we will use in calculations:

$g = 9.80665\ m/s^2$.

The downward force acting on an object is also called the object's weight. Although weight is proportional to mass it is a different quantity.

weight = mass × g

Mass is the amount of matter in an object and therefore does not depend on location. Weight, however, depends on g as well as mass. Because g varies from place to place the weight of the object also varies. The SI unit for weight is the newton (N): $1\ N = 1\ kg\ m/s^2$. The newton is a derived unit made up of three fundamental units: kg (mass), m (length), and s (time).

Example 1.9: Express the weight of a 150-lb student in newtons. ($g = 9.807\ m/s^2$)

Given: 150 lb

Unknown: weight in newtons

Plan: Because weight = m × g, we must know both the mass of the student and g.

Calculation: $g = 9.807\ m/s^2$;

$$\text{mass of student} = 150\ \text{lb} \times \frac{1\ \text{kg}}{2.205\ \text{lb}} = 68.0\ \text{kg}$$

$$\begin{aligned}\text{weight of student} &= (\text{mass of student}) \times g \\ &= 68.0\ \text{kg} \times 9.807\ \text{m/s}^2 \\ &= \underline{667\ \text{N}}\end{aligned}$$

Note: The value used for g in this calculation is $9.807\ m/s^2$ not $9.80665\ m/s^2$. If the latter value had been used, the additional digits would not have affected the answer because the 150-lb measurement limits the answer to only three significant figures. The rule of thumb is to use values that contain one more significant figure than your answer.

The mass of an object can sometimes be obtained by calculating the volume of the object from its dimensions and then multiplying by an average mass per unit volume. The average mass of a small object (like an atom) can be obtained by dividing the mass of a sample by the number of objects in the sample.

Density

The quantity of mass in one unit of volume is the density (d). The density of a sample can be calculated by dividing the mass of the sample by its volume:

$d = m/V$
- d = density of a substance
- m = mass of the sample
- V = volume of the sample

Density of a substance is therefore mass per unit volume. The more mass in a given volume, the greater the density.

Example 1.10: The mass of a 100-mL sample of ethanol is 907 g. Calculate the density of ethanol.

Answer: $d = \frac{m}{V} = \frac{907\ \text{g}}{100\ \text{mL}} = \underline{0.907\ \text{g/mL}}$

Practice Drill on Calculating Density: Complete the following table:

	Sample	Volume	Mass	Density
(a)	Ocean water	2000 mL	2050 g	_____________ g/L
(b)	Air	1.0 L	1.3 g	_____________ mg/mL
(c)	Lead	1.000 mL	11.35 g	_____________ kg/L

Answers: (a) 1.025 g/mL (b) 1.3 mg/mL (c) 11.35 kg/L

Density is a useful property because it allows mass to be calculated from volume and vice versa: rearranging $d = m/V$ gives $V = m/d$ and $m = d \times V$.

Example 1.11: What is the volume of a 1.0-g piece of Fe? The density of Fe is 7.87 g/mL.

Answer: $V = \frac{m}{d} = \frac{1.0\text{ g}}{7.87\text{ g/mL}} = \underline{0.13\text{ mL}}$

You can also treat this problem as a conversion from units of grams into milliliters of Fe. The density provides the needed conversion factor:

$1.0\text{ g Fe} \times \frac{1\text{ mL}}{7.87\text{ g}} = \underline{0.13\text{ mL}}$

Note that the density always gives two conversion factors; for Fe they are: $\frac{1\text{ mL}}{7.87\text{ g}}$ and $\frac{7.87\text{ g}}{1\text{ mL}}$

The conversion factor on the right would be used to find the mass of a sample given its volume.

Practice Drill on Using Densities to Calculate Mass and Volume: Fill in the following table:

	Substance	Volume	Mass	Density
(a)	Fe	_____________	157.4 g	7.87 g/cm^3
(b)	Ice	_____________	2.10 g	0.917 g/mL
(c)	Fat tissue	10.0 mL	_____________	1.6 g/mL

Answers: (a) 20.0 cm^3 (b) 2.29 mL (c) 16 g

Note: Whenever you are given two of the three variables d, m, or V you can calculate the third. In fact, if two of the variables are given in a problem, you will usually need to find the third.

Density depends on the identity of the substance and on the temperature. (See Table 1.7 in your text for some common densities at room temperatures.) Most densities decrease on heating because the substances expand, causing the mass to occupy a larger volume. Water behaves differently; its density has a maximum value of 1.000 g/mL at 3.98°C and decreases with either a rise or fall of temperature from this value. Because of its lower density, ice floats on liquid water.

The specific gravity of a liquid is the ratio of the density of the liquid to the density of water:

$$\text{sp gr} = \frac{\text{density of liquid}}{\text{density of water}}$$

Because the density of water is 1.00 g/mL at room temperature, the specific gravity of a substance is the numerical equivalent of its density in grams per milliliter. Note, however, that specific gravity is unitless while density has units of mass/volume.

Example 1.12: The specific gravity of a 30-cm^3 sample of urine from a dehydrated patient was reported to be 1.035. What is the mass of the sample in grams?

Answer: The density of the sample is 1.035 g/mL or 1.035 g/cm^3. Treating the density as a conversion factor gives

$$30\ \text{cm}^3 \times \frac{1.035\ \text{g}}{1\ \text{cm}^3} = \underline{31\ \text{g}}$$

Work, Heat, and Other Forms of Energy (Section 1.7)

Learning Objectives

1. Give examples of work.
2. State the SI unit of energy.
3. Calculate the kinetic energy, mass, or speed of a moving object, given two of the three quantities.
4. Distinguish between potential energy and kinetic energy.
5. Describe the potential energy changes that occur when charged particles come together or move apart.
6. Distinguish between heat and temperature.
7. Convert joules to calories and vice versa.
8. Convert °F to °C and vice versa.

Review

Work

Work is done when a force moves an object. Work is the product of the force and the distance over which the force acts:

$w = f \times d$ f = force (N)
d = distance over which force acts (m)
w = work (J)

The SI unit for work and energy is the joule (J): 1 J = 1 N m

Example 1.13: A weight lifter uses a force of 1.20×10^3 N to lift a barbell exactly one meter. How much work is done?

Answer: $w = f \times d = 1.20 \times 10^3\ \text{N} \times 1\ \text{m} = \underline{1.20 \times 10^3\ \text{J}}$
(The one meter is exact so that the number of significant figures will depend on the value for the force.)

Energy

Energy is the ability to do work. Moving objects can do work because they possess kinetic energy (E_k):

$E_k = 1/2\ m \times v^2$

m = mass of moving object (kg)
v = speed of object (m/s)
E_k = kinetic energy (J)

Example 1.14: Calculate the kinetic energy in joules of a 120-lb Olympic runner who finishes each mile in 5.00 minutes. (1.609 km = 1 mile; 1 kg = 2.205 lb)

Answer: Because E_k must be calculated in joules (SI unit for energy), the mass and velocity must also be in SI units. The required conversions are lb → kg and mile/min → m/s. Hence,

$$E_k = \tfrac{1}{2}\, m \times v^2 = \frac{1}{2}\left(120 \text{ lb} \times \frac{1 \text{ kg}}{2.205 \text{ lb}}\right) \times \left(\frac{1 \text{ mile}}{5.00 \text{ min}} \times \frac{1609 \text{ m}}{1 \text{ mile}} \times \frac{1 \text{ min}}{60 \text{ s}}\right)^2 = \underline{783 \text{ J}}$$

An object has potential energy because of its position or the positions of its parts relative to each other. An object raised above the surface of the earth has potential energy; two charged particles separated by a distance also have potential energy. To increase the distance between two oppositely charged particles, work has to be done because they attract each other. This work increases the potential energy of the particles. Likewise work must be done to decrease the distance between two particles of like charges because they repel each other. Because the forces that hold atoms and molecules together consist of attractions between charged particles, it is important to understand this type of energy.

Heat and Temperature

Heat is energy that flows from an object of higher temperature to an object of lower temperature. Heat stops flowing when the temperature of the two objects is the same, at which point they are in thermal equilibrium.

Note: Heat and temperature are not the same. Heat is a form of energy; temperature is the property of matter that determines the direction of heat flow. A liter of water at 90°C for example will provide more heat than 500 mL of water at 90°C.

The SI unit for heat is the joule; the calorie (cal) and the large dietary Calorie (Cal) are also used:

1 cal = 4.184 J
1000 cal (1 kcal) = 1 Calorie

One calorie of heat will raise the temperature of one gram of water by approximately one degree Celsius. Hence, the amount of heat absorbed by an object will depend on the mass of the object and the temperature change of the object:

Example 1.15: A chef must boil 2.0 kg of water to cook a pasta dish. How much heat is needed to bring the water to its boiling point if the water is initially at 25°C? Express your answer in (a) cal and (b) joules.

Given: 2.0 kg water; temperature change 25°C to 100°C (boiling point).

Unknown: Heat absorbed.

Plan:
(a) To raise one gram of water by one Celsius degree takes one calorie. To raise one gram of water by 75 Celsius degrees (100° – 25°) takes (1 × 75) calories. To raise 2000 g (2.0 kg) of water by 75°C takes 1.0 cal/g °C × 75°C × 2000 g = $\underline{1.5 \times 10^5 \text{ cal}}$

(b) 1.5×10^5 cal × 4.18 J/cal = $\underline{6.3 \times 10^5 \text{ J}}$

A thermometer contains a substance with a property that varies with temperature. In a mercury thermometer, the volume of liquid mercury expands as the temperature increases and contracts as the temperature decreases.

There are three temperature scales in common use: Celsius, Kelvin (SI system), and Fahrenheit. Celsius and Fahrenheit temperatures can be interconverted by the relationship:

$F = 32 + 1.8\ C$

SELF-TEST

1. Identify which of the following are examples of matter and which are examples of energy.
   (a) heat from a campfire
   (b) resting neutron
   (c) light from a star
   (d) helium in a balloon

Answer questions 2-5 as true or false.

2. Theories are explanations of laws.
3. The fundamental unit of volume in the SI system is the liter.
4. The gram is the SI unit for mass.
5. The joule is a combined unit.
6. A box is 75 cm by 10 cm by 1.5 m. Find its volume in
   (a) cubic meters (m^3)
   (b) liters ($1000\ L = 1\ m^3$)
   (c) mm^3
7. The nitrate ion, a water pollutant, is a component of fertilizer. Consequently, the nitrate ion is often found in high concentration in the drinking water of agricultural regions. It is especially dangerous to infants, since it combines with hemoglobin, decreasing the oxygen-carrying ability of blood. You want to choose a laboratory to provide <u>reliable measurements</u> of nitrate ion concentrations; you provide Laboratory A and Laboratory B with three samples each containing 10.01 mg of nitrate and you receive the following results:

Laboratory A	Laboratory B
10.0	11.5
10.1	11.8
10.3	12.0

   (a) Measurements from which laboratory are more precise? ____________
   (b) Which laboratory is more accurate? ____________
   (c) Which laboratory should you choose? ____________

8. A car burns 2.5 gallons of fuel in travelling 100 miles. How many liters of fuel are burned per kilometer? (1 gal = 4 quarts; 1 L = 1.057 quarts; 1.609 km = 1 mile)
9. Athletes often "carbohydrate load" before competition to supposedly increase their energy reserves. How many grams of carbohydrates would an athlete get from <u>four</u> 2-ounce servings of pasta if there are 320 g of carbohydrates in a one-pound box? (1 pound = 16 ounces).
10. At what speed in meters/second must a 2000 lb car travel to have a kinetic energy of 6.0×10^5 J? (1 kg = 2.205 lb)

11. To raise one gram of ethanol by one Celsius degree takes 1.8 joules of heat. How many kilojoules are needed to increase the temperature of 50 g of ethanol by 25 Celsius degrees?

ANSWERS TO SELF-TEST QUESTIONS

1. (a) energy (b) matter (c) energy (d) matter
2. T
3. F (cubic meters is the SI unit for volume)
4. F (kilogram is the SI unit for mass)
5. T ($1\ J = 1\ kg\ m^2/s^2$)
6. (a) $0.11\ m^3$ (b) 1.1×10^2 L (c) $1.1 \times 10^8\ mm^3$
7. (a) Lab A (b) Lab A (c) Lab A
8. 0.059 L/km
9. 160 g
10. 36 m/s
11. 1.8 J/g°C × 50 g × 25°C = <u>2.2 kJ</u>

CHAPTER 2
ATOMS, MOLECULES, AND IONS

SELF-ASSESSMENT

The Discovery of Electrons (Section 2.1)

1. Which of the following statements about discharge tubes is incorrect?
   (a) The cathode rays in a discharge tube bend toward a positively charged plate.
   (b) Cathode rays are beams of electrons.
   (c) A discharge tube has two electrodes.
   (d) The properties of cathode rays depend on the nature of the cathode material.

Underline the correct word or phrase.

2. The Thomson experiment determined (e/m, e, m).

3.* The Millikan experiment determined (e, m) by measuring the charges acquired by (cathode rays, oil droplets).

The Nuclear Atom (Section 2.2)

4. After bombarding gold foil with alpha particles, Rutherford believed that the positive charge and most of the atom's mass are concentrated in a small nucleus because
   (a) all of the alpha particles passed through the gold atoms.
   (b) all alpha particles showed large deflections.
   (c) most of the alpha particles passed through the gold atoms, some showed large deflections, and few were even scattered backwards.
   (d) all of the alpha particles collided with the nucleus.

The Chemical Elements and the Periodic Table (Section 2.3)

5. Complete the following table using the periodic table when necessary.

Symbol	Number of protons	Number of electrons	Number of neutrons
^{2}H (Deuterium)	____________	____________	____________
$^{4}He^{2+}$	____________	____________	____________
^{235}U	____________	____________	____________
$^{32}S^{2-}$	____________	____________	____________

6. State the mass of a carbon-12 atom in atomic mass units.

7. Given the following mass and abundance data, calculate the atomic weight of magnesium:

Isotope	Mass (u)	Abundance
^{24}Mg	23.985	78.70%
^{25}Mg	24.986	10.13%
^{26}Mg	25.983	11.17%

8. What is a mass spectrometer used for?

Avogadro's Number and the Mole (Section 2.4)

9. Which has more atoms, one mole of sodium or one mole of iron?

10. What is the mass in grams of 6.022×10^{23} carbon-12 atoms?

11. Which is heavier, a sample of iron containing 1.1×10^{19} atoms or a sample of Li with 2.2×10^{19} atoms?

12. How many moles of atoms are in 1.0×10^{5} u of carbon-12?

Elements, Compounds, and Mixtures (Section 2.5)

13. Identify the following as compounds or mixtures:
    (a) sucrose dissolved in water
    (b) table salt
    (c) seawater
    (d) vinegar
    (e) calcium carbonate

14. How could you separate calcium carbonate from table salt? (<u>Hint</u>: calcium carbonate is insoluble in water.)

Atoms in Combination (Section 2.6)

15. Write formulas for
    (a) glycine (an amino acid), whose molecule contains two C atoms, five H atoms, one N atom, and two O atoms
    (b) nitrogen dioxide

16. Give one example each of a molecular and a nonmolecular compound.

Molar Masses of Compounds (Section 2.7)

17. Given a 32.0-g sample of SO_2, find
    (a) the number of molecules of SO_2 in the sample.
    (b) the grams of oxygen in the sample.

18. How many grams of NaCl contain 3.011×10^{23} Na^+ ions?

Ions and Ionic Compounds (Section 2.8) **and Naming Compounds** (Section 2.9)

19. (a) Name the following anions: SO_4^{2-}; Cl^-; PO_4^{3-}
    (b) Write the formula for the ammonium, sodium, and magnesium cation. (Don't forget the charges).

20. Write the formulas for the following ionic compounds: sodium bromide, calcium nitrate, magnesium phosphate.

21. Name the following binary molecular compounds: SiO_2, NH_3, PCl_5.

ANSWERS TO SELF-ASSESSMENT PROBLEMS

If you missed an answer, <u>be sure</u> to study the relevant section in the textbook and study guide.

1. d

2. e/m

3. e; oil droplets

4. c

5. ^{2}H: 1, 1, 1; He^{2+}: 2, 0, 2; ^{235}U: 92, 92, 143; $^{32}S^{2-}$: 16, 18, 16

6. exactly 12 u

7. 24.31 u

8. A mass spectrometer measures the precise masses and relative abundances of atomic and molecular ions.

9. Same; a mole is a mole is a mole.

10. exactly 12 g

11. Fe; its mass is 1.0×10^{-3} g.

12. $1.0 \times 10^5 \text{ u} \times \dfrac{1 \text{ atom}}{12 \text{ u}} \times \dfrac{1 \text{ mol}}{6.02 \times 10^{23} \text{ atoms}} = 1.4 \times 10^{-20} \text{ mol}$

13. mixtures: a, c, d; compounds: b, e

14. Add water to dissolve table salt and then filter to separate solid $CaCO_3$.

15. (a) $C_2H_5NO_2$ (b) NO_2

16. Possible answers are: water (molecular); graphite (nonmolecular)

17. (a) 3.01×10^{23} molecules (b) 16.0 g oxygen

18. 29.22 g

19. (a) sulfate, chloride, phosphate: (b) NH_4^+, Na^+, Mg^{2+}

20. NaBr; $Ca(NO_3)_2$; $Mg_3(PO_4)_2$

21. silicon dioxide; ammonia; phosphorus pentachloride

REVIEW

The Discovery of Electrons (Section 2.1)

Learning Objectives

1. Sketch a typical discharge tube.
2.* Describe how Thomson determined the electron's charge-to-mass ratio.
3.* Describe how Millikan determined the electron charge.

Review

A typical discharge tube is a glass tube containing two metal plates, or electrodes. Most of the gas is removed from the tube. When these plates are connected to a source of voltage, one electrode, the anode, gains a positive charge while the other electrode, the cathode, gains a negative charge. When the charge difference is sufficient, the cathode emits electrons or cathode rays that travel toward the anode. Controlled electron beams are used in TV picture tubes, PC computer monitors, and electron microscopes.

*Thomson used the effect of magnetic and electric fields on cathode rays to calculate the charge-to-mass ratio of the electron. The ratio is

$e/m = 1.759 \times 10^8$ C/g.

(The coulomb, C, is the SI unit of charge.)

*Robert Millikan determined the electron charge by comparing the rates of fall of charged oil droplets in an electric field with their rates of fall in the absence of a field. The charge of one electron is 1.602×10^{-19} C.

The mass of the electron is obtained by combining e/m and e:

$$m = \frac{e}{e/m} = \frac{e}{1.759 \times 10^8\ C/g} = \frac{1.602 \times 10^{-19}\ C}{1.759 \times 10^8\ C/g} = 9.107 \times 10^{-28}\ g$$

The Nuclear Atom (Section 2.2)

Learning Objectives

1. Describe how positive ions form in a discharge tube.
2. State the evidence for Rutherford's conclusion that the positive charge and most of the atom's mass are concentrated in a nucleus that is very small compared to the atom as a whole.
3. Relate the atomic number Z to the numbers of protons and electrons in a neutral atom.
4. Compare electrons, protons, and neutrons in terms of charge and mass.

Review

An ion is an atom or group of atoms bearing an electrical charge. Positive ions form in discharge tubes when cathode rays collide with gas atoms. The simplest positive ion is the proton, formed by the loss of an electron from an atom of hydrogen. Another important ion is the helium ion, which is formed when a helium atom loses two electrons. The helium ion is identical to the alpha particle emitted by certain radioactive atoms.

The Gold Foil Experiment and the Rutherford Atom

Thomson proposed an atomic model consisting of electrons embedded in a jelly-like sphere of positive electricity. This model assumes that most of the atom's mass is uniformly distributed throughout its volume.

Rutherford's bombardment of gold foil with alpha particles showed that the mass of the atom is concentrated in a small core, or nucleus. The nucleus contains all of the protons in the atom; the number of protons in the nucleus of an atom is called the atomic number (Z). In a neutral atom, the number of protons is equal to the number of electrons. The electrons move around the small nucleus so that the atom is mostly empty space. Chadwick showed that the nucleus also contains neutrons.

Masses and Charges of Atomic Particles

electron	9.10939×10^{-28} g	-1.6022×10^{-19} C
proton	1.67262×10^{-24} g	$+1.6022 \times 10^{-19}$ C
neutron	1.67493×10^{-24} g	0

The Chemical Elements and the Periodic Table (Section 2.3)

Learning Objectives

1. State which elements are gases, which are liquids, and which are solids, at room temperature.
2. Given the symbol for an atom or ion, state the number of protons, neutrons, and electrons it contains.
3. Write the symbol for an atom or ion, given its mass number, atomic number, and charge.
4. State the names and write the symbols of the three isotopes of hydrogen.
5. State the exact mass of a carbon-12 atom in atomic mass units.
6. State the approximate masses of a proton and a neutron in atomic mass units.
7. Calculate chemical atomic weights from isotope mass and abundance data.
8.* Describe the operation of a direction focusing mass spectrometer.

Review

An element is a substance whose atoms all have the same atomic number (Z). Elements are represented by one-, two-, or three-letter symbols. The periodic table (inside front cover of your text and study guide) shows the elements arranged in order of atomic number. See Fig 2.10 in your text to locate the gases, liquids, and solids in the periodic table.

Some Important Facts About Elements

Naturally occurring elements: $Z = 1$ to $Z = 94$, except technetium ($Z = 43$) and promethium ($Z = 61$).

Artificially made elements: $Z = 95$ to $Z = 109$ and Tc and Pm.

Ninety-two elements have been found in the earth's crust; only eight elements account for over 98% of the earth's crust. (See Table 2.2 of your text)

Mass Numbers and Atomic Numbers

The mass number (A) of an atom is the sum of the number of protons and the number of neutrons in its nucleus. The atomic number (Z) is equal to the number of protons in its nucleus. All atoms of a given element have the same number of protons. Isotopes are atoms with the same atomic number, but different mass numbers (they have different numbers of neutrons). The number of neutrons in an isotope is equal to the mass number (A) minus the atomic number (Z). Symbols for isotopes are written with the mass number as a left superscript and the atomic number as a left subscript.

Example 2.1: Give the atomic symbol for the carbon atom that contains seven neutrons. (All carbon atoms have six protons)

Answer: The mass number is $6 + 7 = 13$. The symbol is

mass number (A)→ atomic number (Z)→ $^{13}_{6}C$

The symbol for an ion includes the charge as a right superscript. A positive charge equals the number of electrons lost; a negative charge equals the number of electrons gained.

Example 2.2: Give the complete symbol for an oxygen atom ($Z = 8$; $A = 16$) that has gained two electrons

Answer: $^{16}_{8}O^{2-}$

Practice Drill on Symbols of Isotopes: Complete the following chart, using the periodic table when necessary.

Complete Symbol	Mass Number (A)	Atomic Number (Z)	Number of Protons (Z)	Number of Neutrons (A - Z)	Charge	Number of Electrons
$^{23}_{11}Na^+$	23	11	11	12	+1	10
^{35}Cl	35	17	17	18	0	17
^{45}Sc	45	21	21	24	+1	20
^3H (Tritium, T)	______	______	______	______	0	______

Answers: $^{35}_{17}Cl$, 35, 17, 17, 18, 0, 17; $^{45}_{41}Sc^+$, 45, 21, 21, 24, +1, 20; ^3_1H or ^3_1T, 3, 1, 1, 2, 0, 1

Atomic Masses

Atomic masses are most conveniently expressed in atomic mass units (u), where one atomic mass unit is one-twelfth the mass of a carbon-12 atom. One carbon-12 atom therefore has a mass of exactly 12 u.

Example 2.3: Mass spectrometer data indicate that oxygen-16 atoms are 1.3329096 times heavier than carbon-12 atoms. What is the mass of an oxygen-16 atom?

Answer: $12 \text{ u} \times 1.3329096 = 15.994915 \text{ u}$

Chemical Atomic Weights

Elements are found in nature as mixtures of isotopes. Since the fraction of each isotope is fairly constant, an average mass, called the chemical atomic weight, can be calculated for each element by multiplying each isotope mass by its fractional abundance and summing the products.

Example 2.4: Determine the chemical atomic weight of gallium given the abundance of its two isotopes.

Isotope	Atomic Mass	Abundance
^{69}Ga	68.9257 u	60.27%
^{71}Ga	70.9249 u	39.73%

Answer: Chemical atomic weight of Ga

= atomic mass of ^{69}Ga × fractional abundance + atomic mass of ^{71}Ga × fractional abundance

$= (68.9257 \text{ u} \times 0.6027) + (70.9249 \text{ u} \times 0.3973)$

$= 69.72 \text{ u}$

***The Mass Spectrometer**

Isotopic masses and abundances are determined in mass spectrometers, which are modifications of Thomson's apparatus for determining the charge-to-mass ratio of the electron. A direction-focusing mass spectrometer changes a magnetic or electric field to cause the same deflection (r) for all ions. The detector is at a fixed position so that it will indicate the presence of ions at various field strengths.

Avogadro's Number and the Mole (Section 2.4)

Learning Objectives

1. Convert moles of particles to numbers of particles and vice versa.
2. Convert grams to atomic mass units and vice versa.
3. Use a table of atomic weights to find the molar mass of an atom.
4. Convert grams of substance to moles of substance and vice versa.

Review

The mole, an SI counting unit for atoms, molecules, and ions, is an amount of substance that contains the same number of particles as there are in exactly 12 grams of carbon-12. The number of particles in one mole, called Avogadro's number, is 6.0221367×10^{23}.

Example 2.5: Express 3.01×10^{23} atoms of Fe as moles of Fe atoms.

Answer: $3.01 \times 10^{23} \text{ atoms} \times \dfrac{1 \text{ mole}}{6.022 \times 10^{23} \text{ atoms}} = \underline{0.500 \text{ mol}}$

The Molar Mass

The molar mass of a substance is the mass of one mole of its particles. For example, the molar mass of copper is the combined mass of Avogadro's number of copper atoms (6.022×10^{23} atoms) or one mole of copper atoms. For an element, the molar mass in grams is numerically equal to the chemical atomic weight in atomic mass units (u). One gram of an element contains one mole of atomic mass units:

$1 \text{ g} = 6.0221367 \times 10^{23} \text{ u}$

This equality gives rise to conversion factors for converting grams to u, and vice versa.

Example 2.6: Find the mass of an average Cu atom in grams.

Answer: The chemical atomic weight of copper is 63.5 u; therefore, one atom of copper weighs

$$\frac{63.5 \text{ u}}{1 \text{ Cu atom}} \times \frac{1 \text{ g}}{6.022 \times 10^{23} \text{ u}} = \underline{1.05 \times 10^{-22} \text{ g / atom}}$$

Example 2.7: (a) How many moles are in 6.00 grams of carbon? (b) How many atoms? (c) What is the average mass of a carbon atom in grams?

Answer: (a) The mass of one mole of carbon is 12.01 g. The number of moles (n) in 6.00 g of carbon is

$$n = 6.00 \text{ g C} \times \frac{1 \text{ mol C}}{12.01 \text{ g C}} = \underline{0.500 \text{ mol C}}$$

(b) To find the number of atoms, multiply by Avogradro's number:

$$0.500 \text{ mol C} \times \frac{6.022 \times 10^{23} \text{ atoms}}{1 \text{ mol}} = \underline{3.01 \times 10^{23} \text{ atoms}}$$

(c) One mole of carbon weighs 12.01 g and contains 6.022×10^{23} atoms. The average mass per atom is therefore

$$\frac{12.01 \text{ g}}{1 \text{ mol}} \times \frac{1 \text{ mol}}{6.022 \times 10^{23} \text{ atoms}} = \underline{1.994 \times 10^{-23} \text{ g / atom}}$$

Note: Examples 2.6 and 2.7 (c) show that the average mass per atom can be calculated by using the conversion factor 1 g/6.022×10^{23} u (Example 2.6) or by dividing the molar mass by Avogadro's number (Example 2.7c).

Practice Drill on Converting Between Moles, Grams, and Atomic Mass Units: Fill in the blanks. (Avogadro's number = 6.022×10^{23})

Element (Atomic Weight)	Number of Moles	Grams	Number of Atoms	Average Mass of One Atom in Grams
Ag (107.87)	2.000	215.74	1.2044×10^{24}	1.791×10^{-22}
S (32.06)	1×10^{-2}	0.3207	6.022×10^{21}	5.324×10^{-23} g.
Na (22.99)	2.	45.98 g	12.044×10^{23}	3.81×10^{-23}
I (126.88)	1.00	126.88	6.022×10^{23}	2.107×10^{-22}

Answers: Ag: 215.7 g, 12.04×10^{23} atoms, 1.791×10^{-22} g; S: 1.000×10^{-2} mol; 6.022×10^{21} atoms, 5.324×10^{-23} g; Na: 2.000 mol, 45.98 g, 3.818×10^{-23} g; I (126.9): 127 g, 6.02×10^{23} atoms

Elements, Compounds, and Mixtures (Section 2.5)

Learning Objectives

1. Distinguish between elements, compounds, and mixtures, and give examples of each.

Review

An element is a substance composed of atoms that have the same atomic number. A compound is a substance composed of two or more elements combined in a fixed ratio. Examples of compounds and their compositions by mass include
nitrogen dioxide (brown air pollutant) 30.4% nitrogen, 69.6% oxygen
methane (natural gas) 75.0% carbon, 25.0% hydrogen
ethanol (drinking alcohol) 52.2% carbon, 13.0% hydrogen, 34.8% oxygen.

The properties of a compound are different from those of its constituent elements. For example, ethanol is a liquid at room temperature, while oxygen and hydrogen are gases and carbon is a solid.

Mixtures are different from compounds. A mixture is composed of two or more substances (elements or compounds) that (1) retain their properties and (2) are not combined in a fixed ratio. Examples are air, seawater, and soil. The substances in a mixture can usually be separated from each other by physical processes such as evaporation or filtering.

Atoms in Combination (Section 2.6)

Learning Objectives

1. Find the number of atoms of each element in a given formula.
2. Give examples of molecular and nonmolecular elements and compounds.
3. Name the allotropes of oxygen and carbon.

Review

The fact that the properties and composition of a compound are fixed is evidence of a fixed ratio between the different kinds of atoms. Pure water consists of triatomic units in which two hydrogen atoms bond to one oxygen atom. These separate combinations of atoms are called molecules. A molecule is a distinct unit consisting of two or more bonded atoms. The atomic symbols and subscripts in a molecular formula tell us the kind and number of each atom in the molecule. For example:

Compound	Formula	Kinds of Atoms	Number of Each Atom
ozone	O_3	oxygen	3
phosphorus trichloride	PCl_3	phosphorus	1
		chlorine	3
acetic acid	CH_3COOH	carbon	2
		hydrogen	4
		oxygen	2
alanine (an amino acid)	H_2NCHCH_3COOH	carbon	3
		hydrogen	7
		oxygen	2
		nitrogen	1

	Rules for Chemical Formulas	Example: $CO(NH_2)_2$
1.	The symbols tell which elements are present.	A molecule of urea ($CO(NH_2)_2$) contains atoms of carbon (C), oxygen (O), nitrogen (N), and hydrogen (H).
2.	Parentheses may be used to indicate a group of atoms that tends to stay together and behave as a unit.	The urea molecule contains NH_2 groups.
3.	Subscripts give the number of each kind of atom. A subscript, following a parentheses applies to the entire group.	A molecule of urea contains one carbon atom, one oxygen atom and two NH_2 groups. The total is C-1 atom O-1 atom N-2 atoms H-4 atoms
4.	Some formulas can be written more than one way, but all must show the same number of each kind of atom.	The urea formula can also be written as CON_2H_4 and it is sometimes written as H_2NCONH_2.

Molecular and Nonmolecular Substances

Substances that do not contain discrete molecules are said to be nonmolecular: helium, carbon, and sodium chloride are examples.

Helium consists of independent He atoms.

In carbon the atoms are linked in a continuous three-dimensional pattern.

Sodium chloride consists of a continuous array of Na^+ and Cl^- ions in a specific pattern.

The subscripts in a nonmolecular formula give only the relative numbers of atoms; the formula NaCl shows that the ratio of Na^+ to Cl^- in the crystal is 1:1.

Example 2.8: Give the ratio of silicon atoms to oxygen atoms in SiO_2, the main component of sand.

Answer: one Si for every two O or Si:O = 1:2.

Allotropes are different forms of the same element. Ozone (O_3) and ordinary oxygen (O_2) are examples of allotropic forms.

Molar Masses of Compounds (Section 2.7)

Learning Objectives

1. Calculate the number of moles of each element in a given mass of compound or in a given number of moles of compound.
2. Given the mole ratio of the elements in a compound, write the formula of the compound.
3. Calculate the molar mass of a compound from its formula.
4. Use molar masses to convert moles of compound to grams of compound and vice versa.

Review

The subscripts in a formula give the ratio of atoms in one molecule. This atom ratio is the same as the mole ratio. For example, SiO_2 contains two oxygen atoms to one silicon atom; it also contains two moles of oxygen atoms for each mole of silicon atoms.

Example 2.9: For methane (CH_4), give (a) the mole ratio of carbon to methane and (b) the mole ratio of carbon to hydrogen.

Answer: One molecule of methane contains one carbon atom and four hydrogen atoms. The mole ratios are

(a) $\dfrac{1 \text{ mol C}}{1 \text{ mol } CH_4}$ (b) $\dfrac{1 \text{ mol C}}{4 \text{ mol H}}$

Practice Drill on Formulas: Give the required mole ratios in the following compounds:

Compound	Mole Ratio of Compound to Carbon in Compound	Mole Ratio of Carbon to Hydrogen Atoms
$C_6H_{12}O_6$	$\dfrac{1 \text{ mol } C_6H_{12}O_6}{6 \text{ mol C}}$	$\dfrac{6 \text{ mol C}}{12 \text{ mol H}}$ $\dfrac{1 \text{ mol C}}{2 \text{ mol H}}$
(a) CH_3CH_2OH	____________	____________
(b) CH_3COOH	____________	____________
(c) H_2CO_3	____________	____________

Answer:

(a) $\dfrac{1 \text{ mol } CH_3CH_2OH}{2 \text{ mol C}}$; $\dfrac{2 \text{ mol C}}{6 \text{ mol H}}$ or $\dfrac{1 \text{ mol C}}{3 \text{ mol H}}$ (b) $\dfrac{1 \text{ mol } CH_3COOH}{2 \text{ mol C}}$; $\dfrac{2 \text{ mol C}}{4 \text{ mol H}}$ or $\dfrac{1 \text{ mol C}}{2 \text{ mol H}}$

(c) $\dfrac{1 \text{ mol } H_2CO_3}{1 \text{ mol C}}$; $\dfrac{1 \text{ mol C}}{2 \text{ mol H}}$

Using Mole Ratios in Problems

Example 2.10: Given 2.5 mol of glucose, find (a) the number of moles of oxygen atoms, (b) the number of oxygen atoms, and (c) the number of grams of oxygen atoms.

Given: 2.5 mol glucose: formula $C_6H_{12}O_6$

(a) Unknown: moles of oxygen

Plan: (a) We need a relationship between the given (moles of glucose) and the unknown (moles of oxygen atoms). The formula of glucose gives us the mole ratio 6 mol oxygen/1 mol glucose, which we can use to convert moles of glucose into moles of oxygen atoms:

2.5 mol glucose → mol oxygen

Calculation: $2.5 \text{ mol glucose} \times \dfrac{6 \text{ mol oxygen}}{1 \text{ mol glucose}} = \underline{15 \text{ mol oxygen}}$

(b) Unknown: atoms of oxygen

Plan: To convert moles to atoms, multiply the moles of atoms by Avogadro's number:

Calculation: $15 \text{ mol oxygen atom} \times \dfrac{6.02 \times 10^{23} \text{ atoms}}{1 \text{ mol}} = \underline{9.0 \times 10^{24} \text{ oxygen atoms}}$

(c) Unknown: grams of oxygen

Plan: To convert moles of atoms to grams, multiply the moles by the chemical atomic weight in grams:

Calculation: $15 \text{ mol} \times \dfrac{16.00 \text{ g oxygen}}{1 \text{ mol}} = \underline{2.4 \times 10^{2} \text{ g oxygen}}$

Example 2.11: How many moles of glucose ($C_6H_{12}O_6$) contain 3.01×10^{23} hydrogen atoms?

Given: 3.01×10^{23} atoms hydrogen; formula of glucose, $C_6H_{12}O_6$

Unknown: mol glucose

Plan: We can convert atoms of hydrogen to moles of hydrogen and then to moles of glucose:

atoms of hydrogen → mol hydrogen → mol glucose

The conversion from atoms to moles uses Avogadro's number. The conversion from moles of hydrogen to moles of glucose uses the mole ratio in the formula.

Calculation: $3.01 \times 10^{23} \text{ H atoms} \times \dfrac{1 \text{ mol H}}{6.022 \times 10^{23} \text{ H atoms}} \times \dfrac{1 \text{ mol } C_6H_{12}O_6}{12 \text{ mol H}} = \underline{0.0417 \text{ mol } C_6H_{12}O_6}$

Practice Drill on Using Mole Ratios: Fill in the table for octane (C_8H_{18}).
(atomic weights: C = 12.01; H = 1.01)

Moles of Octane	Moles of Carbon	Grams of Carbon	Atoms of Carbon
(a) 1.00 mol	______	______	______
(b) ______	2.0 mol	______	______
(c) ______	______	2.40 g	______
(d) ______	______	______	6.02×10^{23} atoms

Answers: (a) 1.00 mol; 8.00 mol; 96.1 g; 4.82×10^{24} atoms
(b) 0.25 mol; 2.0 mol; 24 g; 12×10^{23} atoms
(c) 0.0250 mol; 0.200 mol; 2.40 g; 1.20×10^{23} atoms
(d) 0.125 mol; 1.00 mol; 12.0 g; 6.02×10^{23} atoms

Molar Mass of a Compound (also called Formula Weight or Molecular Weight)

The molar mass of a compound is the sum of the molar masses of atoms in the formula.

Example 2.12: Find the molar mass of KCl. (atomic weights: K = 39.1; Cl = 35.5)

Answer: Molar mass	=	39.10 g/mol	+	35.45 g/mol	=	74.55 g/mol
		molar mass of K		molar mass of Cl		

Example 2.13: Find the molar mass of glucose ($C_6H_{12}O_6$). (atomic weights: C = 12.01; H = 1.01; 0 = 16.00)

Answer:			
	carbon	= 6 mol × 12.01 g/mol	= 72.06 g
	hydrogen	= 12 mol × 1.008 g/mol	= 12.10 g
	oxygen	= 6 mol × 16.00 g/mol	= 96.00 g
		molar mass of glucose	= 180.16 g/mol

Using Molar Masses in Problems

If you know the molar mass of a compound, you can find

(a) the number of moles in a given mass of the compound.
(b) the mass of a given number of moles of compound.

Note: You should learn to do these conversions quickly because you will need to do many of them in the upcoming chapters. Remember all you require is the formula of the compound and the atomic weights, which you can get from the periodic table.

Example 2.14: How many moles of the amino acid glycine (molar mass = 75.07 g) are in a 7.50-g sample?

Answer: We will convert grams into moles: 7.50 g → mol.

The conversion factor comes from the molar mass.

$$7.50\text{ g} \times \frac{1\text{ mol glycine}}{75.07\text{ g}} = \underline{0.0999\text{ mol glycine}}$$

Example 2.15: What is the mass in grams of 10.0 mol of glycine?

Answer: Here we convert moles into grams: 10.0 mol → g.

Again the conversion factor comes from the molar mass.

$$10.0\text{ mol} \times \frac{75.07\text{ g glycine}}{1\text{ mol}} = \underline{751\text{ g glycine}}$$

Practice Drill on Using Molar Masses: Complete the table:

Formula	Molar Mass	Grams of Sample	Moles of Sample	Molecules in Sample
H_2O	18.02 g/mol	36.0 g	______	______
NO_2	46.01 g/mol	______	0.2000	______
HCOOH	46.02 g/mol	______	______	6.022×10^{23}

Answers: H_2O: 2.00 mol, 12.0×10^{23} molecules; NO_2: 9.202 g, 1.204×10^{23} molecules; HCOOH: 46.02 g, 1.000 mol

PROBLEM-SOLVING TIP

Summary of Strategies for Interconverting Molecules, Moles, and Grams of a Compound and Its Elements

Given		Unknown	Plan
moles of compound	→	grams of compound	Multiply moles by molar mass.
grams of compound	→	moles of compound	Divide grams by molar mass.
molecules of compound	→	grams of compound	1. Divide molecules by Avogadro's number (6.022×10^{23} molecule/mol) to get moles. 2. Multiply moles by molar mass.
grams of compound	→	molecules of compound	1. Divide grams by molar mass to get moles. 2. Multiply moles by Avogadro's number.
moles of compound	→	moles of one of its elements	1. Use subscript in the formula to find the number of moles of the element in one mole of compound. 2. Multiply moles of compound by this number.
grams of compound	→	grams of one of its elements	1. Divide grams of compound by molar mass to find moles of compound. 2. Multiply moles of compound by subscript in the formula to find moles of the element. 3. Multiply moles of element by its atomic weight in grams to find grams of element.
grams of constituent element	→	grams of compound	1. Divide grams of element by atomic weight in grams to find moles of element. 2. Divide moles of element by the subscript of the element in the compound's formula to get moles of compound. 3. Multiply moles of compound by its molar mass to get grams.

Your additions:

Ions and Ionic Compounds (Section 2.8)

Learning Objectives

1. Learn the formulas, charges, and names of the starred anions and cations in Table 2.6 of your text.
2. Learn the ionic charges associated with periodic groups 1A, 2A, 6A, and 7A.
3. Write the formula of an ionic compound given its component ions.

Review

Compounds can be classified as ionic or covalent. Ionic compounds are composed of ions, covalent compounds are not. To be electrically neutral, an ionic compound must contain both positive ions and negative ions; positive ions are called cations, negative ions are called anions.

Metal atoms often lose electrons and become cations. Metals from groups 1A, 2A, and 3A form cations with charges of +1, +2, and +3, respectively.

Some nonmetal atoms can gain electrons and become anions. The nonmetals of groups 7A, 6A, and 5A form anions with charges of -1, -2, and -3 respectively.

Example 2.16: Al and S are in group 3A and group 6A of the periodic table respectively. What are the charges of the aluminum and the sulfide ions? (Answer: +3, -2)

A polyatomic ion is a group of atoms that bears a charge.

Some Polyatomic Ions

Name	Formula
carbonate	CO_3^{2-}
hydroxide	OH^-
nitrate	NO_3^-
phosphate	PO_4^{3-}
sulfate	SO_4^{2-}
ammonium	NH_4^+

Note: Learn the above ions.

Formulas of Ionic Compounds

Ionic compounds are neutral; the sum of the positive charges on the cations must equal the sum of the negative charges on the anions. These charges therefore determine what the formula must be.

Example 2.17: What is the formula of an ionic compound made of sodium ions, Na^+, and oxide ions, O^{2-}?

Answer: The total charge must be zero. Hence, the compound must contain two Na^+ ions for every O^{2-} ion; the formula is Na_2O.

Example 2.18: Give the formula for the ionic compound containing Fe^{3+} and SO_4^{2-} ions.

Answer: We need to find the simplest combination of +3 ions and -2 ions that will add up to zero. Hence, two Fe^{3+} ions ($2 \times (+3) = +6$) giving a +6 charge are cancelled by three SO_4^{2-} anions contributing a -6 charge ($3 \times (-2) = -6$) so that the correct formula is $Fe_2(SO_4)_3$.

Practice Drill on Writing Formulas of Ionic Compounds: Fill in the blanks:

Cation	Anion	Formula of Ionic Compound
K^+	SO_4^{2-}	K_2SO_4
Mn^{2+}	__________	$Mn(NO_3)_2$
Al^{3+}	SO_4^{2-}	__________
ammonium ion	CO_3^{2-}	__________
calcium	hydroxide	__________
NH_4^+	nitrate	__________
__________	PO_4^{3-}	$Fe_3(PO_4)_2$

Answers: NO_3^-; $Al_2(SO_4)_3$; $(NH_4)_2CO_3$; $Ca(OH)_2$; NH_4NO_3; Fe^{2+}

Naming Compounds (Section 2.9)

Learning Objectives

1. Write names for monatomic positive and negative ions.
2. Write names for ionic compounds and binary molecular compounds whose formulas are given.
3. Write formulas for ionic compounds and binary molecular compounds whose names are given.

Review

Because there are so many different chemical compounds, the International Union of Pure and Applied Chemistry (IUPAC) established a Commission on Nomenclature to name compounds using a consistent system. Many compounds, like H_2O and C_2H_5OH, also have common names. You will learn the common names of compounds as you become familiar with them.

Learn the IUPAC rules in the text for naming ions, ionic compounds, and binary molecular compounds. The best way to learn these rules is to practice using them to name compounds..

Example 2.19: Name the following compounds: (a) NH_4Cl (b) Na_2SO_3 (c)$FeCl_3$ (d) NO_2 (e) CO

Answers: We know that (a), (b), and (c) are ionic compounds because they contain ions we have learned. We must first name the cation and then the anion: (a) ammonium chloride; (b) sodium sulfite; (c) iron(III) chloride. Remember that whenever a metal can form more than one positive ion, the charge on the ion is indicated as a Roman numeral.

NO_2 and CO are binary molecular compounds. The rule is to name first the element that is farthest to the left in the periodic table. Also, if the two elements can form more than one binary compound, Greek prefixes (see Table 2.7 in your text, p. 71) are used to indicate the number of each kind of atom present: (d) nitrogen dioxide; (e) carbon monoxide.

SELF-TEST

True or False

1. The properties of an electron beam depend on the cathode material used in the discharge tube.

2. The positive ions formed in a discharge tube move along with the cathode-ray stream.

3. Thomson was not able to determine the mass or charge of an electron separately.

4. Rutherford's gold foil experiment suggests that an atom is mostly empty space.

Complete the following using the periodic table if needed.

5. The number of electrons in $^{40}Ca^{+2}$ is ______________; the number of neutrons ______________; the number of protons ______________.

6. The atomic weight of calcium (Ca) is 40.1. Find (a) the mass of a Ca atom in atomic mass units and in grams, (b) the mass of a mole of Ca atoms, (c) the number of moles in 80.2 g of Ca and (d) the mass in grams of 2.50 mol of Ca.

7. Find the number of grams of oxygen in 12.0×10^{23} molecules of CO_2. (atomic weight of oxygen is 16.00)

8. Rank the following in order of increasing mass: 3.01×10^{23} atoms of Fe; a sample of glucose ($C_6H_{12}O_6$) containing 32.0 g of oxygen; 2.0 mol of helium gas; 18.02 g of water.

9. Which of the following is a compound: soil, blood, air, water?

10. Write the formulas for the following compounds:
    (a) sodium phosphate
    (b) magnesium nitrate
    (c) ammonium chloride
    (d) Co(II) nitrite
    (e) sulfur hexafluoride
    (f) phosphorus trifloride

11. Potassium iodide (K = 39.10; I = 126.9) tablets were dispensed in certain regions to protect against the effect of exposure to high levels of ^{131}I, a radioactive gas emitted from the Chernobyl nuclear power plant in the Soviet Union. If a person ingested a tablet containing 10 mg of KI,
    (a) how many moles of I^- were ingested?
    (b) what mass of I^- was ingested?
    (c) how many I^- ions were ingested?

ANSWERS TO SELF-TEST QUESTIONS

1. F

2. F

3. T

4. T

5. 18e, 20n, 20p

6. (a) 40.1 u, 6.66×10^{-23} g
   (b) 40.1 g
   (c) 2.00 mol
   (d) 100 g

7. $12.0 \times 10^{23} \text{ molecules } CO_2 \times \frac{1 \text{ mol } CO_2}{6.022 \times 10^{23} \text{ molecules } CO_2} \times \frac{2 \text{ mol oxygen}}{1 \text{ mol } CO_2} \times \frac{16.00 \text{ g oxygen}}{1 \text{ mol O}} = \underline{63.8 \text{ g}}$

8. helium, water, Fe, glucose

9. water

10. (a) Na_3PO_4
    (b) $Mg(NO_3)_2$
    (c) NH_4Cl
    (d) $Co(NO_2)_2$
    (e) SF_6
    (f) PF_3

11. (a) 6.0×10^{-5} mol
    (b) 7.6×10^{-3} g
    (c) 3.6×10^{19} ions

CHAPTER 3

CHEMICAL REACTIONS, EQUATIONS, AND STOICHIOMETRY

SELF-ASSESSMENT

Physical and Chemical Change (Section 3.1)

1. State whether each of the following is a physical or chemical change. Identify the reactants in each chemical change.
   (a) NO from supersonic jet exhaust reacts with ozone (O_3) to form NO_2 and O_2. (This process may contribute to a depletion of the protective ozone layer around the earth.)
   (b) Lakes dissolve more air in cold weather than in hot.
   (c) Carbon monoxide combines with hemoglobin, so that the hemoglobin cannot combine with oxygen.
   (d) Liquid nitrogen boils at -196°C.

Laws of Chemical Combination (Section 3.2)

2. A 79.0-g sample of selenium combines with 2.0 g of hydrogen to produce hydrogen selenide, a toxic gas produced by some plants. Calculate the percentage composition of this gas.

Percentage Composition from a Formula (Section 3.3)

3. (a) Calculate the percentage composition of tetraethyl lead ($Pb(C_2H_5)_4$) the antiknock compound in leaded gasoline.
   (b) How much tetraethyl lead was burned in car engines in a year in which approximately 6×10^6 kg of Pb was released into the environment?

Determination of Formulas (Section 3.4)

4. The old type of wooden matches used a phosphorus sulfide that would ignite upon striking.
   (a) Calculate the empirical formula given that 56.3% of the compound is phosphorus.
   (b) The molar mass of this compound is 220.1 g/mol. Give its molecular formula.

5. Nicotine is a compound composed of C, H, and N. The combustion of 162.2 mg of nicotine produces 440.1 mg of CO_2 and 126.1 mg of H_2O. Determine:
   (a) the empirical formula
   (b) the molecular formula, given that the molecular weight is 162.2 u

Chemical Equations (Section 3.5)

6. Balance the following equations

   (a) $C_8H_8(l) + O_2(g) \rightarrow CO_2(g) + H_2O(g)$

   (b) $Pb(C_2H_3O_2)_2(s) + H_2S(g) \rightarrow PbS(s) + HC_2H_3O_2(l)$

   (c) $(NH_4)_2Cr_2O_7(s) \rightarrow Cr_2O_3(s) + H_2O(g) + N_2(g)$

7. Write balanced chemical equations for the following reactions:
   (a) Ammonium nitrate (NH_4NO_3(s)) when heated forms "laughing gas" (N_2O) and water.
   (b) Gaseous dimethyl sulfide ($(CH_3)_2S$) reacts with oxygen to form liquid dimethyl sulfoxide ($(CH_3)_2SO$).
   (c) Copper metal displaces gold from aqueous solutions of $AuCl_3$ to form gold metal and a solution of $CuCl_2$.

Some Chemical Reactions (Section 3.6)

8. Label the reactions in Exercise 7 as formation, decomposition, displacement, or oxidation and reduction.

Reaction Stoichiometry (Section 3.7)

9. Ammonium nitrate, a component of fertilizer, may decompose explosively. The unbalanced equation is:

   $NH_4NO_3(s) \rightarrow N_2(g) + O_2(g) + H_2O(g)$

   (a) If 160.1 g of NH_4NO_3 decomposes, _____ grams of O_2, _____ grams of N_2, and _____ grams of H_2O will be produced.
   (b) To produce 18.0 g of H_2O, _____ moles of NH_4NO_3 must decompose.

10. Ethyl alcohol, also called ethanol, (C_2H_5OH), is produced from the fermentation of sugar in yeast cells:

    $C_6H_{12}O_6(s) \rightarrow CO_2(g) + C_2H_5OH(l)$ (unbalanced)

    (a) How many milliliters of ethanol are produced by the fermentation of 90.1 g of glucose? (The density of ethanol is 0.789 g/mL.)
    (b) How many grams of glucose must ferment to produce 2.00 L of CO_2(g)? (The density of CO_2 is 1.81 g/L under the conditions of the reaction.)

Limiting and Excess Reactants (Section 3.8)

11. Another student mixes 13.8 g of salicylic acid with 2.04 g of acetic anhydride. For this experiment determine:
    (a) the limiting reagent
    (b) the maximum mass of aspirin that can be produced (yield = 100%)
    (c) the mass of excess reagent remaining after the reaction

Theoretical, Actual, and Percentage Yield (Section 3.9)

12. A future medical student synthesizes aspirin in the laboratory using the following reaction:

    salicylic acid acetic anhydride aspirin

    $2C_7H_6O_3 + C_4H_6O_3 \rightarrow 2C_9H_8O_4 + H_2O$

    How much aspirin did she produce if she reported her yield as 66.7% after reacting 13.8 g of salicylic acid with excess acetic anhydride?

ANSWERS TO SELF-ASSESSMENT PROBLEMS

If you missed an answer, be sure to study the relevant section in the textbook and study guide.

1. Physical changes: (b) and (d); chemical changes: (a) NO and O_3 are reactants; (c) CO and hemoglobin are reactants.

2. 2.5% hydrogen; 97.5% selenium

3. (a) 29.71% C; 64.07% Pb; 6.234% H

   (b) $6 \times 10^6 \text{ kg Pb} \times \dfrac{323 \text{ kg Pb}(C_2H_5)_4}{207 \text{ kg Pb}} = 9 \times 10^6 \text{ kg Pb}(C_2H_5)_4$

4. (a) $56.3 \text{ g P} \times \dfrac{1 \text{ mol P}}{30.97 \text{ g P}} = 1.82 \text{ mol P};\ 43.7 \text{ g S} \times \dfrac{1 \text{ mol S}}{32.07 \text{ g S}} = 1.36 \text{ mol S } (P_4S_3)$

   (b) P_4S_3

5. (a) C_5H_7N; (b) $C_{10}H_{14}N_2$

6. (a) $C_8H_8(l) + 10O_2(g) \rightarrow 8CO_2(g) + 4H_2O(g)$

   (b) $Pb(C_2H_3O_2)_2(s) + H_2S(g) \rightarrow PbS(s) + 2HC_2H_3O_2(l)$

   (c) $(NH_4)_2Cr_2O_7(s) \rightarrow Cr_2O_3(s) + 4H_2O(g) + N_2(g)$

7. (a) $NH_4NO_3(s) \rightarrow N_2O(g) + 2H_2O(l)$

   (b) $2(CH_3)_2S(g) + O_2(g) \rightarrow 2(CH_3)_2SO(l)$

   (c) $3Cu(s) + 2AuCl_3(aq) \rightarrow 2Au(s) + 3CuCl_2(aq)$

8. 7(a) decomposition 7(b) oxidation and reduction 7(c) displacement.

9. (a) $160.1 \text{ g } NH_4NO_3 \times \dfrac{1 \text{ mol } NH_4NO_3}{80.05 \text{ g } NH_4NO_3} \times \dfrac{1 \text{ mol } O_2}{2 \text{ mol } NH_4NO_3} \times \dfrac{32.00 \text{ g } O_2}{1 \text{ mol } O_2} = 32.00 \text{ g } O_2;\ 56.04 \text{ g } N_2;\ 72.08 \text{ g } H_2 0$

   (b) 0.500 mol NH_4NO_3

10. $C_6H_{12}O_6(s) \rightarrow 2CO_2(g) + 2C_2H_5OH(l)$

    (a) 58.4 mL ethanol

    (b) $2.00 \text{ L } CO_2 \times \dfrac{1.81 \text{ g } CO_2}{1 \text{ L } CO_2} \times \dfrac{1 \text{ mol } CO_2}{44.01 \text{ g } CO_2} \times \dfrac{1 \text{ mol } C_6H_{12}O_6}{2 \text{ mol } CO_2} \times \dfrac{180.2 \text{ g } C_6H_{12}O_6}{1 \text{ mol } C_6H_{12}O_6} = 7.41 \text{ g glucose}$

11. (a) acetic anhydride (b) 7.20 g aspirin (c) 8.3 g salicylic acid

12. 12.0 g aspirin

REVIEW

Physical and Chemical Change (Section 3.1)

Learning Objectives

1. Identify the reactants and the products in a given chemical reaction.
2. State whether a given change is a physical change or a chemical reaction.
3. State whether a given property is a physical property or a chemical property.

Review

A physical change is one in which the identity of the substance is maintained and no new elements or compounds are formed.

Example 3.1: List some physical changes that water can undergo.

Answer: melting of ice, evaporation of liquid water, condensing of steam, etc.

A chemical change is a chemical reaction; one or more new substances are formed. Substances that change are reactants and the new substances formed are products.

Example 3.2: Give some examples of chemical reactions and identify the reactants and products in each example.

Answer: rusting of iron; reactants: oxygen and iron; products: rust; burning of gasoline in your car engine; reactants: oxygen and the substances in gasoline; products: mostly carbon dioxide and water vapor

The chemical properties of a substance are the reactions that it can undergo.

Example 3.3: List some chemical properties of oxygen.

Answer: reacts with combustible substances, corrodes some metals; combines with hemoglobin in blood

Physical properties are those that can be observed without changing the identity of the substance. Physical properties include odor, color, melting point, boiling point, density, and solubility.

Example 3.4: List three physical properties of oxygen.

Answer: colorless, odorless, slightly soluble in water

Laws of Chemical Combination (Section 3.2)

Learning Objectives

1. State the laws of conservation of mass, constant composition, and multiple proportions and explain these laws in terms of atomic theory.
2. Calculate the percentage composition of a compound given the appropriate data.

Law of Conservation of Mass

The law of conservation of mass states that there is no measurable change in total mass during a chemical reaction: the sum of the masses of reactants equals the sum of the masses of products. This law is explained by assuming that atoms are neither created nor destroyed in a chemical reaction but only rearranged.

Example 3.5: How many grams of carbon dioxide are formed when 12.0 g of carbon combines with 32.0 g of oxygen?

Answer: sum of masses of reactants = sum of masses of products

$m_{carbon} + m_{oxygen} = m_{carbon\ dioxide}$

12.0 g + 32.0 g = 44.0 g

Practice: Use the above information to find out how much carbon will combine with 48.0 g of oxygen to produce 66.0 g of carbon dioxide. Answer: 18.0 g

Law of Constant Composition

The law of constant composition states that every compound has a fixed elemental composition by mass. For example, 44 g of carbon dioxide always contain 12 g of carbon and 32 g of oxygen. These masses can be used to find the percentage of carbon and oxygen in carbon dioxide.

Example 3.6: A 100.0-mg sample of water decomposes to form 88.8 mg of oxygen and 11.2 mg of hydrogen. Find the percentage of hydrogen and oxygen in 10.0 g of water.

Answer: $\%\ \text{hydrogen} = \frac{\text{mass of hydrogen}}{\text{mass of water}} \times 100\% = \frac{11.2\ \text{mg hydrogen}}{100.0\ \text{mg water}} \times 100\% = 11.2\%\ \text{hydrogen by mass}$

$\%\ \text{oxygen} = 100\% - \text{percent hydrogen} = 100\% - 11.2\% = 88.8\%\ \text{oxygen by mass}$

Note: The 10.0-g sample is irrelevant to the problem; the percentage composition of water is the same for all samples of water regardless of mass.

The law of definite proportions makes the same statement as the law of constant composition, but it focuses on the formation of compounds: **When elements combine to form a compound, they do so in definite proportions by mass.** For example, oxygen and hydrogen always react in the ratio of 88.8 g of oxygen to 11.2 g of hydrogen in forming water. Atoms must therefore combine in a fixed ratio to form a given compound.

The law of multiple proportions applies to two or more compounds that contain the same elements but in different fixed ratios: **When compounds contain the same elements, the masses of one element combined with a fixed mass of the other element are in the ratio of small whole numbers.**

Example 3.7: Show that the following experimental data verify the law of multiple proportions: 30.0 g of nitrogen monoxide (NO) will decompose into 14.0 g of nitrogen and 16.0 g of oxygen and 46.0 g of nitrogen dioxide (NO_2) will decompose into 14.0 g of nitrogen and 32.0 g of oxygen.

Answer: The number of grams of oxygen in NO_2 and NO that combines with 14.0 g of nitrogen is 32.0 g and 16.0 g respectively. To verify the law, we must show that these two masses are in a whole-number ratio:

$$\text{Ratio} = \frac{32.0 \text{ g of oxygen in NO}_2}{16.0 \text{ g of oxygen in NO}} = \frac{2}{1}$$

Percentage Composition from a Formula (Section 3.3)

Learning Objectives

1. Calculate percentage composition from a formula.

Review

The formula of a compound gives the number of moles of each element in 1 mol of the compound: in 1 mol of glucose, $C_6H_{12}O_6$, there are 6 mol of carbon atoms, 12 mol of hydrogen atoms, and 6 mol of oxygen atoms. Therefore, if you know the formula and the molar masses (atomic weights) of the elements in the compound, you can calculate the percentage composition:

Example 3.8: DDT is a pesticide whose use in the U.S. is currently restricted due to its toxicity. Its formula is $C_{14}H_9Cl_5$. Determine the percentage composition of DDT.

Answer:

Given: The formula is $C_{14}H_9Cl_5$. The molar masses are C = 12.01 g/mol; H = 1.008 g/mol; Cl = 35.45 g/mol.

Unknown: % composition

Plan:

$$\%\text{ by mass of each element} = \frac{\text{mass of element in 1 mol of compound}}{\text{mass of 1 mol of compound}} \times 100\%$$

Use atomic weights to determine the mass of each element in 1 mol of compound (numerator) and the mass of 1 mol of compound (denominator). Then substitute these values into the expression given above.

Calculations:

$$\begin{aligned} 14 \text{ mol C} \times 12.01 \text{ g/mol} &= 168.1 \text{ g} \\ 9 \text{ mol H} \times 1.008 \text{ g/mol} &= 9.072 \text{ g} \\ 5 \text{ mol Cl} \times 35.45 \text{ g/mol} &= 177.2 \text{ g} \\ \hline \text{molar mass} &= 354.4 \text{ g/mol} \end{aligned}$$

Therefore,

$$\% \text{ H} = \frac{9.072 \text{ g}}{354.4 \text{ g}} \times 100\% = \underline{2.560\%} \qquad \% \text{ Cl} = \frac{177.2 \text{ g}}{354.4 \text{ g}} \times 100\% = \underline{50.00\%}$$

$$\% \text{ C} = \frac{168.1 \text{ g}}{354.4 \text{ g}} \times 100\% = \underline{47.43\%}$$

Note: Sometimes the elemental percentages do not add up to 100.00. This is usually due to the rounding off of the elemental percentages during their computation.

Determination of Formulas (Section 3.4)

Learning Objectives

1. Use atomic and molecular weights to calculate the formulas of simple molecules.
2. Write the empirical formula of a compound given its molecular formula.
3. Use elemental composition data and a table of atomic weights to calculate empirical formulas.
4. Use empirical formulas and molar masses to calculate molecular formulas.
5. Calculate percentage compositions and formulas from combustion data.

Review

Empirical and Molecular Formulas

There are two kinds of formulas used to describe compounds: empirical formulas and molecular formulas. A molecular formula gives the total number of each kind of atom in one molecule of the compound. An empirical formula gives only the simplest whole-number ratio of the atoms in a compound. The formula of a nonmolecular compound can only be empirical.

Example 3.9: The molecular formula for acetylene, the fuel in oxy-acetylene blow torches, is C_2H_2. Give its empirical formula.

Answer: The acetylene molecule contains 2 C atoms and 2 H atoms. The simplest whole-number ratio is one C atom to one H atom; hence, the empirical formula is CH.

> **Practice**: Give the empirical formula of (a) $C_6H_{12}O_6$; (b) $Hg(CH_3)_2$.
> Answer: (a) CH_2O; (b) HgC_2H_6

Determination of Empirical Formulas

The empirical formula of a compound can be determined from elemental composition data. The data can be in the form of grams in a given mass of sample or in the form of percentage composition.

Example 3.10: A 25.0-g sample of a compound consisting of hydrogen and oxygen decomposes to give 22.2 g of oxygen. Give its empirical formula.

Answer:

Given: 22.2 g of oxygen in 25.0 g of sample. The mass of hydrogen is also known because it makes up the difference:

22.2g O + ? g H = 25.0 g

There are 2.8 g H.

Unknown: empirical formula (simplest whole-number mole ratio of oxygen and hydrogen in the compound.)

Plan: (1) Find number of moles of H and O in sample.
(2) Find simplest whole-number mole ratio of these numbers.

Calculations: (1) $\text{mol of H} = 2.8\ \text{g H} \times \dfrac{1\ \text{mol H}}{1.008\ \text{g H}} = 2.8\ \text{mol}$

$$\text{mol of O} = 22.2\ \text{g O} \times \frac{1\ \text{mol O}}{16.00\ \text{g O}} = 1.39\ \text{mol}$$

(2) To find the ratio, divide each number of moles by the smallest:

$$\frac{2.8}{1.39} = 2.0; \quad \frac{1.39}{1.39} = 1.0$$

The ratio is 2 mol H for 1 mol O; formula = H_2O. (Subscripts of 1 are not written.)

Example 3.11: Caffeine contains 49.5% C, 5.20% H, 28.9% N, and 16.5% O. Determine its empirical formula.

Answer: When you are given the percentage composition, you can imagine an arbitrary 100-g sample. Each percentage then equals the number of grams of that element.

Given: 100 g of caffeine = 49.5 g of C + 5.20 g of H + 28.9 g of N + 16.5 g of O

Unknown: empirical formula

Plan: (1) Find the moles of each element in the 100-g sample.
(2) To find the whole-number mole ratio, divide each number of moles by the smallest number of moles found in step (1).

Calculation: (1) (2) Divide by 1.03

$$49.5\ \text{C} \times \frac{1\ \text{mol C atoms}}{12.01\ \text{g C}} = 4.12\ \text{mol C atoms} \qquad \frac{4.12}{1.03} = 4.00 = 4$$

$$5.20\ \text{g H} \times \frac{1\ \text{mol H atoms}}{1.008\ \text{g H}} = 5.16\ \text{mol H atoms} \qquad \frac{5.16}{1.03} = 5.01 = 5$$

$$28.9\ \text{g N} \times \frac{1\ \text{mol N atoms}}{14.01\ \text{g N}} = 2.06\ \text{mol N atoms} \qquad \frac{2.06}{1.03} = 2.00 = 2$$

$$16.5\ \text{g O} \times \frac{1\ \text{mol O atoms}}{16.00\ \text{g O}} = 1.03\ \text{mol O atoms} \qquad \frac{1.03}{1.03} = 1.00 = 1$$

empirical formula = $\underline{C_4H_5N_2O}$

Determination of Molecular Formulas

Case 1: Finding the Molecular Formula of a Simple Compound Given Molecular Weight

The molecular formula can often be determined from the molecular weight of a substance that consists of one or two elements only.

Example 3.12: Determine the formula for ammonia given that it consists of nitrogen and hydrogen and its molecular weight is 17.03 u.

Answer: Since the atomic weights of N and H are 14.01 u and 1.008 u, respectively, there can only be one N atom in the molecule. The additional 3.02 u must therefore belong to H giving the formula NH_3.

> **Practice**: Nitrous oxide is an inhalation anesthetic containing nitrogen and oxygen. Its molar mass is 44.02 g/mol. Find its molecular formula. Answer: N_2O

Case 2: Finding the Molecular Formula from the Empirical Formula and Molar Mass

The molar mass of a compound is a whole-number multiple of the molar mass of the empirical formula. Therefore, if you divide the molar mass of the compound by the molar mass of the empirical formula, you will obtain a whole number. To find the molecular formula, multiply the subscripts in the empirical formula by this whole number.

Example 3.13: Isofluorane is a general inhalation anesthetic used in surgery. Its empirical formula is $C_3H_2ClF_5O$, and its molar mass is 184.5 g. Find the molecular formula of isofluorane.

Answer:

Given: molar mass = 184.5 g/mol; empirical formula = $C_3H_2ClF_5O$

Unknown: molecular formula

Plan:
(1) Determine the molar mass of the empirical formula.
(2) Divide the molecular molar mass by the empirical molar mass.
(3) Multiply the empirical formula by the resulting whole number.

Calculations:

(1)
$$\begin{aligned}
3 \text{ mol C} \times 12.01 \text{ g/mol} &= 36.03 \text{ g} \\
2 \text{ mol H} \times 1.008 \text{ g/mol} &= 2.016 \text{ g} \\
1 \text{ mol Cl} \times 35.45 \text{ g/mol} &= 35.45 \text{ g} \\
5 \text{ mol F} \times 19.00 \text{ g/mol} &= 95.00 \text{ g} \\
1 \text{ mol O} \times 16.00 \text{ g/mol} &= 16.00 \text{ g} \\
\text{empirical formula mass} &= 184.50 \text{ g}
\end{aligned}$$

(2)
$$\frac{184.5 \text{ g}}{184.50 \text{ g}} = 1$$

(3) molecular formula = $C_3H_2ClF_5O$

Case 3: Finding a Molecular Formula from the Molar Mass and the Percentage Composition

1. Find the empirical formula from the percentage composition. (See Example 3.11).
2. Use the empirical formula to determine the molecular formula from the molar molar mass. (See Example 3.13)

Note: An alternative method is to use the percentage composition and the molar mass to get the molecular formula directly. The steps are

1. Take 1 mol of the compound.
2. Multiply the molar mass by the percentage of each element to obtain the number of grams of each element in 1 mol.
3. Divide the number of grams of each element by its atomic weight to get the number of moles of each element in 1 mol of the compound. These numbers are the subscripts in the molecular formula.

Example 3.14: The molar mass and percentage composition for isofluorane are 184.5 g/mol; C = 19.53%; H = 1.09%; Cl = 19.21%; F = 51.49%; O = 8.67%. Find the molecular formula.

Given: molar mass = 184.5g/mol; C = 19.53%, H = 1.09%; Cl = 19.21%; F = 51.49%; O = 8.67%.

Unknown: molecular formula

Plan and Calculations: (1) Take 1 mol, 184.5 g of isofluorane.
(2) Multiply 184.5 g by the percentage of each element:

$$184.5 \text{ g} \times 0.1953 = 36.03 \text{ g C}$$
$$184.5 \text{ g} \times 0.0109 = 2.01 \text{ g H}$$
$$184.5 \text{ g} \times 0.1921 = 35.44 \text{ g Cl}$$
$$184.5 \text{ g} \times 0.5149 = 95.00 \text{ g F}$$
$$184.5 \text{ g} \times 0.0867 = 16.00 \text{ g O}$$

(3) Divide the number of grams of each element by its atomic weight:

$$36.03 \text{ g} \times \frac{1 \text{ mol}}{12.01 \text{ g}} = 3 \text{ mol C}$$

$$2.01 \text{ g} \times \frac{1 \text{ mol}}{1.008 \text{ g}} = 2 \text{ mol H}$$

$$35.44 \text{ g} \times \frac{1 \text{ mol}}{35.45 \text{ g}} = 1 \text{ mol Cl}$$

$$95.00 \text{ g} \times \frac{1 \text{ mol}}{19.00 \text{ g}} = 5 \text{ mol F}$$

$$16.00 \text{ g} \times \frac{1 \text{ mol}}{16.00 \text{ g}} = 1 \text{ mol O}$$

molecular formula = $\underline{C_3H_2ClF_5O}$

Practice: Halothane, another inhalation anesthetic, has a molar mass of 197.4 g/mol, and its percentage composition is 12.17% C; 0.51% H; 40.48% Br; 17.96% Cl; 28.88% F. Find the molecular formula of halothane. Answer: $C_2HBrClF_3$

Case 4: Finding a Molecular Formula from Combustion Analysis Data and the Molar Mass

Combustion analysis is generally confined to compounds that contain C, H, O, S, and N. The masses of C, H, N, and S in the sample are usually calculated from the measured masses of CO_2, H_2O, SO_2, and NO_2 (or N_2) obtained as combustion products. The mass of oxygen in the sample is found by the difference. (If the compound does not contain oxygen, then one of the other masses can be found by the difference.) The number of moles of each element in the sample is calculated from these masses. This composition data can now be converted to an empirical formula (Case 2) or to the molecular formula (Case 3) if the molar mass is given.

Example 3.15: Vitamin C consists of carbon, hydrogen, and oxygen and has a molecular weight of 176.12 u. A 4.99-mg tablet is burned to produce 2.04 mg of H_2O and 7.50 mg of CO_2. Determine the empirical and molecular formulas of vitamin C.

Answer:

Given: Vitamin C + oxygen $\rightarrow$ CO_2 + H_2O
4.99 mg — 7.50 mg — 2.04 mg

Unknown: molecular formula

Plan: (1) Find moles of each element in the sample. (Remember the moles of carbon in the sample are the same as the moles of carbon in 7.50 mg of CO_2 and the moles of hydrogen in the sample are the same as the moles of hydrogen in 2.04 mg of H_2O.)

(2) Determine empirical formula.

(3) Divide molar mass by molar mass of empirical formula to get a whole number; multiply subscripts in empirical formula by this whole number.

Calculations: (1) To find number of moles of C from grams of CO_2, we need the molar mass of CO_2 (44.01 g):

$$\text{moles of C in tablet} = \text{moles of C in } CO_2 = 7.50 \times 10^{-3}\text{ g} \times \frac{1\text{ mol } CO_2}{44.01\text{ g}} \times \frac{1\text{ mol C}}{1\text{ mol } CO_2} = 1.70 \times 10^{-4}\text{ mol C}$$

Similarly to find the moles of H from grams of H_2O

$$\text{moles of H in tablet} = \text{moles of H in } H_2O = 2.04 \times 10^{-3}\text{ g} \times \frac{1\text{ mol } H_2O}{18.02\text{ g}} \times \frac{2\text{ mol H}}{1\text{ mol } CO_2} = 2.26 \times 10^{-4}\text{ mol H}$$

To find moles of O, find mass by difference and then convert to moles:

grams of O = grams of vitamin C – grams of H – grams of C
grams of H = 2.26×10^{-4} mol H × 1.008 g H/1mol H = 2.28×10^{-4} g
grams of C = 1.70×10^{-4} mol C × 12.01 g C/1 mol C = 2.04×10^{-3} g
grams of O = 4.99×10^{-3} g – 2.28×10^{-4} g – 2.04×10^{-3} g = 2.72×10^{-3} g
moles of O = 2.72×10^{-3} g O × 1 mol O/16.00 g = 1.70×10^{-4} mol O

(2) $\text{empirical formula} = C_{\frac{1.70\times10^{-4}}{1.70\times10^{-4}}} H_{\frac{2.26\times10^{-4}}{1.70\times10^{-4}}} O_{\frac{1.70\times10^{-4}}{1.70\times10^{-4}}} = CH_{1.33}O$

To find a whole number, the subscripts can be multiplied by 3 to give $C_3H_4O_3$.

(3) The molar mass of the empirical formula is 88.06 g. The molecular formula can be found by dividing the molar mass by 88.06 g and then multiplying each subscript in the empirical formula by this number:

$$\frac{176.12\text{ g}}{88.06\text{ g}} = 2;\ \text{molecular formula} = C_6H_8O_6$$

Note: An alternative method is to find the percentage composition from the combustion data and then use the molar mass to find the molecular formula as in Example 3.14.

PROBLEM-SOLVING TIP

Summary of Methods for Determining Empirical and Molecular Formulas

Given	Unknown	Plan
elemental composition (grams of each element)	empirical formula	1. Divide grams of each element by atomic weight to find moles. 2. Determine smallest whole-number ratio of moles of elements in compound.
percentage composition (% by mass of each element)	empirical formula	1. Take 100.0 g of compound. 2. Multiply percentage of each element by 100.0 g to find grams of each element. 3. Divide grams of each element by atomic weight to find moles. 4. Determine smallest whole-number ratio of moles of elements in compound.
molar mass and elements in compound (Case 1)	molecular formula	Works only for simple compounds. By inspection you can determine formula.
molar mass and empirical formula (Case 2)	molecular formula	1. Determine molar mass of empirical formula. 2. Divide given molar mass by empirical molar mass. 3. Multiply subscripts in empirical formula by this number to generate molecular formula.
molar mass and percentage composition (Case 3)	molecular formula	1. Take 1 mol. 2. Multiply molar mass by percentage of each element to find grams of each element in 1 mol. 3. Divide grams of each element by its atomic weight to determine number of moles of each element. 4. Numbers of moles of each element are the subscripts in the formula.
molar mass and combustion data (Case 4)	molecular formula	1. Use the masses of combustion products to find masses of C, H, N, and S in compound. Find mass of oxygen by difference. 2. Use these masses to find empirical formula. 3. Divide molar mass of molecular formula by molar mass of empirical formula. 4. Multiply the subscripts of the empirical formula by this number to find the molecular formula.

Chemical Equations (Section 3.5)

Learning Objectives

1. Write balanced equations for chemical reactions.

Review

Chemical reactions are described by balanced equations in which the formulas for reactants are given to the left of the arrow and the products to the right. The state of each substance is also indicated after its formula: s = solid; l = liquid; g = gas; aq = aqueous (dissolved in water). An equation has to be balanced; the number of each kind of atom to the left of the arrow must be the same as the number to the right. (Atoms are never lost or gained in a chemical reaction.) Equations can be balanced only by changing the coefficients (the numbers preceding the formulas).

Caution: Never balance an equation by changing a subscript in a correct formula.

Example 3.16: Write a balanced equation for the reaction between hydrogen and chlorine gas to form hydrogen chloride gas.

Answer: (1) Write reactant formulas → product formulas (indicate the physical state of each substance):

$H_2(g) + Cl_2(g) \rightarrow HCl(g)$

(2) Choose coefficients to balance the atoms: There are two H atoms on the left and only one on the right. To balance them we must place a 2 in front of the HCl:

$H_2(g) + Cl_2(g) \rightarrow 2HCl(g)$

There are two Cl atoms on the left and two on the right; the equation is now balanced.

Example 3.17: Write a balanced equation for the explosion of TNT ($C_7H_5(NO_2)_3(s)$) to produce carbon, carbon monoxide, nitrogen, and water vapor.

Answer: (1) Write reactant formulas → product formulas (indicate physical states):
(2) Choose coefficients to balance the atoms.

Hint: Start by balancing the compounds with the greatest number of elements. If pure elements appear as reactants or products, balance those elements last.

(1) TNT is the starting material; the rest are products of the explosion.

$C_7H_5(NO_2)_3(s) \rightarrow C(s) + CO(g) + N_2(g) + H_2O(g)$

(2) H: Since there are 5 H atoms on the left and only 2 H atoms on the right, the coefficients for TNT and water must be 2 and 5, respectively. Now H is balanced with 10 H atoms on the left and 10 H atoms on the right:

$\underline{2}C_7H_5(NO_2)_3(s) \rightarrow C(s) + CO(g) + N_2(g) + \underline{5}H_2O(g)$

O: There are 12 O atoms on the left and only 6 O atoms on the right. To balance O without changing H, we need a coefficient of 7 for CO. Now there are 12 O atoms on the left and 12 O atoms on the right:

$2C_7H_5(NO_2)_3(s) \rightarrow C(s) + \underline{7}CO(g) + N_2(g) + 5H_2O(g)$

C: We have 14 C atoms on the left and only 7 C atoms on the right. To balance C without changing O we must use a coefficient of 7 for C(s):

$2C_7H_5(NO_2)_3(s) \rightarrow \underline{7}C(s) + 7CO(g) + N_2(g) + 5H_2O(g)$

N: 6 N atoms appear on the left and only 2 N atoms on the right. The coefficient for N_2 must therefore be 3:

$2C_7H_5(NO_2)_3(s) \rightarrow 7C(s) + 7CO(g) + \underline{3}N_2(g) + 5H_2O(g)$

Example 3.18: During photosynthesis green plants use liquid water and carbon dioxide in the atmosphere to produce glucose ($C_6H_{12}O_6$) and oxygen. Write a balanced equation for this reaction.

Answer: (1) The unbalanced equation is $H_2O(l) + CO_2(g) \rightarrow C_6H_{12}O_6(s) + O_2(g)$

(2) As in Example 3.17 we will start by balancing the compound containing the greatest number of elements ($C_6H_{12}O_6$).

C: There are 6 C atoms on the right so we will place a 6 in front of CO_2:

$H_2O(l) + \underline{6}CO_2(g) \rightarrow C_6H_{12}O_6(s) + O_2(g)$

H: There are 12 H atoms on the right so we will also place a 6 in front of H_2O:

$\underline{6}H_2O(l) + 6CO_2(g) \rightarrow C_6H_{12}O_6(s) + O_2(g)$

O: There are 18 O atoms in total on the left and only 8 on the right. We can increase the O atoms on the right by changing the coefficient of $C_6H_{12}O_6$ or O_2. But if we change the coefficient of $C_6H_{12}O_6$ we would alter the already balanced C and H atoms. Therefore, the required O atoms will come from six O_2 molecules:

$6H_2O(l) + 6CO_2(g) \rightarrow C_6H_{12}O_6(s) + \underline{6}O_2(g)$

Practice: Determine the coefficients for the equation ___$FeS_2(s)$ + ___$O_2(g)$ → ___$Fe_2O_3(s)$ + ___$SO_2(g)$

(Hint: Balance the O atoms last using a fractional coefficient for $O_2(g)$. Then multiply all of the coefficients by two.)

Answer: 4, 11, 2, 8

Example 3.19: Write a balanced equation for the reaction between aqueous (water) solutions of lead nitrate and potassium chloride to form solid lead chloride and a solution of potassium nitrate. Use (aq) to indicate substances that are dissolved in aqueous solutions and (s) to indicate a solid substance.

Answer: (1) The unbalanced equation is $Pb(NO_3)_2(aq) + KCl(aq) \rightarrow PbCl_2(s) + KNO_3(aq)$ (You can combine ions listed in Table 2.6 of your text to obtain these formulas.)

(2) NO_3^-: The NO_3^- ion remains intact in this reaction so it can be treated as a unit. Therefore, two NO_3^- ions on the left require two KNO_3 formula units on the right:

$Pb(NO_3)_2(aq) + KCl(aq) \rightarrow PbCl_2(s) + \underline{2}KNO_3(aq)$

Cl^-: There is one Cl^- on the left and two on the right. Therefore,

$Pb(NO_3)_2(aq) + \underline{2}KCl(aq) \rightarrow PbCl_2(s) + 2KNO_3(aq)$

The above equation is now balanced.

Practice: Fill in the coefficients for the following unbalanced equation:

$\_\_Al_2(SO_4)_3(aq) + \_\_Ca(OH)_2(aq) \rightarrow \_\_Al(OH)_3(s) + \_\_CaSO_4(s)$

Answer: 1, 3, 2, 3

Some Chemical Reactions (Section 3.6)

Learning Objectives

1. Write balanced equations for combination and decomposition reactions.
2. Write balanced equations for displacement reactions.
3. Predict the products and write balanced equations for the combustion of organic compounds.
4. Identify the substances oxidized and reduced in an oxidation-reduction reaction involving oxygen.
5. Write balanced equations for the reduction of oxide ores with carbon.

Review

In this section we will discuss the following types of reactions:

1. combination
2. decomposition
3. displacement
4. combustion
5. oxidation-reduction

(Some reactions may fall into more than one category.)

1. A combinatinon reaction is one in which a compound forms from simpler substances:

   $S(s) + O_2(g) \rightarrow SO_2(g)$

2. A decomposition reaction converts a compound into simpler substances. The products may be elements or compounds. Examples are

   $H_2CO_3(aq) \rightarrow H_2O(l) + CO_2(g)$

   $2AsH_3(g) \rightarrow 2As(s) + 3H_2(g)$

3. In a displacement reaction one element takes the place of another in a compound:

   $Zn(s) + 2HCl(aq) \rightarrow H_2(g) + ZnCl_2(aq)$

   Zn displaces hydrogen from HCl.

4. Combustion means "burning." The combustions we study here are those in which something burns in oxygen. A combustion releases heat and light and is self-sustaining (that is, it keeps going by itself):

   $C_8H_{18}(g) + 10O_2(g) \rightarrow 8CO_2(g) + 4H_2O(g) + \text{heat}$
   (octane)

5. Oxidation-reduction: The addition of oxygen to a substance is called oxidation; the removal of oxygen is called reduction. In the reaction

$$C(s) + H_2O(g) \rightarrow CO(g) + H_2(g)$$

carbon is oxidized and water is reduced. The terms oxidation and reduction are also used for many reactions that do not involve oxygen. The general meaning of these terms is discussed later.

Reaction Stoichiometry (Section 3.7)

Learning Objectives

1. Use mole ratios to solve problems relating quantities of reactant and product.

Review

Stoichiometric problems use mole ratios obtained from the coefficients in balanced chemical equations. These mole ratios are conversion factors that enable you to determine the amount of product formed or reactant used up.

Example 3.20: Calculate the number of moles of CO(g) produced from the explosion of 2.5 mol of TNT ($C_7H_5(NO_2)_3(s)$). The balanced equation is

$$2C_7H_5(NO_2)_3(s) \rightarrow 7C(s) + 7CO(g) + 3N_2(g) + 5H_2O(g)$$

Answer:

Given: 2.5 mol TNT

Unknown: mol CO

Plan: We are asked to convert 2.5 mol TNT to mol CO. The required conversion factor is the mole ratio that contains moles of CO in the numerator (desired unit) and moles of TNT in the denominator. From the coefficients in the balanced equation, we get the mole ratio 7 mol CO/2 mol TNT:

$$2.5 \text{ mol TNT} \times \frac{7 \text{ mol CO}}{2 \text{ mol TNT}} = \underline{8.8 \text{ mol CO}}$$

Note: The coefficients are exact and do not affect the number of significant figures in your answers.

Practice Drill on Mole Ratios: In this same reaction, what mole ratios would you use for problems with the following given and unknown quantities?:

Given	Unknown	Mole Ratio
mol C	mol N_2	________
mol TNT	mol H_2O	________
mol N_2	mol TNT	________
mol CO	mol H_2O	________

<u>Answers</u>: $\frac{3 \text{ mol } N_2}{7 \text{ mol C}}$; $\frac{5 \text{ mol } H_2O}{2 \text{ mol TNT}}$; $\frac{2 \text{ mol TNT}}{3 \text{ mol } N_2}$; $\frac{5 \text{ mol } H_5O}{7 \text{ mol CO}}$

If your known quantity is not given in moles, but in other units such as grams, you must convert to moles before you can use the mole ratio.

PROBLEM-SOLVING TIP

Steps in Solving Stoichiometric Problems

The required steps are:

1. Write a balanced chemical equation: identify given and unknown quantities.
2. Convert the given quantity into moles.
3. Determine the mole ratio between the unknown and the given quantity by the equation coefficients. Multiply the given in moles by the mole ratio to find the unknown in moles.
4. Convert moles of unknown to the required quantity.

<u>Example 3.21</u>: The first step in producing arsenic oxide (As_2O_3) (used to make pesticides) is to roast an arsenic ore such as As_2S_3 in air:

$$2As_2S_3(s) + 9O_2(g) \rightarrow 2As_2O_3(s) + 6SO_2(g)$$

How many grams of SO_2 (an air pollutant) are produced by roasting 950 g of As_2S_3?

Answer: (1) $2As_2S_3(s) + 9O_2(g) \rightarrow 2As_2O_3(s) + 6SO_2(g)$
950 g As_2S_3 (given) ? g SO_2 (unknown)

Plan and Calculations: (2) Before we can use a mole ratio from the coefficients in the balanced equation, we must convert grams of As_2S_3 into moles of As_2S_3:

$$950 \text{ g } As_2S_3 \times \frac{1 \text{ mol } As_2S_3}{246.0 \text{ g } As_2S_3} = 3.86 \text{ mol } As_2S_3$$

(3) We can now use the mole ratio

$$\frac{6 \text{ mol } SO_2}{2 \text{ mol } As_2S_3} \text{ to find moles of } SO_2;\ 3.86 \text{ mol } As_2S_3 \times \frac{6 \text{ mol } SO_2}{2 \text{ mol } As_2S_3} = 11.6 \text{ mol } SO_2$$

(4) To find grams of SO_2 we multiply by the molar mass of SO_2:

$$11.6 \text{ mol } SO_2 = \frac{64.06 \text{ g } SO_2}{1 \text{ mol } SO_2} = \underline{743 \text{ g } SO_2}$$

Practice: Balance the following equation, then fill in the boxes:

$CH_3OH(g) + O_2(g) \rightarrow CO_2(g) + H_2O(g)$

	CH_3OH (grams reacted)	O_2 (grams reacted)	CO_2 (grams produced)	H_2O (grams produced)
(a)	32.0 g	______	______	______
(b)	______	64.0 g	______	______
(c)	______	______	22.0 g	______
(d)	______	______	______	36.0 g

Answers: (a) 48.0 g O_2; 44.0 g CO_2; 36.0 g H_2O
(b) 42.7 g CH_3OH; 58.7 g CO_2; 48.1 g H_2O
(c) 16.0 g CH_3OH; 24.0 g O_2; 18.0 g H_2O
(d) 32.0 g CH_3OH; 48.0 g O_2; 44.0 g CO_2

Limiting and Excess Reactants (Section 3.8)

Learning Objectives

1. Identify the limiting and excess reactants in a reaction mixture.
2. Perform stoichiometric calculations involving limiting and excess reactants.

Review

An enormous number of chemical reactions occurs every second on earth. In most of them the reactants are not present in the proportions given by the equation; as a result, not all of every reactant will be used up.

The limiting reactant or limiting reagent is the reactant that is completely consumed; the limiting reactant therefore determines the amount of product(s) formed. Other reactants that are not totally used up are in excess. Some of the excess reactant(s) will remain after the reaction.

Limiting Reactant Problems

In these problems you are given the quantities of two or more reactants and asked to determine one or more of the following:

(1) the limiting reactant, that is, the reactant that is totally consumed
(2) the amount(s) of product(s) obtained from the limiting reactant
(3) the amount of excess reactant(s) left over

In order to answer (2) or (3), we must first determine (1).

Example 3.22: Uranium reacts with fluorine to form gaseous UF_6:

$U(s) + 3F_2(g) \rightarrow UF_6(g)$

Determine (1) the limiting reactant, (2) the moles of product formed, and (3) the grams of excess reactant remaining when 0.50 mol of U is mixed with 1.0 mol of F_2.

(1) Finding the Limiting Reactant

Given: quantities present: 0.50 mol U and 1.0 mol F_2

Unknown: limiting reactant

Plan: Convert all given quantities to moles. Choose one reactant as a reference. Then calculate the number of moles of the other reactant needed for complete reaction with the reference reactant. If the needed quantity of the other reactant is greater than the quantity present, then this other reactant is the limiting reactant. (The reference reactant is in excess.) If the calculated quantity of the other reactant is less than the quantity present, then this other reactant is in excess. (The reference reactant is limiting.)

Calculations: For our problem we choose U as our reference reactant. 0.50 mol U requires:

$$0.50 \text{ mol U} \times \frac{3 \text{ mol F}_2}{1 \text{ mol U}} = 1.5 \text{ mol F}_2$$

We have only 1.0 mol F_2; there is not enough F_2 present to react with all of the U present. As a result U will be in excess and F_2 is the limiting reactant.

Note: If you had chosen F_2 as your reference reactant your result would be the same:

$$1 \text{ mol F}_2 \times \frac{1 \text{ mol U}}{3 \text{ mol F}_2} = 0.33 \text{ mol U}$$

1.0 mol F_2 requires 0.33 mol U to react completely. Since 0.50 mol U are present, all of the F_2 will react; F_2 is the limiting reactant and U is in excess.

Note: An alternative method to finding the limiting reactant is to find the moles of product each given amount of reactant would produce if it reacted completely. The reactant that forms the least amount of product is the limiting reactant.

(2) Finding the Moles of Product Formed

Given:

$$U(s) + 3F_2(g) \rightarrow UF_6(g)$$

1.0 mol (F_2)   Unknown: ? (UF_6)

Plan: Now we have a simple stoichiometric problem like the ones we did before: Given the moles of a reactant, find the moles of product. Again we use the mole ratio from the balanced equation:

Calculation:

$$1.0 \text{ mol F}_2 \times \frac{1 \text{ mol UF}_6}{3 \text{ mol F}_2} = 0.33 \text{ mol UF}_6$$

(3) Finding the Grams of Reactant(s) Remaining

Given:

	U(s) +	F_2(g) → UF_6(g)
moles present:	0.50 mol	1.0 mol
moles reacting:	0.33 mol	1.0 mol

Unknown: grams of U in excess

Plan: After finding the limiting reagent, the remaining moles of the reactant in excess is determined by subtracting the moles of reactant that react from the moles present:

Calculations: moles of U in excess = moles present – moles reacting
= 0.50 mol – 0.33 mol
= 0.17 mol U

$$\text{grams of U in excess} = 0.17 \text{ mol U} \times \frac{238.0 \text{ g U}}{1 \text{ mol U}} = \underline{40 \text{ g}}$$

Practice Drill on Limiting Reactants: Determine the limiting reactant for each of the following reactions:

Reaction	Quantities Present
(a) $2H_2 + O_2 \rightarrow 2H_2O$	2 mol H_2; 2 mol O_2
(b) $2H_2 + O_2 \rightarrow 2H_2O$	1.0 g H_2; 1.0 g O_2
(c) $2C_2H_6 + 7O_2 \rightarrow 4CO_2 + 6H_2O$	4.3 mol C_2H_6; 6.7 mol O_2
(d) $2C_2H_6 + 7O_2 \rightarrow 4CO_2 + 6H_2O$	8.0 g C_2H_6; 32.7 g O_2

Answers: (a) H_2 (b) O_2 (c) O_2 (d) C_2H_6

Theoretical, Actual, and Percent Yield (Section 3.9)

Learning Objectives

1. Calculate the percent yield from the theoretical and actual yields.

Review

The theoretical yield is the amount of product you calculate will be produced from a given amount of reactant. The actual yield is what you collect in a real reaction. The actual yield is less than the theoretical yield due to various factors. Percent yield is

$$\text{percent yield} = \frac{\text{actual yield}}{\text{theoretical yield}} \times 100\%$$

Example 3.23: A student uses the following reaction to synthesize aspirin in the laboratory:

$C_7H_6O_3(aq)$	$+ HC_2H_3O_2(aq)$	$\rightarrow C_9H_8O_4$	$+ H_2O$
(salicylic acid)	(acetic acid)	(aspirin)	

The student reacts 13.8 g of salicylic acid with excess acetic acid and obtains 12.01 g of aspirin. What is the percent yield for the experiment?

Answer:

Given: actual yield = 12.01 g

Unknown: $\text{percent yield} = \dfrac{\text{actual yield}}{\text{theoretical yield}}$

Plan: Actual yield (12.01 g) is given to us. We must therefore determine the theoretical yield to find percent yield.

(1) $C_7H_6O_3(aq) + HC_2H_3O_2(aq) \rightarrow C_9H_8O_4(s) + H_20(l)$
13.8 g ? g (theoretical yield)

$$\text{g salicylic acid} \xrightarrow{(2)} \text{mol salicylic acid} \xrightarrow{(3)} \text{mol asprin} \xrightarrow{(4)} \text{g aspirin (theoretical yield)}$$

Calculation: (2) $13.8 \text{ g salicylic acid} \times \dfrac{1 \text{ mol salicylic acid}}{138.1 \text{ g salicylic acid}} = 0.0999 \text{ mol salicylic acid}$

(3) $0.0999 \text{ mol salicylic acid} \times \dfrac{1 \text{ mol aspirin}}{1 \text{ mol salicylic acid}} = 0.0999 \text{ mol aspirin}$

(4) $0.0.999 \text{ mol aspirin} \times \dfrac{180.2 \text{ g aspirin}}{1 \text{ mol aspirin}} = 18.0 \text{ g aspirin (theoretical yield)}$

Therefore, $\text{percent yield} = \dfrac{12.01 \text{ g}}{18.0 \text{ g}} \times 100\% = \underline{66.7\%}$

SELF-TEST

1. Circle those changes that are chemical:
   (a) dissolving
   (b) burning
   (c) evaporating
   (d) digesting
   (e) rusting

2. Xylocaine is a local anesthetic used in dentistry and medicine. Its formula is $C_{14}H_{22}N_2O$.
   (a) Determine the percentage composition of this compound.
   (b) How many milligrams of carbon are in 1.5 mg of xylocaine?
   (c) If 1.5 mg of xylocaine is completely oxidized (reacted with oxygen) what mass of CO_2 will be produced?

3. Lactic acid is a carbon-, hydrogen-, and oxygen-containing organic compound that accumulates in muscle tissue during vigorous exercise and induces a feeling of tiredness. Its molar mass is 90.08 g/mol. A 50.00 g sample of lactic acid contains 20.00 g of carbon and 3.355 g of hydrogen. Determine its
   (a) percentage composition
   (b) empirical formula
   (c) molecular formula

4. Balance the following equations:
   (a) $NH_4NO_3(s) \rightarrow N_2(g) + O_2(g) + H_2O(g)$
   (b) $CCl_4(g) + HF(g) \rightarrow CF_2Cl_2(g) + HCl(g)$
   (c) $FeS_2(s) + O_2(g) \rightarrow SO_2(g) + Fe(s)$

(d) $Ca(OH)_2(aq) + Na_2SO_3(aq) \rightarrow CaSO_3(s) + NaOH(aq)$
(e) $C_4H_9SH(g) + O_2(g) \rightarrow CO_2(g) + H_2O(g) + SO_2(g)$

5. Label each equation in Question 4 as oxidation, reduction, formation, decomposition, or displacement.

6. Using equation 4(a), complete the following: When 40.0 g NH_4NO_3 decomposes ________ g N_2, _______ g O_2, and _______ g H_2O are formed.

7. Equation 4(b) describes the production of CF_2Cl_2, a refrigerant and aerosol propellant known as Freon-12. How many kilograms of HF are needed to produce 2.00×10^3 kg of Freon-12? (Production of Freons has decreased in the U.S. because of suspected damage to the ozone layer.)

8. FeS_2 (pyrite or fool's gold) is present in a 2.011-g sample. Oxidation of this sample produces 0.542 g SO_2. What percentage of the sample is FeS_2? See Equation (c) of Question 4.

9. Equation 4(d) shows one of the steps in a recycling process for SO_2, an air pollutant emitted from power plants that burn fossil fuels. How many kilograms of NaOH will be generated when 200 kg of $Ca(OH)_2$ are mixed with 200 kg of Na_2SO_3? Which reactant is in excess? How much of it remains after the reaction?

ANSWERS TO SELF-TEST QUESTIONS

1. b, d, e

2. (a) 71.75% C; 9.46% H; 11.96% N; 6.83% O (b) 1.1 mg (c) 3.9 mg

3. (a) 40.00% C; 6.710% H; 53.29% O (b) CH_2O (c) $C_3H_6O_3$

4. (a) $2NH_4NO_3(s) \rightarrow 2N_2(g) + O_2(g) + 4H_2O(g)$
   (b) $CCl_4(g) + 2HF(g) \rightarrow CF_2Cl_2(g) + 2HCl(g)$
   (c) $FeS_2(s) + 2O_2(g) \rightarrow 2SO_2(g) + Fe(s)$
   (d) $Ca(OH)_2(aq) + Na_2SO_3(aq) \rightarrow CaSO_3(s) + 2NaOH(aq)$
   (e) $2C_4H_9SH(g) + 15O_2(g) \rightarrow 8CO_2(g) + 10H_2O(g) + 2SO_2(g)$

5. (a) decomposition (b) displacement (c) oxidation (d) displacement (e) oxidation

6. 14.0 g N_2; 8.00 g O_2; 18.0 g H_2O

7. 662 kg HF

8. 25.2% FeS_2

9. 127 kg NaOH; $Ca(OH)_2$ in excess; 82 kg $Ca(OH)_2$ remaining

CHAPTER 4
SOLUTIONS AND SOLUTION STOICHIOMETRY

SELF-ASSESSMENT

Solutions and Mixtures (Section 4.1)

1. Match column A to column B. Hint: solubility of NaCl is 35.7 g NaCl in 100 g water at 25°C.

A	B
clam chowder	liquid solution
gasoline	saturated solution
air	heterogeneous mixture
35.7 g NaCl dissolved in 100 g water (25°C)	supersaturated solution
35.8 g NaCl dissolved in 100 g water (25°C)	gaseous solution

2. Write equations to represent the ionization, if any, in aqueous solution of the following compounds:
Li_3PO_4 (ionic)
HBr (molecular and a strong electrolyte)
HCOOH (molecular and a weak electrolyte)
CH_3OCH_3 (molecular and a nonelectrolyte)

Metathesis Reactions (Section 4.2)

3. Write balanced net ionic and molecular equations for the following reactions. Use the solubility rules in Table 4.1 of this study guide to predict the products.
(a) $Ca(OH)_2(aq) + Li_2CO_3(aq) \rightarrow$
(b) $AgNO_3(aq) + Na_2SO_4(aq) \rightarrow$
(c) $K_2S(aq) + Pb(NO_3)_2(aq) \rightarrow$

4. A 95.0-g sample of sterling silver was sent to a lab to determine percentage by mass of silver. The sample was dissolved in acid and then precipitated as AgCl. If the precipitate weighed 114.0 g, what is the percent by mass of Ag in the original sample? (Atomic weights: Ag = 107.87; Cl = 35.45)

5. Complete the following equations
(a) $HNO_3(aq) + Ba(OH)_2(aq) \rightarrow$
(b) $H_2S(aq) + KOH(aq) \rightarrow$

6. Write molecular equations for reactions in aqueous solution that produce
   (a) $NH_3(g)$
   (b) $HCN(g)$
   (c) $H_2S(g)$

Ionic Displacement Reactions (Section 4.3)

7. Use the activity series to explain the following:
   (a) A silver "Christmas tree" can be made by immersing tree-shaped copper wire in a $AgNO_3$ solution.
   (b) The solder used by plumbers consists of lead and tin. Recently it has been shown that drinking water that runs through soldered pipes may acquire high Pb^{2+} concentrations especially if the water is somewhat acidic (excess $H^+(aq)$).

Solution Concentration: Percent by Mass and Molarity (Section 4.4)

8. A one-liter sample of seawater contains 3.6% salt by mass and has a density of 1.025 g/mL. If the water evaporates, how many grams of salt will remain?

9. Polychlorinated biphenyls (PCBs) have recently been banned because of their toxicity to human and animal life. However PCBs are widely dispersed in the environment; for example, ducks in Lake Ontario and the Hudson River regions showed PCB contaminations of 7.5 ppm.
   (a) Express this concentration in percent by mass.
   (b) How many micrograms (μg) of PCBs would be present in a 2.0-kg duck? (It was suggested that no more than two meals of duck should be consumed per month.) (1 μg = 10^{-6} g)

10. DDT, a pesticide banned in 1972, is only one of many toxic pollutants present in fresh water. To protect marine life, its level should not exceed 1.0×10^{-3} μg/L.
   (a) Calculate the molarity of DDT in a lake where its concentration is 1.0×10^{-3} μg/L. The molar mass of DDT is 342.0 g/mol.
   (b) What is the percent by mass of DDT in the lake water? Density of lake water is 1.0 g/mL.

11. To correct a low blood potassium level, a patient is given an intravenous KCl solution that contains 0.10% KCl by mass and has a density of 1.0 g/mL. Find the molarity of the KCl in this solution. (atomic weights: K = 39.10; Cl = 35.45)

12. What is the molarity of OH^- in a solution prepared by mixing 100 mL of 0.10 *M* $Ca(OH)_2$ with 50 mL of 0.20 *M* NaOH? Assume volumes are additive.

Preparation of Solutions By Dilution (Section 4.5)

13. Describe how you would prepare 1.00 L of acid that is 38.0% sulfuric acid by mass and has a density of 1.29 g/mL from 15.0 *M* sulfuric acid. The molar mass of H_2SO_4 is 98.08 g/mol.

Solution Stoichiometry (Section 4.6)

14. Acetaldehyde (CH_3CHO) is an example of an organic pollutant that depletes dissolved oxygen and thus destroys aquatic life. The reaction is

$$2CH_3CHO(aq) + 3O_2(aq) \rightarrow 2CO_2(aq) + 4H_2O(l)$$

How many kilograms of O_2 (aq) are consumed when 1.5×10^3 L of a 0.010 *M* solution of acetaldehyde are released into a lake?

15. Exactly 25.0 mL of a 0.100 *M* solution of calcium iodide precipitates all of the silver in a 2.01-g unknown sample. Calculate the percent by mass of silver in the sample. (atomic weight: Ag = 107.9)

ANSWERS TO SELF-ASSESSMENT PROBLEMS

If you missed an answer, be sure to study the relevant section in the textbook and study guide.

1. Clam chowder is a heterogeneous mixture. Gasoline is a liquid solution. Air is a gaseous solution. 35.7 g of NaCl in 100 g of water is a saturated solution while 35.8 g NaCl in 100 g water is a supersaturated solution.

2. $Li_3PO_4(s) \rightarrow 3Li^+(aq) + PO_4^{3-}(aq)$ $\qquad$ $HCOOH(aq) \rightleftharpoons H^+(aq) + HCOO^-(aq)$
   $HBr(aq) \rightarrow H^+(aq) + Br^-(aq)$ $\qquad$ $CH_3OCH_3(aq)$ no ionization

3. The net ionic equations are
   (a) $Ca^{2+}(aq) + CO_3^{2-}(aq) \rightarrow CaCO_3(s)$ $\qquad$ (c) $S^{2-}(aq) + Pb^{2+}(aq) \rightarrow PbS(s)$
   (b) $2Ag^+(aq) + SO_4^{2-}(aq) \rightarrow Ag_2SO_4(s)$

4. $\text{grams of Ag} = 114.0 \text{ g AgCl} \times \frac{107.87 \text{ g Ag}}{143.32 \text{ g AgCl}} = 85.80 \text{ g};\ \% \text{ Ag} = \frac{85.80}{95.0} \times 100 = \underline{90.3\%}$

5. (a) $2HNO_3(aq) + Ba(OH)_2(aq) \rightarrow Ba(NO_3)_2(aq) + 2H_2O(l)$
   (b) $H_2S(aq) + 2KOH(aq) \rightarrow K_2S(aq) + 2H_2O(l)$

6. Possible answers are:
   (a) $NH_4Cl(aq) + KOH(aq) \rightarrow KCl(aq) + H_2O(l) + NH_3(g)$
   (b) $KCN(aq) + HCl(aq) \rightarrow KCl(aq) + HCN(g)$
   (c) $FeS(s) + 2HCl(aq) \rightarrow FeCl_2(aq) + H_2S(g)$

7. (a) Copper is above silver in the activity series. (b) Lead is above hydrogen in the activity series.

8. 37 g

9. (a) $\% \text{ by mass} = \frac{75 \text{ g}}{1.0 \times 10^6 \text{ g}} \times 100\% = \underline{7.5 \times 10^{-4}\%}$

   (b) $\frac{7.5 \text{ g PCB}}{1.0 \times 10^{-6} \text{ g}} \times \frac{10^3 \text{ g}}{1 \text{ kg}} \times \frac{10^6 \ \mu\text{g}}{1 \text{ g}} \times 2.0 \text{ kg} = \underline{1.5 \times 10^4 \ \mu\text{g of PCB}}$

10. (a) 2.9×10^{-12} *M* $\qquad$ (b) 1.0×10^{-10} %

11. For one liter,
    $\text{moles of KCl} = 1.0 \times 10^3 \text{ g soln} \times \frac{0.10 \text{ g KCl}}{100 \text{ g soln}} \times \frac{1 \text{ mol KCl}}{74.55 \text{ g KCl}} = 0.013 \text{ mol};\ \text{molarity} = \frac{0.013 \text{ mol}}{1 \text{ L}} = \underline{0.013\ M}$

12. $\text{moles of OH}^- = \left(2 \times \frac{0.10 \text{ mol}}{1 \text{ L}}\right) \times 0.100 \text{ L} + \frac{0.20 \text{ mol}}{1 \text{ L}} \times 0.050 \text{ L} = \underline{0.030 \text{ mol}};\ \text{molarity} = \frac{0.030}{0.150 \text{ L}} = \underline{0.20\ M}$

13. moles of sulfuric acid in 1.00 L of battery acid =

$$1.00\text{ L} \times \frac{10^3\text{ mL}}{1\text{ L}} \times \frac{1.29\text{ g soln}}{1\text{ mL}} \times \frac{38.0\text{ g H}_2\text{SO}_4}{100\text{ g soln}} \times \frac{1\text{ mol H}_2\text{SO}_4}{98.08\text{ g H}_2\text{SO}_4} = 5.00\text{ mol}$$

$$\text{liters of 15.0 } M \text{ sulfuric acid soln} = 5.00\text{ mol} \times \frac{1\text{ L}}{15.0\text{ mol}} = \underline{0.333\text{ L}}$$

Slowly add 333 mL of the 15.0 M H_2SO_4 solution to about 600 mL of water in a 1-L volumetric flask, then add water to bring the volume up to the 1-L mark. (Remember always add acid to water, not the reverse.)

14. $$\text{grams of O}_2 = 1.5 \times 10^3\text{ L CH}_3\text{CHO} \times \frac{0.010\text{ mol CH}_3\text{CHO}}{1\text{ L CH}_3\text{CHO}} \times \frac{3\text{ mol O}_2}{2\text{ mol CH}_3\text{CHO}} \times \frac{32.00\text{ g O}_2}{1\text{ mol O}_2} = \underline{0.72\text{ kg}}$$

15. $$0.0250\text{ L CaI}_2 \times \frac{0.100\text{ mol CaI}_2}{1\text{ L CaI}_2} \times \frac{2\text{ mol Ag}}{1\text{ mol CaI}_2} \times \frac{107.9\text{ g Ag}}{1\text{ mol Ag}} = 0.540\text{ g Ag};\quad \%\text{ Ag} = 0.540\text{ g}/2.01\text{ g} = \underline{26.9\%}$$

REVIEW

Solutions (Section 4.1)

Learning Objectives

1. Distinguish between a solution and a heterogeneous mixture.
2. Give examples of solid, liquid, and gaseous solutions.
3. Describe the dynamic equilibrium that exists in a saturated solution.
4. Distinguish between unsaturated, saturated, and supersaturated solutions.
5. Write equations for the dissociation of ionic compounds.
6. Write equations for the ionization of strong and weak molecular electrolytes.

Review

A heterogeneous mixture is a combination of two or more components that are not uniformly mixed. A solution is a homogeneous mixture in which the components are uniformly dispersed. The component that determines whether the solution is solid, liquid, or gas is the solvent; the other component(s) are the solute(s). If all components are in the same state before mixing, then the solvent is the component present in largest quantity.

Example 4.1: Identify the solvent in the following solutions:
(a) 1.5 mg CO(g) uniformly mixed with 10 kg N_2 (g)
(b) 20 g of acetone(l) mixed with 150 g of water(l)
(c) sterling silver (92.5% silver and 7.5% copper)

Answer: (a) N_2 (b) water (c) silver Example (b) describes an aqueous solution because water is the solvent.

Solubility

The solubility of a substance is the quantity required to saturate a given amount of solvent. A saturated solution is one in which dissolved solute is in dynamic equilibrium with undissolved solute; solute particles enter the solution and precipitate from the solution at the same rate, so there is no change in concentration. A supersaturated solution holds more solute than a saturated solution. Supersaturated solutions are unstable, but may persist in the absence of "seed" crystals.

Ionic Solutions

An ionic solution contains dissolved ions and can conduct an electric current. The properties of the solution depend only on the kinds of ions present and not on their source. A solution prepared by dissolving sodium nitrate and potassium chloride in water can be made identical to a solution made by dissolving potassium nitrate and sodium chloride in water.

An electrolyte is a substance that forms ions when it is dissolved in water. A strong electrolyte is mostly or completely ionized in aqueous solution. When an ionic compound dissolves, the ions simply dissociate:

$KNO_3(s) \rightarrow K^+(aq) + NO_3^-(aq)$

> **Practice**: Write dissociation equations for the ionic compounds $NaNO_3$, KCl, and K_2SO_4.

Some molecular compounds become completely ionic in aqueous solution, and are strong electrolytes. Examples are HCl, HBr, and HNO_3.

$HBr(aq) \rightarrow H^+(aq) + Br^-(aq)$

A weak electrolyte is a molecular substance that provides only a few ions in aqueous solution. Organic acids such as formic acid are weak electrolytes.

$HCOOH(aq) \rightleftharpoons H^+(aq) + HCOO^-(aq)$
formic acid

The double arrow indicates that the reaction takes place in both directions. A dynamic equilibrium exists in solution between the ions and the undissociated molecules. In aqueous solution most of the formic acid is un-ionized (HCOOH).

> **Practice**: Write equations for the ionization in solution of
> (a) acetic acid (CH_3COOH) a weak electrolyte (b) HI, a strong electrolyte

Compounds that do not form ions in water are called nonelectrolytes. Methanol (CH_3OH) is an example. The dissolving of methanol in water could be represented as

$CH_3OH(l) \rightarrow CH_3OH(aq)$

Metathesis Reactions (Section 4.2)

Learning Objectives

1. Write molecular and net ionic equations for precipitation reactions.
2. Use solubility rules to predict whether a precipitation reaction will occur.
3. Perform the calculations required for a gravimetric analysis.

4. Write equations for complete and stepwise acid-base neutralizations.
5. Find a netralization reaction to produce a given salt.
6. Write equations for metathesis reactions that produce NH_3, HCN, H_2S, CO_2, and SO_2 gases.

Review

Metathesis is a reaction in which atoms or ions exchange partners. When ions are involved it occurs when one of the products is a precipitate or a molecular substance. A metathesis reaction can be described by a net ionic equation or by a molecular equation. A net ionic equation omits the ions that do not participate in the reaction (spectator ions).

Example 4.2: Write the net ionic equation for the reaction in which $AgNO_3$(aq) and KCl(aq) form AgCl, a precipitate, and KNO_3(aq).

Answer: The reactants are completely ionized. Of the products, only KNO_3 is ionized. The reaction is

$Ag^+(aq) + NO_3^-(aq) + K^+(aq) + Cl^-(aq) \rightarrow AgCl(s) + NO_3^-(aq) + K^+(aq)$

Eliminating the ions that remain unchanged gives the net ionic equation:

$Ag^+(aq) + Cl^-(aq) \rightarrow AgCl(s)$

Molecular equations show the complete neutral formulas for all of the substances.

Example 4.3: Write the molecular equation for the above reaction.

Answer: $AgNO_3(aq) + KCl(aq) \rightarrow AgCl(s) + KNO_3(aq)$

Practice: Write the balanced molecular and net ionic equation for the reaction between $Pb(NO_3)_2$(aq) and Na_2S(aq) to form PbS(s) and $NaNO_3$(aq).

How can you determine when a precipitate will form? If you know some general rules of solubility you can predict whether a solid will form in a reaction (see Table 4.1 below).

Table 4.1 General Rules of Solubilities of Ionic Compounds

Ionic compound containing:	Solubility	Principal exceptions
NO_3^-	soluble	____________
Li^+, Na^+, K^+, NH_4^+	soluble	____________
Cl^-, Br^-, I^-	soluble	Sparingly or slightly soluble; halides of Ag^+, Ag^{2+}, Pb^{2+}, Hg_2^+
SO_4^{2-}	soluble	Sparingly or slightly soluble; Ca^{2+}, Sr^{2+}, Ba^{2+}, Ag^+, Ag^{2+}, and Pb^{2+}
OH^-	sparingly or slightly soluble	Soluble; hydroxides of Ba^{2+}, Li^+, Na^+, K^+ and NH_4^+
CO_3^{2-}, PO_4^{3-}, SO_3^{2-}, S^{2-}	sparingly soluble	Soluble; compounds containing Li^+, Na^+, K^+, and NH_4^+

Example 4.4: Use the preceding solubility table to predict whether a precipitate will form when solutions of $Pb(NO_3)_2$ and KI are mixed.

Answer: The possible products are PbI_2 and KNO_3. According to the preceding table on solubilities, PbI_2 is not soluble and therefore it would precipitate out.

Practice: Which of the following pairs of compounds will form a precipitate when their solutions are mixed?

(a) $Na_2S + AgNO_3$
(b) $LiCl + K_2CO_3$
(c) $(NH_4)_3PO_4 + Ca(NO_3)_2$

Answer: (a) and (c)

Gravimetric Analysis

A gravimetric analysis consists of determining the amount of a substance by weighing a reaction product.

Example 4.5: The concentration of chloride ion in a water sample is found by reacting the sample with $AgNO_3$ solution and weighing the product.

$Ag^+(aq) + Cl^-(aq) \rightarrow AgCl(s)$

Excess $AgNO_3$ was added to a one liter sample of ocean water, and 72.0 g of AgCl precipitated. How many grams of Cl^- are in one milliliter of this sample?

Given: $Ag^+(aq) + Cl^-(aq) \rightarrow AgCl(s)$
excess ? 72.0 g

Unknown: grams of Cl^- in one mL of sample.

Plan: We will first find the amount of Cl^- in the one liter using the same method that we used in other stoichiometry problems. We can the divide the amount of chloride ion in one liter by 1000 (or multiply by 1 L/1000 mL) to find Cl^- in one milliliter of sample.
(atomic weights: Ag = 107.87; Cl = 35.45)

$$72.0\text{ g AgCl} \rightarrow \text{mol AgCl} \rightarrow \text{mol Cl}^- \rightarrow \text{g Cl}^- \rightarrow \frac{\text{g Cl}^-}{1\text{ mL}}$$

Calculations:

$$\frac{72.0\text{ g AgCl}}{\text{liter sample}} \times \frac{1\text{ mol AgCl}}{143.32\text{ g AgCl}} \times \frac{1\text{ mol Cl}^-}{1\text{ mol AgCl}} \times \frac{35.45\text{ g Cl}^-}{1\text{ mol Cl}^-} \times \frac{1\text{ L}}{1000\text{ mL}} = \underline{0.0178\text{ g / mL}}$$

Acid-Base Neutralization and Reactions Evolving Gases

Acids produce $H^+(aq)$ ions in aqueous solution; bases produce $OH^-(aq)$ ions. Strong acids and strong bases are completely or almost completely ionized in aqueous solution.

$HCl(aq) \rightarrow H^+(aq) + Cl^-(aq)$
strong acid

$NaOH(s) \rightarrow Na^+(aq) + OH^-(aq)$
strong base

Weak acids and weak bases are only partially ionized in aqueous solution; they remain mostly in molecular form.

$HCOOH(aq) \rightleftharpoons H^+(aq) + HCOO^-(aq)$
weak acid (formic acid)

$NH_3(aq) + H_2O(l) \rightleftharpoons NH_4^+(aq) + OH^-(aq)$
weak base

Salts are ionic compounds whose ions are neither H^+ or OH^-.

$NaNO_3(s) \rightarrow Na^+(aq) + NO_3^-(aq)$

Practice: Use the following equations to identify the underlined substance as a strong acid, strong base, weak acid, weak base, or a salt:

(1) $\underline{Ba(OH)_2(s)} \rightarrow Ba^{+2}(aq) + 2OH^-(aq)$

(2) $\underline{CH_3NH_2(aq)} + H_2O(l) \rightleftharpoons CH_3NH_3^+(aq) + OH^-(aq)$

(3) $\underline{KNO_2(s)} \rightarrow K^+(aq) + NO_2^-(aq)$

(4) $\underline{HBr(aq)} \rightarrow H^-(aq) + Br^-(aq)$

Answers: (1) strong base (2) weak base (3) salt (4) strong acid

Neutralization is the reaction between an acid and a base. Neutralization in aqueous solution has the general form:

acid(aq) + base(aq) $\rightarrow$ salt(aq) + water(l)

An example is

$HNO_3(aq) + KOH(aq) \rightarrow KNO_3(aq) + H_2O(l)$

(Note the above neutralization reaction is written as a molecular equation.)

Acids such as H_3PO_4, which contain more than one acidic hydrogen atom, are neutralized in distinct steps:

(1) $H_3PO_4(aq) + NaOH(aq) \rightarrow NaH_2PO_4(aq) + H_2O(l)$

(2) $NaH_2PO_4(aq) + NaOH(aq) \rightarrow NaHPO_4(aq) + H_2O(l)$

(3) $Na_2HPO_4(aq) + NaOH(aq) \rightarrow Na_3PO_4(aq) + H_2O$

Complete neutralization is the sum of the three steps:

$H_3PO_4(aq) + 3NaOH(aq) \rightarrow Na_3PO_4(aq) + 3H_2O(l)$

When a stronger acid is added to the salt of a weaker acid, the weaker acid forms by metathesis. Weak acids with small molecules such as HCN and H_2S may be evolved as gases. The weak acids H_2CO_3 and H_2SO_3 decompose to give $CO_2(g)$ and $SO_2(g)$, respectively.

Ammonia gas (NH_3) is a weak base that forms when salts containing ammonium ion are treated with a strong base.

Practice: Complete the following reactions.

Reactants		Products
(a) $NaOH(aq) + HCN(aq)$	$\rightarrow$	$? + H_2O(l)$
(b) $NH_3(aq) + ?$	$\rightarrow$	$NH_4^+(aq) + Cl^-(aq)$
(c) $H_2CO_3(aq)$	$\rightarrow$	$H_2O(l) + ?$
(d) $H_2SO_3(aq)$	$\rightarrow$	$H_2O(l) + ?$
(e) $KCN(aq) + HCl(aq)$	$\rightarrow$	$KCl(aq) + ?$
(f) $H_2SO_4(aq) + NaOH(aq)$	$\rightarrow$	$NaHSO_4(aq) + ?$

Answers: (a) NaCN(aq) (b) HCl(aq) (c) CO_2(g) (d) SO_2(g) (e) HCN(g) (f) H_2O(l)

Ionic Displacement Reactions (Section 4.3)

Learning Objectives

1. Write molecular and net ionic equations for displacement reactions.
2. Use the activity series to predict whether or not a displacement reaction will occur.

Review

A more active element will displace a less active element from a compound. The activity series (Table 4.6 in your textbook) lists metals and hydrogen in order of their activity.

Example 4.6: Given the relative abilities of some metals to displace hydrogen from acids, predict the products in the following molecular equations:

(a) $Zn(s) + Pb(NO_3)_2(aq) \rightarrow$? + ?

(b) $Fe(s) + Cu(NO_3)_2(aq) \rightarrow$? + ?

From Table 4.6 in your textbook: Ca > Mg > Zn > Pb > Cr > Fe > Cu > Ag

Answer: (a) $Zn(NO_3)_2(aq) + Pb(s)$ (b) $Fe(NO_3)_2(aq) + Cu(s)$

Practice: A copper penny is dropped into $AgNO_3$ solution. What color will the penny turn?
Answer: silver

Solution Concentration (Section 4.4)

Learning Objectives

1. Calculate the percent by mass of each component in a solution, given the appropriate data.
2. Use parts per thousand, parts per million, parts per billion, and parts per trillion in concentration calculations.
3. Describe how to make solutions of a given molarity.
4. Calculate molarity, solution volume, or number of moles of solute, given two of the three quantities.
5. Calculate molarity, solution volume, or mass of solute, given two of the three quantities.
6. Calculate the molarity of a solution given its density and percent composition by mass.
7. Calculate the molarity of an ion in solution from the molarity of the compound supplying the ion.

Review

The concentration of a solution is the amount of solute in a given quantity of solvent or solution. Although there are several ways of expressing the concentration of a solution, only two are discussed in this chapter: percent by mass and molarity. Concentration units are used widely in all of the sciences so it is especially important that you study this section very carefully.

Percent by Mass

$$\text{percent by mass} = \frac{\text{mass of solute}}{\text{mass of solution}} \times 100\%$$

Example 4.7: 10.0 g of NaCl is added to 100.0 g of water. Find the percent by mass of salt.

Answer: $$\text{percent by mass of solute} = \frac{\text{mass of NaCl}}{\text{mass of solution}} \times 100\%$$

$$\begin{aligned}\text{mass of solution} &= \text{grams of solute} + \text{grams of solvent}\\ &= 10.0\text{ g} + 100.0\text{ g}\\ &= 110.0\text{ g solution}\end{aligned}$$

$$\text{percent by mass} = \frac{10.0\text{ g}}{110.0\text{ g}} \times 100\% = \underline{9.09\%}$$

Practice: Calculate the percent by mass of KCl in a solution prepared by adding 5.0 g NaCl and 5.0 g KCl to 100.0 g of water. Answer: 4.6%

Note: Percent is equivalent to parts per hundred (pph); 1% = 1 part solute per 100 parts solution (1 g per 100 g). Similar concentration units are

ppm = parts per million (1 g per 10^6 g)
ppb = parts per billion (1 g per 10^9 g)
ppt = parts per trillion (1 g per 10^{12} g)

Caution: Sometimes ppt is used for parts per thousand (1 g per 10^3 g).

The concentrations of many air and water pollutants are expressed as ppm or ppb because of their low concentrations.

Example 4.8: A 22-kg tuna fish contains 0.30 ppm mercury. How many grams of mercury are present?

Answer: $$22\text{ kg fish} \times \frac{1000\text{ g fish}}{1\text{ kg fish}} \times \frac{0.30\text{ g mercury}}{1.0 \times 10^6\text{ g fish}} = \underline{6.6 \times 10^{-3}\text{ g mercury}}$$

Molarity

Molarity (*M*) is the number of moles of solute in one liter of solution:

$$\text{molarity} = \frac{\text{number of moles of solute}}{\text{volume of solution in liters}}$$

Example 4.9: Find the molarity of glucose in a 3.0-L solution containing 0.15 mol of glucose.

Answer: $$\text{molarity} = \frac{\text{moles of glucose}}{\text{liters of solution}} = \frac{0.15\text{ mol glucose}}{3.0\text{ L}} = \underline{0.050\ M}$$

Practice: If 0.15 more moles of glucose are added to the above solution with no change in volume, what is the new molarity? Answer: 0.10 *M*

Example 4.10: A 250.0-mL solution contains 18.02 g of glucose. Calculate the molarity of the glucose solution. The molar mass of glucose is 180.2 g/mol.

Answer: $\text{molarity} = \dfrac{\text{mol glucose}}{\text{liters solution}}$

$$\text{mol glucose} = 18.02 \text{ g glucose} \times \frac{1 \text{ mol glucose}}{180.2 \text{ g glucose}} = 0.1000 \text{ mol}$$

$$\text{liters of solution} = 250.0 \text{ mL} \times \frac{1 \text{ L}}{1000 \text{ mL}} = 0.1000 \text{ mL}$$

$$\text{molarity} = \frac{0.1000 \text{ mol glucose}}{0.2500 \text{ L}} = \underline{0.4000\ M}$$

Caution: Be sure to express the quantity of solute in moles and volume of solution in liters before calculating the molarity.

Finding Molarity from Percent by Mass and Density Data

Example 4.11: Find the molarity of a solution that contains 98 % H_2SO_4 and has a density of 1.84 g/mL.

(It is often helpful to define given information and the unknown before you start to solve.)

Answer:

Given: $d = 1.84 \text{ g/mL}; \quad 98\%\ H_2SO_4 \text{ by mass} = \dfrac{98 \text{ g } H_2SO_4}{100 \text{ g solution}}$

Unknown: $\text{molarity} = \dfrac{\text{moles of } H_2SO_4}{\text{liters of acid solution}}$

Plan: Find moles of H_2SO_4 in 1 L of solution as follows:

(1) Convert 1 L into grams using density.
(2) Use percent by mass to find grams of H_2SO_4.
(3) Divide grams of H_2SO_4 by molar mass of H_2SO_4 to find moles in 1 L (or multiply by 1 mol/molar mass). This is the molarity.

In summary: $1 \text{ L} \xrightarrow{(1)} \dfrac{\text{g of solution}}{1 \text{ L solution}} \xrightarrow{(2)} \dfrac{\text{g } H_2SO_4}{1 \text{ L solution}} \xrightarrow{(3)} \dfrac{\text{mol } H_2SO_4}{1 \text{ L solution}}$

Calculations: $1 \text{ L} \times \dfrac{1000 \text{ mL}}{1 \text{ L}} \times \dfrac{1.84 \text{ g solution}}{1 \text{ mL}} \times \dfrac{98 \text{ g } H_2SO_4}{100 \text{ g solution}} \times \dfrac{1 \text{ mol } H_2SO_4}{98.08 \text{ g } H_2SO_4} = \dfrac{18.4 \text{ mol}}{1 \text{ L}} = \underline{18.4\ M}$

Finding Moles, Mass, or Volume from Molarity

Note that the relationship $\text{molarity} = \dfrac{\text{moles}}{\text{volume (L)}}$ involves three quantities, so that given two quantities you can calculate the third: moles = molarity × volume (L); volume (L) = moles/molarity

<u>*Example 4.12*</u>: Given a 0.1000 *M* solution, calculate: (a) moles of solute in 10.00 L of solution (b) milliliters of solution that contain 0.4000 mol of solute.

Answer: (a) 10.00 L → ? moles
(given) (unknown)

Plan: volume (L) × molarity (mol/L) = moles

Calculations: $10.00\ \text{L} \times \dfrac{0.1000\ \text{mol solute}}{1\ \text{L}} = \underline{1.000\ \text{mol solute}}$

(b) 0.4000 mol → liters → milliliters
(given) (unknown)

Plan: $\dfrac{\text{moles}}{\text{molarity}} = \text{volume (L)}$

Calculations: $\dfrac{0.4000\ \text{mol}}{0.1000\ M} = 4.000\ \text{L} = \underline{4000\ \text{mL}}$

Note: Molarity is really a conversion factor. The factor 0.100 mol/1 L changes volume to moles in Example 4.12(a); the factor 1 L/0.100 mol changes moles to volume in Example 4.12(b). The calculations for example (b) using molarity as a conversion factor are 0.4000 mol × 1 L/0.1000 mol × 1000 mL/1 L = <u>4000 mL</u>.

Practice Drill on Molarity: The following table gives information about four different $AgNO_3$ solutions. Fill in the blanks. (The molar mass of $AgNO_3$ is 169.9 g/mol.)

<u>Solution number</u>	<u>*M* (mol/L)</u>	<u>Volume</u>	<u>Moles of $AgNO_3$</u>	<u>Grams of $AgNO_3$</u>
(1)	0.0125	? L	0.0250	?
(2)	?	10.0 mL	?	34.0g
(3)	0.150	? mL	?	3.40g
(4)	0.200	150mL	?	?

<u>Answers</u>: (1) 2.00 L: 4.25 g (2) 20.0 *M*: 0.200 mol (3) 133 mL: 0.0200 mol (4) 0.0300 mol: 5.10 g

Preparing Solutions

<u>*Example 4.13*</u>: Water chemists often use a 0.01000 *M* $AgNO_3$ solution in testing for chloride ion. How would you prepare 500.0 mL of 0.01000 *M* $AgNO_3$? The molar mass of $AgNO_3$ is 169.9 g.

Given: molarity = 0.01000 *M*; volume = 500.0 mL

Unknown: We will make the solution by weighing out the correct mass of $AgNO_3$, then add enough water to make a total volume of 500.0 mL. The <u>unknown</u> is therefore the correct <u>mass (grams) of $AgNO_3$</u>.

Plan: Find the moles of $AgNO_3$ in 500.0 mL (0.5000 L) of solution using the given molarity (0.01000 *M*) and then convert to grams of $AgNO_3$.

molarity (mol/L) × volume (L) → mol $AgNO_3$ → g $AgNO_3$

Calculations: $0.5000\ \text{L} \times \dfrac{0.01000\ \text{mol AgNO}_3}{1\ \text{L}} \times \dfrac{169.9\ \text{g AgNO}_3}{1\ \text{mol AgNO}_3} = 0.8495\ \text{g AgNO}_3$

The 0.8495 g of $AgNO_3$ is weighed out on an analytical balance and carefully transferred to a 500-mL volumetric flask. Enough water is added to bring the volume up to the 500-mL mark on the neck of the flask.

Practice: How would you prepare 250.0 mL of the 0.01000 *M* $AgNO_3$ solution?

Answer: Weigh out 0.4248 g of $AgNO_3$, transfer to a 250.0-mL volumetric flask. Add enough water to reach the 250-mL mark.

Molarities of Ions

Keep in mind that the molarity of an ion in solution may not be the same as the molarity of its compound. Calcium hydroxide, for example, produces 2 mol of OH^- ion per mole of $Ca(OH)_2$:

$$Ca(OH)_2(s) \rightarrow Ca^{2+}(aq) + 2OH^-(aq)$$

The molarity of OH^- is thus twice the molarity of $Ca(OH)_2$ while the molarity of Ca^{2+} is the same as the molarity of $Ca(OH)_2$.

Example 4.14: How many liters of 0.10 *M* Na_3PO_4 contain 0.27 mol of Na^+?

Given: 0.10 *M* Na_3PO_4; 0.27 mol Na^+; molar ratio of $\frac{Na^+}{Na_3PO_4} = \frac{3}{1}$

Unknown: liters of Na_3PO_4 solution containing 0.27 mol Na^+

Plan: (1) Find the number of moles of Na_3PO_4 that contain 0.27 mol Na^+.
(2) Use the molarity to find the volume of solution that contains this number of moles of Na_3PO_4.

$$0.27 \text{ mol Na}^+ \xrightarrow{(1)} \text{mol Na}_3\text{PO}_4 \xrightarrow{(2)} \text{liters of Na}_3\text{PO}_4$$

Calculations: $$0.27 \text{ mol Na}^+ \times \frac{1 \text{ mol Na}_3\text{PO}_4}{3 \text{ mol Na}^+} \times \frac{1 \text{ L}}{0.10 \text{ mol Na}_3\text{PO}_4} = \underline{0.90 \text{ L}}$$

PROBLEM-SOLVING TIP

What to do if you cannot solve a problem:

Very often we can determine the given and unknown information but we cannot come up with a correct plan. It is sometimes helpful to work backwards, that is, start with the unknown and look for given quantities that relate to it. In the preceding problem, for example, your thought process could have gone like this:

Plan: The unknown is volume of Na_3PO_4 solution. The only given quantity that involves the Na_3PO_4 solution is the molarity.

If I know the moles of Na_3PO_4 in the solution, then I can find the volume from the given molarity.

How can I find the moles of Na_3PO_4? Examining the given again I note that moles of Na^+ is known. How do the moles of Na^+ relate to moles of Na_3PO_4? I can convert moles of Na^+ to moles of Na_3PO_4 by the molar ratio 3:1 from the formula.

Hence my plan is mol Na^+ $\rightarrow$ mol Na_3PO_4 $\rightarrow$ volume Na_3PO_4

Practice Drill on Molarity of Ions: Fill in the following chart:

	Molarity of solution	Molarity of ion	Volume of solution	Moles of Ion
(1)	0.10 M Na_2CO_3	? M Na^+	100 mL	? mol Na^+
(2)	? M $AlBr_3$	? M Al^{+3}	100 mL	0.20 mol Br^-
(3)	0.10 M Na_3PO_4	? M Na^+	? L	0.270 mol Na^+
(4)	? M $Ca(OH)_2$	0.800 M OH^-	150 mL	? mol Ca^{2+}

Answers: (1) 0.20 M Na^+; 0.020 mol Na^+ (2) 0.067 M $AlBr_3$; 0.067 M Al^{3+}
(3) 0.30 M Na^+; 0.90 L (4) 0.400 M; 0.0600 mol Ca^{2+}

When solutions that contain different molarities of the same ion are mixed, the new molarity will have an intermediate value. It is usually safe to assume that the final volume is the sum of the volumes that were mixed, so the new molarity is calculated as in Example 4.15.

Example 4.15: What is the molarity of NO_3^- in a solution made by adding 40.0 mL of 0.100 M $Mg(NO_3)_2$ to 10.0 mL of 0.800 M KNO_3? Assume volumes are additive.

Given: 40.0 mL of 0.100 M $Mg(NO_3)_2$; ratio of $NO_3^-/Mg(NO_3)_2 = 2/1$
10.0 mL of 0.800 M KNO_3; ratio of $NO_3^-/KNO_3 = 1/1$

Unknown: molarity of NO_3^-

Plan: (1) Find total moles of NO_3^-. (2) Divide by total volume of solution to find molarity.

Calculations: (1) Find the moles of NO_3^- in each solution and add:

moles of NO_3^- in $Mg(NO_3)_2$ solution =

$$0.0400\ \text{L} \times \frac{0.100\ \text{mol Mg(NO}_3)_2}{1\ \text{L}} \times \frac{2\ \text{mol NO}_3^-}{1\ \text{mol Mg(NO}_3)_2} = 8.00 \times 10^{-3}\ \text{mol}$$

$$\text{moles of NO}_3^-\ \text{in KNO}_3\ \text{solution} = 0.0100\ \text{L} \times \frac{0.800\ \text{mol KNO}_3}{1\ \text{L}} \times \frac{1\ \text{mol NO}_3^-}{1\ \text{mol KNO}_3} = 8.00 \times 10^{-3}\ \text{mol}$$

Total moles of $NO_3^- = 8.00 \times 10^{-3}$ mol + 8.00×10^{-3} mol = 16.00×10^{-3} mol

(2) volume = 40.0 mL + 10.0 mL = 50.0 mL = 5.00×10^{-2} L

$$\text{molarity} = \frac{16.00 \times 10^{-3}\ \text{mol NO}_3^-}{5.00 \times 10^{-2}\ \text{L}} = \underline{0.320\ M}$$

Practice: Compute the molarity of Na^+ in a solution formed by dissolving 42.0 g of $NaHCO_3$ in 1.0 L of water and then adding 100 mL of a 0.100 M Na_2CO_3 solution. Assume volumes are additive. Molar mass of $NaHCO_3$ is 84.02 g/mol. (Answer: 0.473 M Na^+)

PROBLEM-SOLVING TIP

Summary of Strategies for Solving Some Specific Concentration Problems

Given	Unknown	Plan
% by mass and density of solution	molarity	Find moles of solute in 1-L solution. 1. Find mass of 1 L; use mass = d × V. 2. Find mass of solute in 1 L; multiply mass of 1 L (step 1) by percentage by mass (expressed as a decimal). 3. Find moles of solute in 1 L; Use molar mass to convert grams (step 2) to moles.
molarity and density	% by mass	1. Find mass of solute in 1 L; multiply (molarity) mol/L by (molar mass) g/mol. 2. Find mass of 1 L; use mass = V × density. 3. $\% = \frac{\text{g of solute in 1 L (step 1)}}{\text{total g of 1 L (step 2)}} \times 100\%$
% by mass, density and volume of solution	mass of solute in given volume of solution	1. Find mass of given volume of solution; use mass = V × density. 2. Find mass of solute in given volume: multiply mass of solution (step 1) by % by mass (expressed as a decimal).
% by mass, density, and mass of solute	volume of solution containing given mass	1. Find mass of solution; multiply given mass of solute by $\frac{\text{100 g solution}}{\text{\% by mass solute}}$ 2. Find volume of solution: multiply mass of solution (step 1) by the inverse of density (mL/g).
% by mass	ppm, ppb	% by mass is really pph (parts per hundred) or grams of solute per 100 g of solution. To get ppm: $\frac{\text{grams of solute}}{\text{100 grams of solution}} \times \frac{10^6 \text{ grams solution}}{\text{1 gram solute}}$ for ppb: $\frac{\text{grams of solute}}{\text{100 grams solution}} \times \frac{10^9 \text{ grams solution}}{\text{1 gram solute}}$
ppm or ppb and density	molarity	given ppm $= \frac{\text{grams of solute}}{10^6 \text{ grams of solution}}$ 1. Find the number of moles in grams of solute: multiply grams by mol/g (from molar mass). 2. Find liters of solution: multiply by 10^6 g of solution by inverse of density (mL/g). Convert milliliters to liters. 3. $M = \frac{\text{mol}}{\text{L}} = \frac{\text{step 1}}{\text{step 2}}$

Your additions:

Preparation of Solutions by Dilution (Section 4.5)

Learning Objectives

1. Calculate the volume of concentrated solution needed to make a given volume of dilute solution.

Review

A dilute solution is often prepared from a more concentrated one. The required volume of concentrated solution is determined by the following dilution formula:

$$V_C \times M_C = V_D \times M_D$$

V_C = volume of concentrated solution
M_C = molarity of concentrated solution

V_D = volume of dilute solution
M_D = molarity of dilute solution

The moles of solute are the same in both the concentrated and the dilute solutions. The dilution formula simply states this fact (moles = V × *M*).

Example 4.16: How many milliliters of 16 *M* HNO_3 solution are required to prepare 0.50 L of 0.10 *M* HNO_3?

Given: concentrated solution: 16 *M* HNO_3 dilute solution: 0.10 *M*; 0.50 L

Unknown: volume of concentrated solution, V_C

Plan: $V_C \times M_C = V_D \times M_D$

Calculations:

$$V_C \times \frac{16 \text{ mol}}{1 \text{ L}} = 0.50 \text{ L} \times \frac{0.10 \text{ mol}}{1 \text{ L}}$$

$$V_C = \frac{0.50 \text{ L} \times 0.10 \text{ mol/L}}{16 \text{ mol/L}} = 0.0031 \text{ L} = \underline{3.1 \text{ mL}}$$

Practice: A student prepared 1.0 L of 0.10 *M* HCl solution by using 8.3 mL of a concentrated solution. What is the molarity of the concentrated solution? **Answer:** 12 *M*

Solution Stoichiometry (Section 4.6)

Learning Objectives

1. Use molarities and solution volumes in stoichiometric calculations.
2. Describe a titration procedure.
3. Perform typical titration calculations.

Review

The number of moles of a given reactant or product in solution is often calculated from the volume and molarity of the solution.

<u>*Example 4.17*</u>: How many milliliters of 0.0100 *M* $NaHCO_3$ are needed to neutralize a 100-mL sample of stomach acid in which the HCl is 0.0500 *M*?

$$NaHCO_3(aq) + HCl(aq) \rightarrow H_2O(l) + CO_2(g) + NaCl(aq)$$

Given: 0.0100 *M* $NaHCO_3$; HCl: 100 mL; 0.0500 *M*

Unknown: milliliters of $NaHCO_3$ solution

Plan: We will follow the basic procedure for solving stoichiometry problems (Problem-Solving Tip: Steps in Solving Stoichiometry Problems in Chapter 3):

(1) Convert volume and molarity of HCl to moles of HCl. ($M \times V = \text{moles}$)
(2) Use molar ratio to find moles of $NaHCO_3$.
(3) Convert moles of $NaHCO_3$ to volume, using molarity as a conversion factor.

$$\text{volume} \times \text{molarity (of HCl)} \xrightarrow{(1)} \text{mol HCl} \xrightarrow{(2)} \text{mol NaHCO}_3 \xrightarrow{(3)} \text{volume NaHCO}_3$$

Calculations:

$$0.100\text{ L HCl} \times \frac{0.0500\text{ mol HCl}}{1\text{ L HCl}} \times \frac{1\text{ mol NaHCO}_3}{1\text{ mol HCl}} \times \frac{1000\text{ mL NaHCO}_3}{0.0100\text{ mol NaHCO}_3} = \underline{500\text{ mL NaHCO}_3}$$

Practice: How many milliliters of the same $NaHCO_3$ solution would be necessary to neutralize 200 mL of 0.0200 *M* HCl solution? <u>Answer</u>: 400 mL $NaHCO_3$

<u>*Example 4.18*</u>: How many grams of glycerol ($C_3H_5(OH)_3$) will react with 100 mL of 0.250 *M* HNO_3 solution to form nitroglycerin ($C_3H_5(NO_3)_3$)? (Nitroglycerin dilates coronary arteries and lowers blood pressure.)

$$C_3H_5(OH)_3(aq) + 3HNO_3(aq) \rightarrow C_3H_5(NO_3)_3(aq) + 3H_2O(l)$$

Given: HNO_3 solution: 100 mL; 0.250 *M*

Unknown: grams of $C_3H_5(OH)_3$

Plan:

$$V \times M\text{ (of HNO}_3) \xrightarrow{(1)} \text{mol HNO}_3 \xrightarrow{(2)} \text{mol C}_3\text{H}_5(\text{OH})_3 \xrightarrow{(3)} \text{g C}_3\text{H}_5(\text{OH})_3$$

Calculations:

$$0.100\text{ L HNO}_3 \times \frac{0.250\text{ mol HNO}_3}{1\text{ L HNO}_3} \times \frac{1\text{ mol C}_3\text{H}_5(\text{OH})_3}{3\text{ mol HNO}_3} \times \frac{92.09\text{ g C}_3\text{H}_5(\text{OH})_3}{1\text{ mol C}_3\text{H}_5(\text{OH})_3} = \underline{0.767\text{ g C}_3\text{H}_5(\text{OH})_3}$$

Practice: How many milliliters of 0.250 *M* HNO_3 solution are required to produce 0.921 g of $C_3H_5(NO_3)_3$? (See equation in above example.) <u>Answer</u>: 48.7 mL

Volumetric Analysis

A <u>volumetric analysis</u> obtains information by measuring the volume of a reacting solution. This is often done by <u>titration</u>, which is the controlled addition of one solution to another. An indicator may be used; this is a substance that changes color at the <u>endpoint</u> of the titration.

A <u>standard solution</u> is one whose concentration is precisely known.

Example 4.19: A student uses a 0.1256 *M* NaOH solution to determine the molarity of an unlabeled H_3PO_4 solution in the laboratory. She found that 36.25 mL of the NaOH solution neutralized 50.00 mL of the acid solution. Determine the molarity of the H_3PO_4 solution.

$$3NaOH(aq) + H_3PO_4(aq) \rightarrow Na_3PO_4(aq) + 3H_2O(l)$$

Given: NaOH solution: 36.25 mL; 0.1256 *M*; H_3PO_4 solution: 50.00 mL

Unknown: molarity of H_3PO_4 solution

Plan:

$$\text{molarity} = \frac{\text{moles of } H_3PO_4}{\text{liters of } H_3PO_4}$$

We are given liters of H_3PO_4 (0.05000 L). We must calculate moles of H_3PO_4 using the procedure for solving stoichiometry problems:

$$V \times M(\text{of NaOH}) \rightarrow \text{mol NaOH} \rightarrow \text{mol } H_3PO_4$$

Calculations:

$$0.03625 \text{ L} \times \frac{0.1256 \text{ mol NaOH}}{1 \text{ L}} \times \frac{1 \text{ mol } H_3PO_4}{3 \text{ mol NaOH}} = 0.001518 \text{ mol } H_3PO_4$$

$$\text{molarity} = \frac{0.001518 \text{ mol } H_3PO_4}{0.05000 \text{ L}} = \underline{0.03036\ M}$$

SELF-TEST

1. Indicate whether the following statements are true or false.
   (a) A red solution is a homogeneous mixture. ✓
   (b) The solvent is always the component in excess. X
   (c) The solvent is the component that determines whether the solution is gas, liquid, or solid. ✓
   (d) All saturated solutions are concentrated. X
   (e) Solubilities generally change with temperature. ✓

2. Label each of the following compounds as a strong electrolyte (s), weak electrolyte (w), or nonelectrolyte (n):
   (a) $LiNO_3$ (b) HNO_3 (c) CH_3COOH (d) sucrose ($C_{12}H_{22}O_{11}$)
   N S W N

3. Using Table 4.1 in this study guide (Rules of Solubilities), indicate which of the following will precipitate:
   (a) PbI_2 (b) $SrSO_4$ (c) $Ba(NO_3)_2$ (d) $Al(OH)_3$ (e) Li_2S
   ✓

4. Complete the following equations:
   (a) $HCN(aq) + NaOH(aq) \rightarrow\ ? + ?$ H_2O + NaCN
   (b) $H_2CO_3(aq) \rightarrow\ ? + ?$ $2H^+ + CO_3^-$
   (c) $Al(OH)_3 + HCl \rightarrow\ ?$ $AlCl_2$
   weak base + strong acid
   (d) $NaOH(aq) + ? \rightarrow NH_3(g) + NaCl(aq) + ?$
   strong base + salt of weak base

5. Part of the activity series is: Ca>Zn>Pb>H>Cu>Au. Use it to determine which of the following could happen:
   (a) Zinc metal deposits on a copper penny that is placed in $Zn(NO_3)_2$ solution.
   (b) Hydrogen gas evolves when Ca metal is dropped into water.
   (c) A gold ring dissolves when vinegar (acetic acid solution) is accidentally spilled on it.
   (d) An acidic solution dissolves lead out of solder. Solder is a mixture of tin and lead metals.

6. Complete the table for three samples of drinking water contaminated with DDT. The molar mass of DDT is 342.0 g/mol and the density of the drinking water is 1.0 g/mL.

Sample	ppm	% by mass	molarity
(a)	3.0	?	?
(b)	?	?	1.2×10^{-5} *M*
(c)	?	2.0×10^{-4} %	?

7. The blood glucose level of a 70-kg student rises from 90 mg/dL to 175 mg/dL during an exam. Assume that each kilogram of body mass contains 43 mL of blood, and calculate
   (a) the molarity of glucose before and during the exam (the molar mass of glucose is 180.2 g/mol.)
   (b) the increase in the number of moles of glucose in the student's blood stream

8. A potassium depleted patient requires 500 mL of a 0.0100 *M* solution of KCl. How can this solution be prepared from one that is 0.100% KCl by mass and has a density of 1.00 g/mL? The molar mass of KCl is 74.55 g/mol.

9. A scaly layer of $CaCO_3$ may form inside a teakettle if the heated water contains Ca^{2+} and HCO_3^- ions. The reaction is:

   $2HCO_3^-(aq) + Ca^{2+}(aq) \rightarrow CaCO_3(s) + CO_2(g) + H_2O(l)$

   Determine the maximum mass of $CaCO_3$ that could form from 500 mL of water in which $[Ca^{2+}] = 0.00300$ *M* and $[HCO_3^-] = 0.00480$ *M*. The molar mass of $CaCO_3$ is 100.1 g/mol.

10. Patients who do not secrete sufficient gastric acid are advised to drink a 10-mL dose of dilute HCl solution several times a day. Find the molarity of this HCl solution if each milliliter neutralizes 2.7×10^{-4} moles of OH^-.

ANSWERS TO SELF-TEST QUESTIONS

1. (a), (c), and (e) are true; (b) and (d) are false

2. (a) and (b) s (c) w (d) n

3. (a), (b), and (d) will precipitate

4. (a) $NaCN(aq) + H_2O(l)$
   (b) $H_2O(l) + CO_2(g)$
   (c) $Al(OH)_3(s) + 3HCl(aq) \rightarrow AlCl_3(aq) + 3H_2O(l)$
   (d) $NaOH(aq) + NH_4Cl(aq) \rightarrow NH_3(g) + NaCl(aq) + H_2O(l)$

5. (b) and (d)

6. (a) 3.0×10^{-4} %; 3.0 g DDT/10^6 g × 1 mol DDT/342.0 g DDT × 1.0 g/L × 10^3 mL/1L = $\underline{8.8 \times 10^{-6}\ M \text{ DDT}}$
   (b) 4.1 ppm; 1.2×10^{-5} mol/1 L × 342.0 g/1 mol × 1 L/1000 mL × 1 mL/1.0 g × 100% = $\underline{4.1 \times 10^{-4}\%}$
   (c) 2.0 ppm; 5.8×10^{-6} *M*

7. (a) 5.0×10^{-3} *M*; 9.7×10^{-3} *M* (b) 0.014 mol

8. 0.500 L soln × 0.0100 mol KCl/1 L soln × 74.55 g KCl/1 mol soln × 100 g soln/0.100 g KCl × 1 mL soln/1.0 g soln = <u>671 mL</u>

9. $0.500\ \text{L} \times 0.00480\ \text{mol}\ HCO_3^-/1\ \text{L} \times \dfrac{1\ \text{mol}\ CaCO_3}{2\ \text{mol}\ HCO_3^-} \times 100.1\ \text{g}/1\ \text{mol}\ CaCO_3 = \underline{0.120\ \text{g}\ CaCO_3}$

10. 0.27 *M* (10 mL not needed)

8. $.5 \times 0.01\ M = .005$ mol KCl needed

$.1\%$ KCl × 1000 mL = 100g/L

$\dfrac{1.34\ \text{mol/L}}{x\ \text{L}} = \dfrac{.005}{.5\ \text{L}}$ $x = 0.67$

.1g in 100mL.

CHAPTER 5

GASES AND THEIR PROPERTIES

SELF-ASSESSMENT

The Kinetic Theory of Matter (Section 5.1)

Match the terms in Column A to the descriptions in Column B.

1.

A	B
(1) Brownian motion	(a) Matter consists of particles in rapid, continual, and random motion.
(2) diffusion	(b) Random motion of atoms and molecules.
(3) "thermal motion"	(c) Random motion of fine particles dispersed in a fluid medium.
(4) gas pressure	(d) Spontaneous spreading of one substance throughout another.
(5) kinetic theory	(e) The result of molecules or atoms colliding with the walls of a container.

Pressure (Section 5.2)

2. On a day when the atmospheric pressure is 752 torr, the mercury level in the open arm of a manometer is 16.5 cm higher than in the arm connected to a closed flask. Find the pressure in the flask in atmospheres.

The Gas Laws (Section 5.3)

3. A student collects 250 mL of H_2 at 750 torr and 30°C. Calculate the volume of H_2 in milliliters after each of the following changes:
   (a) The pressure doubles while the temperature remains constant.
   (b) The temperature drops to 15°C while the pressure remains at 750 torr.
   (c) The pressure increases to 2.00 atm and the temperature drops to 0°C.

The Ideal Gas Law (Section 5.4)

4. The capacity of a mountain climber's lungs is 3.5 L. How many moles of air do they hold at 37°C (body temperature) and 0.50 atm (the approximate air pressure at 18,000 ft)?

5. Oxygen gas is collected in a 10.0-L tank that can safely contain 30.0 atm pressure. To what temperature can the tank be heated if it contains 160 g of O_2?

6. A 0.39-g sample of an unknown gas occupies 150 mL at 2.0 atm and 300K. Find (a) the molar mass of the gas and (b) its density at STP.

Gas Stoichiometry (Section 5.5)

7. The equation for the combustion of octane (a component of gasoline) is

   $C_8H_{18}(l) + 25/2O_2(g) \rightarrow 8CO_2(g) + 9H_2O(l)$

   Find the volume at STP of the CO_2 produced by burning 1.4 kg of octane.

Mixtures of Gases (Section 5.6)

8. A 5.00-g mixture containing calcium carbide (CaC_2) is dissolved in water producing acetylene gas (C_2H_2):

   $CaC_2(aq) + 2H_2O(l) \rightarrow Ca(OH)_2(aq) + C_2H_2(g)$

   If the C_2H_2 occupies a volume of 250 mL over water at 25°C and 762 torr, what was the percent by mass of CaC_2 in the mixture? The vapor pressure of water at 25°C is 24 torr.

9. A sample of air contains 0.34 mol% CO_2. Find the partial pressure of CO_2 if the atmospheric pressure is 761 torr.

10. Both halothane ($C_2HBrClF_3$) and enflurane ($C_3H_2ClF_5O$) are used as inhalation anesthetics. Which gas will diffuse more rapidly through the lungs and by what approximate factor?

***Kinetic Theory Revisited** (Section 5.7)

Underline the correct word(s) for the following statements:

11. The PV product for an ideal gas is 2/3 of the (kinetic, potential) energy of the molecules.

12. As the temperature of a gas increases, the root mean square speed of the gas molecules (increases, decreases).

13. According to Maxwell-Boltzman distribution curves, the most probable speed of a sample of gas molecules is (always, sometimes, never) less than the root mean square speed.

Deviations from Ideality (Section 5.8)

14. In a real gas attractive forces tend to (lower, raise) the PV product while molecular size tends to (lower, raise) the PV product relative to an ideal gas.

*15. Use the ideal gas law and the van der Waals equation to calculate the pressure in torr of 0.150 mol of O_2 that occupies 20.0 L at 1000°C.

   For O_2: $a = 1.360\ L^2\ atm/mol^2$; $b = 0.03183\ L/mol$

ANSWERS TO SELF-ASSESSMENT PROBLEMS

If you missed an answer be sure to read the relevant section in the textbook and study guide.

1. (1)c (2)d (3)b (4)e (5)a

2. P = 752 torr + 165 torr = 917 torr = 1.21 atm

3. (a) 125 mL
(b) V = 250 mL × (273 + 15)K/(273 + 30)K = 238 mL
(c) V = 250 mL × 0.987 atm/2.00 atm × (273 + 0)K/(273 + 30)K = 111 mL

4. n = 0.50 atm × 3.5 L/(0.0821 L atm/mol K × 310K) = 6.9×10^{-2} mol

5. T = 30.0 atm × 10.0 L/((160/32.0) mol × 0.0821 L atm/mol K) = 731K

6. (a) $\mathfrak{M}$ = 0.39 g × 0.0821 L atm/mol K × 300K/(2.0 atm × 0.150 L) = 32 g/mol
(b) d = 32 g/22.4 L = 1.4 g/L

7. n_{CO_2} = 1400 g C_8H_{18}/114.2 g/mol C_8H_{18} × 8 mol CO_2/1 mol C_8H_{18} = 98.07 mol
V = 98.07 mol × 22.4 L/mol = 2.20×10^3 L

8. $n_{C_2H_2}$ = (762 – 24)/760 atm × 0.250 L/(0.0821 L atm/mol K × 298K) = 9.92×10^{-3} mol = 9.92×10^{-3} mol CaC_2
% CaC_2 = 100% × 9.92×10^{-3} mol × 64.10 g/mol/5.00 g mixture = 12.7%

9. P_{CO_2} = 0.0034 × 761 torr = 2.6 torr

10. enflurane; $\sqrt{197/184}$

11. kinetic

12. increases

13. always

14. lower, raise

15. ideal gas law P = 596 torr
van der Waals P = 597 torr

REVIEW

The Kinetic Theory of Matter (Section 5.1)

Learning Objectives

1. Explain how diffusion and Brownian motion provide support for the kinetic theory of matter.
2. Describe some experimental evidence that indicates that molecular motion increases with temperature.
3. State the assumptions of the kinetic theory of gases.
4. Explain why gases fill their containers and why gases are easily compressed.

Review

The kinetic theory of matter states that matter is composed of particles in continuous, rapid, and random motion, and that the energy and average speed of the moving particles increase with increasing temperature. Evidence for these assumptions is provided by Brownian motion and diffusion:

Brownian motion is the random, spontaneous motion of finely divided particles suspended in a liquid. This motion is caused by random collisions of the liquid molecules with the fine particles.

Diffusion is the spontaneous spreading of one substance through another as a result of random motion. The rate of diffusion increases with increasing temperature.

When applied to gases, kinetic theory states:

(1) Gas molecules or atoms are moving randomly and continuously, filling their containers as they collide with each other and with the container walls.

(2) Gas molecules are much smaller than the distance between them so that most of the container (gas volume) is empty space. As a result, gas can be compressed easily.

(3) The attractive forces between gas molecules are weak and generally negligible. Attractive forces become significant at low temperatures and high densities.

(4) The average kinetic energy of a collection of gas molecules depends on the temperature. This explains the increase in diffusion rate that accompanies an increase in temperature.

Pressure (Section 5.2)

Learning Objectives

1. Calculate the pressure exerted by a given force on a given area.
2. Explain how a mercury barometer is used to measure atmospheric pressure.
3. Convert from one pressure unit to another.
4. Explain how a manometer is used to find the pressure in a closed system.

Review

Pressure is produced by a force acting on a surface and is defined as force per unit area.

$$\text{pressure} = \frac{\text{force}}{\text{area of surface}}$$

The SI unit for pressure is the pascal (Pa) or newton per square meter (N/m^2). The bar is a larger unit.

1 Pa = 1 N/m^2

1 bar = 10^5 Pa

Example 5.1: Find the pressure exerted by a 100-kg man stepping on 49 cm^2 of your foot with all of his weight.

Given: m = 100 kg; area = 49 cm^2

Unknown: pressure

Plan: Use P = F/A. The area, A, in SI units is

$$49\ cm^2 \times \left(\frac{1\ m}{100\ cm}\right)^2 = 4.9 \times 10^{-3}\ m^2$$

The force, F, in this case is the weight of the man.
$w = m \times g = 100\ kg \times 9.80\ m/s^2 = 980\ N$.

Therefore, $P = F/A = 980\ N/4.9 \times 10^{-3}\ m^2 = 20\ N/m^2 = \underline{20\ Pa}$

Atmospheric Pressure

Study the figure of the barometer in your textbook (Figure 5.4, p. 163). The pressure of the atmosphere at sea level will support a column of mercury about 76 cm (760 mm) high. The greater the atmospheric pressure the higher the column. The pressure units most commonly used by chemists are based on this fact.

1 atmosphere (atm) = 760 mm of mercury (760 mm Hg)
1 mm Hg = 1 torr
1 atm = 760 torr

(One atmosphere is also 1.01 bar = 1.01 × 10^5 Pa)

Memorize the relationships between atmospheres, mm Hg, and torr, and be able to convert from one unit to the other.

Example 5.2: At Leadville, Colorado (elevation 10,000 ft), the average atmospheric pressure is about 0.70 bar. Express this pressure in (a) Pa (b) atm (c) torr.

Answers:

(a) $$0.70\ bar \times \frac{1.01 \times 10^5\ Pa}{1.01\ bar} = \underline{7.0 \times 10^4\ Pa}$$

(b) $$0.70\ bar \times \frac{1\ atm}{1.01\ bar} = \underline{0.70\ atm}$$

(c) $$0.70\ bar \times \frac{760\ torr}{1\ atm} = \underline{5.3 \times 10^2\ torr}$$

Practice Drill: Converting Pressure Units

	torr (or mm Hg)	atm	Pa
(a)	750	?	?
(b)	?	0.930	?
(c)	?	?	5.05×10^4

Answers: (a) 0.987 atm; 9.97×10^4 Pa (b) 707 torr; 9.39×10^4 Pa (c) 380 torr; 0.500 atm

The Manometer

Pressure inside a closed system can be measured with a manometer, which gives the difference between the pressure in the system and the outside pressure.

The key part of a simple manometer is a U-tube containing mercury. Study the diagram (Figure 5.5, p. 164) and description in your textbook. The mercury level is lower on the side where the pressure is higher, and the difference in height (mm Hg) between the two sides gives the difference in pressure:

P (closed side) = P (atmosphere) + h (outside minus inside)

Example 5.3: On a day when atmospheric pressure is 770 torr, the mercury level in the outside arm of a manometer is 15 mm higher than the level in the arm that is connected to a closed flask. Find the pressure in the flask.

Given: Outside P = 770 torr; h = 15 mm. The flask pressure is higher.

Answer: P (flask) = 770 torr + 15 torr = 785 torr.

Practice: Gas is released from the flask until the inside mercury level is 20 mm higher than the level open to the atmosphere. Find the new inside pressure. Answer: 750 torr

The Gas Laws (Section 5.3)

Learning Objectives

1. State Boyle's law and use it to calculate the variation of gas volume with pressure and vice versa.
2. Convert temperature readings from degrees Celsius to kelvins and vice versa.
3. State Charles's law and use it to calculate the variation of gas volume with temperature and vice versa.
4. Calculate changes in gas volume and density caused by simultaneous temperature and pressure changes.
5. State the values for standard temperature, standard pressure, and the molar volume of a gas under standard conditions.
6. State Avogadro's law and explain how volume is related to numbers of molecules of gas.

Review

Four properties of a gas sample, pressure (P), volume (V), temperature (T), and number of moles (n), are related by "gas laws."

Boyle's Law: The Relationship Between Pressure and Volume of a Gas

At constant temperature the volume of a fixed quantity of gas is inversely proportional to the pressure:

$P \times V = \text{constant};$

$P_i \times V_i = P_f \times V_f$ (where "i" is initial value and "f" is final value)

Example 5.4: A gas sample is compressed at constant temperature to half of its original volume. What is the change in pressure?

Answer: Pressure and volume changes are reciprocal. If the volume is multiplied by 1/2 (reduced to half), the pressure is multiplied by 2 (doubled).

> **Practice:** If the pressure of a gas sample is increased from 10.0 torr to 30.0 torr at constant T, what is the effect on volume? Answer**:** The volume becomes one-third of the initial value.

Example 5.5: The volume of an air-filled balloon is 2.0 L at 1.0 atm. What would its volume be on a mountain top where the pressure is 0.65 atm? Assume the temperatures are the same.

Given: $P_i = 1.0$ atm $P_f = 0.65$ atm
$V_i = 2.0$ L

Unknown: $V_f = ?$ L

(Always label initial and final values clearly.)

Plan: $P_i \times V_i = P_f \times V_f$ Rearranging gives

$$V_f = V_i \times \frac{P_i}{P_f}$$

Calculation:

$$V_f = 2.0\text{ L} \times \frac{1.0\text{ atm}}{0.65\text{ atm}} = \underline{3.1\text{ L}}$$

Alternate Method: We can also think of calculations like these in terms of "correction factors" that convert the initial value to the final value. In Example 5.5 the initial volume is converted to the final volume by a pressure correction factor.

$$P_f = 2.0\text{ L} \times \frac{1.0\text{ atm}}{0.65\text{ atm}} = \underline{3.1\text{ L}}$$

↑

pressure correction factor

Note: A factor greater than 1 produces an increase; a factor smaller than 1 produces a decrease. Check the factor you use to make sure it produces the effect you expect. In Example 5.5 the pressure decreased, causing an increase in volume. Hence, the correction factor 1.0 atm/0.65 atm is greater than 1. (A factor has the same units in the numerator and the denominator.)

Practice Drill on Boyle's Law: Fill in the blanks for the four gas samples.

Sample	P_i	V_i	P_f	V_f
a	1.0 atm	28.0 mL	7.0 atm	? mL
b	? atm	2.4 L	1.0 atm	1.8 L
c	600 torr	5.00 L	600 atm	? mL
d	1.0 torr	1.00 L	? atm	1.00 mL

Answers: (a) 4.0 mL (b) 0.75 atm (c) 6.58 mL (d) 1.3 atm

Note: Don't just memorize formulas; try to understand the laws by thinking about how they work. Visualize gas samples and imagine how they would be affected by compression, by heating, and so forth.

Charles's Law: The Relationship Between Temperature and Volume of a Gas

At constant pressure the volume of a fixed quantity of gas is directly proportional to the Kelvin temperature.

$$V = \text{constant} \times T; \quad \frac{V_i}{T_i} = \frac{V_f}{T_f}$$

Remember that the Kelvin temperature is obtained by adding 273.15 to the Celsius temperature. A temperature of 273 kelvins is often called "absolute zero."

T (Kelvin) = °C + 273.15

Practice: Complete the following table:

	°C	K
absolute zero	–273	0
water freezes	0	____
typical room temperature	25	____
normal body temperature	37	____
water boils	100	____

Answers: 273K, 298K, 310K, 373K

Example 5.6: If a balloon that occupies 2.0 L at 25°C is put in a freezer at –10°C, what would its volume become? Assume constant pressure.

Given: $T_i = 25°C$; $T_f = -10°C$
$V_i = 2.0$ L

Unknown: $V_f = ?$ L

Plan: Convert temperature into kelvins and use Charles's law.

Calculation: $T_i = 25 + 273 = 298K$; $T_f = -10 + 273 = 263K$

$$V_f = V_i \times \frac{T_f}{T_i} = 2.0\ L \times \frac{263K}{298K} = \underline{1.8\ L}$$

Alternate Method: Use a correction factor: The volume changes by the same factor as the Kelvin temperature.

$$V_f = V_i \times \text{temperature factor} = 2.0\ L \times \frac{263K}{298K} = \underline{1.8\ L}$$

Note: Always make sure your correction factor makes sense. In this case the correction factor must be less than unity because the volume must decrease.

Practice: The 2.0-L balloon is heated from 25°C to 50°C. What is its new volume?
Answer: 2.2 L

The Relationship Between Pressure, Volume, and Temperature

Combining Boyle's and Charles's laws gives

$$\frac{P_i \times V_i}{T_i} = \frac{P_f \times V_f}{T_f}$$

This law can be used to find the final P, V, or T when the other two variables change simultaneously.

Example 5.7: If the balloon that occupies 2.0 L at 1.0 atm and 25°C is carried to an altitude of 5000 ft where the pressure and temperature are 0.83 atm and 10°C, what would its new volume be?

Given: $P_i = 1.0$ atm; $P_f = 0.83$ atm
$T_i = 25°C = 298K$; $T_f = 10°C = 283K$
$V_i = 2.0$ L

Unknown: $V_f = ?$ L

Plan: Substitute these values into the combined gas law:

$$\frac{P_i \times V_i}{T_i} = \frac{P_f \times V_f}{T_f}$$

Calculation: $$\frac{1.0\ \text{atm} \times 2.0\ L}{298K} = \frac{0.83\ \text{atm} \times V_f}{283K};\quad \underline{V_f = 2.3\ L}$$

Alternate Method: Use pressure and temperature correction factors: The pressure decrease acts to increase V; the P factor will be greater than 1. The temperature decrease acts to decrease V; the T factor will be less than 1.

$$V_f = 2.0\ L \times \frac{1.00\ \text{atm}}{0.83\ \text{atm}} \times \frac{283K}{298K} = \underline{2.3\ L}$$

↗ pressure factor chosen to increase volume

↖ temperature factor chosen to decrease volume

Practice Drill on Combining Boyle's and Charles's Laws: Fill in the blanks for the following gas samples:

	V_i	T_i	P_i	T_f	P_f	V_f
(1)	10 L	37°C	2.0 atm	17°C	0.50 atm	?
(2)	10 L	37°C	600 torr	57°C	300 torr	?
(3)	10 L	27°C	0.30 atm	47°C	0.90 atm	?
(4)	10 L	27°C	1.0 atm	–3°C	2.0 atm	?

Answers: (1) 37 L (2) 21 L (3) 3.6 L (4) 4.5 L

Density Changes

$$\text{density} = \frac{\text{mass}}{\text{volume}}$$

If the volume of a sample increases, the density decreases and vice versa. The effect of pressure and temperature change on density is the inverse of the effect on volume, so the correction factors for density are the reciprocals of those for volume.

Example 5.8: The density of an oxygen sample is 1.43 g/L at 0°C and 1.0 atm. Find its density at 25°C and 2.0 atm.

Given: $T_i = 0°C = 273K$ $T_f = 25°C = 298K$
$P_i = 1.0$ atm $P_f = 2.0$ atm
$d_i = 1.43$ g/L

Unknown: $d_f = ?$ g/L

Plan: Use pressure and temperature factors to convert the initial density to the final density.

If the temperature increases, the gas expands to fill more volume. Its density must therefore decrease and the T factor will be less than 1.

The increase in pressure will act to decrease the volume so that the density will increase. Hence, the P factor will be greater than 1.

Calculation: $$d_f = 1.43\ g/L \times \frac{273K}{298K} \times \frac{2.0\ atm}{1.0\ atm} = \underline{2.6\ g/L}$$

Molar Volume and Avogadro's Law

The molar volume, or volume of one mole of a substance, can be obtained by dividing the number of moles (n) into the volume (V) occupied by that number of moles. The molar volume of a gas changes with P and T because V changes.

Molar volumes for different gases are very similar if measured at identical pressures and temperatures. At standard temperature and pressure, which are 0°C and 1 atm (often called "STP"), the molar volumes are approximately 22.4 L.

Example 5.9: Estimate the molar volume of any gas at 298K and 1 atm.

Plan: Multiply the molar volume at STP by a temperature correction factor.

Calculation: $$\text{V (molar) at 298K} = 22.4\ \text{L/mol} \times \frac{298\text{K}}{273\text{K}} = \underline{24.5\ \text{L/mol}}$$

Practice: Estimate the number of moles of gas in 1.0 L at STP. Answer: 0.045 mol

Avogadro's hypothesis: Equal volumes of gases at the same temperature and pressure contain equal numbers of molecules. The law that equal volumes contain equal numbers of moles is a restatement of this hypothesis, which Avogadro published in 1811.

The Ideal Gas Law (Section 5.4)

Learning Objectives

1. State the ideal gas law.
2. Show how the ideal gas law includes Boyle's law, Charles's law, and Avogadro's law.
3. State Amontons' law and use it to calculate the variation of gas pressure with temperature and vice versa.
4. Use experimental data to estimate the value of R.
5. Use the ideal gas law to calculate n, P, V, or T, given values for three of the quantities.
6. Calculate molar masses from mass and volume data for gases.
7. Calculate molar masses from gas densities and vice versa.

Review

The ideal gas law combines all of the other gas laws. Its form is

$PV = nRT$

P = pressure (atm)
V = volume (L)
n = moles of gas (mol)

$$R\ (\text{gas constant}) = 0.0821\ \frac{\text{L atm}}{\text{mol K}}$$

T = Kelvin temperature

Substitutions and rearrangements can bring in other quantities such as

(1) moles per liter: moles per liter = n/V:

$$\frac{P}{RT} = n/V$$

(2) molar mass: substituting $n = m/\mathcal{M}$, where m = mass of sample

$$PV = \frac{m}{\mathcal{M}}RT$$

(3) density: substituting d = m/V:

$$P = \frac{d}{\mathcal{M}}RT$$

Example 5.10: Scientists are trying to influence people and industries worldwide not to use chlorofluorocarbon compounds (such as CCl_2F_2) in refrigerants, insulation, aerosol sprays, and so forth, because of their effect on the protective ozone layer. The following three problems deal with a 0.500-L aerosol can whose contents are under 1.50 atm pressure at 25°C.

(a) (Using PV = nRT.) Find the number of moles of gas in the spray can described above.

Given: $P = 1.50$ atm; $V = 0.50$ L; $T = 25°C = 298K$

Unknown: $n = ?$ mol

Plan: Solve the ideal gas law for n and substitute the given values:

Calculation: $PV = nRT$;
$n = PV/RT$

$$n = \frac{1.50 \text{ atm} \times 0.500 \text{ L}}{0.0821 \text{ L atm/ mol K} \times 298\text{K}} = \underline{0.0307 \text{ mol}}$$

(b) (Calculating molar mass from mass of sample under known conditions.) The gas in the aerosol can weighs 3.71 g. Determine its molar mass, $\mathfrak{M}$.

Plan: Use $PV = \frac{m}{\mathfrak{M}}RT$ and solve for $\mathfrak{M} = mRT / PV$

Calculation:
$$\mathfrak{M} = \frac{3.71 \text{ g} \times 0.0821 \text{ L atm/ mol K} \times 298\text{K}}{1.50 \text{ atm} \times 0.500 \text{ L}} = \underline{121 \text{ g/mol}}$$

(c) (Calculating molar mass from density of sample under known conditions.) The density of the gas in the aerosol can is 7.41 g/L under the given conditions. Find the molar mass.

Plan: Use $P = \frac{d}{\mathfrak{M}}RT$ and solve for $\mathfrak{M}$: $\mathfrak{M} = dRT / P$

Calculation:
$$\mathfrak{M} = \frac{7.41 \text{ g/L} \times 0.821 \text{ L atm/ mol K} \times 298\text{K}}{1.50 \text{ atm}} = \underline{121 \text{ g/mol}}$$

Practice Drill on the Ideal Gas Law: Fill in the missing values.

	P	V	T	n	density	mass	molar mass
(a)	700 torr	500 mL	25°C	?	?	?	16.03 g/mol
(b)	1 atm	22.4 L	0°C	?	?	32.0 g	?
(c)	1.5 atm	0.500 L	25°C	?	7.41 g/L	?	?

Answers: (a) 0.0188 mol, d = 0.602 g/L, m = 0.301 g (b) 1 mol, d = 1.43 g/L, $\mathfrak{M}$= 32.0 g/mol
(c) 0.0307 mol, m = 3.71 g, $\mathfrak{M}$ = 121 g/mol

Gas Stoichiometry (Section 5.5)

Learning Objectives

1. Perform stoichiometric calculations involving gas volumes.

Review

Gas quantities are often given in terms of their volumes under stated conditions. Consequently, when a chemical reaction involves gases, stoichiometry problems may require the calculation of the volume of a gas that is produced or consumed. As in all stoichiometry problems, the mole ratio obtained from the balanced equation is central to the solution of the problem.

Example 5.11: The carbon dioxide concentration in the atmosphere is increasing at an alarming rate of 1 ppm/y. Most of this CO_2 is produced from the burning of fossil fuels such as coal and petroleum. How many grams of isooctane (C_8H_{18}) must burn to release 1000 L CO_2 at 30°C and 750 torr? The equation is

$$2C_8H_{18} + 25O_2 \rightarrow 16CO_2 + 18H_2O$$

Given: $V = 1000$ L
$T = 30°C = 303K$
$P = 750$ torr

Unknown: ? g

Plan: $P, V, T \text{ of } CO_2 \xrightarrow{(1)} \text{mol } CO_2 \xrightarrow{(2)} \text{mol } C_8H_{18} \xrightarrow{(3)} \text{g } C_8H_{18}$

Calculations:

$$(1)\ \text{mol } CO_2 = n = \frac{PV}{RT} = \frac{(750/760)\text{atm} \times 1000 \text{ L}}{0.0821 \text{ L atm/mol K} \times 303\text{K}} = 39.7 \text{ mol}$$

$$(2)\ 39.7 \text{ mol } CO_2 \times \frac{2 \text{ mol } C_8H_{18}}{16 \text{ mol } CO_2} = 4.96 \text{ mol } C_8H_{18}$$

$$(3)\ 4.96 \text{ mol } C_8H_{18} \times \frac{114.2 \text{ g } C_8H_{18}}{1 \text{ mol } C_8H_{18}} = \underline{566 \text{ g } C_8H_{18}}$$

Practice: How many liters of CO_2 measured at STP would be released from the burning of 20.0 L of isooctane at 30°C and 750 torr? <u>Answer</u>: 142 L

Stoichiometry Calculations When All Reactants and Products are Gases

A corollary of Avogadro's law is that the volume ratios of gases at constant temperature and pressure are the same as their mole ratios. This is useful in stoichiometry problems that deal only with gas volumes and not with their masses.

Example 5.12: In an automobile engine some fuel is incompletely oxidized so that CO is produced as well as CO_2. The equation is

$$2C_8H_{18}(g) + 17O_2(g) \rightarrow 16CO(g) + 18H_2O(g)$$

How many liters of O_2 measured at STP is needed to produce 150 L of CO measured at 30°C and 1.0 atm?

Given: $V = 150\ L$
$T = 30°C$
$P = 1.0\ atm$

Unknown: liters at STP

Plan: (1) Use volume ratio. (2) Multiply volume of CO at 30°C and 1.0 atm from (1) by a temperature correction factor.

Calculation: (1) $150\ L\ CO \times \frac{17\ L\ O_2}{16\ L\ CO} = 159\ L\ O_2$ at 30°C and 1.0 atm

(2) $159\ L\ O_2 \times \frac{273K}{303K} = \underline{143\ L\ O_2}$

Mixtures of Gases (Section 5.6)

Learning Objectives

1. State Dalton's law of partial pressures and use it to calculate partial pressures, mole fractions, and mole percents in gas mixtures.
2. Calculate the partial pressure of a gas collected over water.
3. Perform stoichiometric calculations involving gases collected over water.
4. State Graham's law of effusion.
5. Use molar masses to calculate relative effusion rates, and vice versa.

Review

The partial pressure of a gas in a mixture is the pressure the gas would exert if it was the only gas present. Dalton's law of partial pressures states that the total pressure of a gas mixture is the sum of the partial pressures of the gases in the mixture. In a mixture of nitrogen and oxygen, for example,

$$P_{total} = P_{N_2} + P_{O_2}$$

P_{total}	P_{N_2}	P_{O_2}
total pressure	partial pressure of N_2	partial pressure of O_2

Dalton's law applies because the components in a gaseous mixture behave independently.

Mole Fraction

The mole fraction of a component in a mixture is the number of moles of the component divided by the total number of moles. For example, in air the mole fraction of nitrogen (X_{N_2}) is

$$X_{N_2} = \frac{n_{N_2}}{n_T} \quad \text{where } n_T = n_{N_2} + n_{O_2} + n_{\text{other components}}$$

Example 5.13: A typical sample of dry air contains 4.20 mol of oxygen, 15.60 mol of nitrogen, and 0.20 mol of argon. Find the mole fraction of each gas.

Answer: $n_T = 4.20 \text{ mol} + 15.60 \text{ mol} + 0.20 \text{ mol} = 20.00 \text{ mol}$

$$X_{O_2} = \frac{n_{O_2}}{n_T} = \frac{4.20 \text{ mol}}{20.00 \text{ mol}} = \underline{0.210}$$

$$X_{N_2} = \frac{n_{N_2}}{n_T} = \frac{15.60 \text{ mol}}{20.00 \text{ mol}} = \underline{0.7800}$$

$$X_{Ar} = \frac{n_{Ar}}{n_T} = \frac{0.20 \text{ mol}}{20.00 \text{ mol}} = \underline{0.010}$$

Note: Mole fractions are sometimes expressed as mole percents. The mole percents of O_2, N_2, and Ar in dry air are 21%, 78%, and 1.0%, respectively.

Mole Fraction and Pressure

Because the pressure of each gas is proportional to the number of moles, the mole fractions in a gas are equal to the pressure fractions. In air, for example,

$X_{N_2} = \frac{n_{N_2}}{n_{total}} = \frac{P_{N_2}}{P_{total}}$ and so forth, for all of the components of air.

Example 5.14: At high elevations one may experience hypoxia, which is the breathing of air deficient in O_2. Calculate X_{O_2} and P_{O_2} at an elevation of 18,000 ft where the atmospheric pressure is 0.500 atm. The mole fractions are: nitrogen 78%, oxygen 21%.

Answer: The mole fractions in a mixture do not change with conditions, although the pressure may. Hence, the mole fraction is

$X_{O_2} = \underline{0.21}$ and

$P_{O_2} = 0.21 \times 0.500 \text{ atm} = \underline{0.10 \text{ atm}}$

Practice: On a cold, rainy day when the humidity is 100%, the atmospheric pressure is 770 torr, and the temperature is 15°C, the partial pressure of water vapor in the air is 12.8 torr. Find the mole fraction of water vapor. Answer: 0.0166

Collecting Gases Over Water

Equilibrium vapor pressure of water: A confined gas in contact with liquid water will become saturated with water vapor. The partial pressure of the water vapor under these conditions is called the equilibrium vapor pressure, or simply the vapor pressure of water. Values are given for different temperatures in Table 5.3 (p. 185) of your textbook. (Vapor pressure of a liquid depends only on temperature.)

Example 5.15: Air becomes saturated with water vapor in the lungs. Estimate the partial pressure of water vapor in the lungs at a normal body temperature of 37°C to the nearest whole number.

Answer: Table 5.3 in your textbook gives the vapor pressure of water at 35°C and 40°C as 42 torr and 55 torr, respectively. Therefore, the partial pressure of water at 37°C is 47 torr:

$$\Delta VP = 13 \text{ torr}; \quad \Delta T = 5°C, \text{ and } \frac{13 \text{ torr}}{5°} \times 2°C = 5 \text{ torr to be added to 42 torr.}$$

A gas produced in a laboratory reaction is often captured by collecting it over water. The gas becomes saturated with water vapor, which has to be considered in finding the amount of gas from the volume collected.

Example 5.16: A student collects 500 mL of hydrogen over water at 25°C and 1 atm total pressure. Find the partial pressure of hydrogen (P_{H_2} in the collected sample.

Answer: From Table 5.3, (P_{H_2O}) = 24 torr. Therefore P_{H_2} = 760 torr – 24 torr = 736 torr

Partial Pressures in Stoichiometry Calculations

In stoichiometry calculations the molar quantities of gaseous substances may be calculated from P, V, T data.

Example 5.17: Dry ammonium nitrite, like many nitrogen compounds, may decompose explosively. In water, however, the reaction is controllable if carried out with caution.

$$NH_4NO_2(aq) \rightarrow N_2(g) + 2H_2O(l)$$

How many moles of ammonium nitrite must decompose for a student to collect a 300-mL volume of nitrogen over water at 22°C and 765 torr?

Given: V = 300 mL
T = 22°C = 295K
P_{total} = 765 torr

Unknown: moles of NH_4NO_2 decomposing

Plan:
(1) Subtract water vapor pressure from 765 torr to obtain partial pressure of nitrogen.
(2) Use n = PV/RT to obtain moles of nitrogen.
(3) Convert result from (2) to moles of NH_4NO_2.

Calculations: (1) The vapor pressure of water at 22°C is 20 torr (Table 5.3)

$$P_{N_2} = 765 \text{ torr} - 20 \text{ torr} = 745 \text{ torr.}$$

$$\text{(2) moles of nitrogen} = PV/RT = \frac{(745/760)\text{atm} \times 0.300\text{L}}{(0.0821 \text{ L atm/mol K}) \times 295\text{K}} = 0.0121 \text{ mol}$$

$$\text{(3) } 0.0121 \text{ mol } N_2 \times \frac{1 \text{ mol } NH_4NO_2}{1 \text{ mol } N_2} = \underline{0.0121 \text{ mol } NH_4NO_2}$$

Graham's Law of Effusion

Effusion is the passage of gas molecules through very small openings. Light gases effuse more rapidly than heavy gases.

Graham's Law of Effusion: Under identical conditions the rates (r) of effusion of different gases are inversely proportional to the square root of their molar masses ($\mathfrak{M}$).

A useful form of this law for two gases A and B is

$$\frac{r_A}{r_B} = \frac{\sqrt{\mathfrak{M}_B}}{\sqrt{\mathfrak{M}_A}}$$

Example 5.18: (a) Which component of air, N_2 or O_2, will escape more rapidly through minute pores in the material of a balloon? (b) Compare their rates.

Answer: (a) The molar masses are: N_2 = 28.0 g/mol and O_2 = 32.0 g/mol. Nitrogen will effuse faster because it has the lighter molecules. (b) The ratio of the two rates is

$$\frac{r_{N_2}}{r_{O_2}} = \frac{\sqrt{\mathfrak{M}_{O_2}}}{\sqrt{\mathfrak{M}_{N_2}}} = \frac{\sqrt{32.0}}{\sqrt{28.0}} = \underline{1.07}$$

The nitrogen escapes 1.07 times faster than the oxygen.

Practice: Identical balloons are filled with helium and hydrogen. Which gas will escape more rapidly? How much more rapidly? Answer: H_2, 1.41

Gas diffusion is the spontaneous spreading of one gas through another. Like effusion, it depends on the speeds of the molecules and is faster for lighter molecules. Rates of diffusion can be compared in a very approximate way using Graham's law.

Digging Deeper: Kinetic Theory Revisited (Section 5.7)

Learning Objectives

1.* State the relation between the PV product of an ideal gas and its kinetic energy.
2.* Calculate the kinetic energy of a gas from its temperature and vice versa.
3.* State how root mean square speeds vary with temperature and with molar mass.
4.* Calculate the root mean square speed of a collection of gas molecules.
5.* Sketch a typical Maxwell-Boltzmann distribution curve; locate the most probable and root mean square speeds on the curve, and state how the shape of the curve varies with temperature.

Review

This section relates molecular speeds and energies to the PV product. Some quantities it deals with are:

(1) u, the speed of one molecule

(2) $\overline{u^2}$, the average of the squares of all the speeds

(3) u_{rms} (root mean square speed), the square root of $\overline{u^2}$

(4) E_k (molar kinetic energy), the total kinetic energy associated with translation (the motion of molecules from one point to another) in a mole of gas.

Important Formulas:

(1) Since the general formula for kinetic energy is $E_k = 1/2\ mv^2$, it follows that

$$E_k = 1/2\ \mathfrak{M}\overline{u^2} \quad \text{where } \mathfrak{M} = \text{molar mass}$$

(2) A key formula, not proved in the text, relates the PV product of n moles of gas to the total kinetic energy of translation:

$$2/3\ nE_k = PV = nRT$$

(3) Combining (1) and (2) gives

$$u_{rms} = \sqrt{3RT/\mathfrak{M}}$$

Kinetic calculations usually involve some combination of these equations. Here the preferred value for R, the gas constant, is the one that contains energy in joules, R = 8.314 J/mol K.

The Maxwell-Boltzmann distribution curve for a gas shows the fraction of molecules that have a given speed at a given temperature. (See Figure 5.19, p. 193, in text.) The most probable speed is the speed of the greatest number of molecules. It is always less than the root mean square speed. Increasing the temperature results in a greater value for the most probable speed, more molecules with very high speeds, and a flatter distribution curve. Maxwell-Boltzmann curves have been experimentally verified.

Example 5.19: (a) At what Kelvin temperature will 1 mol of helium atoms have a kinetic energy of 3.72×10^3 J? (b) Helium is used in the gaseous mixture inhaled by deep sea divers. Calculate u_{rms} of He atoms in the lungs at normal body temperature (37°C).

Answer: (a) Use formula (2) above and rearrange to find T for 1 mol:

$$T = \frac{2}{3}E_k/R = \frac{2}{3} \times \frac{3.72 \times 10^3\ \text{J/mol}}{8.314\ \text{J/mol°C}} = \underline{298\text{K}}$$

$$\text{(b)}\ u_{rms} = \sqrt{\frac{3RT}{\mathfrak{M}}} = \sqrt{\frac{3 \times 8.314\ \text{J/mol K} \times 310\text{K}}{4.003\ \text{g/mol}}} = \underline{44.0\ \text{m/s}}$$

Deviations From Ideality (Section 5.8)

Learning Objectives

1. State two or three ways in which an ideal gas would differ from a real gas.
2. State the effect of molecular attractions and size on the PV product of a gas.
3. Explain why real gases exhibit almost ideal behavior at low pressures and at high temperatures.
4.* Use the van der Waals equation to calculate the pressure of a confined gas.
5.* Use the van der Waals constants to compare gas molecules with respect to size and attractive forces.

Review

An ideal gas is an imaginary gas consisting of point molecules that have zero volume and exert no forces on each other. Real gases approximate ideal behavior at low pressures and high temperatures, but exhibit substantial deviations under other conditions. Attractive forces tend to lower the PV product of a real gas; molecular size tends to increase the number of collisions and thus raise the PV product. For most gases at ordinary temperatures, the effect of attractive forces predominates up to pressures of several hundred atmospheres. Higher pressures, however, push the molecules closer together, and thus increase the magnitude of the size effect. (See Figure 5.21, p. 96, in text.)

*The van der Waals equation is an equation of state for real gases that contains correction terms to compensate for attractive forces and molecular volume. The van der Waals constants a and b vary from gas to gas: a is a measure of attractive forces and b is a measure of molecular volume. The equation is

$$\left(P + \frac{an^2}{V^2}\right)(V - nb) = nRT$$

P = pressure of gas (atm)
V = volume of container of gas (L)
R = 0.0821 L atm/mol K
n = moles of gas
a and b are constants for each gas

Example 5.20: (a) A chemist wants to collect 50.0 mol of H_2 in a 10.5-L steel vessel at 25°C. Use (1) the van der Waals equation and (2) ideal gas law to determine the pressure inside the steel vessel. (b) Account for the difference in computed values.

Answer: (a) (1) Rearrange the van der Waals equation to solve for P and then substitute given values (a and b are given in Table 5.6 of your textbook):

$$P = \frac{nRT}{V - nb} - \frac{an^2}{V^2} = \frac{50.0 \text{ mol} \times 0.0821 \text{ L atm/mol K} \times 298\text{K}}{10.5 \text{ L} - 50.0 \text{ mol} \times 0.02661 \text{ L/mol}} - \frac{0.2444 \text{ atm L}^2/\text{mol}^2 \times (50.0 \text{ mol})^2}{(10.5 \text{ L})^2}; \quad \underline{P = 128 \text{ atm}}$$

(2) P = nRT/V = 50.0 mol × 0.0821 L atm/mol K × 298K/10.5 L = 117 atm

(b) In hydrogen, the attractive effects are smaller than in most gases (compare a values in Table 5.6), therefore the molecular size effect predominates and raises the pressure above the ideal gas value.

PROBLEM SOLVING TIP

Important Formulas Involving Gases

Formula	Meaning of Variables	Significance of Formula
P = F/A	P = pressure (Pa) F = force (N) A = surface area (m^2)	Pressure results from a force being exerted on a surface area. If two of the variables are known, the third can be calculated.
$P_iV_i = P_fV_f$ (P × V = constant)	V_i, P_i = initial volume and pressure of gas sample V_f, P_f = final volume and pressure Units of V_i and V_f must be the same; units of P_i and P_f must be the same. T must be constant.	Boyle's Law: P and V are inversely related at constant T and n. If three of the quantities are known, the fourth can be calculated.

Formula	Meaning of Variables	Significance of Formula
$\frac{V_i}{T_i} = \frac{V_f}{T_f}$ $\left(\frac{V}{T} = \text{constant}\right)$	T_i, T_f = initial and final temperatures in kelvins. P must be constant.	Charles's law: V and T of a gas are directly related at constant P and n. If three of the quantities are known, the fourth can be calculated.
$\frac{P_i V_i}{T_i} = \frac{P_f V_f}{T_f}$	Same as above	Combined Boyle's and Charles's Law: If five of the quantities are known, the sixth can be calculated.
$\frac{V_i}{n_i} = \frac{V_f}{n_f}$ $\left(\frac{V}{n} = \text{constant}\right)$	n_i, n_f = initial and final number of moles	Avogadro's law: At a given P and T, the volumes of gas samples are proportional to he numbers of moles they contain. Often used for reactions that involve various gases, so that volumes instead of moles can be used in the stoichiometric calculations.
$PV = nRT$	P = pressure (atm) V = volume (L) n = number of moles R = 0.0821 L atm/mol K T = temperature (K)	Ideal gas law: Combination of Boyle's, Charles's and Avogadro's laws. If three of the variables are known, the fourth can be calculated.
$PV = \frac{m}{\mathcal{M}} RT$	m = mass of gas (g) $\mathcal{M}$ = molar mass (g/mol)	Derived from ideal gas law; can be used to calculate $\mathcal{M}$ of an unknown gas when mass of sample and other variables are known.
$P = \frac{d}{\mathcal{M}} RT$	d = density (g/L)	Derived from ideal gas law; can be used to calculate $\mathcal{M}$ or d of a gas when the other variables have been measured.
$P_{total} = P_A + P_B + P_C$	P_{total} = total pressure of a gaseous mixture containing components A, B, and C. P_A, P_B, P_C = partial pressure of gas A, B, and C respectively. (Partial pressure is the pressure the gas would exert if it were the only gas present.) Units of pressure must be the same.	Dalton's law: The total pressure exerted by a gaseous mixture is the sum of the partial pressures of the components.
$X_A = \frac{P_T}{P_T}$	X_A = mole fraction of component A in the mixture	The mole fraction of a component in a mixture is the same as its pressure fraction.

Formula	Meaning of Variables	Significance of Formula
$r_A \times \sqrt{\mathfrak{M}_A} = r_B \times \sqrt{\mathfrak{M}_B}$	r_A, r_B = rates of effusion of gas A and gas B $\mathfrak{M}_A$, $\mathfrak{M}_B$ = molar mass of gas A and gas B	Graham's law of effusion: The rates of effusion (escaping through a very small opening) of gases are inversely proportional to the square root of their molar masses. Relative rates of effusion of different gases can thus be calculated under identical conditions. Also, one effusion rate can be estimated if the other is known.
$\left(P + \frac{an^2}{V^2}\right)(V - nb) = nRT$	a = the van der Waals gas constant that depends on strength of intermolecular forces (Units are atm L^2/mol^2) b = the van der Waals gas constant that depends on molecular size (Units are L/mol)	The van der Waals equation of state for a real gas: In real gases intermolecular attractive forces exist and molecules occupy volume. The van der Waals equation takes this into account.

SELF-TEST

1. True or false?
   (a) Gases are easily compressed.
   (b) The density of a gas increases with temperature.
   (c) A cube of sugar spreads through hot coffee by Brownian motion.
   (d) Attractive forces do not exist between real gas molecules.
   (e) The volume of any gas is the volume of the container.
   (f) Increasing temperature increases the diffusion rates of gases.

2. At an altitude of 10,000 ft, a barometer measures the pressure as 0.688 atm. Convert the pressure to
   (a) torr (b) mm Hg (c) cm Hg (d) Pa (e) bar
   (1 atm = 1.01×10^5 Pa; 10^5 Pa = 1 bar)

3. The net pressure on an eardrum whose diameter is 10 mm is 3.2×10^4 Pa. Find the force on this eardrum in newtons.

4. Indicate the factor(s) by which the initial volume must be multiplied to find the final volume for each of the following changes. Assume the other variables are constant:
   (a) Pressure halved ____________________
   (b) Number of moles doubled ____________________
   (c) Temperature increased from 0°C to 27°C. ____________________
   (d) Temperature decreased from 373K to 273K while pressure increased from 373 torr to 761 torr.
   ____

5. A certain balloon at sea level has a volume of 2.4 L at 1.0 atm and 25°C. When it rises 5000 ft to a pressure of 632 mm Hg its volume becomes 2.0 L. Find the temperature at 5000 ft.

6. The atmospheric pressure during hurricane Gilbert was reported as 663 mm Hg and the temperature was 75°C. Calculate the number of molecules in one milliliter of air.

7. The density of ozone is 2.194 g/L at 298K and 850 torr. Find its molar mass.

8. LiOH is used in space capsules to absorb exhaled CO_2: $2LiOH(s) + CO_2(g) \rightarrow Li_2CO_3(s) + H_2O(l)$

   If 5.20 kg of LiOH are used up during a six day voyage to the moon, what volume of CO_2 at 30°C and 1.0 atm was exhaled by the astronaut? The molar mass of LiOH is 23.95 g.

9. Calculate P_{CO_2} at one atmosphere in alveolar air (air in the lung sacs) given that alveolar air is 5.62 mol% CO_2.

10. An inhalation anesthetic consists of C_3H_6, O_2, and He. Which gas will diffuse more rapidly into the lungs? How many times faster does it diffuse than each of the others?

11. Match the variable in Column A to an applicable statement in Column B.

A	B
(1) a	(a) 8.314 J/ mol K
(2) u_{rms}	(b) Inversely proportional to the square root of the molecular mass.
(3) R	(c) A measure of the magnitude of intermolecular forces.
(4) E_k	(d) Proportional to the PV product.

ANSWERS TO SELF-TEST QUESTIONS

1. (a), (e), and (f) T (b), (c), and (d) F

2. (a) 523 torr (b) 523 mm Hg (c) 52.3 cm Hg (d) 6.95×10^4 Pa (e) 0.695 bar

3. $P = F/A;\ F = P \times A = P \times \pi r^2 = 3.2 \times 10^4 \text{ Pa} \times 3.14 \times \left(\frac{10^{-2}\text{ m}}{2}\right)^2 = \underline{2.5 \text{ N}}$

4. (a) 2 (b) 2 (c) 300K/273K (d) (273K/373K) × (373 torr/761 torr)

5. $T_2 = T_1 \times P_2V_2/P_1V_1 = 298\text{K} \times (0.832 \text{ atm} \times 2.0 \text{ L})/(1.0 \text{ atm} \times 2.4 \text{ L}) = \underline{207\text{K}}$

6. $n = 0.872 \text{ atm} \times 0.00100 \text{ L}/(0.0821 \text{ L atm/mol K} \times 348\text{K}) = 0.0305 \times 10^{-3} \text{ mol} \times 6.022 \times 10^{23} \text{ molecules/mol} = \underline{1.84 \times 10^{19} \text{ molecules}}$

7. $\mathfrak{M} = dRT/P = 2.194 \text{ g/L} \times 0.0821 \text{ L atm/ mol K} \times 298\text{K}/1.12 \text{ atm} = \underline{47.9 \text{ g/mol}}$

8. $n_{CO_2} = 5200 \text{ g LiOH} \times \frac{1 \text{ mol LiOH}}{23.95 \text{ g LiOH}} \times 1 \text{ mol } CO_2 / 2 \text{ mol LiOH} = 108.6 \text{ mol}$

   $V_{CO_2} = n_{CO_2}RT/P = 108.6 \text{ mol} \times 0.0821 \text{ L atm/ mol K} \times 303\text{K}/1.0 \text{ atm} = \underline{2.7 \times 10^3 \text{ L}}$

9. $1.00 \text{ atm} \times 0.0562 = \underline{0.0562 \text{ atm}}$

10. He; $2.8 \times r_{O_2}$; $3.2 \times r_{C_3H_6}$

11. (1) c (2) b (3) a (4) d

CHAPTER 6
THERMOCHEMISTRY

SELF-ASSESSMENT

The First Law of Thermodynamics (Section 6.1)

1. Underline the correct choice in parentheses.
   (a) If the internal energy of a system increased by +50 kJ, and +25 kJ of heat was absorbed by the system, then (–25 kJ, +25 kJ) of work was done (on, by) the system.
   (b) (E, q) is a state function, while (w , ΔE) depends on the pathway for a change.
   (c) When ice melts at 25°C its internal energy (increases, decreases), while the internal energy of its surroundings (increases, decreases) by the same amount.

Heats of Reaction; Standard Enthalpy Changes (Sections 6.2 and 6.3)

2. True or false? If a statement is false, change the underlined word(s) to make it true.
   (a) ΔH is equal to the heat of reaction for a chemical reaction occurring at constant pressure.
   (b) An endothermic process gives off heat.
   (c) ΔH_f° for any element at 1 atm is always 0.
   (d) ΔH° is the enthalpy change accompanying a reaction in which all reactants and products are in their standard states.
   (e) A reaction occurring in a rigid closed vessel is an example of a constant volume reaction.
   (f) ΔH_f° for $H_2O(l)$ is equal to ΔH_f° for $H_2O(g)$.

3. Determine the sign and value for ΔH, q, w, and ΔE for the following events:
   (a) A reacting system releases 45 kJ of heat at constant pressure and does 25 kJ of pressure-volume work.
   q = ____; ΔH = ____; w = ____; ΔE = ____.
   (b) A reacting system open to the atmosphere absorbs 50 kJ of heat with no accompanying change in volume.
   q = ____; ΔH = ____; w = ____; ΔE = ____.

4. Given the following thermochemical equation:

   $C_2H_5OH(l) + 3O_2(g) \rightarrow 2CO_2(g) + 3H_2O(l); \quad \Delta H^{\circ} = -1366.8 \text{ kJ/mol}$

   How many kilojoules of heat are released at 1 atm and 25°C when 500.0 mL of ethanol (C_2H_5OH) undergoes this reaction? The density of ethanol is 0.7893 g/mL.

Hess's Law (Section 6.4)

5. Use Hess's law to calculate $\Delta H°$ for $H_2O(l) \rightarrow H_2O(g)$ from the following thermochemical equations:

$2H_2(g) + O_2(g) \rightarrow 2H_2O(g)$ $\quad \Delta H° = -483.6$ kJ/mol

$2H_2(g) + O_2(g) \rightarrow 2H_2O(l)$ $\quad \Delta H° = -571.6$ kJ/mol

6. Find $\Delta H°$ for the following reactions, using ΔH_f° data from Table 6.5 of your textbook. Indicate whether the reactions are endothermic or exothermic.
   (a) $CO(g) + 1/2O_2(g) \rightarrow CO_2(g)$
   (b) $2SO_2(g) + O_2(g) \rightarrow 2SO_3(g)$
   (c) $O_3(g) + H_2S(g) \rightarrow SO_2(g) + H_2O(g)$

Heat Capacity and Specific Heat (Section 6.5)

7. A 75.6-g copper bracelet is transferred from a table top at 25°C to a person's wrist at 35°C.
   (a) How many joules of heat are absorbed by the bracelet as it warms to 35°C? (molar heat capacity of copper = 24.6 J/mol°C)
   (b) Would more or less heat be absorbed if the 75.6-g bracelet were gold? (molar heat capacity of gold = 25.2 J/mol°C)

Calorimetry (Section 6.6)

8. One crisp, drained, and thinly cut slice of bacon was burned in excess oxygen in a bomb calorimeter with a heat capacity of 6.20 kJ/°C. The temperature increased by 28°C. How many dietary Calories are in a slice of bacon? (1 kcal = 4.18 kJ)

9. The label on a cereal box states that a 10-oz serving of cereal contains 110 dietary Calories. The label also indicates that each 10-oz serving contains 3 g protein, 23 g carbohydrate, and 0 g fat. If the average fuel values of carbohydrate and protein are 4.1 kcal/g and 4.3 kcal/g, respectively, determine if the statement on the box is correct.

ANSWERS TO SELF-ASSESSMENT PROBLEMS

If you missed an answer, be sure to study the relevant section in the text and study guide.

1. (a) +25 kJ; on (b) E, w (c) increases; decreases

2. (a) T (b) exothermic (c) sometimes (element must be in its standard state) (d) T (e) T (f) not equal

3. (a) q = –45 kJ; $\Delta H = -45$ kJ; w = –25 kJ; $\Delta E = -70$ kJ
   (b) q = +50 kJ; $\Delta H = +50$ kJ; w = 0; $\Delta E = +50$ kJ

4. 1.171×10^4 kJ

5. +44.0 kJ/mol

6. (a) –283.0 kJ/mol (exothermic) (b) –197.8 kJ (exothermic) (c) –660.7 kJ/mol (exothermic)

7. (a) $75.6\ \text{g} \times \frac{1\ \text{mol}}{63.55\ \text{g}} \times \frac{24.6\ \text{J}}{1\ \text{mol°C}} \times 10°\text{C} = \underline{293\ \text{J}}$ (b) less (fewer moles of Au than of Cu)

8. $28°C \times \frac{6.20\ kJ}{1°C} \times \frac{1\ kcal}{4.18\ kJ} \times \frac{1\ Calorie}{1\ kcal} = \underline{42\ Cal}$

9. $(3\ g \times 4.3\ kcal/g) + (23\ g \times 4.1\ kcal/g) = \underline{107\ Cal}$

REVIEW

Conservation of Energy (Section 6.1)

Learning Objectives

1. State the first law of thermodynamics.
2. Calculate the value of q, w, or ΔE, given values for two of the three quantities.
3. Distinguish between a state function, such as E, and a path-dependent property, such as q or w.

Review

The law of conservation of energy, also known as the first law of thermodynamics, states that energy cannot be created nor destroyed. Energy can be converted from one form into another.

A system is a portion of the universe selected for consideration while the surroundings are the rest of the universe. In chemistry, the reacting substances are frequently considered to be the system and everything else is the surroundings. According to the first law, energy released by the system is absorbed by the surroundings.

Heat, Work, and Internal Energy

The internal energy (E) of a system is the total energy of the individual particles in the system; the sum of their kinetic energy (E_k) and their potential energy (E_p):

$E = E_k + E_p$

A change in the internal energy of a system is the difference between the final energy (after the change) and the initial energy (before the change):

$\Delta E = E_{final} - E_{initial}$ (Δ or "delta" means change)

Energy changes occur in two ways:
(1) Heat (q) flows into a system (absorbed) or out of a system (released).
(2) Work (w) is done on a system by the surroundings, or by a system on the surroundings.

The sum of the two is the energy change.

$\Delta E = q + w$

ΔE = change in internal energy of system
q = heat
w = work

This equation is the mathematical statement of the first law of thermodynamics. Signs are very significant in thermodynamics. The correct signs will be easier to remember if you understand how they relate to energy changes.

Make sure you use them consistently as follows:

+ q = heat absorbed by the system, which adds to its energy
– q = heat released by the system, which takes away from its energy
+ w = work done on the system, which adds to its energy
– w = work done by the system, which takes away from its energy
– ΔE = energy lost by system
+ ΔE = energy gained by system

(From the system's point of view, "+" signifies gain and "–" signifies loss.)

Example 6.1: A sample of air absorbs 1000 kJ of heat and expands, doing 2500 kJ of work. What are the values of q, w, and ΔE for the air sample?

Answer: (1) The air sample is the system. It absorbs 1000 kJ of heat, q = +1000 kJ.
(2) The air sample does 2500 kJ of work; w = –2500 kJ.
Therefore ΔE = q + w = +1000 kJ + (–2500 kJ) = –1500 kJ

The internal energy of the air sample decreased.

Practice Drill on the First Law: Each row gives q, w, and ΔE for a given change. Fill in the missing values.

	q	w	ΔE
(a)	+ 50 kJ	+100 kJ	?
(b)	?	–20 kJ	+150 kJ
(c)	–200 kJ	?	+150 kJ

Answers: (a) +150 kJ (b) +170 kJ (c) +350 kJ

The preceding practice drill shows that a system can undergo the same change in internal energy although the change follows different pathways; the values for q and w are different although ΔE is the same for (a), (b), and (c). This means that internal energy is a state function: ΔE depends only on the initial and final states of the system, not the pathway taken during the change. Note that q and w depend on the actual pathway and so are not state functions.

Heats of Reaction (Section 6.2)

Learning Objectives

1. Distinguish between constant volume reactions and constant pressure reactions, and give examples of each.
2. Given the value of ΔH, state whether a reaction is exothermic or endothermic.
3. Use thermochemical equations to calculate the amount of heat given off or absorbed during a chemical reaction.

Review

A constant volume reaction is one in which the volume does not change, although the pressure may. A constant pressure reaction is one in which the pressure does not change, although the volume may. Constant pressure reactions, such as those we carry out in open beakers and flasks, are the most common.

Example 6.2: Identify the following as constant pressure or constant volume reactions:
(a) A chemical reaction carried out in an open test tube.
(b) A reaction in a sealed container with rigid walls.
(c) A biochemical reaction in the body.

Answer: (a) and (c) are constant pressure reactions (pressure of the atmosphere); (b) is a constant volume reaction because the container walls keep the volume constant.

Enthalpy

The enthalpy (H) is a state function. The enthalpy of a system is defined as the sum of its internal energy and its pressure-volume product:

$H = E + PV$ — E = internal energy, P = pressure, V = volume

Therefore at constant pressure the enthalpy change, ΔH, is

$\Delta H = \Delta E + P\Delta V$.

In spite of its odd name, enthalpy is an important and even familiar thermodynamic property. When a system undergoes a change at constant pressure–a burning log or a baking cake–the heat absorbed or released is an enthalpy change.

For Chemical Reactions at Constant Pressure

(1) $\Delta H = q_p$

For a reaction at constant pressure, the enthalpy change, ΔH, equals the heat, q, exchanged with the surroundings.

ΔH is positive for an endothermic reaction (heat absorbed).
ΔH is negative for an exothermic reaction (heat released).

Example 6.3: A chemical reaction releases 50 kJ of heat at 1 atm. What is the enthalpy change?

Answer: $\Delta H = q_p = -50$ kJ

(2) $w = -P\Delta V$

When a system undergoes a volume change, (ΔV), at constant pressure, (P), the work that it exchanges with the surroundings is $w = -P\Delta V$.

w is positive when ΔV is negative (contracting system); work is done on the system.
w is negative when ΔV is positive (expanding system); work is done by the system.

Example 6.4: A gas forms from two reacting liquids. What is the sign of the work term?

Answer: The volume of the gas is greater than that of the liquids, therefore the volume increases, ΔV is positive, and $-P\Delta V$ is <u>negative</u>; w is negative. Work is done by the system in pushing back the atmosphere.

Thermochemical Equations

A thermochemical equation includes the value of the enthalpy change (ΔH) under specified conditions (usually 1 atm and 298K). The magnitude and sign of ΔH apply to the molar quantities in the equation as written.

Example 6.5: How many kilojoules are released when one gram of methane (CH_4) reacts with oxygen at 25°C?

$$CH_4(g) + 2O_2(g) \rightarrow CO_2(g) + 2H_2O(g) \quad \Delta H_{298} = -896.0 \text{ kJ}$$

Answer: One mole of CH_4 releases 896.0 kJ of heat and weighs 16.04 g. We can combine these two facts to find the heat released by one gram:

$$\frac{-896.0 \text{ kJ}}{1 \text{ mol}} \times \frac{1 \text{ mol}}{16.04 \text{ g}} = \underline{-55.86 \text{ kJ / g}}$$

Practice: A camper has a 20-L tank of propane gas (C_3H_8) under 30 atm pressure at 25°C. How much heat at 25°C and 1 atm can be released by the combustion of the propane?

$$C_3H_8(g) + 5O_2(g) \rightarrow 3CO_2(g) + 4H_2O(l) \quad \Delta H_{298} = -2219.9 \text{ kJ}$$

Hint: use $PV = nRT$ to find moles of propane gas. Answer: 5.4×10^4 kJ

Standard Enthalpy Changes (Section 6.3)

Learning Objectives

1. Distinguish between a standard enthalpy change and other enthalpy changes.
2. Write thermochemical equations for combustion reactions and formation reactions.
3. Use standard enthalpies of combustion to compare the efficiencies of various fuels and to calculate the amount of heat given off during a combustion reaction.
4. Use standard enthalpies of formation to calculate the quantity of heat absorbed or given off during a formation reaction.

Review

A solid, liquid, or gas is in its standard state when its pressure or the pressure on it is one atmosphere. A null mark (°) is used to label standard values, those that apply to substances in their standard states. $\Delta H°$ symbolizes a standard enthalpy change for a reaction in which reactants and products are in their standard states. The temperature can be specified as a subscript; e.g., ΔH°_{273}, ΔH°_{298}. If no temperature is given, we can assume the value is the one for 298K.

Example 6.6: For which of the following reactions would the enthalpy change be standard?:

(a) $1/2N_2$ (g, 1.5 atm) + O_2 (g, 1 atm) $\rightarrow$ NO_2 (g, 1 atm) at 25°C
(b) $1/2N_2$ (g, 1 atm) + O_2 (g, 1 atm) $\rightarrow$ NO_2 (g, 1 atm) at 25°C
(c) $1/2N_2$ (g, 1 atm) + O_2 (g, 1 atm) $\rightarrow$ NO_2 (g, 1 atm) at 50°C.

Answers: (b) and (c); ΔH for (b) is ΔH°_{298}, ΔH for (c) is ΔH°_{323}. (The N_2 in (a) is not in its standard state.)

Standard Enthalpies of Combustion and Formation

A standard enthalpy of combustion is the standard enthalpy change associated with the combustion of 1 mol of a substance in oxygen. For example, ΔH in the following thermochemical equation is the standard enthalpy of combustion of glucose.

$C_6H_{12}O_6(s) + 6O_2$ (g, 1 atm) → $6H_2O(l) + 6CO_2$(g, 1 atm) $\Delta H° = -2816$ kJ

A standard enthalpy of formation (ΔH_f°) is the standard enthalpy change associated with the formation of 1 mol of a substance from the most stable forms of its elements. ΔH_f° for the most stable form of an element is therefore zero. For example, the tabulated ΔH_f° value of –285.8 kJ/mol for liquid H_2O implies the thermochemical equation:

H_2 (g, 1 atm) + $1/2O_2$ (g, 1 atm) → $H_2O(l)$ $\Delta H_f^\circ = -285.8$ kJ

> **Practice:** Write the thermochemical equation for the formation of 1 mol of $CO_2(g)$ from its elements under standard conditions. Use Table 6.5 (p. 221) in your textbook: Standard enthalpies of Formation at 25°C.
>
> Answer: C (graphite) + $O_2(g) \rightarrow CO_2(g)$ $\Delta H_f^\circ = -393.5$ kJ

Example 6.7: How much heat is released when 100.0 g of liquid methanol (CH_3OH) burns at 25°C and 1 atm? For methanol $\Delta H^\circ_{combustion} = -726.5$ kJ/mol (Table 6.2 in your textbook).

Answer: The combustion of 1 mol of CH_3OH will release 726.5 kJ of heat. Therefore:

100.0 g CH_3OH → mol CH_3OH → kJ

$$100.0\ \text{g} \times \frac{1\ \text{mol}}{32.04\ \text{g}} \times \frac{-726.5\ \text{kJ}}{1\ \text{mol}} = \underline{-2.267 \times 10^3\ \text{kJ}}$$

> **Practice:** How many kilojoules of heat are absorbed by the formation of 96.0 g of $O_3(g)$ from O_2 at 25°C and 1 atm? (From Table 6.5: ΔH_f° for $O_3(g) = +142.7$ kJ/mol)
>
> Answer: +285 kJ

Hess's Law (Section 6.4)

Learning Objectives

1. State Hess's law and use it to find unknown enthalpy changes.
2. Calculate reaction enthalpies from enthalpies of formation.

Review

Hess's law states that the enthalpy change for a specific reaction is the same whether the reaction occurs in one or several steps.

Hess's law can be used to determine ΔH values that might be difficult to measure directly. Keep in mind two rules when using Hess's law:

1. Reversing an equation changes the sign of ΔH.
2. If you multiply an equation by a factor, you must also multiply ΔH by that same factor.

Example 6.8: Use the thermochemical equations (a) and (b) to find ΔH° for equation (c):

(c) $SO_2(g) + 1/2O_2(g) \rightarrow SO_3(g)$ $\quad \Delta H° = ?$

(a) $S(s) + O_2(g) \rightarrow SO_2(g)$ $\quad \Delta H_f^\circ = -297$ kJ

(b) $S(s) + 3/2O_2(g) \rightarrow SO_3(g)$ $\quad \Delta H_f^\circ = -396$ kJ

Plan: Find a combination of (a) and (b) that adds up to the desired (c).

Compare reactants and products in (c) with those in (a) and (b).

$SO_2(g)$ is a reactant in equation (c) so you need to add in an equation in which $SO_2(g)$ appears as a reactant. Use equation (a) in reverse, changing the sign of its ΔH°:

(1) $SO_2(g) \rightarrow S(s) + O_2(g)$ $\quad \Delta H_1^\circ = 297$ kJ

$SO_3(g)$ is a product in your unknown equation (c) so that equation (b) can be used as is;

(2) $S(s) + 3/2O_2(g) \rightarrow SO_3(g)$ $\quad \Delta H_2^\circ = -396$ kJ

Sum of (1) and (2) gives (c):

$SO_2(g) + S(s) + 3/2O_2(g) \rightarrow S(s) + O_2(g) + SO_3(g)$ $\quad \Delta H° = 297 \text{ kJ} + (-396 \text{ kJ})$

(c) $SO_2(g) + 1/2O_2(g) \rightarrow SO_3(g)$ $\quad \Delta H° = -99$ kJ

Using ΔH_f° Values

ΔH° for any reaction can be computed from $\Delta H°_f$ values; the formula is

$$\Delta H^\circ_{reaction} = \sum \Delta H_f^\circ \text{ (products)} - \sum \Delta H_f^\circ \text{ (reactants)}$$

(Σ means "sum of")

Example 6.9: Calculate ΔH° for the decomposition of nitroglycerin. Use ΔH_f° for nitroglycerine = –58.1 kcal/mol and Table 6.5 in your textbook. The equation is

$$4C_3H_5(NO_3)_3(l) \rightarrow 12CO_2(g) + 10H_2O(g) + 6N_2(g) + O_2(g)$$

Answer: Look up the ΔH_f° value for each reactant and product. Multiply each one by its coefficient in the equation:

	$4C_3H_5(NO_3)_3(l)$	→	$12CO_2(g)$	+	$10H_2O(g)$	+	$6N_2(g)$	+	$O_2(g)$
ΔH_f°, kJ	4 × (–243)		12 × (–393.5)		10 × (–241.8)		0		0
	–972		–4722		–2418				

Add the ΔH_f° values for the products and subtract the ΔH_f° values for the reactants to get ΔH°:

ΔH° = –4722 – 2418 – (–972) = <u>-6.168×10^3 kJ</u>

Note: If the ΔH° is known, this formula can be used to find a missing ΔH_f° value.

Heat Capacity and Specific Heat (Section 6.5)

Learning Objectives

1. Calculate molar heat capacities from specific heats and vice versa.
2. Calculate the quantity of heat absorbed or released by a given mass of substance during some temperature change.

Review

The heat capacity of an object is the amount of heat required to raise its temperature by 1°C. The units are usually J/°C. Heat capacity depends on size; for example, it takes more heat per degree to warm a lake than to warm a glass of water.

Example 6.10: The heat capacity of a calorimeter is 600 J/°C. How much heat is needed to raise its temperature by 10°C?

Answer: 10°C × 600 J/°C = 6000 J

Specific heat is heat capacity per gram; its units are usually J/g°C. The specific heat of a given substance is the same for all samples regardless of size.

Example 6.11: The specific heat of liquid water is 4.18 J/g°C. How much heat is needed to raise the temperature of 6.00 g water by 15.0°C?

Answer: 4.18 J/g°C × 6.00 g × 15.0°C = 376 J

Note: The general formula for all specific heat problems is:

q = specific heat × mass × ΔT

Given three of the above quantities the fourth can be calculated.

Molar heat capacity is heat capacity per mole; its units are usually J/mol°C. The molar heat capacity of a given substance is the same for all samples regardless of size.

Example 6.12: The molar heat capacity of copper is 24.6 J/mol°C. How many joules are released by 0.10 mol of copper cooling from 85.0°C to 82.0°C?

Answer: The change in temperature is $\Delta T = T_{final} - T_{initial}$ = 82.0°C – 85.0°C = –3.0°C.

The heat released is q = 24.6 J/mol°C × 0.10 mol × (–3.0°C) = –7.4 J

Note: When the temperature of the system increases, ΔT is positive, heat is absorbed, and q is positive. When the temperature decreases, ΔT is negative, heat is released, and q is negative.

The relationship between specific heat and molar heat capacity is:

molar heat capacity = specific heat × molar mass

Example 6.13: The specific heat of aluminum is 0.900 J/g°C and its molar mass is 26.98 g/mol. What is its molar heat capacity?

Answer: 0.900 J/g°C × 26.98 g/mol = 24.3 J/mol°C.

Calorimetry (Section 6.6)

Learning Objectives

1. Sketch a bomb calorimeter and describe its operation.
2. Sketch a "coffee cup" calorimeter and describe its operation.
3. Calculate heats of reaction from calorimetric data.
4. Given the reaction equation, state whether the constant volume heat of reaction is equal to the enthalpy change.

5.* Calculate enthalpy changes from constant volume heats of reaction and vice versa.

Review

A calorimeter is used in the laboratory to measure heats of reaction (ΔH at constant pressure). A bomb calorimeter consists of an explosion proof reaction vessel (called a "bomb"), surrounded by a liquid. The heat evolved by a reaction in the reaction vessel raises the temperature of both the vessel and the surrounding liquid. The amount of heat is calculated from the temperature rise and the heat capacity of the calorimeter:

heat of reaction = heat absorbed by calorimeter = heat capacity of calorimeter × ΔT

Example 6.14: The combustion of a 0.180-g sample of glucose raises the temperature of a bomb calorimeter from 25.01°C to 27.13°C. The heat capacity of the calorimeter is 1.46×10^3 J/°C. Find
(a) the measured heat of combustion of glucose in kilojoules per mole
(b) the number of dietary Calories (kilocalories) associated with one gram of glucose

Given: 0.180 g glucose; $\Delta T = 27.13°C - 25.01°C = 2.12°C$

heat capacity = 1.46×10^3 J/°C

Unknown: (a) heat of reaction (kJ/mol)
(b) dietary Calories (kcal) per gram of glucose

Plan: heat evolved by 0.180 g glucose = heat capacity × ΔT

Calculations: $1.46 \times 10^3\ \text{J/°C} \times 2.12°C = 3.10 \times 10^3\ \text{J} = 3.10\ \text{kJ}$

We are not finished, because we need heat of reaction in <u>kilojoules per mole</u> of glucose:

(a) $$\frac{3.10\ \text{kJ}}{0.180\ \text{g glucose}} \times \frac{180.2\ \text{g glucose}}{1\ \text{mol glucose}} = 3.10 \times 10^3\ \text{kJ / mol}$$

(b) Since 1 kcal = 4.184 kJ and 1 kcal = 1 dietary Calorie, then

$$\frac{3.10\ \text{kJ}}{0.180\ \text{g}} \times \frac{1\ \text{kcal}}{4.184\ \text{kJ}} = 4.12\ \text{kcal / g} = \underline{4.12\ \text{Cal / g}}$$

Note: The measured heat of reaction is also the ΔH only if the pressure was constant. However, a bomb calorimeter maintains the system at constant volume and the pressure is not necessarily constant. The pressure remains constant in the bomb calorimeter only if:

(1) Gases are not involved in the reaction, or
(2) Gases are involved, but the number of moles of gas is not changed by the reaction.

Example 6.15: In which of the following reactions is the constant volume heat of reaction the same as the enthalpy change?

(a) $C_3H_8(g) + 5O_2(g) \rightarrow 3CO_2(g) + 4H_2O(l)$
(b) $C_6H_{12}O_6(s) + 6O_2(g) \rightarrow 6CO_2(g) + 6H_2O(l)$
(c) $HCl(aq) + NaOH(aq) \rightarrow NaCl(aq) + H_2O(l)$

Answer: (b) and (c). Equation (b) shows the same number of moles of gas on both sides. Equation (c) does not contain gases. In equation (a) the products contain fewer moles of gas than the reactants, so there is a change in pressure at constant volume.

Styrofoam "coffee cup" calorimeters are sometimes used for reactions in aqueous solution. Here the solution itself is the fluid that absorbs the heat.

*An enthalpy change can be estimated from a heat of reaction at constant volume (ΔE) using the formula:

$\Delta H = \Delta E + RT\Delta n$

where Δn is the change in moles of gas shown in the equation.

PROBLEM-SOLVING TIP

Summary of Some Important Thermodynamic Formulas

Formula	Meaning of each variable in the formula	Significance
$\Delta E = q + w$	ΔE = change in internal energy. q = heat: q is positive when heat is absorbed; q is negative when heat is lost. w = work: w is positive when work is done on the system; w is negative when work is done by the system.	Statement of the first law of thermodynamics; the internal energy can change by heat exchange with surroundings and by work.
$w = -P\Delta V$	w = work P = pressure ΔV = change in volume Work is positive when the system contracts: work is negative when the system expands.	The system does work when it expands; work is done on the system when it contracts.

Formula	Meaning of each variable in the formula	Significance
(a)* $\Delta H = \Delta E + P\Delta V$ (b) $= \Delta E + RT\Delta n$	ΔH = change in enthalpy P = pressure during change ΔV = change in volume ΔE = change in internal energy Δn = moles of gas of products – moles of gas of reactants $R = 8.314 \times 10^3$ kJ/mol K	(a) The enthalpy change at constant pressure is the sum of the internal energy change and the $P\Delta V$ product. (b) For reactions involving gases, ΔH can be found by adding the heat of reaction at constant volume (ΔE) to $RT\Delta n$.
$\Delta H = q$ (at constant P)	q = heat, released or absorbed ΔH = enthalpy change: ΔH is negative when heat is released; ΔH is positive when heat is gained.	ΔH is the heat of a reaction at constant pressure.
$\Delta H° = \Sigma\Delta H_f^{o}$ (products) – $\Sigma\Delta H_f^{o}$ (reactants)	$\Delta H°$ = standard enthalpy change ΔH_f^{o} = standard enthalpy of formation (Values listed in Table 6.5 of your textbook.)	Tabulated $\Delta H°_f$ values can be used to calculate $\Delta H°$ for a reaction. If $\Delta H°$ is given then an unknown ΔH_f^{o} value can be found.
q = heat capacity $\times \Delta T$	q = heat absorbed or released; heat capacity of an object is the heat needed to increase its temperature by 1°C. ΔT = change in temperature.	Given two of the quantities, the third can be calculated.
$q = m \times$ specific heat $\times \Delta T$	q = heat absorbed or released; m = mass of the object specific heat = heat capacity per gram. (units are usually J/g°C)	Given three of the quantities, the fourth can be calculated.

Your additions:

SELF-TEST

1. A gas in a balloon absorbs heat and expands. Underline the correct word(s).
   (a) Work is done (on, by) the gas. The work term is therefore (negative, positive).
   (b) The sign of q is (negative, positive).
   (c) If the amount of heat absorbed is greater than the amount of work done, then ΔE for the gas is (positive, negative).
   (d) If the pressure of the gas is 1 atm during the expansion, then ΔH is (greater than, less than, equal to) the heat absorbed.

2. A gas absorbs 200 J of heat and expands from 6.0 L to 9.0 L under a constant pressure of 1 atm. Calculate q, w, ΔE, and ΔH. (1 L atm = 101.3 J)

3. ΔH°_{298} for the combustion of $CH_4(g)$ with liquid water as a product is –890.4 kJ/mol.
   (a) Write the thermochemical equation.
   (b) How many kilojoules of heat are produced from the combustion of 500.0 L of methane at 1 atm and 298K? (R = 0.0821 L atm/mol K)
   (c) Determine ΔH°_f for $CH_4(g)$, given that $\Delta H^\circ_f(CO_2(g)) = -393.5$ kJ/mol; $\Delta H^\circ_f(H_2O(l)) = -285.8$ kJ/mol

4. Determine the ΔH° for $2CO(g) + O_2(g) \rightarrow 2CO_2(g)$ given that $\Delta H^\circ_f(CO_2(g)) = -393.5$ kJ/mol; $\Delta H^\circ_f(CO(g)) = -110.5$ kJ/mol

5. A chemistry student adds 25.0 g of milk at 12.0°C to 150 g of coffee at 75.0°C in a styrofoam cup. If the specific heat of milk is 3.80 J/g°C, estimate the final temperature. Since coffee is mostly water, use the specific heat of water, 4.18 J/g°C. (Hint: the heat lost by the coffee is gained by the milk.)

6. Which liberates more heat, the combustion of benzene ($C_6H_6(l)$), to form $CO_2(g)$ and $H_2O(g)$ in a closed calorimeter or the same reaction open to the atmosphere?

7. True or false? If a statement is false, change the underlined word(s) to make it true.
   (a) The standard state of carbon is a <u>diamond</u>.
   (b) ΔH° for a combustion reaction is a <u>smaller</u> negative quantity if water vapor rather than liquid water is formed.
   (c) When a gas forms from two reacting liquids in a calorimeter the heat of reaction is <u>never</u> equal to the enthalpy change.
   (d) When a gas condenses work is done <u>on</u> the gas.

8. A student mixes 50.0 mL of 1.00 M $NH_3(aq)$ with 200.0 mL of a solution containing excess HCl. The thermochemical equation for the neutralization reaction is

   $NH_3(aq) + HCl(aq) \rightarrow NH_4Cl(aq)$ $\qquad$ $\Delta H = -57.0$ kJ

   Estimate the resulting increase in temperature. The specific heat and density of the solution are the same as for water, 4.18 J/g°C and 1.00 g/mL. Assume volumes are additive.

ANSWERS TO SELF-TEST QUESTIONS

1. (a) by; negative (b) positive (c) positive (d) equal to

2. $q = +200$ J; $w = -3.0$ L atm $= -3.0 \times 10^2$ J; $\Delta E = -1.0 \times 10^2$ J; $\Delta H = +200$ J

3. (a) $CH_4(g) + 2O_2(g) \rightarrow CO_2(g) + 2H_2O(l)$ $\Delta H° = -890.4$ kJ/mol
(b) (Use $PV = nRT$) 18.2 kJ
(c) −74.7 kJ/mol

4. −566 kJ

5. Heat lost by coffee = heat gained by milk

$-150 \text{ g} \times 4.18 \text{ J/g°C} \times (T_f - 75°C) = 25.0 \text{ g} \times 3.8 \text{ J/g°C} \times (T_f - 12°C)$; $T_f = \underline{66.7°C}$

6. Closed calorimeter (constant V)

7. (a) graphite; (b), (c), and (d) true

8. $$0.0500 \text{ L } NH_3 \times \frac{1.00 \text{ mol } NH_3}{1 \text{ L}} \times \frac{-57.0 \text{ kJ}}{1 \text{ mol reaction}} = -2.85 \text{ kJ}$$

$$\Delta T = \frac{2.85 \times 10^3 \text{ J}}{250 \text{ g} \times 4.18 \text{ J / g°C}} = 2.73°C$$

CHAPTER 7
QUANTUM THEORY AND THE HYDROGEN ATOM

SELF-ASSESSMENT

Useful data: $c = 3.00 \times 10^8$m/s; $h = 6.626 \times 10^{-34}$ J s; 1 mi = 1.609 km. Energy of ground-state hydrogen electron is $E_1 = -2.179 \times 10^{-18}$ J

Electromagnetic Radiation and Spectra (Section 7.1)

1. Complete the following table:

Spectral region in which wave is found	Frequency (Hz)	Wavelength
X-rays	2.0×10^{16}	14 ? nm
Radio	2.0×10^{8}	1.499? m
Infrared	2.0×10^{14}	1.499? µm
Visible (Blue)	?6.5×10^{14}	460 nm
?infrared	? 2.998×10^{15}	100 nm
? microwaves	?9.993×10^{10}	0.3 cm

384 551 km

2. The moon is about 239,000 miles from the Earth. How many seconds did it take for an electromagnetic signal from the Apollo mission on the moon to reach earth? 1.28 s

The Quantum Theory of Radiation (Section 7.2)

3. Calculate the energy of the highest energy and the lowest energy photon in the table of Question 1.

4. DNA molecules are damaged by 180 to 320 nm ultraviolet radiation. Calculate the energy in kilojoules of 1 mol of 280-nm photons.

1.85528×10^{-40}

5.* The minimum energy needed to remove an electron from an aluminum surface is 6.7×10^{-19} J. Find the kinetic energy of an electron that is ejected from an aluminum surface by 200-nm radiation.

3.23×10^{-19} J

The Hydrogen Spectrum and the Bohr Atom (Section 7.3)

6. Complete the following statements for a hydrogen electron that drops directly from the n = 3 level to the n = 1 level.
   (a) The energy of the emitted photon is 1.94×10^{-18} J.
   (b) The wavelength of the emitted photon is 102 nm and is in the ______ spectral series of hydrogen.
   (c) If the electron dropped from the n = 3 to the n = 2 level and then dropped from the n = 2 to the n = 1 level, the total energy released would be ________.

7. True or false? If a statement is false, change the underlined word(s) to make it true.
   (a) As n becomes higher the energy of the level becomes <u>lower</u>. Higher
   (b) The difference in energy between the n = 2 and n = 1 levels is <u>greater</u> than the difference in energy between the n = 3 and n = 2 levels.
   (c) The wavelength of the photon absorbed by the transition n = 2 → n = 3 is <u>greater</u> than the wavelength of the photon absorbed by the transition n = 1 → n = 2. T
   (d) The Bohr equation for He^+ is

$$E_n = \frac{-4 \times 2.179 \times 10^{-18} \text{ J}}{n^2}.$$ The energy of its ground state is $\underline{-8.716 \times 10^{-18} \text{ J}}$ T

Matter Waves (Section 7.4)

8. Underline the correct answers:
   (a) As the speed of the beam of electrons in an electron microscope is increased, the associated wavelength (increases, decreases, remains the same).
   (b) The mass of an electron is 9.11×10^{-31} kg. The wavelength of electrons moving at 1.06×10^6 m/s is 6.86×10^{-10} cm, 6.86×10^{-6} mm.

The Uncertainty Principle (Section 7.5)

9. (a) As we come to know the position of a very small particle more precisely we come to know (more, less) about its speed.
   (b) An attempt to measure position and speed has a (greater, smaller) effect on a proton than on a lead atom.

Wave Mechanics (Section 7.6)

10. True or false?. If a statement is false, change the underlined word(s) to make it true.
   (a) Bohr's planetary model for the atom has been replaced by a wave mechanical model that uses <u>probability density</u> to describe the positions of the electrons.
   (b) Three quantum numbers, <u>n, l, and m_l</u> are required to specify an orbital.
   (c) For all 2p orbitals <u>n = 2 and l = 0</u>.
   (d) There are <u>32</u> possible states for electrons in the n = 4 shell.
   (e) Each orbital in a subshell has a different <u>m</u> quantum number.
   (f) The <u>n = 2</u> shell consists of s, p, and d subshells.

$c = \lambda \nu$

$\lambda = \frac{c}{\nu}$

11. Circle the electron states that are not allowed:

(a)	n = 0;	l = 0;	m_l = −1;	m_s = +1/2
(b)	n = 2;	l = 0;	m_l = 0;	m_s = +1/2
(c)	n = 2;	l = 2;	m_l = 2;	m_s = −1/2
(d)	n = 1;	l = 2;	m_l = 2;	m_s = −1/2

$E = h\nu$

Electron Densities (Section 7.7)

12. Draw balloon pictures depicting the shapes and spatial orientations of the s, p_z, and d_{xz} orbitals.

13. True or false? If a statement is false, change the underlined word(s) to make it true.
    (a) The mathematical description of an electron state is its wave function, ψ.
    (b) The probability of finding an electron at a particular point in space is proportional to the value of ψ.
    (c) ψ^2 is the electron density.
    (d) A balloon diagram of the p_z orbital indicates a uniform electron density lying on the surface of a sphere.
    (e) The electron density is zero at a node.
    (f) The p_x, p_y, and p_z orbitals in the n = 3 shell have the same energy, shape, and spatial orientation.

ANSWERS TO SELF-ASSESSMENT PROBLEMS

If you missed an answer, be sure to study the relevant section in the textbook and study guide.

1. X-rays: 15 nm; radio: 1.5 m; infrared: 1.5 μm; visible: 6.5×10^{14} Hz; ultraviolet: 3.0×10^{15} Hz; microwaves: 1×10^{11} Hz

2. 2.39×10^5 mi $\times$ 1.609 km/1 mi $\times$ 10^3 m/1 km $\times$ 1 s/2.998×10^8 m = 1.28 s

3. X-rays: 1.3×10^{-17} J; radio: 1.3×10^{-25} J

4. 7.1×10^{-19} J/photon $\times$ 6.02×10^{23} photons/mol $\times$ 1 kJ/10^3 J = 427 kJ

5. $E = hc/\lambda = 6.626 \times 10^{-34}$ J s $\times$ 2.998×10^8 m/2.00×10^{-7} m = 9.93×10^{-19} J;
   KE = 9.93×10^{-19} J – 6.7×10^{-19} J = 3.2×10^{-19} J

6. (a) 1.937×10^{-18} J (b) 102.6 nm; Lyman (c) 1.937×10^{-18} J released

7. (a) higher (b) T (c) T (d) T

8. (a) decreases (b) 6.86×10^{-10} cm

9. (a) less (b) greater

10. (a) T (b) T (c) n = 2, l = 1 (d) T (2^4 = 32) (e) T (f) n = 3 shell

11. (a), (c), and (d) are not allowed.

12. See balloon pictures within this chapter of the study guide.

13. (a) T (b) ψ^2 (c) T (d) s (e) T (f) same energy and shape but different spatial orientations.

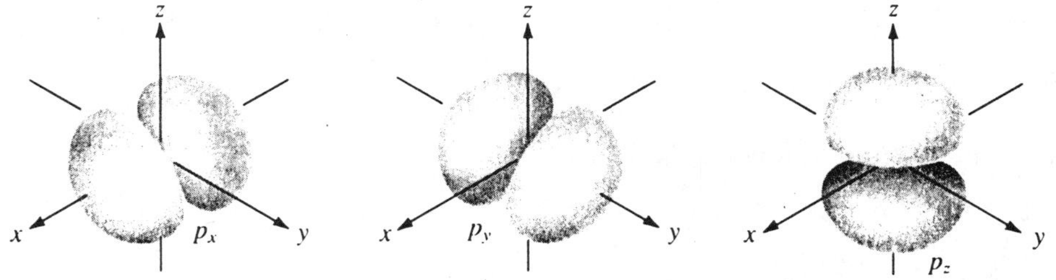

REVIEW

Introduction

An atom is usually in an state of minimum energy (the ground state), but it can absorb energy and exist in a higher energy or excited state. Excited states are temporary; atoms tend to return to their ground states by emitting radiation. The study of the absorption and emission of radiation by matter is called spectroscopy.

Electromagnetic Radiation and Spectra (Section 7.1)

Learning Objectives

1. Calculate the frequency of electromagnetic radiation from its wavelength, and vice versa.
2. List the regions of the electomagnetic spectrum in order of increasing frequency or decreasing wavelength.
3. List the colors of the visable spectrum in order of increasing frequency or decreasing wavelength.
4. Distinguish between a continuous spectrum and a line spectrum.
5. Distinguish between an emission spectrum and an absorbtion spectrum.

Review

Electromagnetic radiation consists of radio, microwaves, infrared, visable, ultraviolet, x-rays, and gamma radiation, in order of increasing frequency. Each of these forms has a range of wavelengths and frequencies. Radiation is chracterized by its wavelength (λ) and frequency (υ) such that

$\lambda \times \upsilon = c$

λ = wavelength (m)
υ = frequency (s^{-1} or Hz)
c = speed of light = 2.998×10^8 m/s

The units of wavelength are any multiple or fraction of a meter. Nanometers (1 nm = 10^{-9} m) are the most frequently used unit in chemical spectroscopy.

Example 7.1: Ozone very strongly absorbs harmful ultraviolet radiation with wavelengths shorter than 330 nm. Find the frequency of this wavelength.

Answer: $\lambda \times \upsilon = c$

$$\upsilon = \frac{c}{\lambda} = \frac{2.998 \times 10^{8}\ \text{m/s}}{330\ \times 10^{-9}\ \text{m}} = \underline{9.08 \times 10^{14}\ \text{s}^{-1}}$$ The answer can also be reported as 9.08×10^{14} Hz

Practice Drill: Complete the following chart:

1 micrometer (μm) = 10^{-6}m
1 nanometer (nm) = 10^{-9} m
1 picometer (pm) = 10^{-12} m

Spectral region	Approximate frequency range (Hz)	Wavelength range
Radio	$0 - 10^{9}$	∞ – ? cm
Microwaves	$10^{9} - 5 \times 10^{11}$	? cm – ? cm
Infrared	$5 \times 10^{11} - 4 \times 10^{14}$	? μm – ? μm
Visible	$4 \times 10^{14} - 8 \times 10^{14}$	? nm – ? nm
Ultraviolet	$8 \times 10^{14} - 10^{17}$	? nm – ? nm
X-rays	$10^{17} - 10^{19}$	? nm – ? nm
Gamma Rays	10^{19} and higher	? pm – ? pm

Answers: Radio: ∞ to 30 cm; microwaves: 30 cm to 0.06 cm; infrared: 600 μm to 0.75 μm; visible: 750 nm to 375 nm; ultraviolet: 375 nm to 3 nm; x-rays: 3 nm to 0.03 nm; gamma rays: 30 pm and shorter

Emission and Absorption Spectra

White light contains all the visible wavelengths. The colors of light are

red, orange, yellow, green, blue, indigo, violet

increasing υ → → → → → → →
decreasing λ → → → → → → →

A beam of light can be separated into its component wavelengths by passing it through a spectroscope; the resulting spectrum consists of the waves spread out in order of wavelength or frequency. Glowing sources give rise to emission spectra: liquids and solids produce continuous emission spectra: gases produce line spectra containing ony certain wavelengths. An absorption spectrum is obtained when light is passed through a substance that absorbs only certain wavelengths of light.

Quantum Theory of Radiation (Section 7.2)

Learning Objectives

1. Calculate the frequency, wavelength, or photon energy, given one of the quantities.
2.* Explain how the value of Planck's constant can be obtained from a plot of photoelectron kinetic energy versus photon frequency.

3.* Calculate W from υ_0 and vice versa.
4.* Use W (or υ_0) and υ to calculate the kinetic energy of a photoelectron.

Review

Light has a dual nature. It exhibits wave-like properties when it forms interference patterns, but it delivers energy in the form of particle-like quanta called photons.

The energy of a photon is related to its frequency by Planck's law:

$E = h\upsilon$

E = energy of photon (J)
h = Planck's constant = 6.626×10^{-34} J s
υ = frequency (s^{-1})

Example 7.2: Calculate the energy in joules of a 240-nm ultraviolet photon arriving from the sun.

Answer: $E = h\upsilon$ and $\upsilon = c/\lambda$ (from $c = \lambda\upsilon$) so that

$$E = \frac{hc}{\lambda} = \frac{6.626 \times 10^{-34}\ \text{J s} \times 2.998 \times 10^{8}\ \text{m/s}}{240 \times 10^{-9}\ \text{m}} = 8.28 \times 10^{-19}\ \text{J}$$

Practice Drill on Planck's law: Complete the following chart:

	Energy of photon	Wavelength	Frequency
(a)	? J	600 nm	? Hz
(b)	? J	? nm	1.05×10^{17} Hz
(c)	2.7×10^{-15} J	? pm	? Hz

Answers: (a) 3.31×10^{-19}J; 5.00×10^{14} Hz (b) 6.96×10^{-17} J; 2.86 nm (c) 74 pm; 4.1×10^{18} Hz

✧The Photoelectric Effect

The light-induced emission of electrons from a metal surface is called the photoelectric effect.

The kinetic energy of photoelectrons is $KE = h\upsilon - W$, where υ is the frequency of the incident light and W is the energy required to dislodge the electron from the metal's surface.

The slope of a kinetic energy versus frequency plot is the same for all metals and is equal to Planck's constant.

The threshold frequency (υ_o), the frequency below which electrons will not be ejected from the metal surface, varies from metal to metal and is given by $\upsilon_o = W/h$.

The Spectrum of Hydrogen (Section 7.3)

Learning Objectives

1. Calculate excited and ground-state energies for the hydrogen electron, given the quantum number n.
2. Calculate the energy, frequency, and wavelength of a photon emitted or absorbed during an electron transition in the hydrogen atom.

3. Identify the electron transitions that give rise to the Lyman Balmer series of spectral lines (see Figure 7.16 (p. 254) of your text).

Review

The emission spectrum of hydrogen consists of several series of lines in the visible, ultraviolet, and infrared regions. Bohr explained this spectrum by assuming that the electron could travel in various planetary-like orbits around the nucleus. Each orbit is characterized by a quantum number, n, such that the energy of the electron orbit is

$E_n = 2.179 \times 10^{-18}\ \text{J}/n^2$

The lowest energy orbit corresponds to n = 1 and is called the ground state; it is the state normally occupied by the hydrogen electron. Higher energy orbits are called excited states. When the value of n is greater than 1, the electron has more energy, is less tightly bound, and is farther from the nucleus than in the ground state.

Example 7.3: Calculate the energy of the state corresponding to n = 3 for the hydrogen atom.

Answer: $E_n = \dfrac{-2.179 \times 10^{-18}\ \text{J}}{n^2} = \dfrac{-2.179 \times 10^{-18}\ \text{J}}{9} = \underline{-2.421 \times 10^{-19}\ \text{J}}$

The energies of different states can be shown on an energy level diagram as seen below. An electron transition from one energy level to another gives off or absorbs a single photon of energy equal to the difference between the two levels:

$E_{photon} = E_{higher} - E_{lower}$

Absorption of Photons

$\_\_\_\_\_\_\_\_\_\_\_\_\_\_ E_{higher}$

$\uparrow \qquad \xleftarrow{h\nu}$

$\_\_\_\_\_\_\_\_\_\_\_\_\_\_ E_{lower}$

Emission of Photons

$\_\_\_\_\_\_\_\_\_\_\_\_\_\_ E_{higher}$

$\downarrow \qquad \xrightarrow{h\nu}$

$\_\_\_\_\_\_\_\_\_\_\_\_\_\_ E_{lower}$

Example 7.4: Calculate (a) the energy in joules and (b) the wavelength in nanometers of the photon emitted when an electron drops from the n = 3 to n = 1 energy level.

Answer: (a) $E_{photon} = E_{higher} - E_{lower} = E_3 - E_1 = \dfrac{-2.179 \times 10^{-18}\ \text{J}}{9} - \dfrac{(-2.179 \times 10^{-18}\ \text{J})}{1} = \underline{1.937 \times 10^{-18}\ \text{J}}$

(b) $E_{photon} = hc/\lambda$ so that $\lambda = hc/E$. Substituting appropriate values,

$$E_{photon} = \frac{6.626 \times 10^{-34}\ \text{J s} \times 2.998 \times 10^{8}\ \text{m/s}}{1.937 \times 10^{-18}\ \text{J}} = 10.26 \times 10^{-8}\ \text{m} = \underline{102.6\ \text{nm}}$$

Practice: Electromagnetic radiation is absorbed by a hydrogen atom in the ground state, and the electron is excited into the n = 3 level. Calculate the (a) energy absorbed in joules and (b) wavelength of this radiation in nanometers. Answers: (a) 1.937×10^{-18} J (b) 102.6 nm

The ultraviolet Lyman series of spectral lines results from transitions from high energy levels to the n = 1 level; the visible Balmer series is caused by transitions to the n = 2 level; and the infrared series results from transitions to n = 3 or higher levels.

The Bohr model cannot be extended to molecules, or to atoms other than hydrogen; it can only be successfully applied to one-electron systems. Bohr's concept of energy states has withstood the test of time, but the planetary model of the atom with well-defined orbits has been abandoned.

Matter Waves (Section 7.4)

Learning Objectives

1. Calculate the de Broglie wavelength with a beam of moving particles.
2. State how the de Broglie wavelength varies with increasing particle mass and speed.

Review

Moving particles have associated matter waves. The wavelength of a matter wave can be calculated using de Broglie's relationship;

$\lambda = h/mv$

λ = wavelength of matter waves (m)
h = Planck's constant = 6.6262×10^{-34} J s
m = mass of moving particle (kg)
v = speed of particles (m/s)

Matter waves are also called de Broglie waves.

Example 7.5: High-energy neutrons are being used to shrink malignant tumors. Given the neutron mass as 1.67×10^{-27} kg, calculate the de Broglie wavelength in picometers associated with a neutron beam moving at 2.00×10^5 m/s. ($1 \text{ m} = 10^{12}$ pm)

Answer: Substitute given values into the de Broglie relationship:

$$\lambda = \frac{h}{mv} = \frac{6.626 \times 10^{-34}\ \text{J s}}{1.67 \times 10^{-27}\ \text{kg} \times 2.00 \times 10^5\ \text{m/s}} = 1.98 \times 10^{-12}\ \text{m} = \underline{1.98\ \text{pm}}$$

Practice: Would the wavelength associated with protons moving at this same speed be larger or smaller than the neutrons? Neutrons are slightly more massive than protons. Answer: larger

Note that the wavelengths associated with everyday, massive, slow-moving objects are extremely small and therefore cannot be detected. Measurable wavelengths are associated with particles with small masses like electrons, protons, and neutrons.

The Uncertainty Principle (Section 7.5)

Learning Objectives

1. State the uncertainty principle and explain why we cannot predict precise paths or orbits for very small particles such as electrons.

Review

The act of measuring a particle's position can alter both its position and momentum. Furthermore, a more precise position measurement will introduce greater uncertainty into the momentum measurement, and vice versa. The uncertainties are inherent in the measuring process. The Heisenberg uncertainty principle states that it is impossible to make simultaneous exact measurements of both the position and momentum of a particle. The uncertainties are appreciable for very small particles, so there is no chance of predicting precise orbits for electrons and other subatomic particles. The uncertainty principle does not, however, rule out predictions based on probabilities.

Wave Mechanics (Section 7.6)

Learning Objectives

1. Write symbols for the orbital and spin quantum numbers and give their range of values.
2. State the number of orbitals in each shell and each subshell.
3. State the total number of states available to an electron in each shell or subshell.
4.* Describe some of the evidence for electron spin.

Review

Wave mechanics provides a model of the hydrogen atom that is consistent with the uncertainty principle. Each state of the atom is associated with a wave function or orbital from which the properties on an electron in that state can be calculated. Many of these properties are expressed in terms of probabilities rather than certainties.

Electron Orbitals and Quantum Numbers

Orbitals are characterized by three quantum numbers;

Principal quantum number (n)	$n = 0, 1, 2, \ldots.$
Azimuthal quantum number (l)	$l = 0, 1, 2, \ldots, n - 1$
Magnetic quantum number (m_l)	$m_l = -1, \ldots, +1$

The principal quantum number, n, is the quantum number discovered by Bohr. It can have only integer values starting with n = 1. Orbitals with the same n value are in the same shell.

The total number of orbitals in a shell is equal to n^2. The energy of an electron in a hydrogen orbital is determined by the principal quantum number:

$$E_n = \frac{-2.179 \times 10^{-18}\ \text{J}}{n^2}$$

The azimuthal quantum number, l, gives the shape of the orbital; orbitals with the same l value are in the same subshell. l can have only integer values from $l = 0$ up to $n - 1$, thus the maximum number of subshells in the shell is equal to the principal quantum number, n. The n = 1 energy shell has one subshell; the n = 2 energy shell has two subshells and so on.

Each subshell has a letter symbol that corresponds to its l value; each subshell also contains $2l + 1$ orbitals. This is summarized in the following table:

l value	0	1	2	3
subshell letter symbol	s	p	d	f
number of orbitals in subshell	1	3	5	7

The symbol for an orbital consists of the n value followed by the letter symbol for the subshell:

Example 7.6: What are the n and l quantum numbers for a 2s, 3p, and 4d orbital?

Answer:			
	2s:	n = 2,	$l = 0$
	3p:	n = 3,	$l = 1$
	4d:	n = 4,	$l = 2$

Example 7.7: List the subshells in the n = 3 shell and the number of orbitals in each subshell.

Answer: The number of subshells is equal to n. The three n = 3 subshells have l values of 0, 1, and 2, and therefore are the s, p, and d subshells with one, three, and five orbitals, respectively.

Practice Drill: Complete the following table:

Electron	n	l
4s	4	0
2p	2	1
5f	5	3
3d	3	2

Answers: 4s: n = 4, $l = 0$; 2p; 5f: n = 5, $l = 3$; 3d

The magnetic quantum number, m_l, determines the behavior of an electron in a magnetic field. The values of m depend on l; m_l has positive and negative integral values ranging from $-l, \ldots, +l$ including 0. It is important to remember that each orbital in a subshell has a different magnetic quantum number.

Example 7.8: What are the magnetic quantum numbers for the three orbitals in a p subshell?

Answer: Since $m_l = -l, \ldots, +l$, and l for a p subshell is 1, then $m_l = -1, 0, +1$.

Practice: Give the magnetic quantum numbers for the orbitals in an s and in a d subshell.
Answers: s: $m_l = 0$; d: $m_l = -2, -1, 0, +1, +2$

An electron in an orbital will be in one of two spin states given by the spin quantum number, m_s: $m_s = +1/2$ or $m_s = -1/2$.

*Evidence for electron spin is provided by the Stern-Gerlach experiment and the Zeeman effect.

Example 7.9: Give four possible quantum numbers for a 2p hydrogen electron.

Answer:

n = 2
$l = 1$ (p electron)
$m_l = 0$ ($m_l = +1$ or $m_l = -1$ are also correct)
$m_s = +1/2$ ($m_s = -1/2$ is also correct)

Practice Drill on Quantum Numbers: Fill in the following chart;

State of hydrogen electron	Four possible quantum values			
	n	l	m_l	m_s
1 s	____	____	____	____
2 p	____	____	____	____
____	3	2	–2	+1/2
4 f	____	____	____	____

Answers: 1s: n = 1, $l = 0$, $m_l = 0$, $m_s^* = 1/2$; 2p: n = 2, $l = 1$, $m_l^* = -1$, $m_s^* = +1/2$; 3d; 4f: n = 4, $l = 3$, $m_l^* = +2$, $m_s^* = -1/2$. *These are not the only correct answers for m_l and m_s.

Electron Densities (Section 7.7)

Learning Objectives

1. Sketch balloon pictures for s, p, and d orbitals.
2. Distingish between p_x, p_y, and p_z orbitals.
3. Distinguish between electron density and radial electron density.
4. State how the most probable density of an electron from the nucleus varies with increasing n and, within a shell, with increasing l.

Review

The value of the wave function at any point in space is symbolized by ψ. (Remember that ψ is a mathematical description of an electron state. Each orbital has different ψ expression.) The probability of finding an electron in a small volume of space centered about some point is proportional to the value of ψ^2 at that point. Probabilities can be visualized by plotting ψ^2 versus the distance from the nucleus or by means of electron cloud (dot) diagrams. ψ^2 is called the electron density because as ψ^2 increases, the density of the electron cloud also increases. Electron density diagrams contain nodal regions, in which the electron density falls to zero. The electron density in an s orbital is spherically symmetrical.

A balloon picture of an orbital depicts a surface of uniform electron density chosen so that most of the electron density is inside it. The balloon picture of an s orbital shows a sphere (often shown only in cross-section). A balloon picture for a p orbital shows two lobes of density separated by a nodal plane. The three p orbitals have the same shape, but are oriented differently in space. The d and f orbitals are more elaborate, usually containing several lobes and two or more nodal planes.

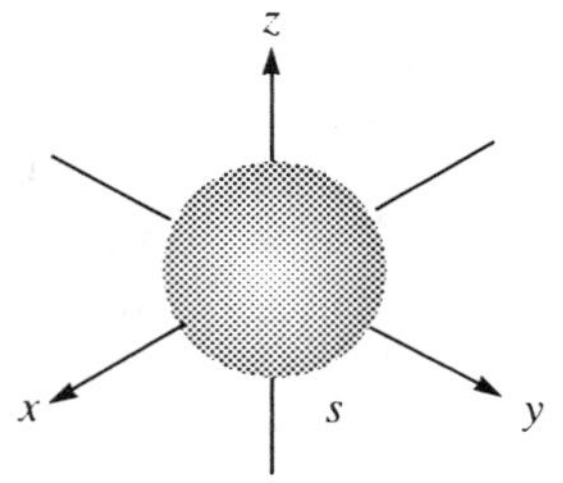

(a)

Balloon picture for s orbital

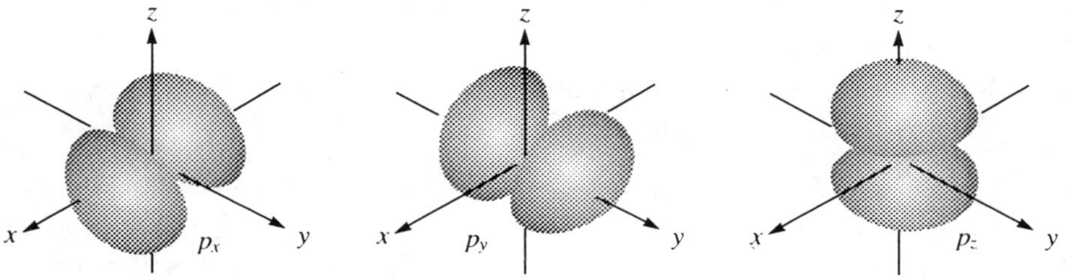

(b)

Balloon picture for the three p orbitals

*Radial electron density plots show the probability of finding an electron at a given distance from the nucleus regardless of direction. The most probable distance for the electron is where the radial density is a maximum. The most probable distance increases with increasing n; it is equal to 52.9 pm for a ls hydrogen electron. Within a shell, the most probable distance decreases with increasing *l*.

PROBLEM-SOLVING TIP

Summary of Important Relationships in Quantum Theory

Relationship	Meaning and units of variables	Significance
$c = \lambda\nu$	c = speed of electromagnetic radiation = 2.998×10^{8} m/s λ (lambda) = wavelength (m) ν (nu) = frequency (s^{-1} or Hz)	Given λ you can calculate ν and vice versa. λ and ν are inversely proportional; greater ν, shorter λ.
$E = h\nu$	E = energy of photon (J) h = Planck's constant = 6.6261×10^{-34} J s ν = frequency of wave (s^{-1})	Light energy is carried in packets called photons. Equation shows dual nature of light; photon from particle-like nature, ν from wave-like nature. Given ν you can calculate E and vice versa.
$^*KE = h\nu - W$	KE = kinetic energy of ejected electron (J) $h\nu$ = energy of photon striking the surface (J) W = work function (J), the minimum energy needed to eject an electron.	Photoelectric effect; energy of photons ejects electrons (photoelectrons) from metals and imparts KE to them.
$^*\nu_o = W/h$	ν_o = threshold frequency (s^{-1}) above which an electron will be ejected with KE.	Not all light has enough energy to eject electrons. ν_o is threshold frequency below which no electrons will be emitted.
$E_n = \dfrac{-2.179 \times 10^{-18}\ \text{J}}{n^2}$	E_n = energy (J) of the electron in the nth energy shell of the hydrogen atom. n = principal quantum number of the shell.	Bohr's model of H atom. Electron energy depends on the principle quantum number, n. Electrons can have only those energies for which n is an integer quantity, starting with n = 1.
$E_{photon} = -2.179 \times 10^{-18}\ \text{J}\left(\dfrac{1}{n_H^2} - \dfrac{1}{n_L^2}\right)$	E_{photon} = energy of photon emitted or absorbed by the hydrogen atom n_H = principal quantum number of excited state n_L = principal quantum number of lower energy state	Light is emitted or absorbed when electrons change energy shells. Use to relate E_{photon} to the energy shells involved in the transition.
$\lambda = h/mv$	λ = de Broglie wavelength (m) for matter wave m = mass of moving particles (kg) v = speed of moving particle (m/s) h = 6.626×10^{-34} J s	Moving particles have an associated wavelength, which is greater for smaller and faster particles.

Your additions:

SELF-TEST

1. The frequency of a 530-nm green light is 5.66×10^{14} Hz and the energy of its photon is 3.75×10^{-19} J.

2. Rank the following in order of decreasing wavelength (increasing energy): x-rays, ultraviolet, microwaves, blue light, and red light. x rays microwaves ultraviolet blue red

3. For He^+, $E_n = \dfrac{-4 \times 2.179 \times 10^{-18}\ J}{n^2}$, calculate the following:
   (a) energy of the ground state -8.716^{-18}
   (b) energy of the photon released when electron drops from $n = 4 \rightarrow n = 1$ 9.26×10^{-18}
   (c) wavelength associated with photon absorbed when electron jumps from $n = 1 \rightarrow n = 4$ 2.1 nm

4. A 60-kg marathon runner who averages 4 m/s has an associated de Broglie wavelength of 2.76×10^{-27} nm; the wavelength of an electron (mass 9.11×10^{-31} kg) traveling at 1.5×10^8 m/s is 4.85×10^{-3} nm.

5. Match each statement in Column A to the formula that supports it in Column B.

A	B
(1) The energy of the hydrogen electron increases as its principal quantum number increases.	(a) $\lambda = \dfrac{h}{mv}$
(2) The kinetic energy of a photoelectron is the difference between the striking photon energy and the work function of the metal.	(b) $E = h\upsilon$
(3) Moving particles have an associated wavelength.	(c) $E_n = \dfrac{-2.179 \times 10^{-18}\ J}{n^2}$
(4) The energy of light is proportional to its frequency.	(d) $E_{photon} = E_H - E_L$
(5) The energy of a photon released when an electron jumps from a higher energy state to a lower energy state is the difference between the two energy states.	(e)* $KE = h\upsilon - W$

6. Fill in the following:
   (a) The shape of an orbital is related to the ℓ quantum number.
   (b) The values of quantum numbers l and m_l are both 0 for a 4s orbital.
   (c) All electrons in the same shell have the same quantum number n.
   All electrons in the same subshell have the same quantum number ℓ.
   All electrons with parallel spins have the same quantum number m_s.

7. In each of the following, change one quantum number to make it an allowed electron state. (There is more than one right answer.)

(a)	$n = 1$;	$l = 1$; → L=0	$m_l = 0$;	$m_s = 1/2$
(b)	$n = 3$;	$l = 2$;	$m_l = 3$; $m_l=2$	$m_s = -1/2$
(c)	$n = 3$;	$l = 0$;	$m_l = 0$;	$m_s = 1$ $m_s=1/2$
(d)	$n = 1$; n=3	$l = 2$;	$m_l = 2$;	$m_s = -1/2$

8. Match Column A to the best answer in Column B:

A		B	
(1)	balloon picture of an s orbital	(a)	electron density
(2)	wave function, ψ	(b)	zero electron density
(3)	ψ^2	(c)	sphere
(4)	dot diagram	(d)	one way to represent the electron cloud
(5)	node	(e)	mathematical function for an electron state

ANSWERS TO SELF-TEST QUESTIONS

1. 5.66×10^{14} Hz; 3.75×10^{-19} J

2. microwaves > red light > blue light > ultraviolet > x-rays

3. (a) -8.716×10^{-18} J (b) 8.171×10^{-18} J (c) 0.2431 nm

4. 2.8×10^{-27} nm; 4.8×10^{-3} nm

5. (1) c (2) e (3) a (4) b (5) d

6. (a) azimuthal or l (b) 0 (c) n; l; m_s

7. (a) $l = 0^*$ (b) $m_l = 2^*$ (c) $m_s = +1/2^*$ (d) $n = 3^*$ (*Not the only correct answer.)

8. (1) c (2) e (3) (a) (4) d (5) b

CHAPTER 8
MANY ELECTRON ATOMS AND THE
PERIODIC TABLE

$1s$
$4p$ $2p$

SELF-ASSESSMENT

Electron Energies in Many-Electron Atoms (Section 8.1)

Underline the correct word.

1. The 3p electron in chlorine has a (higher, lower) energy than the 3p electron of bromine. The (bromine, chlorine) 3p electron is more tightly bound to the atom.

2. Rank the following electrons of uranium (a seventh period element) in order of increasing energy: 4s, 4d, 4p, and the 4f.

The Pauli Exclusion Principle (Section 8.2)

3. True or false? If the statement is false, change the underlined word(s) to make it true.
 (a) Four is the maximum number of quantum numbers that can be the same for two electrons in the same atom.
 (b) The maximum number of electrons that can occupy the n = 3 shell is <u>8</u>. (18)
 (c) Paired electrons have <u>the same</u> values for the spin quantum number.
 opposite

Electron Configurations (Section 8.3)

4. Write the following for Co (Z = 27):
 (a) expanded and abbreviated electron configuration $1s^2 2s^2 2p^6 3s^2 3p^6 4s^2 3d^7$ $[Ar]3d^7 4s^2$
 (b) orbital diagram for the 3d subshell _____ ↑↓ ↑↓ ↑ ↑ ↑
 (c) number of unpaired electrons ____3____

The Periodic Table (Section 8.4)

5. Use the periodic table to help answer the following:
 (a) State the number of periodic groups in the s block ____2____, p block ____6____, and the d block ____10____.

(b) State which block contains (1) the halogens __________; (2) Group 5A __________; (3) Sc __________; (4) the actinides __________.

(c) Write the abbreviated electron configuration of Hg (Z = 80). __________.

(d) Write the valence shell configuration for the Group 4A elements. __________.

(e) Write the abbreviated electron configuration of S^{2-} __________; Mn^{3+} __________; Sr^{2+} __________; Br^{-} __________.

(f) State the number of valence electrons in Rb __________; Ga __________; Po __________.

6. Fill in the following chart, using the periodic table when necessary:

	Element	Group	Period	Valence Shell Configuration
(a)	________	________	________	$6s^2$
(b)	________	4A	4	________
(c)	At	________	________	________
(d)	________	8A	________	$5s^2\ 5p^6$

Atomic Properties (Section 8.5)

Use the periodic table to help answer Questions 7-9.

7. Rank the following atoms in order of increasing degree of paramagnetism: Li, Ar, Mn and Fe.

8. Which of the following contain three unpaired electrons? N, N^{3-}, V^{3+}, V, Co^{2+}, O^{2-}

9. Rank Cl, S, Sc, and Ca in order of
   (a) increasing atomic radius
   (b) decreasing ionization energy
   (c) increasing metallic character

10. The first ionization energy of B is less than the first ionization energy of Be although B is to the right of Be in the first period. Offer an explanation based on atomic structure. Be electron config.

Underline the correct word(s).

11. The greater the tendency of an atom to gain an electron the more (positive, negative) the atom's electron affinity.

Descriptive Chemistry and the Periodic Table (Section 8.6)

12. Metals tend to (gain, lose) electrons in combining with other elements. Metals are found toward the (left, right) and the (top, bottom) of the periodic table.

13. Nonmetals in general have (higher, lower) ionization energies than metals.

ANSWERS TO SELF-ASSESSMENT PROBLEMS

If you missed an answer, be sure to study the relevant section in the textbook and study guide.

1. higher, bromine

2. 4s < 4p < 4d < 4f

3. (a) three (b) 18 (c) different
4. (a) $1s^2\ 2s^2\ 2p^6\ 3s^2\ 3p^6\ 4s^2\ 3d^7$; $[Ar]\ 4s^2\ 3d^7$ (b) ↑↓ ↑↓ ↑ ↑ ↑ (c) 3 unpaired
5. (a) 2; 6; 10
   (b) (1) p (2) p (3) d (4) f
   (c) $[Xe]\ 6s^2\ 4f^{14}\ 5d^{10}$
   (d) $s^2\ p^2$
   (e) [Ar]; $[Ar]\ 3d^4$, [Kr]; [Kr]
   (f) 1; 3; 6
6. (a) Ba: 2A, 6, $6s^2$
   (b) Ge: 4A, 4, $4s^2\ 4p^2$
   (c) At: 7A, 6, $6s^2\ 6p^5$
   (d) Xe: 8A, 5, $5s^2\ 5p^6$
7. Ar < Li < Fe < Mn
8. N, V, Co^{2+}
9. (a) (b) (c) Cl, S, Sc, Ca
10. The 2p electron of boron ($2s^2\ 2p^1$) experiences a smaller effective charge and is therefore less tightly bound than the two 2s electrons of beryllium ($2s^2$).
11. negative
12. lose, left and bottom
13. higher

REVIEW

Introduction

The orbitals in many-electron atoms have the same quantum numbers and shapes as hydrogen orbitals, but differ in their electron energies and in the most probable distances of their electrons from the nucleus.

Electron Energies in Many-Electron Atoms (Section 8.1)

Learning Objectives

1. List the similarities and differences between the orbitals of a many-electron atom and those of hydrogen.
2. List the orbitals within a shell in order of increasing electron energy.
3. State how the energy of an electron in a given orbital varies with increasing atomic number.
4. Describe the shielding effect, and explain why the energy of an electron increases with increasing *l* within a shell.

Review

Two new facts are important in dealing with many-electron atoms:
1. All orbital energies decrease with increasing atomic number.
2. The energies within a given shell increase with increasing *l*.

Example 8.1: How does the energy of the 2p orbital in Na differ from that of the 2p orbital in hydrogen?

Answer: The energy for any given orbital decreases with increasing atomic number so that the Na 2p orbital has a lower energy than the H 2p orbital.

> **Practice:** Compare the energy of the 3d orbitals of Sb and Te.
> Answer: 3d of Te < 3d of Sb

The effective nuclear charge experienced by an electron is less than the actual nuclear charge because of the shielding effect of inner electrons. Within a given shell the shielding effect is greater on subshells with higher *l* values.

Example 8.2: Rank electrons in the n = 4 subshell in decreasing order of the effective nuclear charge they experience

Answer: $\xrightarrow[\text{decreasing effective nuclear charge}]{4s \; > \; 4p \; > \; 4d \; > \; 4f}$

> **Practice:** Rank the subshells in decreasing order of shielding by inner electrons.
> Answer: 4f > 4d > 4p > 4s

The difference in effective nuclear charge causes the increase in energy with increasing *l* value. For example, the energies of electrons in an n = 4 shell increase in the order 4s < 4p < 4d < 4f.

> **Practice:** For each of the following pairs, indicate which orbital would have the higher energy in Cs, a many-electron atom. (a) 2s or 2p (b) 3d or 3s (c) 1s or 2s
> Answers: (a) 2p (b) 3d (c) 2s

The Pauli Exclusion Principle (Section 8.2)

Learning Objectives

1. State and apply the Pauli exclusion principle.

Review

The Pauli exclusion principle states that only two electrons can occupy the same orbital, and these two electrons have opposite spins.

Example 8.3: Give four possible quantum numbers for each of two electrons in a 2p orbital.

Answer: For both 2p electrons, n = 2 and $l = 1$. Because m_l can range from $-l$ to $+l$, m_l can be -1, 0, or +1 for a 2p orbital and will be the same for both electrons. (Let us choose $m_l = 1$.) The Pauli principle tells us that two electrons in one orbital must be paired and therefore must differ in their spin quantum number: $m_s = +1/2$ for one electron and $m_s = -1/2$ for the other electron. The four possible quantum numbers are

n = 2	n = 2
$l = 1$	$l = 1$
$m_l = 1$	$m_l = 1$
$m_s = +1/2$	$m_s = -1/2$

Electron Configurations (Section 8.3)

Learning Objectives

1. State the orbital filling order for the first 36 elements.
2. Write ground-state electron configurations in both abbreviated and expanded forms for the first 36 elements.
3. Write orbital diagrams for the first 36 elements.

Review

The distribution of electrons in the orbitals of an atom is called the electron configuration. The ground-state electron configuration is the configuration that provides the lowest energy for the atom as a whole. Ground-state configurations are obtained by the aufbau procedure which adds electrons one at a time into available orbitals according to an experimentally determined orbital filling order. For the first 36 elements ($Z = 1 \rightarrow Z = 36$) the order is:

1s 2s 2p 3s 3p 4s 3d 4p →
orbital filling order

Each orbital holds no more than two electrons (Pauli exclusion principle). Electrons in a partially filled subshell remain unpaired with parallel spins until each orbital has one electron in it (Hund's rule).

Hund's rule: Electrons occupy orbitals of a subshell singly and with parallel spin (same spin quantum numbers) until each orbital has one electron in it. Hund's rule can be explained by the tendency of electrons to reduce repulsion by staying apart. Pairing occurs when each orbital in the subshell already contains one electron. Except for chromium and copper, the above filling order gives correct configurations for the first 36 elements.

Example 8.4: Apply the orbital filling order, the Pauli exclusion principle, and Hund's rule to determine the ground-state electron configuration and orbital diagram for N (Z = 7).

Answer: The orbital filling order puts two electrons in the 1s orbital ($1s^2$). The next orbital to fill up is the 2s orbital ($2s^2$). The remaining three electrons are in the 2p subshell ($2p^3$). The three electrons occupy different 2p orbitals and have parallel spins

$1s^2\ 2s^2\ 2p^3$ (electron configuration)

⇅	⇅	↑	↑	↑
1s	2s		2p	

(orbital diagram)

Practice: Show that P (Z = 15) has three unpaired electrons.

Answer: $1s^2\ 2s^2\ 2p^6\ 3s^2\ 3p^3$ ↑ ↑ ↑ (3p)

Example 8.5: Give the ground-state electron configuration for Fe (Z = 26) using the orbital filling order.

Answer: $1s^2\ 2s^2\ 2p^6\ 3s^2\ 3p^6\ 4s^2\ 3d^6$

Example 8.6: Give the orbital diagram for Fe.

Answer:	1s	2s	2p	3s	3p	4s	3d
	↑↓	↑↓	↑↓ ↑↓ ↑↓	↑↓	↑↓ ↑↓ ↑↓	↑↓	↑↓ ↑ ↑ ↑ ↑

Practice: Give the orbital diagram for the 3d subshell of Mn (Z = 25) and Co (Z = 27).

Answer: Mn: ↑ ↑ ↑ ↑ ↑ (3d); Co: ↑↓ ↑↓ ↑ ↑ ↑ (3d)

An abbreviated form based on noble gas configurations is often used to write long configurations:

$1s^2 = [He]$

$1s^2\ 2s^2\ 2p^6 = [Ne]$

$1s^2\ 2s^2\ 2p^6\ 3s^2\ 3p^6 = [Ar]$

$1s^2\ 2s^2\ 2p^6\ 3s^2\ 3p^6\ 4s^2\ 3d^{10}\ 4p^6 = [Kr]$

The abbreviated configurations for C (Z = 6) and P (Z = 15) are

Expanded Form	Abbreviated Form
$C = \underbrace{1s^2}_{He}\ 2s^2\ 2p^2$	$[He]\ 2s^2\ 2p^2$
$P = \underbrace{1s^2\ 2s^2\ 2p^6}_{Ne}\ 3s^2\ 3p^3$	$[Ne]\ 3s^2\ 3p^3$

Practice: Give the expanded and abbreviated electron configuration for Se (Z = 34).
Answer: abbreviated: $[Ar]\ 4s^2\ 3d^{10}\ 4p^4$

PROBLEM-SOLVING TIP

Rules for Determining Ground-State Electron Configurations for Atoms ($Z = 1 \rightarrow Z = 36$)

1. Determine the number of electrons in the atom. (Same as the atomic number of the atom.)
2. Fill each subshell in the orbital filling order until all of the electrons are accounted for. The order is:

   $\xrightarrow{\text{1s 2s 2p 3s 3p 4s 3d}}$

   The number of orbitals and electrons in each subshell are:

Subshell	Number of orbitals	Maximum number of electrons
s	1	2
p	3	6
d	5	10
f	7	14

3. If a subshell is incomplete, assign one electron to each orbital and give them parallel spins. Assign the remaining electrons, if any, to the already occupied orbitals showing antiparallel spins. (Make sure you have the right number of electrons.)

4. To write the configuration in abbreviated form, substitute the noble gas symbol in brackets for the part that is the noble gas configuration.

The Periodic Table (Section 8.4)

Learning Objectives

1. Identify the s, p, d, and f blocks in the periodic table.
2. Given a periodic table, identify the representative elements, the transition elements, the lanthanides, and the actinides.
3. Give the name of each of the representative groups.
4. Explain why the elements in a periodic group exhibit similar chemical properties.
5. Use group numbers to state the number of valence electrons in a representative element.
6. Use the periodic table to write ground-state electron configurations of atoms and monatomic ions.

Review

The vertical groups in the periodic table contain elements with similar chemical properties. Each horizontal row, or period, begins with the addition of an s electron to an unoccupied shell and ends with a noble gas. The period number is the principal quantum number, n, of the shell that begins to fill and that acquires the noble gas configuration. A long period ($n > 3$) contains transition elements in which $(n-1)$d orbitals fill.

Example 8.7: State the order in which the subshells fill in the third period (Na to Ar).

Answer: The n = 3 shell begins to fill with the 3s followed by the 3p. The 3p is filled with Ar.

Practice: State the order of filling in the 4th period. Answer: 4s, 3d, 4p

The divisions in the periodic table are (see Figure 8.4, in text):

Representative elements (A Groups)
s block (Groups 1A, 2A)
p block (Groups 3A-8A)

Transition elements (B Groups)
d block (Groups 3B-8B, followed by 1B and 2B)
f block (inner transition elements)
lanthanides (Z = 58-71, rare earth elements)
actinides (Z = 90-103)

See Table 8.2 in your textbook for the names of the representative groups.

Example 8.8: In which block, s, p, d or f, do the following elements appear? (Use the periodic table): Na, Mn (manganese), Rn (radon), U, Ce (cerium). For representative elements also give the name of the group.

Answer:
Na (Group 1A) s-block ; alkali metals
Mn (transition element, Group 7B) d-block
Rn (Group 8A) p-block; noble gases
U (actinide, Z = 92) f-block
Ce (lanthanide, Z = 58) f-block

Practice: Label the following elements as representative, transition metal, lanthanide, or actinide: Lu (lutetium), P, Co, Th (thorium); Answer: lanthanide; representative; transition metal, actinide

Finding Electron Configurations from the Periodic Table

The ground-state configuration of an element can usually be determined from its position in the periodic table. The procedure is:

1. Find the noble gas in the period preceding the atom of interest. This noble gas symbol begins the abbreviated form of the electron configuration.
2. Count the number of spaces across the period in which the element occurs. This gives the number of additional electrons in the orbitals that follow the noble gas configuration.

Example 8.9: Use the periodic table to determine the electron configuration for selenium, Se (Z = 34).

Answer:

1. Se is in the fourth period, Group 6A. The preceding noble gas is argon, Ar, in the third period. Therefore the configuration starts with [Ar].
2. It takes 16 spaces in the fourth period to get to Se: 2 electrons from the s block, 10 electrons from the d block, and 4 electrons from the p block. The configuration of Se is therefore

   $[Ar]4s^2\ 3d^{10}\ 4p^4$. (The d block begins in the 4th period, so these are the first level of d electrons, 3d.)

Practice: Use the periodic table to write the electron configuration for bismuth, Bi (Z = 83). Answer: $[Xe]\ 6s^2\ 4f^{14}\ 5d^{10}\ 6p^3$

Valence Electrons

Elements from the same group have similar chemical properties because their atoms have the same outer shell configurations. The outermost electron shell (shell with the highest n value) is called the valence shell. Except for He the number of valence electrons in an atom of a representative element is the same as its group number. For d-block and f-block elements, valence electrons include those from the n – 1 and n – 2 shell respectively

Example 8.10: Give the valence shell configurations for selenium and phosphorus.

Answer: The periodic table shows
P: Group 5A – so 5 valence electrons, $s^2\ p^3$; 3rd period – so $3s^2\ 3p^3$
Se: Group 6A – so 6 valence electrons, $s^2\ p^4$; 4th period – so $4s^2\ 4p^4$

Practice: Give the valence shell configurations for Rb and Te.
Answer: $5s^1$; $5s^2\ 5p^4$

Electron Configurations of Monatomic Ions

When an atom forms a positive ion, its electrons leave in order of decreasing principal quantum number. (This is not always the reverse of the filling order.) Within a shell (same n value), d electrons are lost before p electrons, and p electrons are lost before s electrons.

Example 8.11: Write ground-state electron configurations of Na^+, Al^3, and Co^{3+}.

Answer: Start by writing the configuration for the neutral atoms:

Na (Z = 11)	[Ne] $3s^1$
Al (Z = 13)	[Ne] $3s^2\ 3p^1$
Co (Z = 27)	[Ar] $3d^7\ 4s^2$

Now remove the electrons in order of decreasing n and *l*; Na will lose its 3s electrons, Al will lose one 3p and two 3s electrons, and Co will lose two 4s electrons and one 3d electron. The electron configurations are

Na^+ [Ne]

Al^{3+} [Ne]

Co^{3+} [Ar] $3d^6$

Practice: Give the configurations for Mn^{7+} and Sr^{2+}.
Answer: Mn^{7+}: [Ar]; Sr^{2+}: [Kr]

The ground-state configuration of a stable negative monatomic ion is the same as that of the noble gas at the end of its period.

Example 8.12: Give the ground-state configuration for S^{2-}.

Answer: Find the number of electrons in S and then add two more to account for the charge:

total electrons = 16 (from S atom) + 2 (from charge) = 18 electrons

The configuration will be the same as the atom with 18 electrons, namely Ar.

Practice: Which noble gases have the configurations of P^{3-}, Br^{-}, and I^{-}?
Answer: Ar, Kr, Xe

Atomic Properties (Section 8.5)

Learning Objectives

1. Use electron configurations to predict whether atoms and ions are paramagnetic or diamagnetic.
2. Use periodic trends to compare atomic and ionic radii.
3. Use periodic trends to compare ionization energies of atoms and ions.
4. Use periodic trends to compare electron affinities of atoms.
5. Explain why the second electron affinity for an atom is always positive.

Review

Many of an element's properties depend on its atomic number and electron configuration, and can therefore be predicted from its position in the periodic table.

Paramagnetism: Atoms or ions with unpaired electrons are attracted toward a magnetic field and are said to be paramagnetic. Paramagnetic behavior increases with the number of unpaired electrons. If all electrons are paired then the atom or ion is diamagnetic, and is not attracted by a magnetic field.

Example 8.13: Rank Sc (Z = 21), Ti (Z = 22), and V (Z = 23) in order of increasing paramagnetism.

Answer: We must first find the electron configurations and then count the number of unpaired electrons:

Sc [Ar] $4s^2\ 3d^1$ (1 _ _ _ _) one unpaired electron

Ti [Ar] $4s^2\ 3d^2$ (1 1 _ _ _) two unpaired electrons

V [Ar] $4s^2\ 3d^3$ (1 1 1 _ _) three unpaired electrons

Sc < Ti < V
→ increasing paramagnetic behavior

Practice: Rank the following in order of decreasing paramagnetism:
Ca^{2+}, Mn^{2+}, Fe^{2+} Answer: $Mn^{2+} > Fe^{2+} > Ca^{2+}$

Atomic size: Atomic size is measured by the atomic radius, which is one-half the average distance between centers of identical touching atoms.

Atomic radii generally increase from top to bottom within a group (because the number of electron shells increases).

Atomic radii generally decrease from left to right across a period (because the nuclear charge increases while electrons are being added to already occupied shells).

Positive ions are smaller than their parent atoms: negative ions are larger.

The decrease in atomic radius across the lanthanide series is called the lanthanide contraction.

Ionization energy (IE): IE is the energy required to remove an electron from a gaseous atom or ion. The more tightly held the electron, the more energy it takes to remove it and the higher the IE.

IE's generally decrease going down a group (as atomic size increases the outer electrons are less tightly held).

IE's generally increase going across a period from left to right (as atomic size decreases the outer electrons are more tightly held).

The first ionization energy, which removes the first or highest energy electron from the neutral atom, is always less than the second, which removes a second electron from the positive ion. The second IE is less than the third, and so forth.

An atom with a low ionization energy will form positive ions more easily. Metals especially those toward the left and bottom of the table, readily form stable positive ions. Nonmetals, except for hydrogen, do not form stable positive ions.

Electron affinity (EA): EA is the enthalpy change that occurs when a gaseous ground-state atom or ion gains an electron. Group 2A elements and noble gases have positive EA's. All other atoms have electron affinities; that is, they release energy when they gain an electron.

The halogens have the largest negative electron affinities so that they form very stable negative ions. Metal atoms do not form stable negative ions.

Summary of Periodic Trends

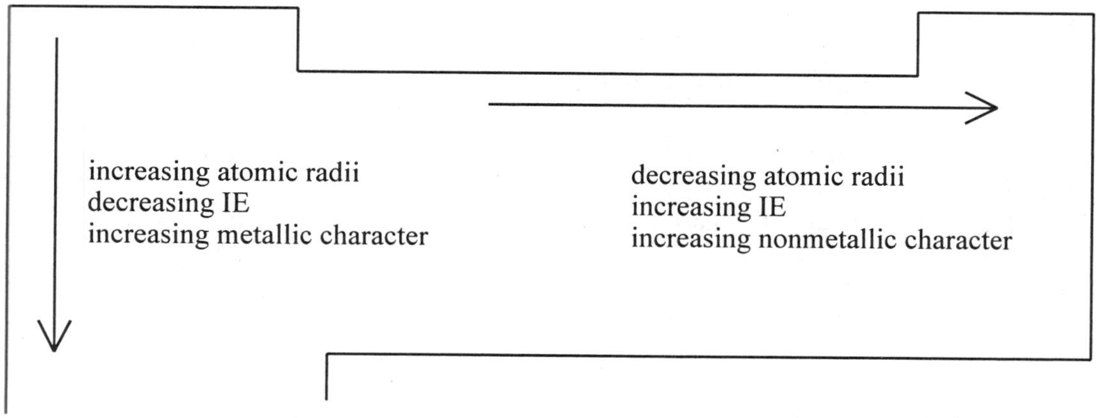

Electron affinities exhibit many irregularities. Elements with large negative values are found in the upper right portion of the table.

Example 8.14: Use the periodic table to predict which atom in each pair has the smaller atomic size and also the greater first IE: (a) C or O (b) N or P (c) Na or Mg

Answer: (a) O (b) N (c) Mg

Practice: Which of the following pairs has a more negative electron affinity? (a) Na or Cl (b) Mg or S (c) O or Te Answers: (a) Cl (b) S (c) O

Descriptive Chemistry and the Periodic Table (Section 8.6)

Learning Objectives

1. Distinguish between metals and nonmetals on the basis of their physical and chemical properties.

Review

Summary of the General Properties of Metals and Nonmetals

	Physical Properties	Chemical Properties	+Location in Periodic Table
Metals	Lustrous, conductors of electricity and heat, malleable and ductile.	Low IE, many form positive ions. Little electron affinity.	All of the s block except for H. All of the d and f blocks. The p block to the left of the +stepwise line.
Nonmetals	Not lustrous, poor conductors of electricity and heat Solids are brittle.	High IE. Very negative electron affinity. Many form negative ions.	Right of the +stepwise line (except for H).

+ The periodic table is divided by a stepwise line into metallic and nonmetallic elements. Next to the stepwise line are found semimetals, which share some of the characteristics of both metals and nonmetals.

SELF-TEST

You may use the periodic table for the self-test.

1. Choose the highest energy electron of each pair given:
   (a) Li 1s, Na 1s (c) K 2s, K 2p
   (b) Na 1s, Na 2s (d) K 2s, Ca 2s

2. A 4p electron has the quantum numbers: n = 4, $l = 1$, $m_l = 1$, $m_s = 1/2$. Give the four quantum numbers for
   (a) another electron in the same 4p orbital $n = 4$ $l = 1$, $m_l = 0$ $m_s = -1/2$
   (b) a 4p electron in a different orbital but with a parallel spin

3. Give the expanded electron configuration for
   (a) P^{3-} (Z = 15) $1s^2 2s^2 2p^6 3s^2 3p^6$
   (b) Sr (Z = 38) $1s^2 2s^2 2p^6 3s^2 3p^6 4s^2 3d^{10} 4p^6 5s^2$
   (c) Cr^{3+} (Z = 24) $1s^2 2s^2 2p^6 3s^2 3p^6 4s^2 3d^1$

4. Match the atom or ion in Column A with its abbreviated configuration in Column B.

A	B
(1) Li^+ (Z = 3) C	(a) [Ar] $3d^5$
(2) Br^- (Z = 35) d	(b) [Ar] $4s^2$
(3) Mn^{2+} (Z = 25) A	(c) [He]
(4) Ca (Z = 20) b	(d) [Kr]

5. Use the periodic table to find the
   (a) number of valence electrons in P
   (b) abbreviated electron configuration of Cd
   (c) outer shell configuration of radon, Rn

6. For the following pairs state which atom (1) has the higher IE, (2) has more metallic character, and (3) is more paramagnetic.
   (a) Al, S
   (b) O, B
   (c) S, Se
   (d) Ba, O

7. Fill each blank with increases, decreases, or remains the same.
   (a) As a nonmetal gains electrons its radius decreases x.
   (b) As you go down Group 2A the degree of paramagnetism remains the same.
   (c) As the positive charge on a manganese ion increases from +2 to +3, the number of unpaired electrons increase. x

ANSWERS TO SELF-TEST QUESTIONS

1. (a) Li 1s (b) Na 2s (c) K 2p (d) K 2s

2. (a) $n = 4, l = 1, m_l = 1, m_s = -1/2$; (b) $n = 4, l = 1, m_l = -1, m_s = 1/2$ ($m_l = 0$ is also correct)

3. (a) P^{3-}: $1s^2\ 2s^2\ 2p^6\ 3s^2\ 3p^6$
   (b) Sr: $1s^2\ 2s^2\ 2p^6\ 3s^2\ 3p^6\ 4s^2\ 3d^{10}\ 4p^6\ 5s^2$
   (c) Cr^{3+}: $1s^2\ 2s^2\ 2p^6\ 3s^2\ 3p^6\ 3d^3$

4. (1) c (2) d (3) a (4) b

5. (a) 5 (b) [Kr] $5s^2\ 4d^{10}$ (c) $6s^2\ 6p^6$

6. (a) S higher IE and more paramagnetic; Al more metallic
   (b) O higher IE and more paramagnetic; B more metallic
   (c) S higher IE; S and Se have the same number of unpaired electrons; Se more metallic
   (d) O higher IE and more paramagnetic; Ba more metallic

7. (a) increases (b) remains the same (c) decreases

SURVEY OF THE ELEMENTS 1
GROUPS 1A, 2A, AND 8A

SELF-ASSESSMENT

Hydrogen (Section S1.1)

1. Write equations for two reactions used to prepare hydrogen gas.

2. Write a balanced chemical equation for the reaction of hydrogen gas with
   (a) nitrogen gas
   (b) sodium
   (c) WO_3

The Alkali Metals and the Akaline Earth Metals (Sections S1.2 and S1.3)

3. Compare Na and Mg with respect to
   (a) reactivity
   (b) hardness
   (c) density
   (d) melting point

4. Which of the following statements are true?
   (a) Each Group 1A metal has the lowest ionization energy and the largest atoms of any element in its period.
   (b) All Group 1A and Group 2A metals react with water to form H_2 and OH^-.
   (c) All Group 1A and Group 2A metals react with F_2, Cl_2, Br_2, and I_2 to form ionic halides.
   (d) All hydrides of Group 1A and Group 2A react with water to liberate H_2.
   (e) When oxygen is in abundant supply, all alkali metals form normal oxides (oxides that contain O^{2-} ions.)

5. Complete the following equations:
   (a) $Sr(s) + 2H_2O(l) \rightarrow\ ?\ +\ ?$
   (b) $3Ba(s) + N_2(g) \xrightarrow{\text{heat}}\ ?$
   (c) $Ba(s) + O_2(g) \rightarrow\ ?$
   (d) $2Cs(s) + Cl_2(g) \rightarrow\ ?$
   (e) $6Li(s) + N_2(g) \rightarrow\ ?$
   (f) $4Li(s) + O_2(g) \rightarrow\ ?$

6. Write an equation for the
   (a) electrolysis of brine [concentrated NaCl(aq)]
   (b) heating of hard water (contains Ca^{2+} and HCO_3^- ions)
   (c) reaction between stomach acid (HCl) and bicarbonate of soda ($NaHCO_3$, ingredient in some antacids)
   (d) conversion of limestone ($CaCO_3$) to slaked lime ($Ca(OH)_2$). (Hint: Show two reactions.)

The Noble Gases (Section S1.4)

7. Which noble gas is described in each of the following statements?
   (a) It is the second most abundant element in the universe. It is formed on earth by nuclear reactions underground.
   (b) It forms compounds with fluorine and oxygen.
   (c) It is formed from the decay of radioactive radium in the soil.
   (d) It represents an anomaly in that it has a greater atomic weight than the element following it in the periodic table.

ANSWERS TO SELF-ASSESSMENT PROBLEMS

1. From coke: $C(s) + H_2O(g) \rightarrow CO(g) + H_2(g)$
   From active metals and aqueous acid: $Zn(s) + 2HCl(aq) \rightarrow H_2(g) + ZnCl_2(aq)$

2. (a) $3H_2(g) + N_2(g) \xrightarrow[\text{catalyst}]{\text{500 atm, 450°C}} 2NH_3(g)$
   (b) $H_2(g) + 2Na(s) \rightarrow 2NaH(s)$
   (c) $3H_2(g) + WO_3(s) \rightarrow W(s) + 3H_2O(g)$

3. (a) Mg is less reactive than Na.
   (b), (c) Mg is harder and denser.
   (d) Mg has a higher melting point.

4. (a), (c), and (d) are true.
   (b) Be does not react with water.
   (e) Only Li in Group 1A forms the normal oxide; the other metals form peroxides (containing O_2^{2-}) or superoxides (containing O_2^-).

5. (a) $Sr(s) + 2H_2O(l) \rightarrow Sr(OH)_2(aq) + H_2(g)$
   (b) $3Ba(s) + N_2(g) \xrightarrow{\text{heat}} Ba_3N_2(s)$
   (c) $Ba(s) + O_2(g) \xrightarrow{\text{heat}} BaO_2(s)$
   (d) $2Cs(s) + Cl_2(g) \rightarrow 2CsCl(s)$
   (e) $6Li(s) + N_2(g) \xrightarrow{\text{heat}} 2Li_3N(s)$
   (f) $4Li(s) + O_2(g) \rightarrow 2Li_2O(s)$

6. (a) $2NaCl(aq) + 2H_2O(l) \rightarrow 2NaOH(aq) + H_2(g) + Cl_2(g)$
   (b) $Ca^{2+}(aq) + 2HCO_3^-(aq) \rightarrow CaCO_3(s) + CO_2(g) + H_2O(l)$
   (c) $NaHCO_3(s) + HCl(aq) \rightarrow NaCl(aq) + H_2O(l) + CO_2(g)$
   (d) $CaCO_3(s) \xrightarrow{\text{heat}} CaO(s) + CO_2(g)$; $CaO(s) + H_2O(l) \rightarrow Ca(OH)_2(s)$
   (lime)

7. (a) He (b) Xe (c) Rn (d) Ar

REVIEW

Hydrogen (Section S1.1)

Learning Objectives

1. Describe the occurrence, properties, preparation, and uses of hydrogen.
2. Write equations for the reactions of hydrogen with oxygen, nitrogen, active metals, and halogens.

Review

Occurrence
- most abundant element in universe.
- on earth found mostly combined with heavier elements (H_2O, hydrocarbons are examples).

Physical Properties
- lightest element, gas under ordinary conditions
- colorless, odorless
- very low solubility in water

Chemical Properties
- flammable
- reducing agent
- combines with nonmetals to form covalent compounds
- combines with Group 1A and 2A metals to form ionic hydrides

Some Important Equations

$O_2(g) + 2H_2(g) \rightarrow 2H_2O(g)$

$N_2(g) + 3H_2(g) \xrightarrow[\text{catalyst}]{\text{500 atm, 45°C}} 2NH_3(g)$ (Haber Process)

$Cl_2(g) + H_2(g) \rightarrow 2HCl(g)$

$2M(s) + H_2(g) \rightarrow 2MH(s)$ (M = any Group 1A metal)

$M(s) + H_2(g) \xrightarrow{\text{heat}} MH_2(s)$ (M = Ca, Sr, Ba)

$WO_3(s) + 3H_2(g) \rightarrow W(s) + 3H_2O(g)$
(Used as a reducing agent)

Ions and Compounds
- H^+ (produced by acids)
- H^- (in ionic hydrides, unstable)
- hydrocarbons (organic molecules)

Preparation
- methane with steam: $CH_4(g) + H_2O(g) \xrightarrow[\text{900° C}]{\text{Ni}} CO(g) + 3H_2(g)$
- carbon monoxide with steam: $CO(g) + H_2O(g) \xrightarrow[\text{iron oxide}]{\text{450°C}} CO_2(g) + H_2(g)$
- coke with steam: $C(s) + H_2O(g) \xrightarrow{\text{about 1000°C}} CO(g) + H_2(g)$
- cracking of hydrocarbons
- electrolysis of water or brine
- displacement reactions between active metals and aqueous acid solutions

Uses
- fuel
- reducing agent
- reagent in the synthesis of NH_3, HCl, and CH_3OH

The Alkali Metals (Section S1.2)

Learning Objectives

1. List the alkali metals and describe their occurrence, preparation, properties, and uses.
2. Write equations for the reactions of alkali metals with water, oxygen, halogens, hydrogen, and nitrogen.
3. Write equations for the reactions of alkali metal oxides and hydrides with water.
4. State the formulas, names, common names, and uses for the more important alkali metal compounds.

Review

Alkali metals

lithium (Li) → most abundant
sodium (Na) → most abundant
potassium (K)
rubidium (Rb)
cesium (Cs)
francium (Fr) – radioactive

Occurrence

- Found as 1+ ions of soluble salts in oceans, salt lakes, brine wells, and salt deposits.

Physical Properties

- See Table S1.3 in text.

Chemical Properties

- Very reactive; alkali metals do not exist in nature in the elemental form, but only as ions (see below).

Equations for Some Important Reactions of Alkali Metals

$\underline{2H_2O(l)} + 2M(s) \rightarrow H_2(g) + 2MOH(aq)$ (M = any Group 1A metal)
$\underline{X_2(g)} + 2M(s) \rightarrow 2MX(s)$ (heat and light evolved) (X = F, Cl, Br, I and M = any Group 1A metal)
$\underline{H_2(g)} + 2M(s) \rightarrow 2MH(s)$ (M = any Group 1A metal)
$\underline{N_2(g)} + 6Li(s) \xrightarrow{\text{heat}} 2Li_3N(s)$ (Only Li reacts with N_2)

Alkali metals react with oxygen to form various types of oxides:
$\underline{O_2(g)} + 4Li(s) \rightarrow 2Li_2O(s)$
$\underline{O_2(g)} + 2Na(s) \rightarrow Na_2O_2(s)$ sodium peroxide
$\underline{O_2(g)} + Rb(s) \rightarrow RbO_2(s)$ rubidium superoxide

Preparation

An electric current is passed through molten alkali chlorides or hydroxides:

$$2NaCl(l) \xrightarrow{\text{electrolysis}} 2Na(l) + Cl_2(g)$$

Uses

- Na used as a liquid heat exchanger in some nuclear reactors
- Na used in vapor lamps
- Cs used in solar cells

Alkali Metal Compounds

- Ionic and soluble in water.
- Some properties of alkali metal compounds:
  1. chlorides; not very reactive
  2. hydroxides; soluble, basic, react with acids: $NaOH(aq) + HCl(aq) \rightarrow H_2O(l) + NaCl(aq)$
  3. oxides; react with water to form hydroxide: $K_2O(aq) + H_2O(l) \rightarrow 2KOH(aq)$
  4. hydrides; react with water to form hydrogen: $LiH(aq) + H_2O(l) \rightarrow LiOH(aq) + H_2(g)$

Group 2A: The Alkaline Earth Metals (Section S1.3)

Learning Objectives

1. List the alkaline earth metals and describe their occurrence, preparation, properties, and uses.
2. Compare Group 1A and Group 2A metals with respect to atomic radius, ionization energy, hardness, density, and melting point.
3. Write equations for the reactions of Group 2A metals with water, oxygen, hydrogen, halogens, and nitrogen.
4. Write equations for the reactions of Group 2A oxides, nitrides, and hydrides with water.
5. State the formulas, names, common names, and uses for the more important Group 2A compounds.
6. Write equations for the preparation of calcium oxide (lime) and calcium hydroxide from limestone.

Review

Alkaline Earth Metals

beryllium (Be)
magnesium (Mg)
calcium (Ca)
strontium (Sr)
barium (Ba)
radium (Ra) – radioactive

Occurrence

- found as 2+ ions in various minerals

Physical Properties

- Group 2A metals are harder, denser, and have higher melting points than their neighbors in Group 1A (see Table S1.5 in your text).

Chemical Properties

- form ionic compounds (see below)

Equations for Some Important Reactions of Alkaline Earth Metals

$\underline{2H_2O(l)} + M(s) \rightarrow M(OH)_2(aq) + H_2(g)$ (M = Mg [requires heat], Ca, Sr, Ba)

$\underline{X_2(g)} + M(s) \rightarrow MX_2(s)$ (M = any Group 2A metal; X = F, Cl, Br, I)

$\underline{H_2(g)} + M(s) \xrightarrow{\text{heat}} MH_2(s)$ (M = Ca, Sr, Ba)

$\underline{N_2(g)} + 3M(s) \xrightarrow{\text{heat}} M_3N_2(s)$ (M = any Group 2A metal)

$\underline{O_2(g)} + 2M(s) \rightarrow 2MO(s)$ (M = any Group 2A metal: Be and Mg require heat)

$O_2(g) + Ba(s) \xrightarrow{\text{heat}} BaO_2(s)$

Preparation

- All except Be are prepared by forming the chloride from a naturally occurring compound and then electrolyzing the molten chloride:
  $MgCl_2(l) \xrightarrow{\text{electrolysis}} Mg(l) + Cl_2(g)$
- Be is prepared from the reaction between BeF and Mg.

Uses

- Mg used in flash bulbs and flares
- Be and Mg used in lightweight structural alloys

Alkaline Earth Metal Compounds

- Some important reactions of alkaline earth metal compounds with water are
  $MO(s) + H_2O(l) \rightarrow M(OH)_2(s)$ (M = Ca, Sr, Ba)
  $M_3N_2(s) + 6H_2O(l) \rightarrow 2NH_3(g) + 3M(OH)_2(aq)$ (M = Ca, Sr, Ba)
  $MH_2(s) + 2H_2O(l) \rightarrow 2H_2(g) + M(OH)_2(aq)$ (M = any Group 2A metal)
- Lime (CaO) is prepared from limestone ($CaCO_3$): $CaCO_3(s) \xrightarrow{\text{heat}} CaO(s) + CO_2(g)$
- Slaked lime ($Ca(OH)_2$) is prepared from lime: $CaO(s) + H_2O(l) \rightarrow Ca(OH)_2\,(aq)$
- Lime is used in the preparation of mortar, cement, and plaster.
- Calcium sulfate occurs naturally as gypsum ($CaSO_4 \bullet 2H_2O$) and is used to make wallboard and other plaster-containing materials.

Group 8A: The Noble Gases (Section S1.4)

Learning Objectives

1. List the noble gases, and describe their occurrence, preparation, properties, and uses.
2.* Write formulas for some noble gas compounds and relate their formation to group trends in atomic properties.

Review

Noble Gases

helium (He)
neon (Ne)
argon (Ar)
krypton (Kr)
xenon (Xe)
radon (Rn) – radioactive

Occurrence

- Helium is the second most abundant element in the universe. On earth, it is produced underground from the decay of naturally occurring radioactive elements.
- Argon constitutes 1% of the atmosphere.
- Radon is a decay product of radium that is present in soil. It is therefore always present in air. Both radon and its decay products are radioactive and can be health hazards when allowed to accumulate in enclosed spaces such as in uranium mines. Radon levels in houses built over soil containing large deposits of radium can also contain high concentrations of radon.

Properties
- odorless, colorless, tasteless gases
- chemically inert, for the most part (Xe and Kr form a few compounds with F, O, and N)

Preparation
- He is obtained mostly from natural gas deposits.
- All noble gases except Rn can be obtained from liquid air by fractional distillation; the temperature of liquid air is raised slowly so that each component can be collected individually as it boils off.
- Rn can be obtained from the gases emitted by radium salts.

Uses
- Ar and Kr fill electric light bulbs.
- He forms a nonreactive atmosphere for high temperature work such as in welding. It is used to fill balloons and blimps, is a constituent in breathing mixtures for divers, and is used in low temperature experiments as a liquid (4.2K).
- Measurement of Rn emission from the ground is currently being used to predict earthquakes.
- The noble gases (especially neon) are used in electric signs because they emit characteristic colors when electric currents are passed through them (neon lights).

SELF-TEST

Choose the correct answer to the following:

1. Hydrogen is
   (a) the most abundant element on earth
   (b) inert
   (c) used in the synthesis of important industrial chemicals
   (d) found on earth mostly in its elemental form

2. Alkali metals
   (a) are present as 1+ ions in nature
   (b) exist in their elemental form
   (c) are prepared by passing an electric current through aqueous salt solutions
   (d) have no practical uses

3. An insoluble compound is
   (a) NaBr
   (b) CsOH
   (c) $Mg(OH)_2$
   (d) $KHCO_3$

4. Pick out the false statement.
   (a) Group 2A metals are harder, denser, and have a higher melting point than their neighbors in Group 1A.
   (b) All of the alkali and alkaline metals react with water.
   (c) All of the alkali and alkaline metals react with F_2, Cl_2, Br_2, and I_2.
   (d) All alkaline earth metals react with O_2 to form an oxide.

5. Match the substance listed in column A to one of its uses listed in column B.

A	B
(1) CaO (lime)	(a) used to fill blimps and balloons
(2) H_2	(b) used in lightweight alloys
(3) Be	(c) used in vapor lamps
(4) Na	(d) preparation of mortar, cement, and plaster
(5) He	(e) used as a reducing agent

6. Complete and balance the following equations:

(a) $CH_4(g) + H_2O(g) \rightarrow\ ? + ?$

(b) $H_2(g) + 2K(s) \rightarrow\ ?$

(c) $2H_2O(l) + 2Cs(s) \rightarrow\ ? + ?$

(d) $N_2(g) + 6Li(s) \rightarrow\ ?$

(e) $\underset{\text{(excess)}}{O_2(g)} + Rb(s) \rightarrow\ ?$

(f) $2Mg(s) + O_2(g) \xrightarrow{\text{heat}}\ ?$

(g) $3Ba(s) + N_2(g) \xrightarrow{\text{heat}}\ ?$

(h) $Sr(s) + H_2(g) \xrightarrow{\text{heat}}\ ?$

(i) $K_2O(s) + H_2O(l) \rightarrow\ ?$

ANSWERS TO SELF-TEST QUESTIONS

1. (c)

2. (a)

3. (c)

4. (b) Be does not

5. (1) d (2) e (3) b (4) c (5) a

6. (a) $CH_4(g) + H_2O(g) \rightarrow CO(g) + 3H_2(g)$
(b) $H_2(g) + 2K(s) \rightarrow 2KH(s)$
(c) $2H_2O(l) + 2Cs(s) \rightarrow H_2(g) + 2CsOH(aq)$
(d) $N_2(g) + 6Li(s) \rightarrow 2Li_3N(s)$
(e) $O_2(g) + Rb(s) \rightarrow RbO_2(s)$
(f) $2Mg(s) + O_2(g) \xrightarrow{\text{heat}} 2MgO(s)$
(g) $3Ba(s) + N_2(g) \xrightarrow{\text{heat}} Ba_3N_2(s)$
(h) $Sr(s) + H_2(g) \xrightarrow{\text{heat}} SrH_2(s)$
(i) $K_2O(s) + H_2O(l) \rightarrow 2KOH(aq)$

CHAPTER 9
THE CHEMICAL BOND

SELF-ASSESSMENT

Use the periodic table when necessary.

Bond Types; Lewis Symbols (Introduction, Section 9.1)

1. Match the substance in Column A with its bond type (or types) in Column B.

A	B
(1) H_2S	(a) ionic
(2) KNO_3	(b) covalent
(3) NaCl	(c) metallic
(4) Ag	(d) covalent and ionic

2. Give the Lewis symbol for (a) C, (b) O, (c) Sn, (d) S.

The Ionic Bond (Section 9.2)

3. Circle the formulas that are not correct for stable ions, and then write them correctly.

   Ca^{+}, Na^{+}, Cl^{-}, Fe^{2+}, Zn^{2+}, S^{2-}, Mg^{2-}, F^{+}, Sn^{4+}

4. Rank the following ionic compounds in order of increasing lattice energy: KNO_3, $Ca(NO_3)_2$, $CsNO_3$.

The Covalent Bond (Section 9.3)

5. Rank the C–F, C–Cl, and C–Br bonds in order of (a) increasing bond strength and (b) increasing bond length.

6. Use bond energy data from Table 9.3 in your text to estimate ΔH° for the formation of 1 mol of HCl from its elements:

   $Cl_2(g) + H_2(g) \rightarrow 2HCl(g)$

Drawing Lewis Structures (Section 9.4)

7. Determine which of the following Lewis structures are incorrect and then draw the correct structure.

 (a) N$_2$:N̈=N̈:

 (b) HCN H=C=N̈: H—C≡N

 (c) CHCl$_3$ (chloroform) (d) NO$_3^-$

 :Cl̈:
 ‖
 :C̈l-C-C̈l:
 │
 H

 $$\left[\ :\overset{..}{\underset{..}{O}}-\overset{\overset{:O:}{\|}}{N}-\overset{..}{\underset{..}{O}}: \ \right]^-$$

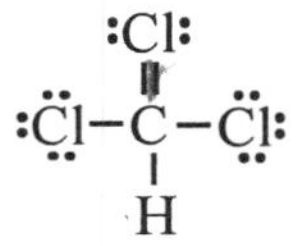

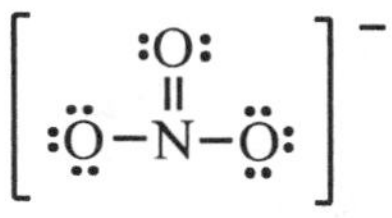

8. Label the formal charges in :C≡O:

9. Draw Lewis structures for H$_2$SO$_4$ and H$_3$PO$_4$ in which there are no formal charges. (Hint: Remember that third period elements can expand their octets.)

 O
 ‖
 O=S—OH
 │
 O
 H

 H-O—P=O-H
 │
 O
 H

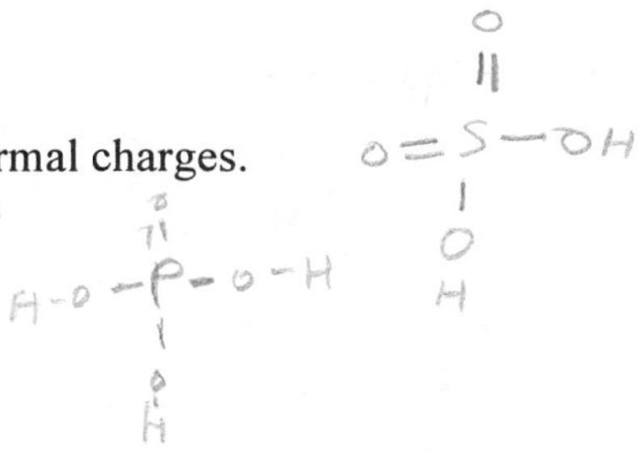

Resonance Structures (Section 9.5)

10. Draw resonance structures for

 (a) CO$_3^{2-}$ (carbonate ion) (b) HCOO$^-$ (formate ion) (c) HCOOH (formic acid)

11. Use formal charges to determine which resonance structure, if any, makes the greater contribution to each hybrid in Exercise 10.

12. Draw the Kekulé structures for bromobenzene (C$_6$H$_5$Br).

Exceptions to the Octet Rule (Section 9.6)

13. Draw Lewis structures for (a) SF$_6$, (b) BeH$_2$, and (c) POCl$_3$.

 F F
 \ /
 F–S–F
 /
 F F

 H–Be–H

 O
 ‖
 P
 Cl Cl Cl

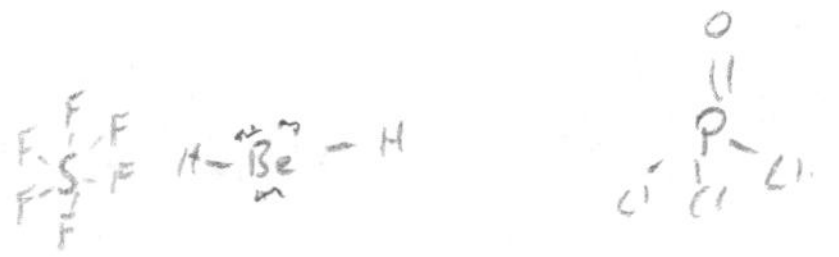

14. Which is the preferred Lewis structure for BF$_3$? Why?

 :F: :F: :F:
 ‖ (+1) │ │
 (a) :F̈-B-F̈: (b) :F̈-B-F̈: (c) (+1) F̈=B-F̈:
 (-1) (-1)

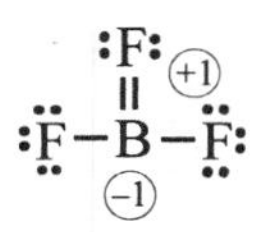

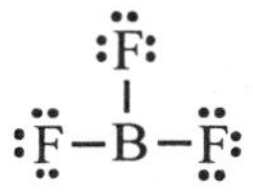

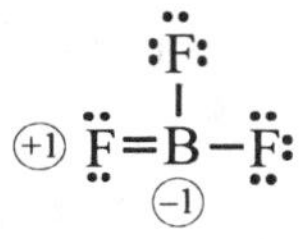

Polar and Nonpolar Covalent Bonds (Section 9.7)

15. Rank the following bonds in order of increasing polarity: I–I, C–I, C–Cl, C–F, C–H.

16. Which of the following are predominantly ionic compounds: O$_2$, CsBr, BeH$_2$, CH$_4$, AlF$_3$, CCl$_4$?

17. Draw Lewis structures for the following molecules and indicate the polarity of each bond by a crossed arrow: (a) BeCl$_2$ (b) CO$_2$ (c) BF$_3$

ANSWERS TO SELF-ASSESSMENT PROBLEMS

1. (1) b (2) d (3) a (4) c

2. ·Ċ· ·Ö· ·Sn· ·S̈·

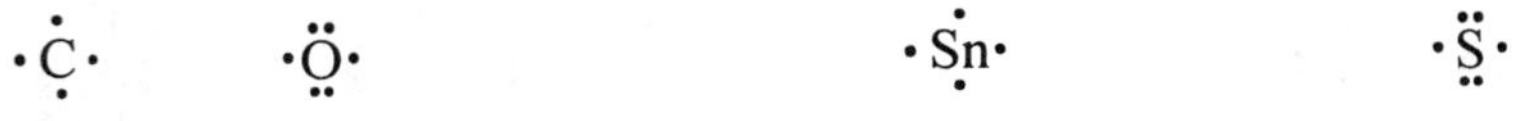

3. Ca^{2+}, Mg^{2+}, F^-

4. $CsNO_3 < KNO_3 < Ca(NO_3)_2$

5. (a) C–Br < C–Cl < C–F (b) C–F < C–Cl < C–Br

6. $\Delta H° = (435.9 \text{ kJ} + 243.4 \text{ kJ} - 2 \times 432.0 \text{ kJ})/2 = \underline{-92.35 \text{ kJ}}$

7. :N≡N: H–C≡N:

 :C̈l:
 |
 :C̈l–C–C̈l:
 |
 H

8. :C≡O: (−1 on C, +1 on O)

9.
 :O:
 ‖
 H–Ö–S–Ö–H
 :O:

 :O:
 ‖
 H–Ö–P–Ö–H
 :O:
 |
 H

10,11. (a)
$$\left[\begin{array}{c} :O: \\ \| \\ :Ö–C–Ö: \\ _{(-1)} \quad _{(-1)} \end{array} \right]^{2-} \longleftrightarrow \left[\begin{array}{c} :Ö:^{(-1)} \\ | \\ :Ö–C=O \\ _{(-1)} \end{array} \right]^{2-} \longleftrightarrow \left[\begin{array}{c} :Ö:^{(-1)} \\ | \\ Ö=C–Ö:^{(-1)} \end{array} \right]^{2-}$$

(b)
$$\left[\begin{array}{c} :O: \\ \| \\ H–C–Ö: \\ _{(-1)} \end{array} \right]^{-} \longleftrightarrow \left[\begin{array}{c} :Ö:^{(-1)} \\ | \\ H–C=Ö \end{array} \right]^{-}$$

(c)
 :O: :Ö:⁻¹
 ‖ |
 H–C–Ö–H ⟷ H–C=Ö–H
 ₊₁

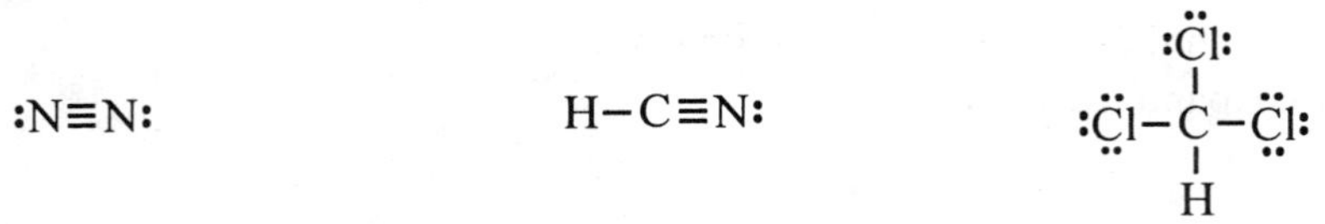

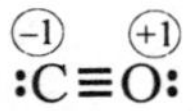

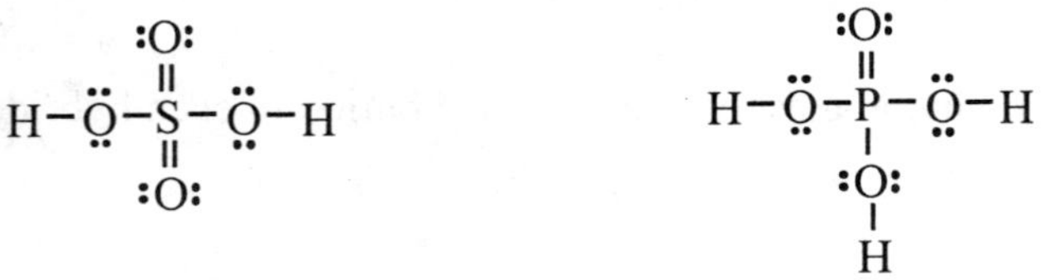

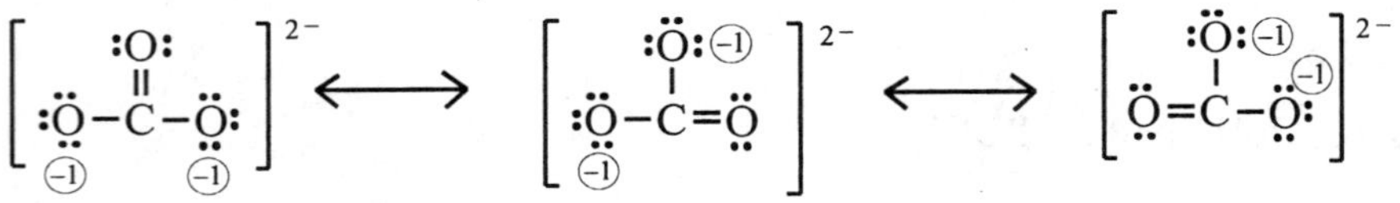

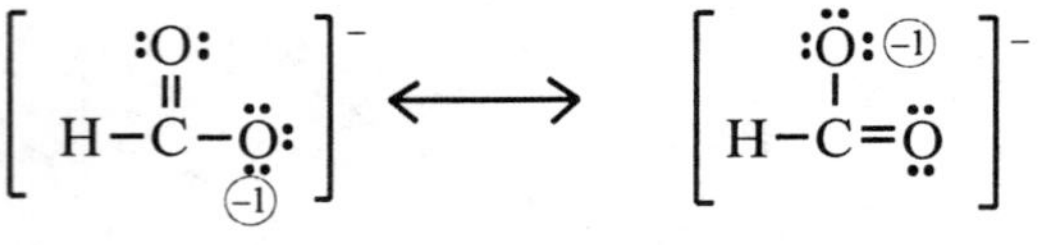

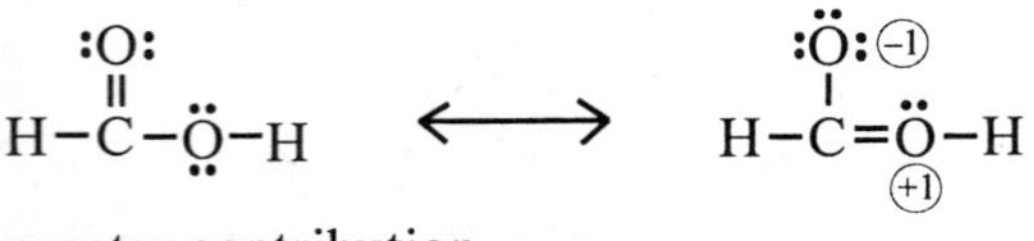

 greater contribution
 (no formal charge)

(a) and (b) Structures are equivalent.

12.

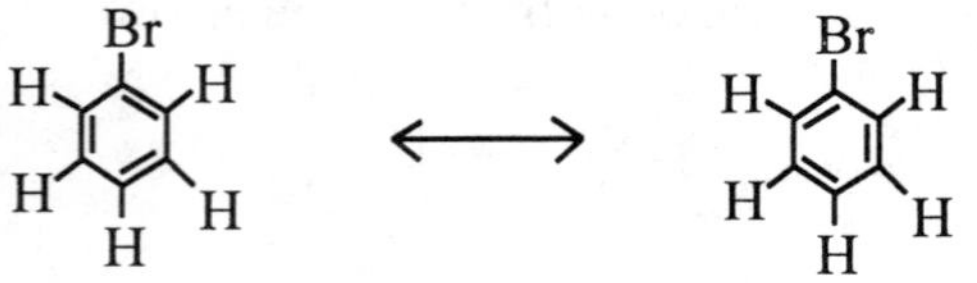

13. (a)
```
      :F:
   :F:|:F:
      >S<
   :F:|:F:
      :F:
```
(b) H–Be–H

(c)
```
       :O:
        ‖
   :Cl– P –Cl:
        |
       :Cl:
```

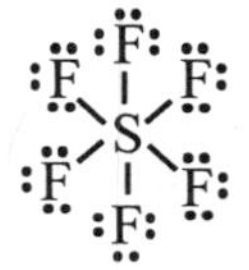

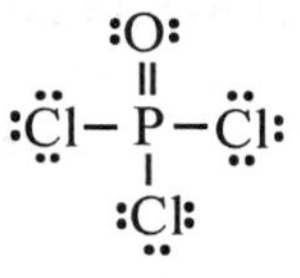

14. (b); no formal charges

15. I–I < C–H < C–I < C–Cl < C–F

16. CsBr, AlF$_3$

17. (a) :Cl– Be –Cl: (b) O̤ = C = O̤ (c)
```
     ←   →
   :F̈ – B – F̈:
        |
       :F̈:↕
```

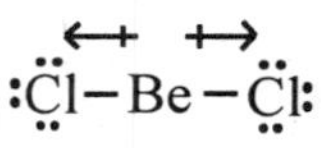

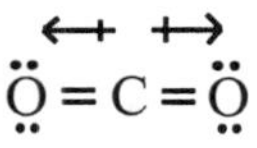

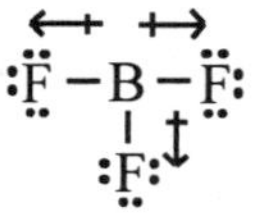

REVIEW

Introduction

The three principal types of chemical bonds are <u>ionic</u>, <u>covalent</u>, and <u>metallic</u>.

<u>Ionic bonds</u> form when electrons are transferred from one atom to another, usually from metal atoms to nonmetal atoms, and the resulting ions attract each other.

<u>Covalent bonds</u> form when atoms share one, two, or three pairs of electrons. The bond derives its strength from a build-up of electron density between the bonded nuclei.

<u>Metallic bonds</u> form when valence electrons are not confined to individual atoms or bonds, but flow freely through the crystal.

Lewis Symbols (Section 9.1)

Learning Objectives

1. Describe the three principal types of chemical bonds (see above).
2. Write Lewis symbols for the main group atoms.

Review

The <u>Lewis symbol</u> for an atom consists of its atomic symbol surrounded by a number of dots equal to the number of valence electrons.

Example 9.1: Give the Lewis symbol for S.

Answer: Sulfur is in Group 6A of the periodic table; it has six valence electrons: ·S̈·

Practice Drill on Lewis Symbols: Complete the following table:

Element	Group in periodic table	Lewis symbol
P		
Te		
Rn		
Mg		

Answers: P: Group 5A; Te: Group 6A; Rn: Group 8A; Mg: Group 2A

·P̈· ·T̈e· :R̈n: ·Mg·

The Ionic Bond (Section 9.2)

Learning Objectives

1. Use Lewis symbols and electron configurations to represent the formation of monatomic ions.
2. Know which groups of ions generally have noble gas configurations and which groups do not.
3. Describe how lattice energy varies with increasing ionic charge and decreasing ionic radii.
4. List some crystal properties associated with high lattice energy.
5.* Diagram the Born-Haber cycle for the formation of an ionic solid and use it to calculate lattice energies from experimental data.
6.* Given five of the six enthalpy changes in the Born-Haber cycle, calculate the missing enthalpy change.

Review

Atoms and ions with the same electron configurations are said to be isoelectronic.

Monatomic cations of the Group 1A and Group 2A metals are isoelectronic with the noble gas atoms preceding their periods. Monatomic anions are isoelectronic with the noble gas atoms at the end of their periods.

Example 9.2: Use Lewis symbols and electron configurations to show the formation of Mg^{2+} from Mg.

Answer: ·Mg· $\rightarrow Mg^{2+} + 2e^-$

↑ ↑

[Ne]$3s^2$ [Ne]

Practice: Complete the following:

(a) :Ṅ· $+ 3e^- \rightarrow$? K· $\rightarrow$? + e^-

[He]$2s^22p^3$ [Ar]$4s^1$

Answer: (a) :N̈: [Ne] (b) K^+ [Ar]

The octet rule states that bonded atoms often have noble gas configurations. This rule is obeyed when the metals of Group 1A and Group 2A combine with the nonmetals of Group 5A, 6A, and 7A. (The octet rule is not obeyed by transition metal atoms.)

Example 9.3: Use Lewis formulas to show the formation of K_2O, an ionic compound.

Answer: 2K• + •Ö• → $2K^+$ + :Ö: $^{2-}$

Each K• donates an electron and each •Ö• accepts two electrons.

The Strength of Ionic Bonding: Lattice Energy

The ions in an ionic crystal are held together by the attractive force between unlike charges. The lattice energy is the energy required to break up 1 mol of the crystal into independent gaseous ions. Lattice energies increase with increasing ionic charge and decreasing ionic radii. High lattice energies are associated with hardness and high melting points.

Example 9.4: Rank the following ionic crystals in order of increasing lattice energy: KF, KBr, KCl, KI

Answer: The ionic radius decreases in the order: $I^- > Br^- > Cl^- > F^-$

The lattice energy increases in the same order: KI < KBr < KCl < KF

> **Practice:** Which of the following pairs has a greater lattice energy?:
> (a) $MgCl_2$ or KCl (b) K_2O or K_2S
>
> Answers: (a) $MgCl_2$ (b) K_2O

*The lattice energy of an ionic compound is determined by the Born-Haber method. The known enthalpy of formation (ΔH_f°) is the sum of the known enthalpies of sublimation and ionization of the metal, the known enthalpies of dissociation and ionization of the nonmetal, and the unknown lattice energy.

The Covalent Bond (Section 9.3)

Learning Objectives

1. Distinguish between single, double, and triple bonds in Lewis structures.
2. State how bond energies vary with multiple-bond character and with atom size.
3. Use bond energy data to estimate reaction enthalpies.

Review

The vast majority of compounds contain nonmetal atoms held together by covalent bonds. A covalent bond is the sharing of electrons between bonded atoms.

Lewis Structures

Covalent molecules and polyatomic ions can be represented by Lewis structures in which each shared electron pair is shown as two dots or as one line. All of the valence electrons of all the individual atoms must be shown.

The Lewis structures of most molecules satisfy the octet rule. The two bonds in H_2S, for example, are electron pairs that form when sulfur shares its two unpaired electrons with the two hydrogen atoms. As a result the total number of valence electrons around the sulfur satisfies the octet rule.

H· + ·S̤̈· + ·H → H:S̤̈:H, or H–S̤̈–H

Valence electrons that are not involved in bonding are called lone pairs or nonbonding pairs; sulfur has two lone pairs in H_2S.

The Strength of Covalent Bonds: Bond Energies and Bond Lengths

The bond length is the equilibrium distance between the centers of two bonded atoms.

The bond energy (D) is the average energy required to break a bond in a gaseous molecule. For diatomic molecules such as Cl_2, the bond energy is also the energy needed to dissociate 1 mol of gaseous molecules into gaseous atoms:

$Cl_2(g) \rightarrow 2Cl(g)$ $\qquad$ $D(Cl–Cl) = \Delta H° = +243.4$ kJ

Bond breaking is an endothermic process. The reverse, bond formation, is exothermic:

$2Cl(g) \rightarrow Cl_2(g)$ $\qquad$ $\Delta H° = -243.4$ kJ

Keep these two facts in mind:

(1) Bonds between small atoms tend to be stronger and shorter than bonds between large atoms.
(2) Increasing multiple-bond character makes for stronger and shorter bonds. Turn to Table 9.3 in your text to get a feel for the different bond energies and to verify the trends.

Example 9.5: Rank the following bonds in order of increasing bond energy: H–Cl, H–F, H–I, H–Br

Answer: H–I < H–Br < H–Cl < H–F. Substantiate the order with Table 9.3 in your text.

Practice: Rank the above bonds in order of decreasing bond length.
Answer: H–I > H–Br > H–Cl > H–F

Example 9.6: Which molecule,

acetone, $CH_3–C(=\ddot{O}:)–CH_3$, or ethanol, $CH_3–CH_2–\ddot{O}–H$ (with C bearing H above and H below), has the shorter and the stronger carbon-oxygen bond?

Answer: The carbon-oxygen double bond in acetone is stronger and shorter than the carbon-oxygen single bond in ethanol.

Bond energies can be used to estimate reaction enthalpies.

Example 9.7: Estimate ΔH° for the hydrogenation of acetone.

$$
\begin{array}{ccc}
\text{H} & :\!\ddot{\text{O}}\!: & \text{H} \\
| & \| & | \\
\text{H}-\text{C}-\text{C}-\text{C}-\text{H} \\
| & & | \\
\text{H} & & \text{H}
\end{array}
\quad + \quad \text{H}_2 \quad \longrightarrow \quad
\begin{array}{c}
\text{H}\diagdown \\
\text{H} \; \ddot{\text{O}}: \; \text{H} \\
| \;\; | \;\; | \\
\text{H}-\text{C}-\text{C}-\text{C}-\text{H} \\
| \;\; | \;\; | \\
\text{H} \; \text{H} \; \text{H}
\end{array}
$$

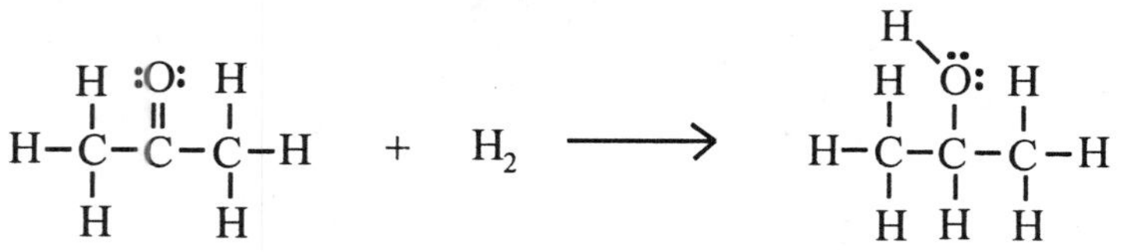

Answer: The steps include breaking a C=O bond, and a H–H bond, and forming a C–O bond, a C–H bond, and an O–H bond. (The other 6 C–H bonds remain intact.) ΔH° is positive for the bond breaking, negative for the bond forming. From Table 9.3 in your text;

Breaking 1 mol of C=O bonds:	ΔH° = 802 kJ
Breaking 1 mol of H–H bonds:	ΔH° = 436 kJ
Forming 1 mol of C–O bonds:	ΔH° = -360 kJ
Forming 1 mol of O–H bonds:	ΔH° = -463 kJ
Forming 1 mol of C–H bonds:	ΔH° = -413 kJ
Answer:	ΔH° = 2 kJ

Drawing Lewis Structures (Section 9.4)

Learning Objectives

1. Draw Lewis structures for covalently bonded molecules and ions.
2. Find the formal charge on each atom in a Lewis structure.

Review

Drawing Lewis Structures

The three steps for drawing Lewis structures are

1. Find the number of valence electrons.
2. Draw the skeleton structure with single bonds.
3. Distribute the remaining valence electrons.

These steps are illustrated in the following examples.

Example 9.8 (Molecule with Lone Pair): Draw the Lewis structure of PH₃.

Answer:

Step 1: **Count the valence electrons.**

Atoms	Group	Number of electrons
P	5	5
3H	1	3 × 1 = 3
		Total = 8 electrons, 4 pairs

Step 2: **Draw the skeleton structure with single bonds.** A hydrogen atom can have only one bond, therefore the only possible arrangement is

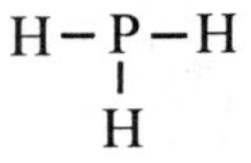

Each single bond is a pair of electrons. Three pairs of electrons have been used up.

Step 3: **Distribute the remaining electrons.** There are four pairs of electrons; one pair remains. Each H atom already has two electrons and can have no more. The P atom needs a pair to complete its octet:

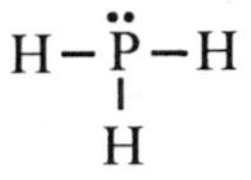

Example 9.9 (An Ion): Draw the Lewis structure of NH_4^+.

Answer:

Step 1: **Count the valence electrons.**

Atoms	Group	Number of electrons
N	5	5
4H	1	4

Total = 9 electrons – 1 electron = 8 electrons, 4 pairs

A +1 charge on an ion means that an electron must be subtracted.

Step 2: **Draw the skeleton with single bonds.** Since each H can have only one bond, the only possible arrangement is

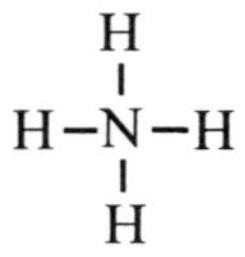

The four single bonds have used the four pairs of electrons, so the above structure is correct.

Example 9.10 (A Molecule With a Double Bond): Draw the Lewis structure of C_2H_4.

Answer:

Step 1: **Count the valence electrons.**

Atoms	Group	Number of electrons
2C	4	8
4H	1	4

Total = 12 electrons, 6 pairs

Step 2: **Draw the skeleton with single bonds.** Since each H can have only one bond, the only possible arrangement is

$$
\begin{array}{c}
\text{H} \diagdown \qquad \diagup \text{H} \\
\qquad \text{C}-\text{C} \\
\text{H} \diagup \qquad \diagdown \text{H}
\end{array}
$$

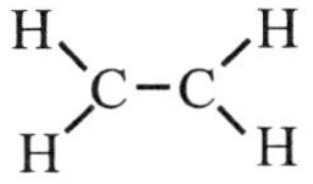

The five single bonds have used five pairs of electrons.

Step 3: **Distribute the remaining electrons.** There are six pairs of electrons; one pair remains.

Both carbon atoms need a pair to complete their octets, therefore they will have to share the remaining pair. The result is a double bond.

$$
\begin{array}{c}
\text{H} \diagdown \qquad \diagup \text{H} \\
\qquad \text{C}-\text{C} \\
\text{H} \diagup \uparrow \ \ \uparrow \diagdown \text{H}
\end{array}
\qquad\qquad
\begin{array}{c}
\text{H} \diagdown \qquad \diagup \text{H} \\
\qquad \text{C}=\text{C} \\
\text{H} \diagup \ \ \uparrow \ \ \diagdown \text{H}
\end{array}
$$

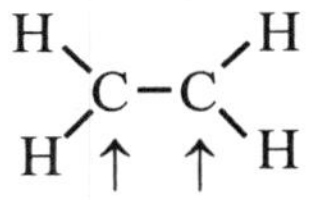

needs one more pair The one remaining pair is shared.

Example 9.11 (An Acid that Contains Oxygen): Write the Lewis structure of HNO_3.

Answer:

Step 1: **Count the valence electrons.**

Atoms	Group	Number of electrons
H	1	1
N	5	5
3O	6	18

Total = 24 electrons, 12 pairs

Step 2: **Draw the skeleton with single bonds.** The skeletons of most oxygen-containing acids can be drawn as follows:

1. Bond each oxygen to the central atom (in this case N).
2. Bond each hydrogen to a different oxygen. The arrangement for HNO_3 is therefore:

$$
\begin{array}{c}
\text{H}-\text{O} \diagdown \quad \diagup \text{O} \\
\qquad\qquad \text{N} \\
\qquad\qquad | \\
\qquad\qquad \text{O}
\end{array}
$$

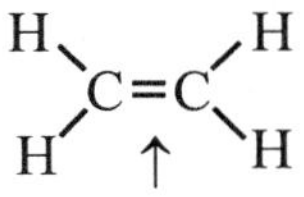

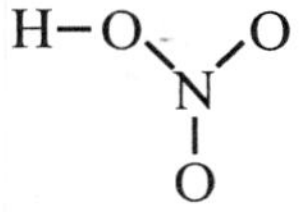

The four single bonds have used four pairs of electrons.

Step 3: **Distribute the remaining electrons.** Four pairs of electrons were used; eight pairs remain. Nine pairs would be needed to complete all the octets with lone pairs; therefore, one pair must be shared between nitrogen and one of the oxygens.

needs two more pairs needs three more pairs
 ↓ ↓

$$
\begin{array}{c}
\text{H}-\text{O} \diagdown \quad \diagup \text{O} \\
\text{needs one more pair} \rightarrow \text{N} \\
| \\
\text{O} \leftarrow \text{needs three more pairs}
\end{array}
\qquad
\begin{array}{c}
\text{H}-\ddot{\text{O}} \diagdown \quad :\ddot{\text{O}}: \\
\qquad\qquad \text{N} \diagup\!\!\!\!= \\
| \\
:\ddot{\text{O}}:
\end{array}
$$

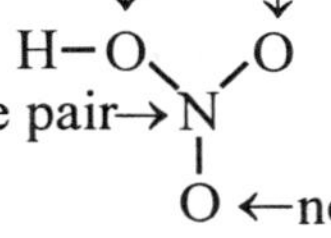

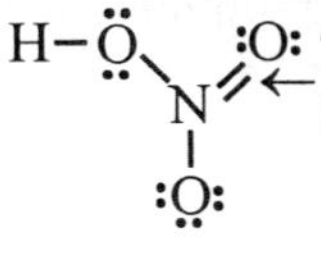

The eighth remaining pair is shared to make a double bond.

Practice Drill on Lewis Structures: Complete the table:

Compound	Number of valence electrons (Step 1)	Skeletons (Step 2)	Lewis structure (Step 3)
CO_2	:O=C=O:		
HCN	H–C–N	H-C-N	H-C≡N
$COCl_2$			:Cl - C(=O) - Cl:
H_2CO_3		H-O-C(=O)-O-H	

Answers: CO_2: 16, O–C–O, Ö=C=Ö ;   HCN: 10, H–C–N, H−C≡N:

$COCl_2$: 24, Cl−C(O)−Cl , :C̈l−C(=:O:)−C̈l: ; H_2CO_3: 24, H−O−C(O)−O−H , H−Ö−C(=:O:)−Ö−H

Formal Charge

Sometimes the total number of electrons assigned to an atom in a Lewis structure is different from the number of electrons it started with. Such an atom is said to have a formal charge.

To find a formal charge:

1. Count the valence electrons on the independent atom.
2. Count the electrons assigned to it in the Lewis structure.
3. Subtract (2) from (1).

To count the electrons assigned to an atom in a Lewis structure, add together:

one-half of its bonding electrons + all of its nonbonding electrons

To check: the formal charges in a molecule must add up to zero. The formal charges in an ion must add up to the ionic charge.

Example 9.12: Find the formal charge on O in CO_2.

Answer:

Step 1: Oxygen has six valence electrons.

Step 2: The Lewis structure for CO_2 is

:Ö=C=Ö:

Each oxygen atom has two bonds (four bonding electrons). The number of assigned electrons is

bonding: $1/2 \times 4 = 2$

nonbonding: 4

Total = 6

Step 3: The number of assigned electrons (6) equals the number of valence electrons (6). The formal charge on the oxygen atom is therefore zero.

Example 9.13: Find the formal charge on each atom in HNO_3.

Answer: The Lewis structure is (see Example 9.11 of this study guide)

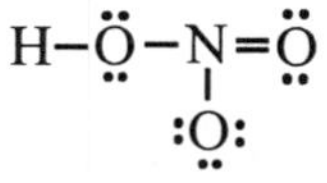

Atom	Valence electrons	Assigned electrons	Charge
H	1	1/2 × 2 = 1	0
O, left	6	(1/2 × 4) + 4 = 6	0
O, right	6	(1/2 × 4) + 4 = 6	0
O, bottom	6	(1/2 × 2) + 6 = 7	-1
N	5	1/2 × 8 = 4	+1

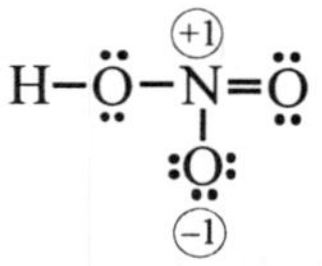

Note: The formal charges on a neutral molecule must add up to zero.

Example 9.14: Find the formal charges in OH^-.

Answer: The Lewis structure is :Ö–H

H: 1 valence electron, 1/2 × 2 = 1 assigned electron; formal charge = 0

O: 6 valence electrons, (1/2 × 2) + 6 = 7 assigned electrons; formal charge = -1

Note: The total formal charge equals the ionic charge.

Practice Drill on Formal Charge: Complete the following for the oxygen atom(s) (Group 6A) in each of the given Lewis structures:

Lewis Structure	One-half bonding electrons	Number of lone pair electrons	Formal charge
(a) :C≡O:			
(b) :N̈=N=Ö:			
(c) [:Ö–N̈=Ö:]⁻			
(d) :C̈l–Ö–C̈l:			

Answers: (a) 3, 2, +1 (b) 2, 4, 0 (c) 1, 6, -1; 2, 4, 0 (d) 2, 4, 0

Resonance Structures (Section 9.5)

Learning Objectives

1. Draw resonance structures for molecules and ions.
2. Use formal charge to determine which resonance structures make the greatest contribution to the resonance hybrid.
3. Draw the important resonance structures of benzene.

Review

In some cases two or more Lewis structures can be drawn for a molecule or ion. These Lewis structures must have the same atomic arrangement and the same number of electron pairs and are called <u>resonance structures</u>.

Resonance structures differ <u>only</u> in the positions of their electron pairs. The electron distribution in the actual molecule is intermediate to those depicted in the resonance structures. Therefore, the actual molecule is called a resonance hybrid. The most stable resonance structures have the least formal charge on their atoms and contribute most heavily to the hybrid. Structures in which adjacent atoms have formal charges of the same sign contribute very little to the hybrid.

Example 9.15: Write resonance structures for CO_3^{2-}. State which one, if any, makes the greatest contribution to the hybrid.

Answer: Follow the steps for writing Lewis structures (see Example 9.8 in this study guide). One double bond is necessary in each structure, however the double bond could be placed in any one of the three possible positions. The resonance structures are:

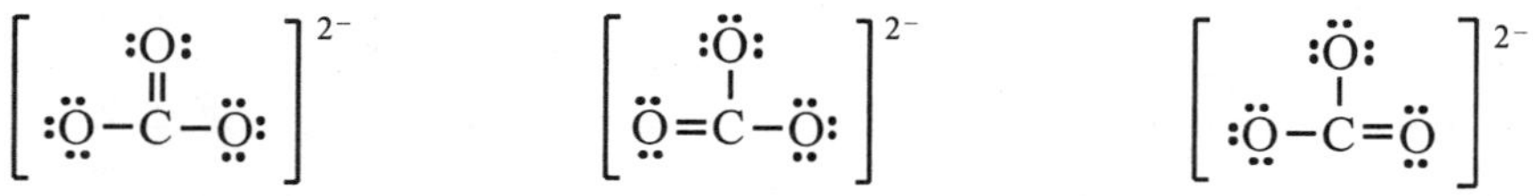

Since these structures all have the same amount of formal charge on their atoms, they contribute equally to the hybrid.

Example 9.16: Which of the following resonance structures for acetic acid (CH_3COOH) contributes more to the resonance hybrid?

I. 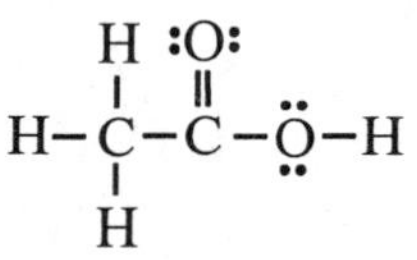

II. 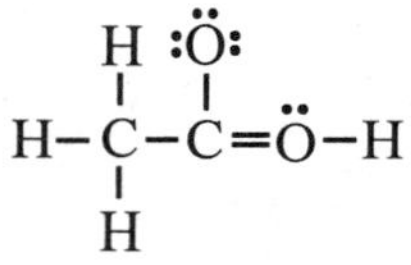

Answer: Assign formal charges to atoms in the structures:

I. 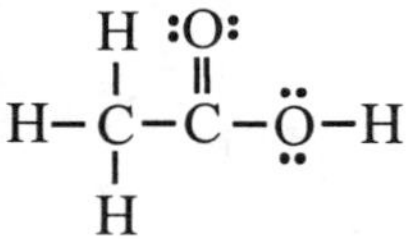

II. 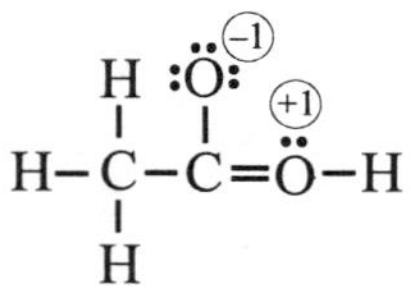

I. All atoms have 0 formal charge.

Structure II has more formal charge, and is therefore less stable; structure I contributes more to the actual molecule.

Benzene: The benzene molecule, C_6H_6, is very important in organic chemistry. Two of its important resonance structures are:

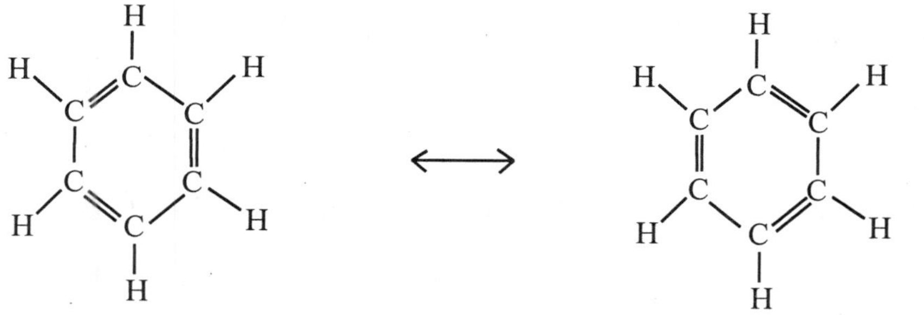

A two-headed arrow is often used to indicate contributions from two resonance structures.

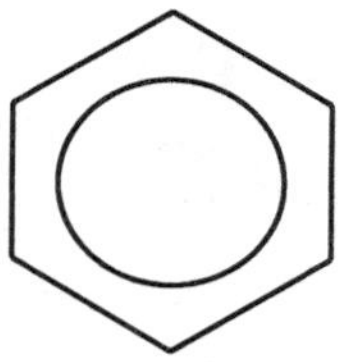

The benzene molecule is usually written this way.

Exceptions to the Octet Rule (Section 9.6)

Learning Objectives

1. State which elements can exhibit octet expansion and draw Lewis structures for molecules containing these elements.
2. State which elements are likely to have incomplete octets and draw Lewis structures for molecules containing such elements.
3. Give examples of free radicals and their Lewis structures.
4. State some of the properties of free radicals.

Review

More Than Four Pairs

Some Lewis structures do not satisfy the octet rule. Nonmetal atoms in the third period and beyond sometimes have expanded octets; that is, more than four electrons pairs. Important examples include sulfur, phosphorus, some halogens, and some noble gases.

Example 9.17: Draw a Lewis structure for IF_7.

Answer:

Step 1: All eight atoms are in Group 7A and have seven valence electrons: $8 \times 7 = 56$ electrons = 28 pairs.

Step 2: A fluorine atom can form only one covalent bond; therefore, the only possible arrangement is

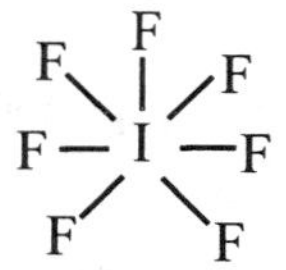

Step 3: Seven pairs are used in forming the single bonds; 21 pairs remain. The Lewis structure is

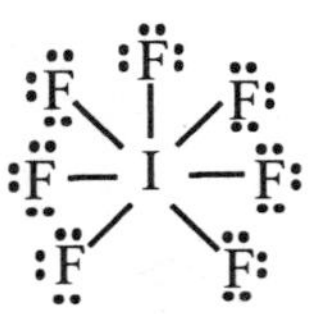

Less Than Four Pairs:

To avoid formal charges, atoms from Groups 2A, 3A, and 4A are sometimes given incomplete octets (less than four electron pairs). Some examples are:

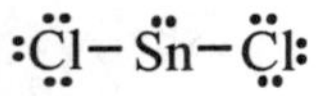

(Sn is in Group 4A)

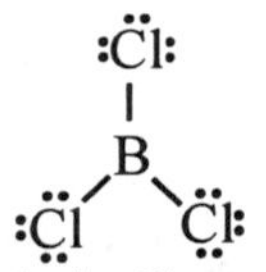

(B is in Group 3A)

(Be is in Group 2A)

Note that Lewis structures could be drawn for the above compounds that satisfy the octet rule, but these would all show positive formal charges on chlorine. Such charges are unlikely, given the known electron- attracting ability of chlorine.

Free Radicals

A free radical is an atom or molecule containing at least one unpaired electron. Such species can never satisfy the octet rule. Most free radicals are very reactive and short-lived; many possess color. NO and NO_2 are well-known examples of stable free radicals, and so is oxygen (O_2) with two unpaired electrons.

Polar and Nonpolar Covalent Bonds (Section 9.7)

Learning Objectives

1. Give examples of polar and nonpolar molecules.
2. Used crossed arrows to represent dipoles in diatomic molecules.
3. State how electronegativities vary across a period and down a group.
4. Use electronegativities to predict whether a bond will be ionic, polar covalent, or nonpolar.

Review

Covalent bonds may be polar or nonpolar. In a polar covalent bond the electron density is not equally shared; the electrons spend more time on one of the bonded atoms than on the other. In a nonpolar bond the electron density is equally shared by the bonded atoms.

Electronegativity

Electronegativity is the tendency of an atom or group of atoms to attract shared electrons. The Pauling electronegativity scale rates elements on a scale of 0 to 4 according to their tendency to attract bonded electrons.

Electronegativities tend to increase from left to right across a period and decrease down a group. Nonmetals tend to be more electronegative than metals. The polarity of a bond increases with the electronegativity difference between two bonded atoms. Most bonds with differences above 1.7 are predominantly ionic, most bonds with differences below 1.7 are predominantly covalent; bonds with no electronegativity differences are nonpolar covalent.

Example 9.18: Use Pauling electronegativities to predict the nature of the bonds in NaBr and HI. From Figure 9.11 in your textbook electronegativities are Na = 0.9; H = 2.1; Br = 2.8; I = 2.5

Answer:	For NaBr the difference is:	2.8 – 0.9 = 1.9 (ionic)
	For HI the difference is:	2.5 – 2.1 = 0.4 (covalent)

Practice: Rank the following bonds in order of increasing polarity: Cl–F, H–F, Br–F.
Answer: Cl–F < Br–F < H–F

Bond Polarities

Any polar molecule is also a molecular dipole. A molecular dipole is shown with a crossed arrow directed toward the negative atom (most electronegative atom) with the cross at the positive end. For example

H⟼F

The dipole moment usually given in units of debyes for molecules, is a measure of the polarity of the molecule.

See Table 9.5 in your text for the dipole moments of some diatomic molecules.

Practice Drill on Molecular Dipoles: Indicate the polarity of each bond for the given molecules.

Molecule	Lewis Structure	Bond Polarities
BrCl	:B̈r–C̈l:	Br⟼Cl
HBr	H–B̈r:	H⟼Cl
NH_3	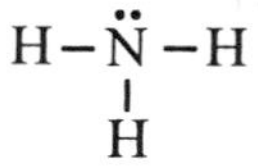	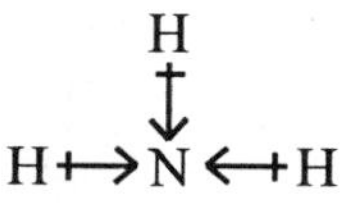
H_2CO_3	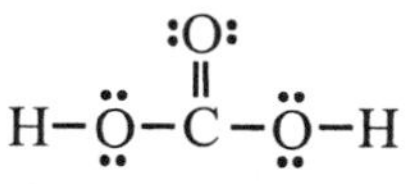	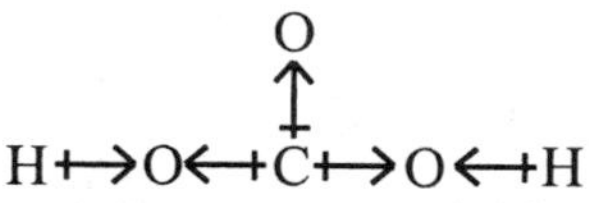

SELF-TEST

1. Give an example of (a) an ionic bond, (b) a nonpolar covalent bond, and (c) a polar covalent bond.
   Na Cl I_2 CO O_3
2. Give the Lewis symbols for Mg, S, and Al. Write the formula for the monatomic ion formed by each one.
   Mg: ·S:
3. On the basis of atomic properties, predict how the following ionic compounds should be ranked in order of increasing melting points: CsCl, $MgCl_2$, NaCl, KCl, CsBr.
4. Draw Lewis structures for $CHCl_3$, HPO_4^{2-}, HCO_3^-, and C_3H_8.
5. Draw three equivalent resonance structures for HPO_4^{2-}.

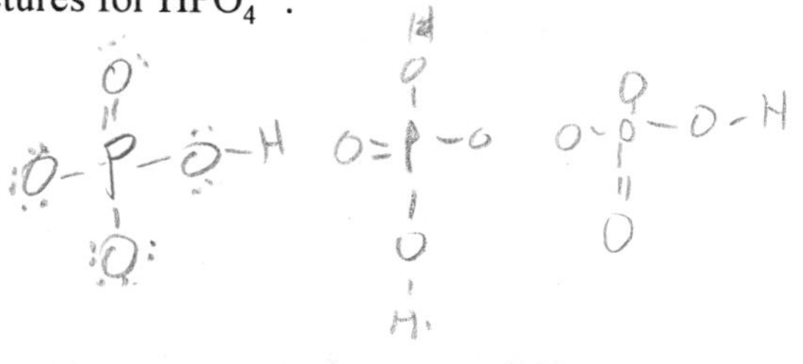

6. Which one of each pair of the following resonance structures should contribute most to the actual molecule?

(a) (1) $H-\overset{\overset{\displaystyle :O:}{\|}}{C}-\overset{\cdot\cdot}{\underset{\cdot\cdot}{O}}-H$ \longleftrightarrow (2) $H-\overset{\overset{\displaystyle :\overset{\cdot\cdot}{O}:}{|}}{C}=\overset{\cdot\cdot}{O}-H$

(b) (1) $H-C\equiv N:$ \longleftrightarrow (2) $H-\overset{\cdot\cdot}{C}=N:$

(c) (1) $H-\overset{\overset{\displaystyle H}{|}}{C}=\overset{\overset{\displaystyle H}{|}}{C}-\overset{\cdot\cdot}{\underset{\cdot\cdot}{C}l:$ \longleftrightarrow (2) $H-\overset{\overset{\displaystyle H}{|}}{C}-\overset{\overset{\displaystyle H}{|}}{C}=\overset{\cdot\cdot}{C}l:$

(d) (1) $H-\overset{\cdot\cdot}{\underset{\cdot\cdot}{O}}-\overset{\overset{\displaystyle :\overset{\cdot\cdot}{O}:}{|}}{\underset{\underset{\displaystyle :\overset{\cdot\cdot}{O}:}{|}}{S}}-\overset{\cdot\cdot}{\underset{\cdot\cdot}{O}}-H$ \longleftrightarrow (2) $H-\overset{\cdot\cdot}{\underset{\cdot\cdot}{O}}-\overset{\overset{\displaystyle :O:}{\|}}{\underset{\underset{\displaystyle :O:}{\|}}{S}}-\overset{\cdot\cdot}{\underset{\cdot\cdot}{O}}-H$

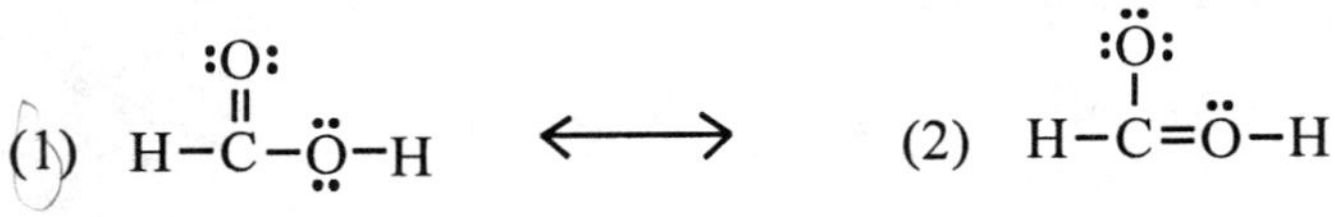

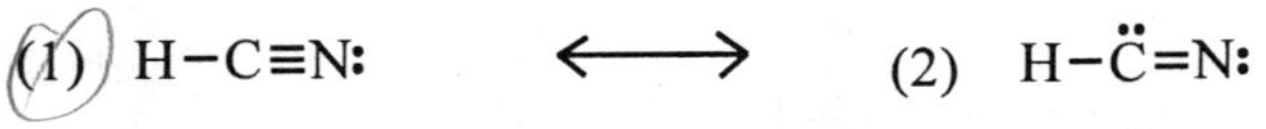

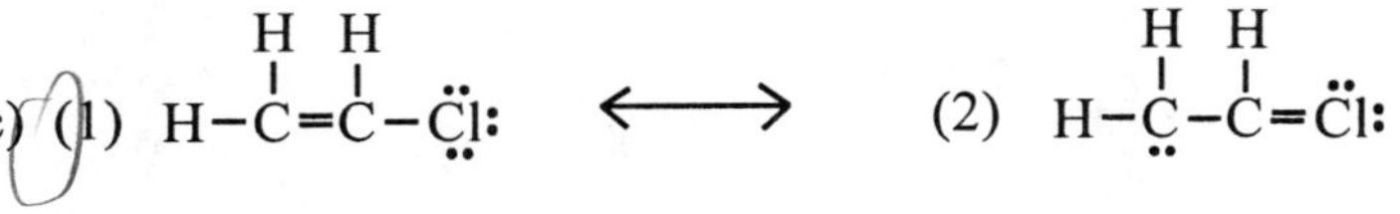

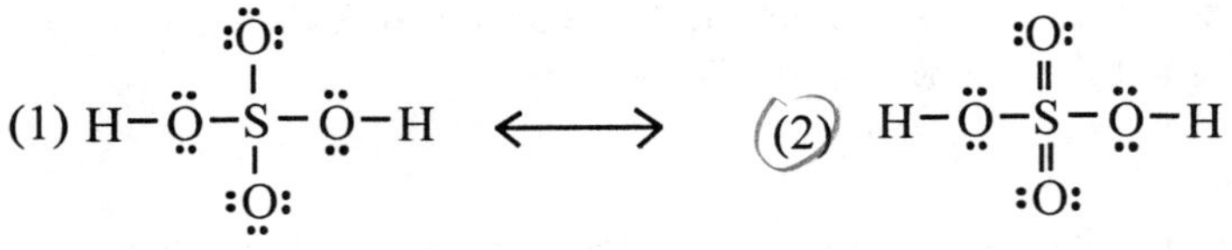

7. The $\Delta H°$ of formation of H_2O (g) is -242 kJ/mol. The H–H bond and O=O bond energies are 435.9 kJ/mol and 498.3 kJ/mol, respectively. Estimate the average O–H bond energy.

8. Use electronegativities to predict which of the following compounds are predominately covalent: F_2, NaF, $MgCl_2$, H_2Se, CH_4.

9. Identify the most polar bond in each molecule:

(a) $Cl-\overset{\overset{\displaystyle Cl}{|}}{\underset{\underset{\displaystyle F}{|}}{C}}-F$

(b) $H-\overset{\overset{\displaystyle H}{|}}{\underset{\underset{\displaystyle H}{|}}{C}}-\overset{\overset{\displaystyle O}{\|}}{C}-\overset{\overset{\displaystyle H}{|}}{\underset{\underset{\displaystyle H}{|}}{C}}-H$

(b) $Cl-\overset{\overset{\displaystyle H}{|}}{\underset{\underset{\displaystyle H}{|}}{C}}-Cl$

(d) $H-\overset{\overset{\displaystyle H}{|}}{\underset{\underset{\displaystyle H}{|}}{C}}-S-H$

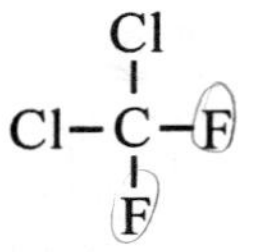

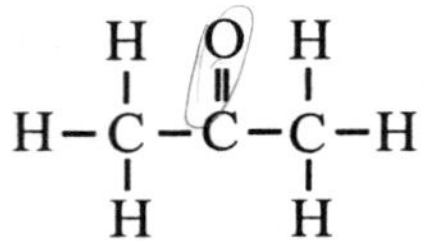

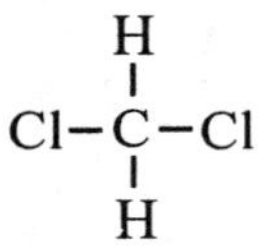

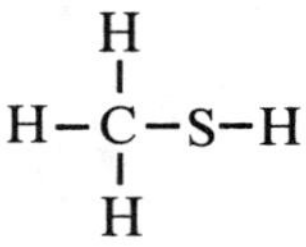

10. Rank the following in order of increasing polarity: NH_3, N_2, NO, AlN.

ANSWERS TO SELF-TEST QUESTIONS

1. (possible answers) ionic: CsBr; covalent: Br_2; polar covalent: HBr

2. ·Mg·, Mg^{2+}; ·S·, S^{2-}; ·Al·, Al^{3+}

3. CsBr < CsCl < KCl < NaCl < $MgCl_2$

4. 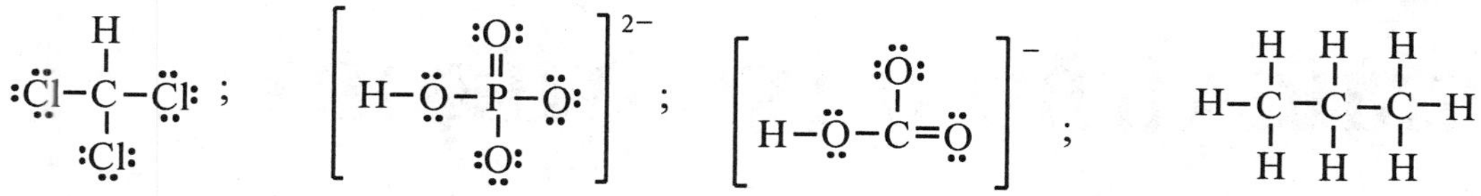

5. 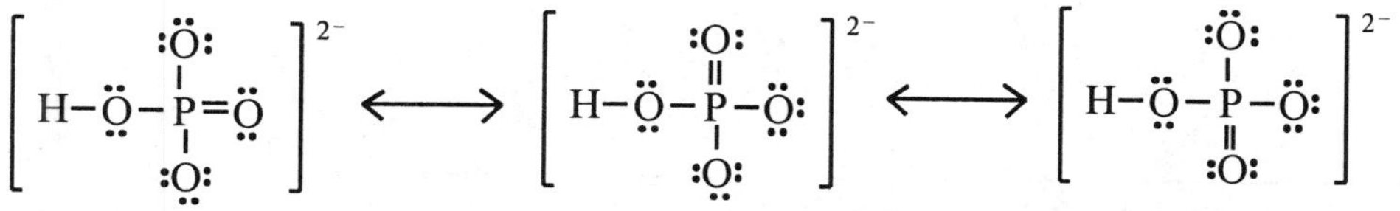

6. (a) 1 (b) 1 (c) 1 (d) 2

7. 463 kJ/mol

8. F_2, H_2Se, CH_4

9. (a) C–F (b) C–Cl (c) C=O (d) C–S

10. N_2, NO, NH_3, AlN

CHAPTER 10

MOLECULAR GEOMETRY AND CHEMICAL BONDING THEORY

SELF-ASSESSMENT

Dipole Moments and Molecular Geometry (Section 10.1)

1. Predict which molecules have net dipole moments:

(a) 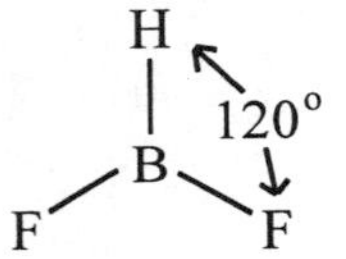

(b) CO

(c) 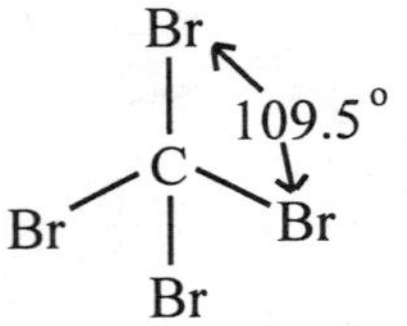

(d) 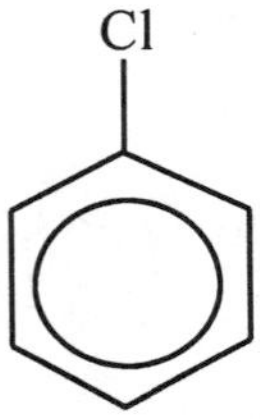

(e) 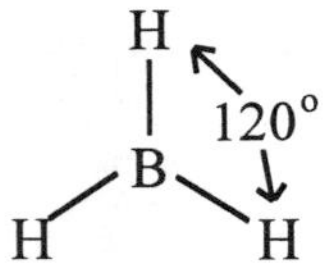

The VSEPR Model (Section 10.2)

True or false? If a statement is false, change the underlined word(s) to make it true. Use the periodic table if necessary.

2. AsH_3 is trigonal planar.

3. Lone-pair electrons tend to compress bond angles.

4. All of the bond angles in SF_6 are 90°.

5. CCl_4 is a polar molecule.

6. The H–C=O bond angle in HCOOH is exactly 120°.

The Shapes of Some Hydrocarbon Molecules (Section 10.3)

7. Draw and name all of the isomers of butene. Label the isomers as structural or cis-trans.

8. Describe the geometry of the bonds around the carbon atoms in
   (a) ethane
   (b) ethylene
   (c) acetylene

Valence Bond Theory (Section 10.4)

9. Match Column A to Column B.

(1) σ bond in H_2	(a) formed by the mixing of two p orbitals with the s orbital in the same energy shell of an atom.
(2) π bond in C_2H_4	(b) formed by the overlap of two sp^2 hybrid orbitals
(3) sp^2 hybrid orbital	(c) formed by the overlap of a hybrid orbital and an s orbital
(4) σ bond between the carbon and hydrogen atom in CH_4	(d) formed by the sideways overlap of two p orbitals
(5) σ bond between the carbon atoms in C_2H_4	(e) formed by the overlap of two s orbitals

10. The Lewis structure for glycine, the simplest amino acid, is

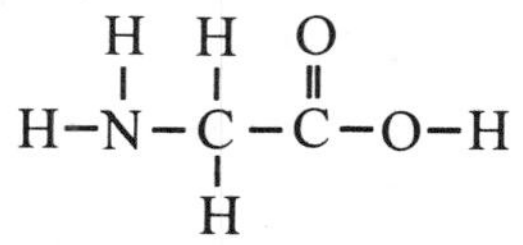

Determine the hybridization of each central atom and estimate all bond angles. Consider the effect of lone-pair electrons and multiple bonds.

Molecular Orbital Theory for Diatomic Molecules (Section 10.5)

Underline the correct word.

11. The sigma bonding orbital provides for (more, less) electron density between the nuclei and (higher, lower) energy than does the sigma antibonding orbital.

12. Rank the following species in order of (a) increasing bond energy, (b) increasing bond length, and (c) increasing paramagnetism:

    N_2^{2+}, N_2^{+}, N_2

A Combined Model for Polyatomic Molecules (Section 10.6)

13. Describe the bonding in NO_3^- in terms of localized sigma bonds and a delocalized pi electron system.

ANSWERS TO SELF-ASSESSMENT PROBLEMS

1. a, b, d

2. Pyramidal

3. T

4. T

5. F

$$Cl-\overset{\overset{\displaystyle Cl}{|}}{\underset{\underset{\displaystyle Cl}{|}}{C}}-Cl$$

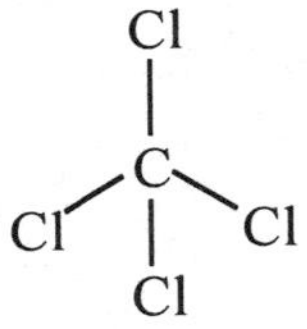

6. F; greater than

7. Structural Isomers

 $CH_2=CH-CH_2-CH_3$

 1-butene

 $CH_3-CH=CH-CH_3$

 2-butene

 $CH_2=\overset{\overset{\displaystyle CH_3}{|}}{C}-CH_3$

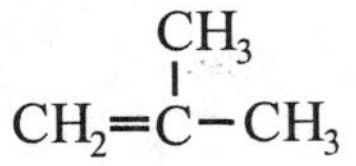

 isobutene

 Cis-Trans Isomers

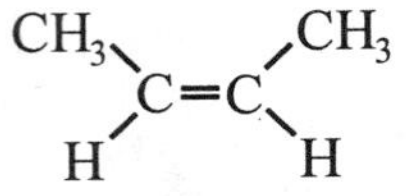

 cis-2-butene trans-2-butene

8. (a) Bond angles are 109.5° to form a tetrahedron.
 (b) Bond angles are about 120°: angles including double bond are greater than 120°, other two bond angles are less than 120°.
 (c) Bond angles are 180° making the molecule linear.

9. (1) e (2) d (3) a (4) c (5) b

10. From left to right; N: sp^3; C: sp^3; C: sp^2; O: sp^3

 ∡ HNH, ∡ HNC = slightly less than 109.5°

 ∡ HCH, ∡ HCC = 109.5°; ∡CC=O, ∡ O=CO > 120°

 ∡CC–O < 120°; ∡ COH < 109.5°

11. more, lower

12. (a) $N_2^{2+} < N_2^{+} < N_2$ (b) $N_2^{2+} < N_2^{+} < N_2$ (c) $N_2 = N_2^{2+} < N_2^{+}$

13. N–O sigma bonds form by the overlap of nitrogen sp^2 orbitals with oxygen orbitals. The pi molecular orbitals are formed from the overlap of the nitrogen p orbital and the oxygen p orbitals that are perpendicular to the plane of the molecule. A pair of electrons occupies the lowest energy pi orbital.

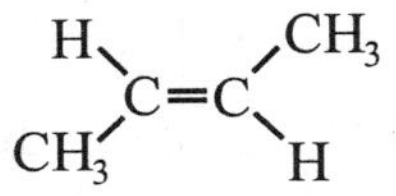

REVIEW

Introduction

Lewis structures, VSEPR configurations, valence bond theory, and molecular orbital theory are models of chemical bonding. Each of them is useful for specific purposes.

Dipole Moments and Molecular Geometry (Section 10.1)

Learning Objectives

1. Use dipole moments to predict the shapes of simple molecules.

Review

The net dipole moment of a molecule is the vector sum of the individual bond moments and therefore depends on the shape of the molecule. In some cases the individual bond moments cancel to give a nonpolar molecule.

The following bond moments will cancel to give a nonpolar molecule.

1. Two equal bonds pointing in opposite direction. Example: CO_2, a linear molecule

   O ⟻ C ⟼ O

2. Three equal bonds at 120° angles. Example: BF_3, a flat triangular molecule

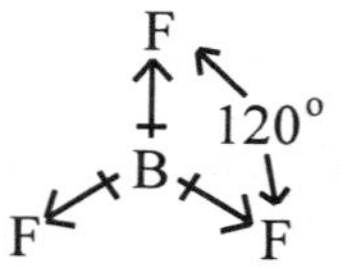

3. Four equal bonds that point toward the corners of a tetrahedron (109.5° angles). Example: CCl_4, a tetrahedral molecule

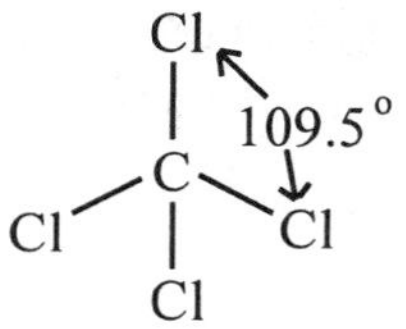

Note: Most other bond arrangements provide some net polarity to the molecule.

Practice Drill on Dipole Moments: Complete the following chart.

Molecule	Shape	Draw the bond dipole moments	Polar or nonpolar molecule
BCl_3	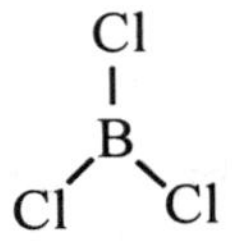		
H_2S	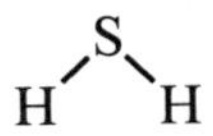		
CS_2	S=C=S		
COS	O=C=S		

Answers:

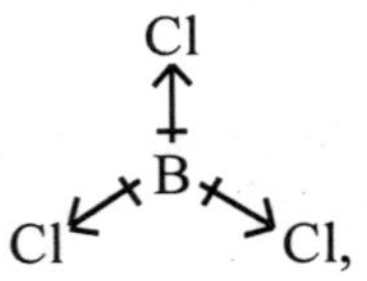 nonpolar; 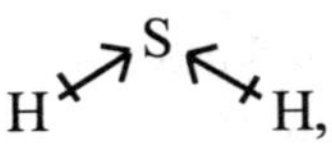 polar; S⟵+C+⟶S, nonpolar; O⟵+C+⟶S, polar

The VSEPR Model (Section 10.2)

Learning Objectives

1. Determine the number of VSEPR electron pairs around a central atom.
2. Draw linear, trigonal planar, tetrahedral, trigonal bipyramidal, and octahedral molecules.
3. Explain deviations from ideal geometry in terms of lone-pair repulsion and multiple-bond character and draw molecules exhibiting such deviations.
4. Use VSEPR to predict bond angles, molecular shapes, and polarities.

Review

The VSEPR model for predicting molecular geometry assumes that Valence Shell Electron Pairs repel each other and stay as far apart as possible. To find the number of VSEPR pairs around a central atom, refer to the Lewis structure and

(a) count each bond to another atom as one pair (whether it is single or multiple)
(b) count each lone pair of electrons as one pair

VSEPR pairs arrange themselves to minimize repulsion as follows:

Table 10.1

Number of VSEPR electron pairs	Electron pair Arrangement	Angle between electron pairs	Example
2	linear	180°	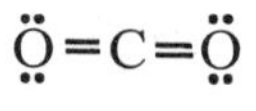
3	trigonal planar	120°	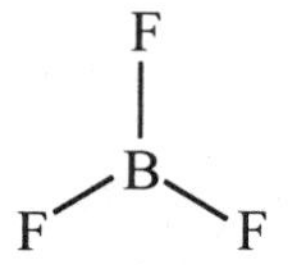
4	tetrahedral	109.5°	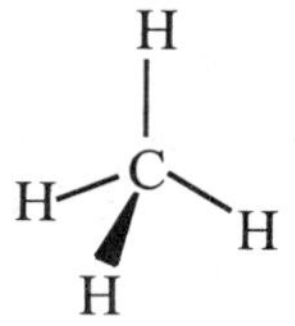
5	trigonal bypyramid	120°, 90°	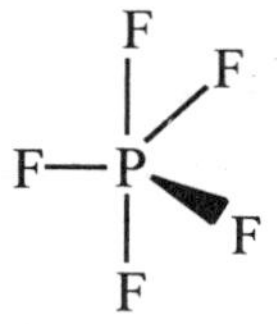
6	octahedral	90°	

When all VSEPR pairs are bond pairs, the molecular shape will be the same as the electron arrangement. The shape of a molecule with lone pairs will be what remains after the lone pair corners are removed.

A bond angle tends to be larger than the ideal value if it includes a multiple bond. A bond angle tends to be smaller if the <u>central</u> atom has lone pair electrons.

PROBLEM-SOLVING TIP

How to Predict the Shape of a Molecule

To predict the shape of a molecule:

1. Draw the Lewis structure.
2. Determine the number of VSEPR electron pairs around the <u>central</u> atom:
   number of VSEPR electron pairs around central atom = number of lone pairs + bond pairs
   (multiple bonds count only as one bond pair).
3. Choose the arrangement and bond angles associated with the number of VSEPR pairs determined in (2). The arrangements and angles in Table 10.1 should be memorized.
4. Remove any corner inhabited by a lone pair. The shape of the molecule corresponds to what is left.
5. Consider deviations from ideal angles:
   (a) The bond angles will be slightly compressed if there are lone-pair electrons on the central atom.
   (b) A bond angle that includes a multiple bond is slightly larger; as a result the other angles are slightly smaller.

Example 10.1: Draw CH_2O (C is the central atom).

Answer: We will follow the five steps given above.

1. 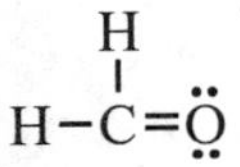
2. Number of VSEPR pairs = 3 bond pairs = 3 VSEPR pairs.
3. Electron pair arrangement = trigonal planar; bond angle is approximately 120°.
4. No lone pairs.
5. Double bond increases < HCO above 120° and decreases < HCH below 120°.

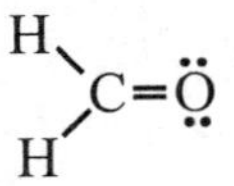

Example 10.2: Draw NH_2^-

Answer:

1. 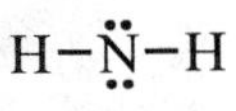
2. Number of VSEPR electron pairs = 2 bond pairs + 2 lone pairs = 4 VSEPR pairs.
3. Electron pair arrangement = tetrahedral; bond angle is approximately 109.5°.
4. Two lone pairs occupy two corners of a tetrahedron. The central N and the two remaining corners make a V-shape.

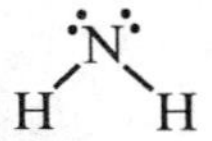

5. (a) Bond angle is less than 109.5° due to the lone pairs.
   (b) No multiple bonds involved.

Practice Drill on Drawing Molecules: Complete the following chart:

Molecule	Lewis structure	Number of VSEPR pairs	Electron Pair Arrangement	Molecular shape	Effect of lone pairs or multiple bonds	Sketch the molecule
PF_3					bond angles < 109.5°	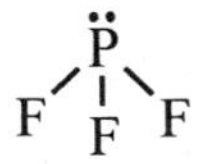
CCl_4						
BrF_3						

Answers: PF_3: 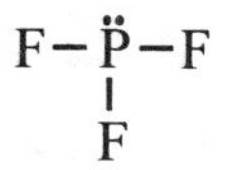

Four VSEPR electron pairs, tetrahedral, pyramid, lone pair decreases bond angle below 109.5°:

CCl_4: 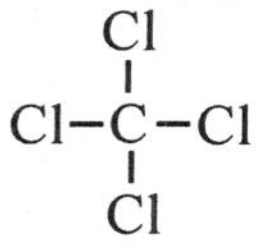

Four VSEPR electron pairs, tetrahedral, tetrahedral, no effect: 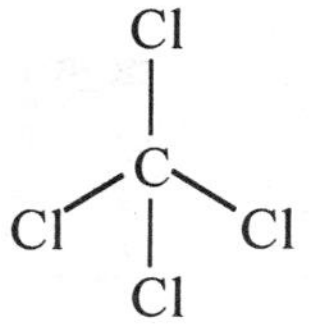

BrF_3: 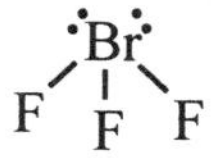

Five VSEPR electron pairs, trigonal bypyramid, distorted T, lone pair compresses bond angle below 90°; (lone-pair electrons occupy equatorial positions):

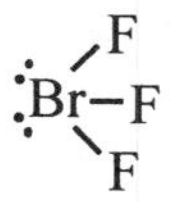 (distorted T-shape)

The Shapes of Some Hydrocarbon Molecules (Section 10.3)

Learning Objectives

1. Write structural formulas for simple alkanes, alkenes and alkynes.
2. Explain why there is free rotation around carbon-carbon single bonds, but not around carbon-carbon double bonds.
3. Draw and name the structural isomers of butane, pentane, and butene.
4. State whether a given alkene could have <u>cis-trans</u> isomers and draw the isomers.
5. Write the structural formulas of benzene, napthalene, and anthracene.

Review

The large number and variety of carbon compounds are due to three factors:

(1) the stability of the carbon-carbon bond so that long chains can form,
(2) the four valence electrons of carbon so that branched chains and bonds to other elements can form,
(3) the small size of the carbon atom so that multiple bonds are possible.

In this section we will apply what we have learned about bonding and molecular geometry to the study of hydrocarbons, compounds containing only carbon and hydrogen atoms.

Alkanes

Alkanes are hydrocarbons that contain only single bonds. Methane, CH_4, the simplest alkane had bond angles of 109.5° with each C–H bond pointing to the corners of a tetrahedron. Bond angles in all of the alkanes are 109.5° so that chains with more than three carbon atoms have a puckered conformation due to rotation about the C–C single bonds.

Structural Isomers

Alkanes with four or more carbon atoms occur as a straight chain (called normal) or as a branched molecule. For example

$CH_3-CH_2-CH_2-CH_3$

 n-butane

$$CH_3-\underset{\underset{\displaystyle CH_3}{|}}{CH}-CH_3$$

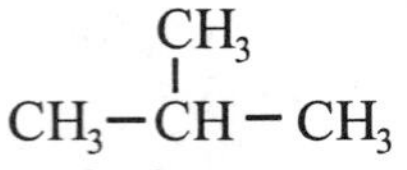

 isobutane

The above are structural isomers because the sequence of bonds is different.

Caution: In drawing structural isomers remember that they can only be interconverted by breaking bonds and not by rotation about a C–C single bond. Therefore the two formulas given below are for the same molecule and are not representations of structural isomers.

$$\underset{\underset{\displaystyle CH_3}{|}}{CH_2}-CH_2-\overset{\overset{\displaystyle CH_3}{|}}{CH}-CH_3$$

$$CH_3-CH_2-CH_2-\underset{\underset{\displaystyle CH_3}{|}}{CH}-CH_3$$

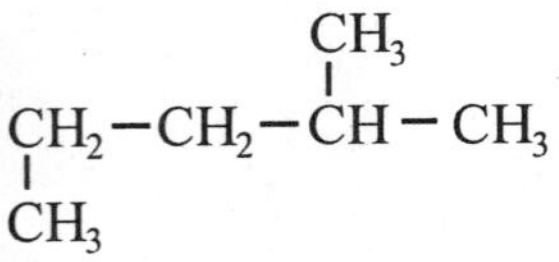

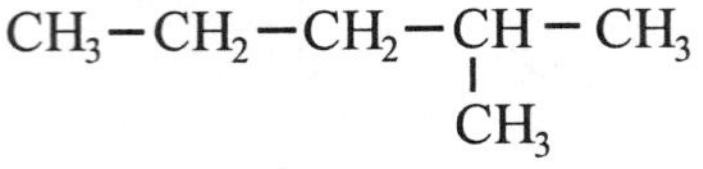

In both formulas there are five carbon atoms in a continuous chain and a branched carbon atom bonded to a second carbon atom from one of the ends. Therefore the two formulas represent the same molecule.

Alkenes and Alkynes

Alkenes are hydrocarbons with double bonds. The simplest is ethene, $CH_2=CH_2$ or ethylene. All bond angles in alkenes which include one or both of the doubly bonded carbon atoms is approximately 120° giving a flat, rigid structure around the double bond.

$$\underset{H}{\overset{H}{\diagdown}}C=C\underset{\diagdown H}{\overset{\diagup H}{}}$$

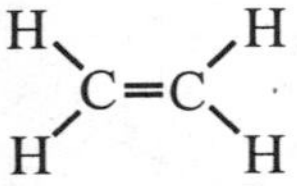

All of the atoms in ethylene are in the same plane and are unable to rotate due to the strong double bond.

Many naturally occurring substances such as β-carotene, recently shown to decrease risk of some cancers, contains alternate double and single bonds and is referred to as a conjugated molecule.

Alkynes contain triple bonds. The simplest alkyne, ethyne or acetylene, $HC\equiv CH$, is a linear molecule because of the 180° bond angles about the triple bond. The appearance of a triple bond in a hydrocarbon means that four atoms will be in a straight line.

Structural Isomers

The appearance of multiple bonds gives rise to isomers. For example, C_4H_6, has two structural isomers which differ in the position of the triple bond.

$HC\equiv C-CH_2-CH_3$

 1-butyne

$CH_3-C\equiv C-CH_3$

 2-butyne

Butene, C_4H_8, has three structural isomers.

Cis-Trans Isomers

These isomers are encountered in alkenes in which each carbon in the double bond carries different substituents. For example

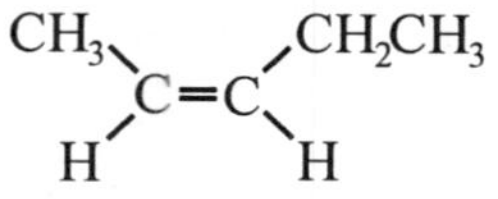

cis isomer

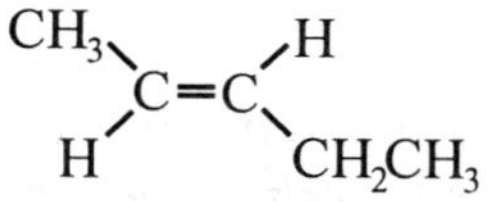

trans isomer

2- pentene

Note that the cis isomer has the substituents on the same side of the double bond while the substituents appear on opposite sides in the trans isomer.

Aromatic Hydrocarbons

We have already discussed benzene in Section 9.5 in which the delocalized pi bonds are repreresented as a circle in a hexagon.

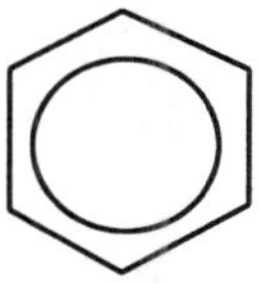

Hydrocarbons which contain benzene rings are called aromatic. Some examples are given in Table 10.1 of your textbook.

Practice: Draw two isomers for
(a) 2-pentene
(b) C_5H_{12}
(c) pentene

Answers:

(a)

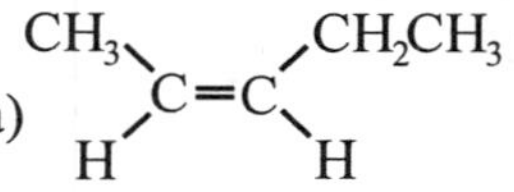

cis

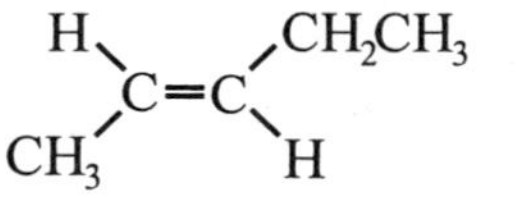

trans

(b) $CH_3–CH_2–CH_2–CH_2–CH_3$ $\qquad$ $CH_3-CH_2-\overset{\overset{\displaystyle CH_3}{|}}{C}H-CH_3$

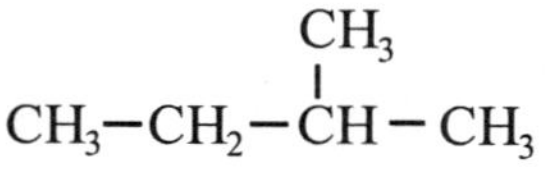

structural isomers

(c) $CH_2{=}CH–CH_2–CH_2–CH_3$ $\qquad$ $CH_3–CH{=}CH–CH_2–CH_3$

structural isomers

Valence Bond Theory (Section 10.4)

Learning Objectives

1. Describe the formation of a covalent bond in terms of valence bond theory.
2. Sketch the ways s and p orbitals overlap to form sigma bonds.
3. State the roles of orbital hybridization and electron promotion in valence bond theory.
4. Sketch an s orbital, a p orbital, and the general shape of a hybrid orbital.
5. Sketch the sigma bonds formed by overlap of hybrid orbitals with s orbitals and with p orbitals.
6. Sketch the sigma bonds and lone pairs in molecules such as CH_4, NH_3, and H_2O.
7. Sketch the formation of a pi bond from two p orbitals.
8. Sketch the sigma and pi bonds in molecules with two or more central atoms.
9. Estimate the angles in molecules with two or more central atoms.

Review

According to valence bond theory, a bond is formed by the overlap of half-filled atomic orbitals on adjacent atoms. The electron density is highest in the overlap region.

The observed bond angles around a central atom are nearly always different from the angles between its s, p, and d atomic orbitals. Valence bond theory explains the observed geometry of the molecule by assuming that atomic orbitals on the central atom have hybridized; that is, they have combined to form a new set of orbitals with the requisite angles.

To determine the type of hydridization:

1. Write the Lewis structure.
2. Count the VSEPR pairs.
3. Consult Table 10.2. (The material in this table should be memorized and understood.)

Table 10.2:

Number of electron pairs	Hybridization	Arrangement of electron pairs
2	sp	linear
3	sp^2	trigonal planar
4	sp^3	tetrahedron
5	sp^3d	trigonal bipyramid
6	sp^3d^2	octahedron

Each VSEPR electron pair around a central atom is either a lone pair in a hybrid orbital or a bonded pair in a sigma bond formed by the overlap of a hybrid orbital with an orbital on another atom.

A sigma (σ) bond is one in which the electron density is greatest along the axis that passes through both nuclei. All single bonds are sigma bonds.

Example 10.3: Determine the hybridization around carbon in

(a) CO_2
(b) CH_4
(c) C_2H_2

Answer:

(a) $\ddot{\underset{..}{O}}=C=\ddot{\underset{..}{O}}$ Two VSEPR electron pairs around carbon requiring sp hybridization.

(b) 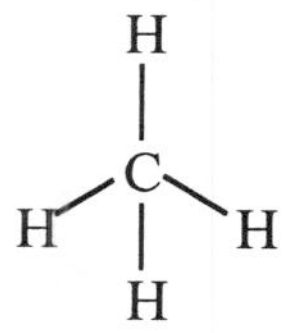 Four VSEPR electron pairs around carbon requiring sp^3 hybridization.

(c) H−C≡C−H Two VSEPR electron pairs around carbon requiring sp hybridization.

Practice: Determine the hybridization for the central atom in PF_3, CCl_4, BrF_3, and SF_6.
Answers: sp^3; sp^3; sp^3d; sp^3d^2

The additional electron pairs in a multiple bond are accommodated in pi bonds.

A pi bond is formed by the overlap of unhybridized p orbitals. It has lobes of electron density above and below the line of atomic centers, and zero density along the line. Carefully study Figures 10.14, 10.17, and 10.23 in your text showing sigma and pi bonds, and sketch them.

Example 10.4:

(a) What is the hybridization around each carbon in acetone ($(CH_3)_2CO$)?
(b) Give the approximate C–C=O, H–C–C, and H–C–H bond angles.

Answer:

(a) The structural formula for acetone is: 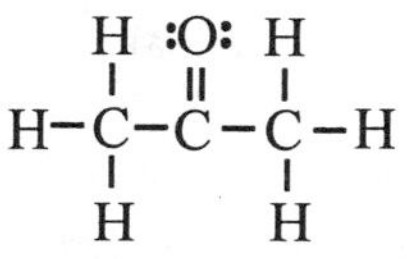

The left-hand carbon with four VSEPR electron pairs forms four sigma bonds. The arrangement is tetrahedral. The hybridization is sp^3.

The middle carbon atom with three VSEPR pairs participates in three sigma bonds; these are trigonal planar with sp^2 hybridization.

The right-hand carbon is the same as the left-hand carbon.

(b) The bond angles can be determined from the number of VSEPR (see Table 10.1 in study guide) or from the hybridization (see Table 10.2 in this study guide) around the central atom. Memorize and understand the information in the tables.

∡C–C=O; 120°
∡ H–C–C; 109.5°
∡ H–C–H; 109.5°

Practice Drill on Hybridization and Bond Angles: Complete the following chart:

Lewis structure	VSEPR pairs around atom 1 atom 2		Hybridization around atom 1 atom 2		Approximate bond angles around atom 1 atom 2	
O ‖ H−C−Ö−H 　　1　2	3	4	sp²	sp³	120	109.5
H \| H−C−N̈−H 　1\|　\|2 　H　H	4	4	sp³	sp³	109.5	109.5
H \| H−C−C≡N: 　1\|　2 　H	4	2	sp³	sp	109.5	180

Answers: CH_3NH_2: 4, 4 sp³, sp³, 109.5°, 109.5°; CH_3CN: 4, 2, sp³, sp, 109.5°, 180°

Molecular Orbital Theory for Diatomic Molecules (Section 10.5)

Learning Objectives

1. Sketch the bonding and antibonding molecular orbitals obtained by combining s orbitals, p_x orbitals, p_y orbitals, and p_z orbitals.
2. Draw an approximate energy level diagram showing the relative energies of the first ten homonuclear diatomic molecular orbitals and their parent atomic orbitals.
3. Write ground-state, molecular orbital configurations and draw orbital occupancy diagrams for first- and second-period homonuclear diatomic molecules.
4. Use molecular orbital configurations to predict whether homonuclear diatomic molecules are stable and paramagnetic.
5. Calculate bond orders from molecular orbital configurations.
6. Use bond orders to compare relative bond energies and bond lengths for similar species.
7. Write molecular orbital configurations for diatomic molecules containing different atoms from the same period.

Review

Molecular orbital theory is a chemical bonding theory which assumes that atomic orbitals overlap and form <u>bonding</u> and <u>antibonding molecular orbitals</u> with shapes, energies, and electron density distributions different from those of the parent orbitals. <u>The number of molecular orbitals is equal to the original number of atomic orbitals.</u>

Electrons in bonding orbitals have lower energies than electrons in the parent orbitals; electrons in antibonding orbitals have higher energies. For example, when the H_{1s} orbital overlaps with another H_{1s} orbital to form a sigma bond, two molecular orbitals are formed: the σ_{1s} (bonding) and the σ^*_{1s} (antibonding).

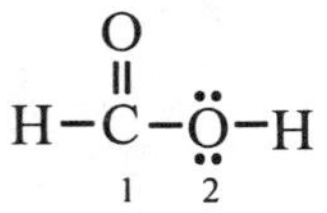

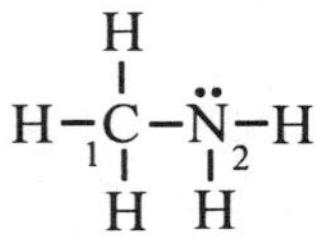

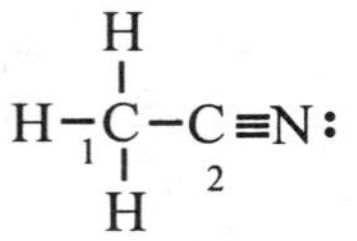

Energy level diagrams are used to show the relative energies of the atomic orbitals and the molecular orbitals:

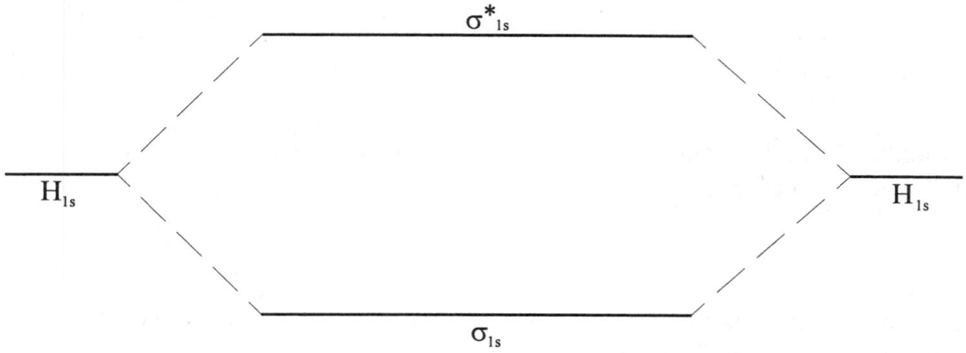

Electrons will fill up molecular orbitals just as they do atomic orbitals, occupying the lower energy orbitals. The ground-state molecular orbital configuration for H_2 is therefore $(\sigma_{1s})^2$. To obtain the ground-state electron configuration of a molecule, (1) list its molecular orbitals in order of increasing energy and (2) fill them according to the aufbau procedure, the Pauli exclusion principle, and Hund's rule.

Practice: Give the ground-state configuration for H_2^-.
Answer: $(\sigma_{1s})^2\ (\sigma^*_{1s})^1$

The sigma (σ) molecular orbitals of homonuclear diatomic molecules that have valence electrons in the n = 2 shell are formed by the overlap of 2s orbitals and by the overlap of the $2p_z$ orbitals. Pi (π) molecular orbitals are formed by the overlap of the $2p_x$ and the $2p_y$ atomic orbitals. The order of increasing energy of these molecular orbitals can be shown with an approximate energy level diagram:

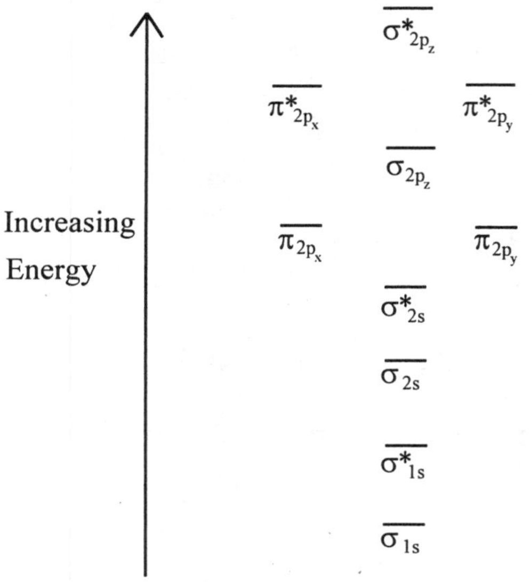

Note: Some diatomic molecules have different energy level diagrams. For example, in O_2 and F_2 the σ_{2p_z} orbital is below the π_{2p_x} and π_{2p_y} orbitals.

A molecule is stable if the total number of electrons in bonding molecular orbitals (bonding electrons) exceeds the total number of electrons in antibonding molecular orbitals (antibonding electrons). Relative stabilities are often compared in terms of bond orders:

Bond order = 1/2 × (number of bonding electrons – number of antibonding electrons)

In general, the higher the bond order, the greater the bond energy (bond less easily broken), and the shorter the bond length.

Example 10.5:

(a) Give the electronic configuration of F_2 and F_2^+ using the approximate energy level diagram.
(b) Predict which species has the higher bond energy.
(c) Which one is paramagnetic?

Answer:

(a) F_2 has 18 electrons. To determine the configuration, we will put a maximum of two electrons in each molecular orbital, filling the lower energy orbitals first (Aufbau principle):

$(\sigma_{1s})^2\ (\sigma^*_{1s})^2\ (\sigma_{2s})^2\ (\sigma^*_{2s})^2\ (\pi_{2p_x})^2\ (\pi_{2p_y})^2\ (\sigma_{2p_z})^2\ (\pi^*_{2p_x})^2\ (\pi^*_{2p_y})^2$

For F_2^+ remove one electron from the highest energy orbital.

In this case, $\pi^*_{2p_x}$ and $\pi^*_{2p_y}$ have the same energy so the answer can be written as either $(\pi^*_{2p_x})^1\ (\pi^*_{2p_y})^2$ or $(\pi^*_{2p_x})^2\ (\pi^*_{2p_y})^1$.

(b) Bond order of $F_2 = 1/2\ (10 - 8) = 1$; $F_2^+ = 1/2\ (10 - 7) = 3/2$

F_2^+ will therefore have the greatest bond energy.

(c) F_2^+ has one unpaired electron, so it is paramagnetic.

Practice Drill on Molecular Orbital Configurations: Complete the following chart using the approximate energy level diagram.

Diatomic	Total number of electrons in molecule	Electronic configuration	Bond order	Paramagnetic? (Yes or No)
N_2^+				
F_2^-				
O_2				

Answers: N_2^+: 13, $(\sigma_{1s})^2\ (\sigma_{1s}^*)^2\ (\sigma_{2s})^2\ (\sigma_{2s}^*)^2\ (\pi_{2px})^2\ (\pi_{2py})^2\ (\sigma_{2pz})^1$, B.O. = 5/2, yes

F_2^-: 19, $(\sigma_{1s})^2\ (\sigma_{1s}^*)^2\ (\sigma_{2s})^2\ (\sigma_{2s}^*)^2\ (\pi_{2px})^2\ (\pi_{2py})^2\ (\sigma_{2pz})^2\ (\pi_{2px}^*)^2\ (\pi_{2py}^*)^2\ (\sigma_{2pz}^*)^1$, B.O. = 1/2, yes

O_2: 16, $(\sigma_{1s})^2\ (\sigma_{1s}^*)^2\ (\sigma_{2s})^2\ (\sigma_{2s}^*)^2\ (\pi_{2px})^2\ (\pi_{2py})^2\ (\sigma_{2pz})^2\ (\pi_{2px}^*)^1\ (\pi_{2py}^*)^1$, B.O. = 2, yes

Note: The O_2 configuration obeys Hund's Rule.

The energy level diagram for a heteronuclear molecule composed of two atoms from the same period is a skewed version of the homonuclear diagram. However, when the atoms come from different periods as in HF or are far apart in the same period the energy level diagram differs significantly.

A Combined Model for Polyatomic Molecules (Section 10.6)

Learning Objectives

1. Use valence bond theory to describe the sigma-bonded framework of a polyatomic molecule and molecular orbital theory to describe its pi electron system.
2. Give examples showing how modern bonding theory eliminates the need for resonance structures.

Review

Chemists often find it convenient to consider the sigma electron system of a polyatomic molecule in terms of localized sigma valence bonds and the pi electron system in terms of delocalized molecular orbitals. The use of molecular orbitals for the pi bonding eliminates the need for different resonance structures. The very important molecule benzene is usually treated in this manner (see Figure 10.38 and 10.39 of your textbook).

Example 10.6: Describe the bonding in CO_3^{2-} in terms of localized bonds and delocalized pi bonding.

Answer: The sigma bonds are formed by the overlap of carbon sp^2 hybrid orbitals with 2p atomic orbitals on oxygen.

Resonance structures indicate that the double bond is distributed over three sites:

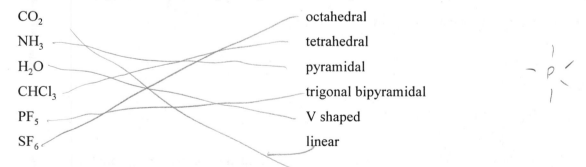

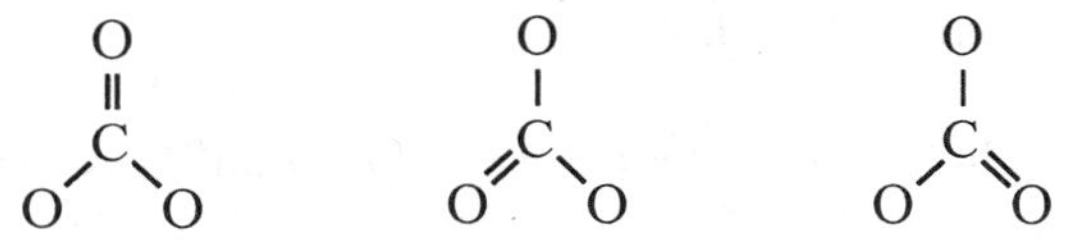

Each of the four atoms contributes an unhybridized p orbital to the π structure; these four parallel p orbitals overlap to form four delocalized molecular orbitals. The lowest energy bonding π molecular orbital accommodates the two electrons.

SELF-TEST

1. Match the compound to its corresponding molecular shape.

CO_2	octahedral
NH_3	tetrahedral
H_2O	pyramidal
$CHCl_3$	trigonal bipyramidal
PF_5	V shaped
SF_6	linear

2. Which of the molecules in question 1 are polar? CHCl3

True or False

3. All bond angles in PF_5 are the same. F

4. N_{sp^3} hybrid orbitals overlap with H_{1s} orbitals to form N–H sigma bonds in NH_3. T

5. According to molecular orbital theory, p atomic orbitals can only form pi molecular orbitals. f

6. The bond angles in NH_3 are greater than 109.5°. E

7. 1-butene and 2-butene are cis-trans isomers. f

8. Circle all paramagnetic species (the relative energies of molecular orbitals of CO are similar to those of N_2):

 F_2^+, O_2^-, N_2, CO, CO^+, CO^-

9. Rank N_2^+, N_2^-, and N_2 in order of increasing ionization energy. (Hint: Electron removal requires less energy if it produces a more stable species and requires more energy if it produces a less stable species.)

10. Describe the bonding in the formate ion in terms of localized sigma bonds and a delocalized pi electron system.

$$\left[\begin{array}{c} :\ddot{O}: \\ | \\ H-C=\ddot{O} \\ \ddot{} \end{array} \right]^{-} \qquad \left[\begin{array}{c} :O: \\ || \\ H-C-\ddot{O}: \\ \ddot{} \end{array} \right]^{-}$$

formate ion
(resonance structures)

ANSWERS TO SELF-TEST QUESTIONS

1. CO_2: linear; NH_3: pyramidal; H_2O: bent or V shaped; $CHCl_3$: tetrahedral; PF_5: trigonal bipyramidal; SF_6: octahedral

2. NH_3, H_2O, $CHCl_3$

3. F; 90° and 120°

4. T

5. F; p orbitals can also form sigma molecular orbitals (σ_{p_z})

6. F; lone-pair repulsion compresses bond angles to less than 109.5°.

7. F

8. F_2^+, O_2^-, CO^+, CO^-

9. $N_2^- < N_2 < N_2^+$

10. Sigma bonds: C–O sigma bonds formed by overlapping sp^2 orbitals of carbon with 2p atomic orbitals of oxygen.

 C–H sigma bond formed by overlapping of a carbon sp^2 hybrid orbital with a hydrogen 1s orbital.

 Pi bonds: overlap of one carbon p orbital with two oxygen p orbitals to form three pi molecular orbitals. The two electrons are in the lowest energy bonding molecular orbital.

CHAPTER 11
OXIDATION-REDUCTION REACTIONS

SELF-ASSESSMENT

Oxidation - Reduction Reactions (Section 11.1)

1. Lead metal will reduce Cu^{2+}(aq) to form copper metal and Pb^{2+}(aq). Write the equations for the half-reactions, the net reaction, and identify the oxidizing and reducing agents. $Pb \longrightarrow Pb^{2+} + 2e^{-}$ $Pb + Cu^{2+} \longrightarrow Pb^{2+} + Cu$.
 $Cu^{2+} + 2e^{-} \longrightarrow Cu$

2. In a redox reaction, the oxidizing agent is (oxidized, reduced) while the reducing agent is (oxidized, reduced).

Oxidation Numbers (Section 11.2)

3. Give the oxidation numbers of manganese in each of the following compounds: MnO_2; $KMnO_4$; Mn_2O_3; $MnCl_2$.
 $4+$ $7+$ $3+$ $2+$

4. Identify the
 (a) substance oxidized and the substance reduced
 (b) oxidizing agent and the reducing agent in the reaction

 $C_3H_6O_3$(aq) + $2IO_4^{-}$(aq) → $2CH_2O$(aq) + CO_2(g) + $2IO_3^{-}$(aq) + H_2O(l)

Writing and Balancing Redox Reactions (Section 11.3)

5. The number of moles of electrons lost by one mole of Cl_2(g) to form ClO_3^{-}(aq) is _____10_____ .

6. Write the half-reactions and balance the net ionic equation for each of the following redox reactions:
 (a) Cu(s) + H^{+}(aq) + HSO_4^{-}(aq) → Cu^{2+}(aq) + SO_2(g) + H_2O(l)
 (b) The reaction between Zn(s) and NO_3^{-}(aq) to produce NH_3(g) and $Zn(OH)_4^{2-}$(aq) in basic solution. [This reaction is used to determine the NO_3^{-}(aq) concentration. Nitrate-containing fertilizers often pollute water in agricultural regions; high concentrations of NO_3^{-}(aq) can be toxic especially to infants and small children.]

Redox Stoichiometry (Section 11.4)

7. A 100.0-mL sample of well water from a farm community was tested for NO_3^{-}(aq) using the reaction described in 6(b) and produced 1.7 mg of NH_3(g).
 (a) Find the molarity of the NO_3^{-}(aq) in the water sample.
 (b) How many milligrams of NO_3^{-} would a child ingest with each liter of well water she drinks?

8. Dissolved oxygen in natural waters is necessary for survival of aquatic life. The concentration of dissolved oxygen can be determined by allowing it to oxidize Mn^{2+} to MnO_2 in basic solution. The $O_2(aq)$ in a certain 1.00-L water sample produces 6.8×10^{-4} mol of MnO_2. Calculate the concentration of dissolved $O_2(aq)$ in milligrams per liter.

ANSWERS TO SELF-ASSESSMENT PROBLEMS

If you missed an answer, be sure to study the relevant section in the textbook and study guide.

1. $Pb(s) \rightarrow Pb^{2+}(aq) + 2e^-$ (oxidation)
   $2e^- + Cu^{2+}(aq) \rightarrow Cu(s)$ (reduction)
   $\overline{Cu^{2+}(aq) + Pb(s) \rightarrow Pb^{2+}(aq) + Cu(s)}$

   Cu^{2+} is the oxidizing agent; Pb is the reducing agent.

2. reduced, oxidized

3. +4; +7; +3; +2

4. (a) $C_3H_6O_3$ is oxidized (the oxidation number of carbon is raised from 0 to +4 in CO_2.); IO_4^- is reduced.
   (b) IO_4^- is the oxidizing agent; $C_3H_6O_3$ is the reducing agent.

5. One way of finding the answer is to write the half-reaction. In basic solution it is:

   $12OH^-(aq) + Cl_2(g) \rightarrow 2ClO_3^-(aq) + 6H_2O(l) + \underline{10e^-}$

6. (a) $Cu(s) \rightarrow Cu^{2+}(aq) + 2e^-$
   $2e^- + 3H^+(aq) + HSO_4^-(aq) \rightarrow SO_2(g) + 2H_2O(l)$
   $\overline{Cu(s) + 3H^+(aq) + HSO_4^-(aq) \rightarrow Cu^{2+}(aq) + SO_2(g) + 2H_2O(l)}$

   (b) $4[4OH^-(aq) + Zn(s) \rightarrow Zn(OH)_4^{2-}(aq) + 2e^-]$
   $8e^- + 6H_2O(l) + NO_3^-(aq) \rightarrow NH_3(g) + 9OH^-(aq)$
   $\overline{7OH^-(aq) + 4Zn(s) + 6H_2O(l) + NO_3^-(aq) \rightarrow 4Zn(OH)_4^{2-}(aq) + NH_3(g)}$

7. (a) $\text{mol } NO_3^- = 1.7 \times 10^{-3} \text{ g } NH_3 \times \frac{1 \text{ mol } NH_3}{17.0 \text{ g } NH_3} \times \frac{1 \text{ mol } NO_3^-}{1 \text{ mol } NH_3} = 1.0 \times 10^{-4} \text{ mol}$

   molarity = 1.0×10^{-4} mol/0.100 L = $\underline{1.0 \times 10^{-3}\text{ M}}$

   (b) $\text{mg } NO_3^- = \frac{1.0 \times 10^{-3} \text{ mol}}{1 \text{ L}} \times \frac{62.0 \text{ g}}{1 \text{ mol}} \times \frac{10^3 \text{ mg}}{1 \text{ g}} = \underline{62 \text{ mg/L}}$

8. $2Mn^{2+}(aq) + 4OH^-(aq) + O_2(aq) \rightarrow 2MnO_2(s) + 2H_2O(l)$

   $$\frac{\text{mg } O_2}{1 \text{ L}} = \frac{6.8 \times 10^{-4} \text{ mol } MnO_2}{1 \text{ L}} \times \frac{1 \text{ mol } O_2}{2 \text{ mol } MnO_2} \times \frac{32.0 \text{ g } O_2}{1 \text{ mol } O_2} \times \frac{10^3 \text{ mg}}{1 \text{ g}} = \underline{11 \text{ mg/L}}$$

REVIEW

Oxidation-Reduction Reactions (Section 11.1)

Learning Objectives

1. Write equations for the half-reactions that occur during oxidation-reduction reactions involving ions.
2. Identify the oxidizing and reducing agents in an oxidation-reduction reaction.

Review

In an oxidation-reduction reaction (also called a redox reaction) electrons are transferred from one substance to another. Oxidation is a loss of electrons; the substance accepting the electrons is the oxidizing agent. Reduction is a gain of electrons; the substance giving up the electrons is the reducing agent. As the reaction proceeds, the oxidizing agent is reduced and the reducing agent is oxidized. The equation for an oxidation-reduction reaction is the sum of the equations for the oxidation and reduction half-reactions.

Example 11.1: Zn metal will dissolve in acidic solution to form $Zn^{2+}(aq)$ and $H_2(g)$ according to the equation:

$Zn(s) + 2H^+(aq) \rightarrow Zn^{2+}(aq) + H_2(g)$

Write equations for the half-reactions and identify the oxidizing and reducing agents.

Answer: $Zn(s) \rightarrow Zn^{2+}(aq) + 2e^-$ (oxidation)
$2e^- + 2H^+(aq) \rightarrow H_2(g)$ (reduction)

Note: The equation for an oxidation half-reaction has electrons on the right side. The equation for a reduction half-reaction has electrons on the left side.

The oxidizing agent is H^+ because it gains electrons and is reduced; the reducing agent is Zn because it loses electrons and is oxidized.

Practice Drill on Redox Reactions: Write the balanced half-reactions for each equation and identify the oxidizing and reducing agents.

(a) $Cu^{2+}(aq) + Zn(s) \rightarrow Zn^{2+}(aq) + Cu(s)$

(b) $2K(s) + Cl_2(g) \rightarrow 2KCl(s)$

(c) $4Al(s) + 3O_2(g) \rightarrow 2Al_2O_3(s)$

Answers:

(a) $Zn(s) \rightarrow Zn^{2+}(aq) + 2e^-$ — Reducing agent is Zn
$2e^- + Cu^{2+}(aq) \rightarrow Cu(s)$ — Oxidizing agent is Cu^{2+}

(b) $K(s) \rightarrow K^+ + e^-$ — Reducing agent is K
$2e^- + Cl_2(g) \rightarrow 2Cl^-$ — Oxidizing agent is Cl_2

(c) $Al(s) \rightarrow Al^{3+} + 3e^-$ — Reducing agent is Al
$4e^- + O_2(g) \rightarrow 2O^{2-}$ — Oxidizing agent is O_2

Oxidation Number (Section 11.2)

Learning Objectives

1. Calculate the oxidation number of each atom in a molecule or ion.
2. State the maximum and minimum oxidation numbers associated with each group of representative elements.
3. Use oxidation numbers to determine whether a reaction is a redox reaction.
4. Use oxidation numbers to predict whether a substance can be an oxidizing agent, a reducing agent, or both.

Review

The oxidation number or oxidation state of a bonded atom is the charge calculated for the atom when the electrons in each bond are assigned to the more electronegative atom.

The oxidation number of an atom in a free element is always zero. The oxidation number of a main group element is never higher than the group number or lower than the group number minus eight. Oxidation states lower than -4 are not observed. You can assign oxidation states by applying the rules in Section 11.2 of your textbook.

Example 11.2: Find the oxidation number of phosphorus in K_3PO_4.

Answer: The sum of the oxidation numbers is zero (Rule 2). The oxygen has an oxidation number of -2 (Rule 4). The oxidation number of potassium is +1 (Rule 3). Using x for the oxidation number of phosphorus gives

$$\underset{K}{3(+1)} + \underset{O}{4(-2)} + \underset{P}{x} = 0; \quad x = +5$$

Practice Drill on Oxidation Numbers: Give the oxidation number of the element indicated.

(a) in	Carbon oxidation number	(b) in	Vanadium oxidation number	(c) in	Sulfur oxidation number
CO_2	+4	V_2O_5	________	S_8	________
CH_4	+4	VI_2	________	SO_2	________
CH_3COOH	________	$VOCl_2$	________	H_2SO_4	________
$C_2O_4^{2-}$	________	$CaVO_3$	________	K_2S	________
C_6H_5F	________	$V(OH)^{2+}$	________	$SOCl_2$	________
$KHCO_3$	________	VF_6^{3-}	________	SF_6	________

Answers:	(a) Carbon	(b) Vanadium	(c) Sulfur
	+4	+5	0
	-4	+2	+4
	0	+4	+6
	+3	+4	-2
	-2/3	+3	+4
	+4	+3	+6

A substance is oxidized if at least one of its elements undergoes an increase in oxidation number; a substance is reduced if at least one of its elements undergoes a decrease in oxidation number. At least two oxidation numbers must change in every redox reaction; one number (or more) will increase and one number (or more) will decrease. Reactions in which all oxidation numbers remain unchanged are not redox reactions.

Example 11.3: Identify the substances oxidized and reduced in the reaction

$3ReO_4^-(aq) + 7NO(g) + 2H_2O(l) \rightarrow 3Re(s) + 7NO_3^-(aq) + 4H^+(aq)$

Answer: The oxidation number of rhenium changes from +7 in ReO_4^- to 0 in Re. A lower oxidation number indicates that electrons have been gained so that rhenium is reduced. Nitrogen changes from +2 in NO to +5 in NO_3^- indicating a loss of electrons; nitrogen is therefore oxidized. We can also say that ReO_4^- is reduced and NO is oxidized.

Practice Drill: Use oxidation numbers to identify the substances oxidized and reduced, and the oxidizing and reducing agents in each of the following reactions.

(a) $I_2(aq) + 2S_2O_3^{2-}(aq) \rightarrow S_4O_6^{2-}(aq) + 2I^-(aq)$

(b) $2CrO_4^{2-}(aq) + 5H_2O(l) + I^-(aq) \rightarrow 2Cr(OH)_4^-(aq) + 2OH^-(aq) + IO_3^-(aq)$

(c) $2MnO_4^-(aq) + 6H^+(aq) + 5SO_3^{2-}(aq) \rightarrow 2Mn^{2+}(aq) + 5SO_4^{2-}(aq) + 3H_2O(l)$

Answers: (a) $S_2O_3^{2-}$ is oxidized and is the reducing agent (oxidation number of S changes from +2 to + 2 1/2); I_2 is reduced and is the oxidizing agent (oxidation number of I changes from 0 to -1).

(b) CrO_4^{2-} is reduced and is the oxidizing agent (Cr changes from +6 to +3); I^- is oxidized and is the reducing agent (I changes from -1 to +5).

(c) SO_3^{2-} is oxidized and is the reducing agent (S changes from +4 to +6); MnO_4^- is reduced and is the oxidizing agent (Mn changes from +7 to +2).

An oxidizing agent must contain at least one element in one of its higher oxidation states so that it can accept electrons. A reducing agent on the other hand must have an element in one of its lower oxidation states so that it can donate electrons. See Figure 11.1 in your text which lists the oxidation states of the elements.

Example 11.4: Can I^- or IO_3^- be used as an oxidizing agent?

Answer: Iodine has a -1 oxidation state in I^- and a +5 oxidation state in IO_3^-. Only IO_3^- can act as an oxidizing agent because -1 is the lowest possible oxidation state for iodine.

Writing and Balancing Redox Reactions (Section 11.3)

Learning Objectives

1. Balance redox equations by the half-reaction method.
2. Balance equations for disproportionation reactions.
3.* Balance redox equations by the oxidation number method.

Review

The half-reaction method of balancing oxidation-reduction reactions consists of writing equations for the half-reactions and adding them so as to equalize electron gain and loss. Water and hydrogen ions are used to balance half reactions in acidic solutions; water and hydroxide ions are used to balance half-reactions in basic solutions. In the following examples we will follow the steps given in Section 11.3 of your text.

Balancing Redox Reactions in Acidic Solution

Example 11.5: A mixture of hydrochloric and nitric acid will dissolve metallic gold to produce $AuCl_4^-$ and NO_2 gas. Write the net ionic equation for the reaction.

Answer:

Step 1: **Write an unbalanced ionic equation. (Eliminate any spectator ions.)**

$Au(s) + Cl^-(aq) + NO_3^-(aq) + H^+(aq) \rightarrow AuCl_4^-(aq) + NO_2(g)$

Step 2: **Identify the oxidized and reduced substances. Write unbalanced equations for the half-reactions.**
Gold is oxidized; the unbalanced half-reaction is

$Au(s) + Cl^-(aq) \rightarrow AuCl_4^-(aq)$

Nitrogen is reduced. The unbalanced half-reaction is

$NO_3^-(aq) \rightarrow NO_2(g)$

Step 3: **Balance each half-reaction.**

Oxidation: (a) **First balance the atoms.**

$Au(s) + \underline{4}Cl^-(aq) \rightarrow AuCl_4^-(aq)$

(b) **Then balance the charge; add electrons to the side with less negative charge:**

$Au(s) + 4Cl^-(aq) \rightarrow AuCl_4^-(aq) + \underline{3e^-}$ (balanced)

The reduction half-reaction is more complicated than the oxidation half-reaction because oxygen is involved. Below we describe the steps you can use to balance oxygen atoms in both basic and acidic solutions.

Reduction: (a) **Balance the atoms that undergo a change in oxidation state.**

$NO_3^-(aq) \rightarrow NO_2(g)$

(b) **Find the change in oxidation state and add electrons to the product side if the oxidation state increases or to the reactant side if the oxidation state decreases.**

$e^- + NO_3^-(aq) \rightarrow NO_2(g)$

Nitrogen decreases from +5 to +4 gaining one electron.

(c) **Balance the charge by adding H^+ for acidic solutions or OH^- for basic solutions.**

$e^- + 2H^+(aq) + NO_3^-(aq) \rightarrow NO_2(g)$

(d) **Now use H_2O to balance hydrogen and oxygen atoms.**

$e^- + 2H^+(aq) + NO_3^-(aq) \rightarrow NO_2(g) + H_20(l)$

Before you go on, look at your two half-reactions. Check the following:

(i) *Does one equation have electrons on the left (reduction) and the other on the right (oxidation)? If not, check over steps 1-3 because you have made an error.*

(ii) Does the number of electrons in each half-reaction equal the change in oxidation number in that half-reaction? If not, check over steps 1-3.

Step 4: **In the complete reaction the same number of electrons are gained as lost.** Since our oxidation half-reaction shows three electrons while our reduction half-reaction shows only one electron, we must multiply the latter reaction by three:

$3[e^- + 2H^+(aq) + NO_3^-(aq) \rightarrow NO_2(g) + H_2O(l)]$

Reduction: $3e^- + 6H^+(aq) + 3NO_3^-(aq) \rightarrow 3NO_2(g) + 3H_2O(l)$

Oxidation: $Au(s) + 4Cl^-(aq) \rightarrow AuCl_4^-(aq) + 3e^-$

Step 5: **Now add the half-reactions.** The electrons will cancel.

$$3e^- + 6H^+(aq) + 3NO_3^-(aq) \rightarrow 3NO_2(g) + 3H_2O(l)$$
$$Au(s) + 4Cl^-(aq) \rightarrow AuCl_4^-(aq) + 3e^-$$
$$\overline{Au(s) + 6H^+(aq) + 3NO_3^-(aq) + 4Cl^-(aq) \rightarrow 3NO_2(g) + 3H_2O(l) + AuCl_4^-(aq)} \text{ (balanced)}$$

Check your equation:

(i) Have the electrons canceled? No electrons should appear.
(ii) Is the total charge the same on both sides?
(iii) Are the atoms balanced?

Now try the practice problem below.

Practice: Write the balanced net ionic equation for the reaction in which $KMnO_4$, $H_2C_2O_4$, and HCl are the reactants and CO_2, KCl, and $MnCl_2$ are the products. ($KMnO_4$, HCl, KCl, and $MnCl_2$ are ionic.)

Step 1: Write the unbalanced net ionic equation (eliminate spectator ions):

$MnO_4^- + C_2O_4^{2-} + 2H^+ + [illegible]^- \rightarrow Mn^{2+} + CO_2 + H_2O$

Step 2: Write the unbalanced reduction half-reaction: $H^+ + C_2O_4^{2-} + 2e^- \rightarrow 2CO_2 + H_2O$

Write the unbalanced oxidation half-reaction: $Mn^{7+} \rightarrow$

Step 3: Balance half reactions with regard to atoms and charge.

Reduction: If oxygen is involved follow substeps (a)-(d)

(a) Balance atoms involved in an oxidation state change.
(b) Add electrons.
(c) Balance charge by adding H^+.
(d) Add H_2O to balance H and O atoms.

Oxidation: Follow substeps (a)-(d) given above.

Step 4: Multiply your oxidation and reduction half-reactions by appropriate factors (if necessary) so that the half-reactions show the same number of electrons:

Reduction:

Oxidation:

Step 5: Add half-reactions to give the balanced net ionic equation.

Answer: $16H^+(aq) + 2MnO_4^-(aq) + 5C_2O_4^{2-}(aq) \rightarrow 2Mn^{2+}(aq) + 8H_2O(l) + 10CO_2(g)$

Practice: Write the balanced net ionic equation for each of the following reactions:

(a) $KMnO_4(aq) + H_2SO_3(aq) \rightarrow MnSO_4(aq) + KHSO_4(aq) + H_2SO_4(aq) + H_2O(l)$

(b) $FeSO_4(aq) + HNO_3(aq) + H_2SO_4(aq) \rightarrow Fe_2(SO_4)_3(aq) + NO(g) + H_2O(l)$

Answers: (a) $2MnO_4^-(aq) + 6H^+(aq) + 5SO_3^{2-}(aq) \rightarrow 2Mn^{2+}(aq) + 5SO_4^{2-}(aq) + 3H_2O(l)$

(b) $3Fe^{2+}(aq) + 4H^+(aq) + NO_3^-(aq) \rightarrow 3Fe^{3+}(aq) + NO(g) + 2H_2O(l)$

Balancing Redox Reactions in Basic Solution

To balance a redox reaction in basic solution follow the same procedure as in acidic solution, but use OH^- ions and H_2O instead of H^+ ions and H_2O.

Example 11.6: A mixture of NaOH and Al will react with water to generate H_2 and $NaAl(OH)_4$. (This mixture is used to clear clogged drains. The NaOH reacts with grease and the generated gas dislodges dirt.) Write the balanced net ionic equation.

Step 1: **Write the unbalanced equation:**

$H_2O(l) + Al(s) + Na^+(aq) + OH^-(aq) \rightarrow Al(OH)_4^-(aq) + H_2(g) + Na^+(aq)$

Step 2: **Write the unbalanced equations for half-reactions:**

Oxidation: $Al(s) \rightarrow Al(OH)_4^-(aq)$

Reduction: $H_2O(l) \rightarrow H_2(g)$

Step 3: **Balance the half reactions with regard to atoms and charge:**

Oxidation: The oxidation half-reaction can be simply balanced by inspection adding OH^- ions to the reactant side and balancing the charge with electrons.

$Al(s) + 4OH^-(aq) \rightarrow Al(OH)_4^-(aq) + 3e^-$

Reduction: Because oxygen is involved we must use the four substeps.

(a) Balance atoms involved in oxidation state change.

$H_2O(l) \rightarrow H_2(g)$

Note hydrogen changes from +1 to 0.

(b) Add electrons to account for oxidation state change.

$2e^- + H_2O(l) \rightarrow H_2(g)$

Two electrons are needed because two H atoms undergo the oxidation state change.

(c) Add OH^- to balance charge.

$2e^- + H_2O(l) \rightarrow H_2(g) + 2OH^-(aq)$

(d) Balance H and O atoms by adding H_2O.

$2e^- + 2H_2O(l) \rightarrow H_2(g) + 2OH^-(aq)$

Step 4: **Equalize electrons:**

Oxidation: $2[Al(s) + 4OH^-(aq) \rightarrow [Al(OH)_4^-(aq) + 3e^-]$

$2Al(s) + 8OH^-(aq) \rightarrow 2Al(OH)_4^-(aq) + 6e^-$

Reduction: $3[2e^- + 2H_2O(l) \rightarrow H_2(g) + 2OH^-(aq)]$

$6e^- + 6H_2O(l) \rightarrow 3H_2(g) + 6OH^-(aq)$

Step 5: **Add half reactions to find balanced net ionic equation:**

$2Al(s) + 8OH^-(aq) + 6H_2O(l) \rightarrow 2Al(OH)_4^-(aq) + 3H_2(g) + 6OH^-(aq)$

Net: $2Al(s) + 2OH^-(aq) + 6H_2O(l) \rightarrow 2Al(OH)_4^-(aq) + 3H_2(g)$

Example 11.7: Write the balanced molecular equation for the reaction in Example 11.6:

Answer: Add enough spectator ions to produce neutral formulas. The spectator ion is Na^+. Two Na^+ ions are needed on each side:

$2Al(s) + 2NaOH(aq) + 6H_2O(l) \rightarrow 2NaAl(OH)_4(aq) + 3H_2(g)$

Practice Drill on Balancing Half-Reactions in Basic Solution: Balance the following half-reactions with regard to atoms and charge:

(a) $MnO_4^-(aq) \rightarrow MnO_2(s)$

(b) $O_2(g) \rightarrow OH^-(aq)$

(c) $NO_3^-(aq) \rightarrow NH_3(g)$

(d) Reduction of $H_2O(l)$ to $H_2(g)$

Answers: (a) $3e^- + 2H_2O(l) + MnO_4^-(aq) \rightarrow MnO_2(s) + 4OH^-(aq)$
(b) $4e^- + 2H_2O(l) + O_2(g) \rightarrow 4OH^-(aq)$
(c) $8e^- + 6H_2O(l) + NO_3^-(aq) \rightarrow NH_3(g) + 9OH^-(aq)$
(d) $2e^- + 2H_2O(l) \rightarrow H_2(g) + 2OH^-(aq)$

Note: A redox reaction in which the same substance is oxidized and reduced is called a disproportionation reaction. In such a reaction both half-reactions start with the same reactant.

For example, chlorine is simultaneously oxidized and reduced when it is bubbled through a sodium hydroxide solution:

$Cl_2(g) + 2OH^-(aq) \rightarrow ClO^-(aq) + Cl^-(aq) + H_2O(l)$

(The above solution is a household bleach and disinfectant.) The balanced half reactions are

oxidation: $Cl_2(g) + 4OH^-(aq) \rightarrow 2ClO^-(aq) + 2H_2O(l) + 2e^-$
reduction: $2e^- + Cl_2(g) \rightarrow 2Cl^-(aq)$

***The Oxidation Number Method for Balancing Redox Equations**

The oxidation number method consists of equalizing oxidation number changes so that the total change is zero.

Redox Stoichiometry (Section 11.4)

Learning Objectives

1. Perform stoichiometric calculations that involve redox reactions.

Review

A redox stoichiometry problem is solved by the usual stoichiometric procedure, using a balanced equation from which molar ratios can be derived. Net ionic equations are often used.

Example 11.8: Silver metal is oxidized by hydrochloric acid. The products are $H_2(g)$ and $Ag^+(aq)$. Find the molarity of the HCl(aq) solution if 10.8 g of Ag(s) is dissolved by 500 mL of the unknown acid solution.

Answer: The first step is to write a balanced equation:

$2Ag(s) + 2H^+(aq) \rightarrow 2Ag^+(aq) + H_2(g)$

The problem is then solved by following the same steps as in other stoichiometry problems:

$10.8 \text{ g Ag} \rightarrow \text{mol Ag} \rightarrow \text{mol HCl} \rightarrow ?\ \text{M HCl}$

$$10.8 \text{ g Ag} \times \frac{1 \text{ mol Ag}}{107.9 \text{ g Ag}} \times \frac{2 \text{ mol H}^+}{2 \text{ mol Ag}} = 0.100 \text{ mol H}^+$$

The 0.100 mol H^+ is in 500 mL (0.500 L) of solution. The molarity is therefore

$$\text{Molarity} = \frac{0.100 \text{ mol H}^+}{0.500 \text{ L}} = \underline{0.200\ M}$$

SELF-TEST

1. $I_2(s)$ is oxidized by thiosulfate, $S_2O_3^{2-}(aq)$, to produce $I^-(aq)$ and $S_4O_6^{2-}(aq)$. Write
   (a) equations for the half-reactions and
   (b) the net ionic equation.
   (c) Identify the oxidizing and reducing agents.

2. True or False: A substance that gains oxygen atoms is oxidized and a substance that gains hydrogen atoms is reduced.

3. Give the oxidation number of sulfur in $Na_2S_2O_3$, SO_2Cl_2, SF_6, and H_2S.

4. Identify the redox reactions:
   (a) $I_3^-(aq) + SO_3^{2-}(aq) + H_2O(l) \rightarrow 3I^-(aq) + SO_4^{2-}(aq) + 2H^+(aq)$
   (b) $3CO(g) ++ Fe_2O_3(s) \rightarrow 2Fe(s) + 3CO_2(g)$

(c) $Al_2(SO_4)_3(aq) + 3Ca(OH)_2(aq) \rightarrow 2Al(OH)_3(s) + 3CaSO_4(s)$
(d) $H_2SO_4(l) + NaNO_3(s) \rightarrow HNO_3(g) + NaHSO_4(s)$

5. Complete each pair of half reactions and write the balanced net ionic equation.
(a) $Mn(OH)_2(s) \rightarrow MnO_2(s)$ (basic medium)
$O_2(g) \rightarrow OH^-(aq)$

(b) $MnO_4^-(aq) \rightarrow Mn^{2+}(aq)$ (acidic solution)
$HSO_3^-(aq) \rightarrow HSO_4^-(aq)$

(c) $Pb(OH)_3^-(aq) \rightarrow PbO_2(s)$ (basic solution)
$OCl^-(aq) \rightarrow Cl^-(aq)$

(d) $Se(s) \rightarrow Se^{2-}(aq)$ (basic solution)
$Se(s) \rightarrow SeO_3^{2-}(aq)$

6. $NO_3^-(aq)$ can be removed from water by methyl alcohol, $CH_3OH(aq)$, which reacts to form $N_2(g)$, $CO_2(g)$ and $H_2O(l)$ in acid.
(a) Write the balanced net ionic equation.
(b) Calculate the volume of $CO_2(g)$ produced at STP when 124.0 g of NO_3^- are reduced by CH_3OH.
(c) Calculate the moles of N_2 gas produced when all of the CH_3OH in 150.0 mL of a 0.0100 *M* CH_3OH solution is oxidized.

ANSWERS TO SELF-TEST QUESTIONS

1. (a) $2S_2O_3^{2-}(aq) \rightarrow S_4O_6^{2-}(aq) + 2e^-$ (oxidation)
$2e^- + I_2(s) \rightarrow 2I^-(aq)$ (reduction)
(b) $2S_2O_3^{2-}(aq) + I_2(s) \rightarrow S_4O_6^{2-}(aq) + 2I^-(aq)$
(c) reducing agent: $2S_2O_3^{2-}(aq)$; oxidizing agent: $I_2(s)$

2. T

3. +2; +6; +6; -2

4. (a) and (b)

5. (a) $2[Mn(OH)_2(s) + 2OH^-(aq) \rightarrow MnO_2(s) + 2H_2O(l) + 2e^-]$
$4e^- + 2H_2O(l) + O_2(g) \rightarrow 4OH^-(aq)$
$2Mn(OH)_2(s) + O_2(aq) \rightarrow 2MnO_2(s) + 2H_2O(l)$

(b) $2[5e^- + MnO_4^-(aq) + 8H^+(aq) \rightarrow Mn^{2+}(aq) + 4H_2O(l)]$
$5[HSO_3^-(aq) + H_2O(l) \rightarrow HSO_4^-(aq) + 2H^+(aq) + 2e^-]$
$2MnO_4^-(aq) + 6H^+(aq) + 5HSO_3^-(aq) \rightarrow 5HSO_4^-(aq) + 2Mn^{2+}(aq) + 3H_2O(l)$

(c) $Pb(OH)_3^-(s) + OH^-(aq) \rightarrow PbO_2(s) + 2H_2O(l) + 2e^-$
$2e^- + OCl^-(aq) + H_2O(l) \rightarrow Cl^-(aq) + 2OH^-(aq)$
$Pb(OH)_3^-(s) + OCl^-(aq) \rightarrow PbO_2(s) + Cl^-(aq) + H_2O(l) + OH^-(aq)$

(d) $2[2e^- + Se(s) \rightarrow Se^{2-}(aq)]$

$6OH^-(aq) + Se(s) \rightarrow SeO_3^{2-}(aq) + 3H_2O(l) + 4e^-$

$3Se(s) + 6OH^-(aq) \rightarrow 2Se^{2-}(aq) + SeO_3^{2-}(aq) + 3H_2O(l)$

6. (a) $6NO_3^-(aq) + 5CH_3OH(aq) + 6H^+(aq) \rightarrow 3N_2(g) + 5CO_2(g) + 13H_2O(l)$

(b) $$V_{CO_2} = 124.0 \text{ g } NO_3^- \times \frac{1 \text{ mol } NO_3^-}{62.0 \text{ g } NO_3^-} \times \frac{5 \text{ mol } CO_2}{6 \text{ mol } NO_3^-} \times \frac{22.4 \text{ L}}{1 \text{ mol}} = \underline{37.3 \text{ L}}$$

(c) $$\text{mol } CH_3OH = 0.1500 \text{ L} \times \frac{0.0100 \text{ mol}}{1 \text{ L}} = 1.50 \times 10^{-3} \text{ mol } CH_3OH$$

$$\text{mol } N_2 = 1.50 \times 10^{-3} \text{ mol } CH_3OH \times \frac{3 \text{ mol } N_2}{5 \text{ mol } CH_3OH} = \underline{9.0 \times 10^{-4} \text{ mol } N_2}$$

SURVEY OF THE ELEMENTS 2
OXYGEN, NITROGEN, AND THE HALOGENS

SELF-ASSESSMENT

Oxygen (Section S2.1)

1. Write equations for the formation of $O_2(g)$ by
   (a) photosynthesis
   (b) photolysis of water

2. Write balanced chemical equations for the following reactions:
   (a) heating of potassium chlorate in the presence of manganese dioxide
   (b) addition of water to barium peroxide
   (c) oxidation of sulfur dioxide
   (d) combustion of methanol (CH_3OH)
   (e) generation of ozone from oxygen
   (f) potassium oxide and water
   (g) dehydration of aluminum hydroxide
   (h) hydration of carbon dioxide

3. Complete the following equations:
   (a) $SO_2(g) + H_2O(l) \rightarrow$?
   (b) $2H_2O_2(l) \rightarrow 2H_2O(l) +$?
   (c) $4NaO_2(s) + 2CO_2(g) \rightarrow$? + ?
   (d) $2Na_2O_2(s) + 2H_2O(l) \rightarrow$? + ?

4. Give molecular orbital configurations for O_2, O_2^-, and O_2^{2-}. Explain in terms of these configurations why O_2^- and O_2^{2-} are more reactive than O_2.

Nitrogen (Section S2.2)

5. (a) Describe the two principal ways that nitrogen is fixed in nature.
   (b) Write equations for one of the above.

6. What is the final nitrogen-containing compound at the end of each of the following sets of reactions?
   (a) Nitrogen reacts with H_2. The product reacts with sulfuric acid and the product of this reaction is gently heated with base.
   (b) N_2 reacts with O_2 during a lightning storm. The product is further oxidized and reacts with moisture.

7. Match the compound in column A with one of its reactions in column B.

A	B
(a) HNO_3	(1) reacts with O_2 to form a brown gas
(b) NO_3^-	(2) disproportionates at room temperature to form NO and NO_2
(c) N_2O_3	(3) its sodium salt reacts with coke to form a nitrite
(d) HNO_2	(4) at room temperature it disproportionates into NO and HNO_3
(e) NO	(5) copper reduces it to NO or NO_2

8. Use the method of half-reactions to write a balanced equation for the dissolving of gold in a mixture of nitric and hydrochloric acids.

9. Give an example of a compound in which nitrogen is in the -3 oxidation state and write a reaction in which it is a reactant.

The Halogens (Section S2.3)

10. (a) Describe how free halogens are prepared.
    (b) Why is fluorine prepared by electrolysis?

11. Give three examples of uses for halogens that are based on their oxidizing ability.

12. Write a balanced chemical equation for the reaction of Br_2 with
    (a) Zn
    (b) H_2
    (c) H_2O
    (d) NaI
    (e) F_2

13. (a) Write equations for the reaction between NaCl and H_2SO_4 (18 M), and between HBr and H_2SO_4 (18 M).
    (b) Explain in terms of your answer in (a) why sulfuric acid cannot be used to prepare HBr and HI from their salts?

14. Match the compound in column A to its formula in column B.

A	B
(a) chloric acid	(1) HIO_4
(b) sodium hypobromite	(2) $NaBrO_3$
(c) periodic acid	(3) $HClO_3$
(d) sodium bromate	(4) NaOBr

15. Hypochlorates, chlorates, and perchlorates are prepared commercially by electrolysis with stirring to mix the products. Write the electrolysis equations.

ANSWERS TO SELF-ASSESSMENT PROBLEMS

1. (a) $6CO_2(g) + 6H_2O(l) \xrightarrow{hv} C_6H_{12}O_6(s) + 6O_2(g)$
   (b) $2H_2O(g) \xrightarrow{hv} 2H_2(g) + O_2(g)$

2. (a) $2KClO_3(s) \xrightarrow[heat]{MnO_2} 2KCl(s) + 3O_2(g)$
   (b) $2BaO_2(s) + 2H_2O(l) \rightarrow 2Ba(OH)_2(aq) + O_2(g)$
   (c) $2SO_2(g) + O_2(g) \rightarrow 2SO_3(g)$
   (d) $2CH_3OH(g) + 3O_2(g) \rightarrow 2CO_2(g) + 4H_2O(g)$
   (e) $3O_2(g) \rightarrow 2O_3(g)$
   (f) $K_2O(s) + H_2O(l) \rightarrow 2KOH(aq)$
   (g) $2Al(OH)_3(s) \xrightarrow{heat} Al_2O_3(s) + 3H_2O(g)$
   (h) $CO_2(g) + H_2O(l) \rightleftharpoons H_2CO_3(aq)$

3. (a) $SO_2(g) + H_2O(l) \rightarrow H_2SO_3(aq)$
   (b) $2H_2O_2(l) \rightarrow 2H_2O(l) + O_2(g)$
   (c) $4NaO_2(s) + 2CO_2(g) \rightarrow 2Na_2CO_3(s) + 3O_2(g)$
   (d) $2Na_2O_2(s) + 2H_2O(l) \rightarrow 4NaOH(aq) + O_2(g)$

4. O_2; $(\sigma_{1s})^2 (\sigma^*_{1s})^2 (\sigma_{2s})^2 (\sigma^*_{2s})^2 (\pi_{2py})^2 (\pi_{2px})^2 (\sigma_{2pz})^2 (\pi^*_{2py})^1 (\pi^*_{2px})^1$
   O_2^-; $(\pi^*_{2py})^2 (\pi^*_{2px})^1$
   O_2^{2-}; $(\pi^*_{2py})^2 (\pi^*_{2px})^2$
   The anions contain more antibonding electrons.

5. (a) N_2 and O_2 combine during lightning storms. Bacteria on the roots of legumes such as beans, peas, and alfalfa fix N_2 from the air.
   (b) $N_2(g) + O_2(g) \rightarrow 2NO(g)$

6. (a) NH_3; $N_2(g) + 3H_2(g) \rightarrow 2NH_3(g)$; $2NH_3(aq) + H_2SO_4(aq) \rightarrow (NH_4)_2SO_4(aq)$
   $NH_4^+(aq) + OH^-(aq) \xrightarrow{heat} NH_3(g) + H_2O(l)$
   (b) HNO_3; $N_2(g) + O_2(g) \rightarrow 2NO(g)$; $2NO(g) + O_2(g) \rightarrow 2NO_2(g)$
   $2NO_2(g) + H_2O(l) \xrightarrow[dilute]{cold} HNO_3(aq) + HNO_2(aq)$

7. (a) 5 (b) 3 (c) 2 (d) 4 (e) 1

8. $Au(s) + 4Cl^-(aq) \rightarrow AuCl_4^-(aq) + 3e^-$
   $3e^- + NO_3^-(aq) + 4H^+(aq) \rightarrow NO(g) + 2H_2O(l)$
   $Au(s) + NO_3^-(aq) + 4H^+(aq) + 4Cl^-(aq) \rightarrow AuCl_4^-(aq) + NO(g) + 2H_2O(l)$

9. $HCl(aq) + NaCN(aq) \rightarrow NaCl(aq) + HCN(g)$

10. Cl_2, Br_2, and I_2 are prepared by oxidizing halide salts; F_2 is prepared by electrolysis because chemical agents are not strong enough to oxidize F^-.

11. Chlorine water is a bleaching agent; the halogens are used as bactericides; chlorine is used to sterilize public water supplies while iodine is used for treatment of drinking water.

12. (a) $Zn(s) + Br_2(l) \rightarrow ZnBr_2(s)$
(b) $H_2(g) + Br_2(l) \rightarrow 2HBr(g)$
(c) $Br_2(aq) + H_2O(l) \rightleftharpoons H^+(aq) + Br^-(aq) + HOBr(aq)$
(d) $2I^-(aq) + Br_2(aq) \rightarrow 2Br^-(aq) + I_2(aq)$
(e) $3F_2(g) + Br_2(l) \rightarrow 2BrF_3(l)$

13. (a) $NaCl(s) + H_2SO_4(18\ M) \rightarrow NaHSO_4(s) + HCl(g)$
$2HBr(aq) + H_2SO_4(18\ M) \rightarrow Br_2(l) + SO_2(g) + 2H_2O(l)$
(b) Sulfuric acid would oxidize HBr and HI to Br_2 and I_2.

14. (a) 3 (b) 4 (c) 1 (d) 2

15. $NaCl(aq) + H_2O(l) \xrightarrow{\text{electrolysis; cold}} NaOCl(aq) + H_2(g)$
$NaCl(aq) + 3H_2O(l) \xrightarrow{\text{electrolysis; hot}} NaClO_3(aq) + 3H_2(g)$
Prolonged electrolysis oxidizes ClO_3^- to ClO_4^-;
$NaClO_3(aq) + H_2O(l) \xrightarrow{\text{electrolysis}} NaClO_4(aq) + H_2(g)$

REVIEW

Oxygen (Section S2.1)

Oxygen is the most abundant element in the earth's crust, and the second most abundant element in the atmosphere.

Learning Objectives

1. Using equations whenever possible, describe how oxygen is produced in nature, in industry, and in the laboratory.
2. Write equations for the preparation of oxygen from $KClO_3$, HgO, BaO_2, Na_2O_2, and H_2O.
3. Write an equation for the formation of O_3 from O_2.
4. Compare the properties of oxygen and ozone, and list the principal uses of each allotrope.
5. Give examples of basic oxides and write equations for their reactions with water and $H^+(aq)$.
6. Give examples of acidic oxides and write equations for their reactions with water and $OH^-(aq)$.
7. Describe the structure, properties, and uses of hydrogen peroxide.
8. Write equations for the reaction of BaO_2 with cold H_2SO_4, the decomposition of H_2O_2, and the oxidation and reduction half-reactions of H_2O_2.
9. Write equations for the reactions of oxide, peroxide, and superoxide ions with water, $H^+(aq)$, and CO_2.
10. List the principal uses of peroxides and superoxides.

Review

Occurrence

- Molecular oxygen (O_2) is 20.9% of the atmosphere in terms of number of molecules.
- In nature it is produced mainly by <u>photosynthesis</u>; chlorophyll in green plants uses solar energy to convert carbon dioxide and water into glucose ($C_6H_{12}O_6$) and oxygen.

  $6CO_2(g) + 6H_2O(l) \xrightarrow{h\nu} C_6H_{12}O_6(aq) + 6O_2(g)$
- O_2 is also produced by photolysis of water molecules: $2H_2O(g) \xrightarrow{h\nu} 2H_2(g) + O_2(g)$
- O_2 is removed from the atmosphere by the metabolism, combustion, and decay of organic compounds that form CO_2 and H_2O.

Physical Properties
- colorless, odorless, gas at ordinary temperatures and pressures
- slightly soluble in water

Chemical Properties
- Combines exothermically with most elements: $Cu(s) + 1/2O_2(g) \rightarrow CuO(s)$
- Combines with most compounds that are not completely oxidized: $NO(g) + 1/2O_2(g) \rightarrow NO_2(g)$
- Organic compounds burn in oxygen to produce oxygen-containing compounds:

$$CH_3SH(g) + 3O_2(g) \rightarrow CO_2(g) + 2H_2O(g) + SO_2(g)$$

Preparation
- Industry: fractional distillation of liquid air
- Laboratory: electrolysis of water, heating of potassium chlorate with a manganese dioxide catalyst:

$$2KClO_3(s) \xrightarrow{MnO_2(s)} 2KCl(s) + 3O_2(g)$$

- Heating of mercury and silver oxides will also evolve oxygen (and the metal) as will the heating or adding of water to ionic peroxides.

$$2K_2O_2(s) + 2H_2O(l) \rightarrow O_2(g) + 4KOH(aq)$$

Uses
- oxidizes and removes impurities from steel
- reacts to form important oxygen-containing compounds
- produces the high temperature in metal fabrication
- used in the hyperbaric oxygen chamber for patients suffering from CO poisoning, gangrene, and bone infection

Ozone (O_3)

Ozone and molecular oxygen are <u>allotropes</u>: different forms of an element that exist in the same physical state.

Properties
- pale blue gas
- condenses into a deep blue liquid
- in the upper atmosphere it absorbs much of the harmful UV radiation from the sun
- in the lower atmosphere it is a pollutant
- ozone is a better oxidizing agent than oxygen

Preparation
- Ozone is formed when O_2 or air is subjected to an electric discharge (as in an ozonizer).

$$3O_2(g) \rightarrow 2O_3(g) \quad \Delta H° = 285.4 \text{ kJ}$$

Uses
- Ozone can be used to destroy harmful microorganisms and oxidize pollutant organic molecules present in public water supplies.

Oxides

The oxidation state of oxygen in oxides is -2. Active metals form ionic oxides. Nonmetals form covalent oxides.

Reactions of Metal Oxides

- Metal oxides are generally basic:
  $K_2O(s) + 2H^+(aq) \rightarrow 2K^+(aq) + H_2O(l)$;
  $SrO(s) + 2H^+(aq) \rightarrow Sr^{2+}(aq) + H_2O(l)$
- Oxides of active metals react with water to form hydroxides: $K_2O(s) + H_2O(l) \rightarrow 2KOH(aq)$
  The hydroxides of sparingly soluble metal oxides can be prepared by metathesis. Heating these hydroxides yields the oxide.
- Ionic metal oxides react with CO_2 to form carbonates: $MgO(s) + CO_2(g) \rightarrow MgCO_3(s)$

Reactions of Nonmetal Oxides

- Most nonmetal oxides are acidic: $2NO_2(g) + 2OH^-(aq) \rightarrow NO_3^-(aq) + NO_2^-(aq) + H_2O(l)$
- Nonmetal oxides react with water to form acids: $2NO_2(g) + H_2O(l) \rightarrow HNO_3(aq) + HNO_2(aq)$
  (The reaction of an acidic oxide with base forms the same products as the reaction of the corresponding acid with base.)

Peroxides and Superoxides

Peroxides contain O–O bonds in which the oxidation state of oxygen is -1. Hydrogen peroxide can be formed from barium peroxide and cold H_2SO_4:

$$BaO_2(s) + H_2SO_4(aq) \xrightarrow{0°C} BaSO_4(s) + H_2O_2(aq)$$

Properties

- pure H_2O_2 is a colorless, syrup-like liquid
- the usual form is a 3% or 6% solution
- it can decompose violently: $2H_2O_2(l) \rightarrow 2H_2O(l) + O_2(g) \quad \Delta H° = -196$ kJ
- H_2O_2 is both an oxidizing and a reducing agent:
  in acid solution: $HNO_2(aq) + H_2O_2(aq) \rightarrow HNO_3(aq) + H_2O(aq)$
  in base solution: $2I^-(aq) + H_2O_2(aq) \rightarrow I_2(aq) + 2OH^-(aq)$

Uses

- Aqueous solutions of varying concentrations are used in water purification (30%)
- as a household antiseptic (3%)
- as a hair bleach (6%)

Ionic peroxides and superoxides

Peroxide (O_2^{2-}) and superoxide ions (O_2^-) have more antibonding electrons than O_2; hence, their bond energies are lower. As a result, they are stronger oxidizing agents than O_2.

Important reactions of O_2^{2-} and O_2^- include:

- with water: $2O_2^{2-} + 2H_2O(l) \rightarrow 4OH^-(aq) + O_2(g)$
- with CO_2: $2O_2^{2-} + 2CO_2(g) \rightarrow 2CO_3^{2-} + O_2(g)$
- with acid: $2O_2^{2-} + 4H^+(aq) \rightarrow 2H_2O(l) + O_2(g)$

- with water: $4O_2^- + 2H_2O(l) \rightarrow 4OH^-(aq) + 3O_2(g)$
- with CO_2: $4O_2^- + 2CO_2(g) \rightarrow 2CO_3^{2-} + 3O_2(g)$
- with acid: $4O_2^- + 4H^+(aq) \rightarrow 2H_2O(l) + 3O_2(g)$

All of the above reactions release oxygen. (KO_2 is used in emergency breathing masks because it absorbs CO_2 and releases O_2.)

Nitrogen (Section S2.2)

Learning Objectives

1. Describe the occurrence, preparation, properties, and uses of nitrogen.
2. Write equations for nitrogen fixation in the atmosphere and by the Haber process.
3. Describe the properties and uses of ammonia and its salts.
4. Write equations for the reactions of NH_3 with water, NH_3 with aqueous acids, and ammonium salts with aqueous bases.
5. List the principal oxides, oxo acids, and oxo anions of nitrogen in order of increasing nitrogen oxidation number.
6. Describe the preparation and uses of N_2O.
7. Write equations for the formation in the atmosphere of NO, NO_2, and N_2O_4.
8. Write equations for the formation of nitrites and nitrous acid, and list their uses.
9. Using equations, describe the Ostwald process for the synthesis of nitric acid.
10. Write half-reaction equations for the reduction of HNO_3 to NO_2, NO, N_2O, N_2, and NH_4^+.
11. List the principal uses of nitric acid and nitrates.
12. Write equations for the reactions of nitrides with water, cyanides with aqueous acids, and hydrazine with oxygen.

Review

Nitrogen is the lightest element in Group 5 and is the most abundant element in the atmosphere. Its oxidation states range from -3 to +5.

Molecular Nitrogen (N_2)

Occurrence

- The atmosphere consists of about 78 mol % N_2.
- Nitrogen is removed from the atmosphere when it forms other compounds (nitrogen fixation) by lightning ($N_2(g) + O_2(g) \rightarrow 2NO(g)$) and by bacterial activity.
- Nitrogen re-enters the atmosphere through combustion, metabolism, and decay of organic compounds.

Properties

- Colorless, odorless gas at ordinary temperatures and pressures.
- Because of the strong triple bond (N≡N) N_2 molecules are extremely stable.

Preparation

- Fractional distillation is used to obtain N_2 from liquid air.

Uses

- Production of ammonia is the largest commercial use (Haber process):

$$N_2(g) + 3H_2(g) \xrightarrow[\text{catalyst}]{400^\circ C,\ 250\ \text{atm}} 2NH_3(g)$$

- Liquid nitrogen (-196°C) is used to freeze food and to maintain a low temperature for reactions.
- Used to provide an inert atmosphere for reactive chemicals and processing food.

Ammonia

Physical Properties

- colorless gas, pungent odor
- high solubility in water (saturated aqueous solution ≈ 15 M)

Chemical Properties

- reacts with water to form a weakly basic solution of "ammonium hydroxide":

$$NH_3(aq) + H_2O(l) \rightleftharpoons NH_4^+(aq) + OH^-(aq)$$

- reacts with acids to form ammonium salts:

$$NH_3(aq) + H^+(aq) \rightarrow NH_4^+(aq)$$

$$NH_3(aq) + HNO_3(aq) \rightarrow NH_4NO_3(aq)$$

Uses

- most goes into the manufacture of fertilizers
- used to produce fibers and plastics, and explosives

Ammonium Salts

Ammonia gas is formed when a compound containing the NH_4^+ ion is gently warmed with aqueous base:

$$NH_4^+(aq) + OH^-(aq) \rightarrow NH_3(g) + H_2O(l)$$

Oxides, Oxo Acids, and Oxo Anions of Nitrogen

The oxidation states of nitrogen in oxides and in oxo acids are positive. In an oxo acid the central atom is bound to oxygen (H_2SO_4, H_3PO_4, HNO_3 are examples): the anion of an oxo acid is called an oxo anion (HSO_4^-, $H_2PO_4^-$, NO_3^- are examples).

Some facts about these compounds are given below.

Properties	Preparation	Reactions	Uses
Oxidation State: +1 Important compound: dinitrogen oxide (nitrous oxide) (N_2O)			
• colorless gas • paramagnetic	$NH_4NO_3(s) \xrightarrow{\text{heat}} N_2O(g) + 2H_2O(g)$	$N_2O(g) \xrightarrow{\text{heat}} N_2(g) + 1/2O_2(g)$	anesthetic in dentistry propellant in cans of whipped cream
Oxidation State: +2 Important compound: nitrogen monoxide (NO)			
• colorless gas, • paramagnetic	$N_2(g) + O_2(g) \xrightarrow[\text{discharge}]{\text{heat or electric}} 2NO(g)$	$2NO(g) + O_2(g) \rightarrow 2NO_2(g)$	
Oxidation State: +3 Important compound: dinitrogen trioxide (N_2O_3)			
• unstable gas; • disproportionates at room temperature: $N_2O_3(g) \rightarrow NO(g) + NO_2(g)$	$NO(g) + NO_2(g) \rightarrow N_2O_3(l)$		

Important compound: nitrous acid (HNO_2)

Properties	**Preparation**	**Reactions**	**Uses**
• pale blue • unstable acid; exists only in cold aqueous solution	$H^+(aq) + NO_2^-(aq) \rightarrow HNO_2(aq)$	$3HNO_2(aq) \xrightarrow[\text{temperature}]{\text{at room}} HNO_3(aq) + 2NO(g) + H_2O(l)$	

Important compounds: nitrites (compounds containing the NO_2^- ion)

$NO(g) + NO_2(g) + 2OH^-(aq) \rightarrow 2NO_2^-(aq) + H_2O(l)$

$NaNO_3(l) + C(s) \xrightarrow{\text{heat}} NaNO_2(l) + CO(g)$

Uses: an additive in cured meats to inhibit growth of botulinis bacteria

Oxidation State: +4 Important compound: nitrogen dioxide (NO_2)

• brown gas
• paramagnetic
• contributes to smog and acid rain

$2NO(g) + O_2(g) \rightarrow 2NO_2(g)$

$2NO_2(g) \rightleftharpoons N_2O_4(g)$

Oxidation State: +5 Important compound: dinitrogen pentoxide (N_2O_5)

• colorless
• crystalline solid compound of NO_2^+ and NO_3^- ions.

unstable vapor decomposes;
$2N_2O_5(g) \rightarrow 4NO_2(g) + O_2(g)$

Important compoound: nitric acid (HNO_3)

• colorless liquid;
• strong acid and oxidizing agent

$NaNO_3(s) + H_2SO_4(18\ M) \xrightarrow{\text{heat}} HNO_3(g) + NaHSO_4(aq)$

Uses: synthesis of
NH_4NO_3 (fertilizer)
plastics
drugs
explosives

Commercially prepared from the Ostwald process;

(1) $4NH_3(g) + 5O_2(g) \xrightarrow[900°\ C]{\text{Pt catalyst}} 4NO(g) + 6H_2O(g)$

(2) $2NO(g) + O_2(g) \xrightarrow{\text{cool}} 2NO_2(g)$

(3) $3NO_2(g) + H_2O(l) \rightarrow 2HNO_3(aq) + NO(g)$
(NO is recycled)

Concentrated HNO_3 oxidizes C, S, and I to form CO_2, H_2SO_4, and HIO_3, respectively.
$S(s) + 6HNO_3(16\ M) \rightarrow H_2SO_4(aq) + 6NO_2(g) + 2H_2O(l)$

Moderately active metals are oxidized by nitric acid:
$4Zn(s) + 10HNO_3(6\ M) \rightarrow NH_4NO_3(aq) + 4Zn(NO_3)_2(aq) + 3H_2O(l)$

Properties	Preparation	Reactions	Uses
		Metals less active than hydrogen reduce HNO_3: $3Cu(s) + 8HNO_3(6M) \rightarrow 3Cu(NO_3)_2(aq) + 2NO(g) + 4H_2O(l)$	
		$Cu(s) + 4HNO_3(12M) \rightarrow Cu(NO_3)_2(aq) + 2NO_2(g) + 2H_2O(l)$	
		Noble metals dissolve in aqua regia (HNO_3 and HCl): $Au(s) + NO_3^-(aq) + 4Cl^-(aq) + 4H^+(aq) \rightarrow AuCl_4^-(aq) + NO(g) + 2H_2O(l)$	

Important compounds: nitrates (NO_3^-)

Properties	Preparation	Reactions	Uses
• soluble salts		Nitrates are potentially explosive: $NH_4NO_3(s) \rightarrow N_2O(g) + 2H_2O(g)$	fertilizers

Other Nitrogen Compounds

Oxidation State: -3 Important compounds: Nitrides (N^{3-})

Properties	Preparation	Reactions	Uses
• white crystalline solids	$N_2(g) + 6Li(s) \rightarrow 2Li_3N(s)$ (Also Al and most Group 2A metals)	$N^{3-}(aq) + 3H_2O(l) \rightarrow NH_3(g) + 3OH^-(aq)$	

Important compounds: cyanides (CN^-)

Properties	Preparation	Reactions	Uses
• cyanides and HCN are toxic		$CN^-(aq) + H^+(aq) \rightarrow HCN(g)$	

Oxidation State: -2 Important compound: hydrazine (N_2H_4)

Properties	Preparation	Reactions	Uses
• colorless liquid with toxic fumes	$2NH_3(l) + OCl^-(aq) \rightarrow N_2H_4(aq) + Cl^-(aq) + H_2O(l)$	$N_2H_4(l) + O_2(l)$ (or H_2O_2) $\rightarrow N_2(g) + 2H_2O(g)$	rocket fuel (see reactions)

Halogens (Section S2.3)

Learning Objectives

1. Describe the occurrence, preparation (including equations), physical properties, and principal uses of the halogens.
2. Write equations for the reactions of the halogens with metals, hydrogen, and water.
3. List the halogens in order of decreasing oxidizing strength, and write equations for halogen-halide displacement reactions.
4. Write equations for the preparation of hydrogen halides from their salts.
5. List the principal uses of the hydrogen halides and their salts.
6. Given the name of a halogen oxo acid or oxo anion, write its formula; write the name when the formula is given.
7. List the halogen oxo acids in order of increasing oxidation number, increasing acid strength, and increasing ability to act as an oxidizing agent.
8. Write equations for the preparation of hypochlorites, chlorates, and perchlorates, and list their principal uses.

Review

Halogens

fluorine (F); most electronegative of all elements
chlorine (Cl)
bromine (Br)
iodine (I)
astatine (At); radioactive

Occurrence

- The halogens exist in nature as halide ions (X^-). The most abundant halogen-containing compound is NaCl.
- F^- ions are in solid minerals such as fluorspar (CaF_2), cryolite (Na_3AlF_6), and fluorapatite ($Ca_5(PO_4)_3F$).
- Cl^-, Br^-, and I^- ions are present in seawater, brine wells, and salt deposits.
- I^- concentrates in seaweed and some shellfish.
- Cl^- is the major anion in the body. Thyroid glands contain iodine, and F^- is necessary for the development of strong bones and teeth.

Properties

At room temperature

- F_2 is a pale yellow gas
- Cl_2 a greenish yellow gas
- Br_2 a fuming red liquid
- I_2 a gray solid with a violet vapor

Reactions of the Halogens

Halogens can gain an electron to form an anion or share electrons with other nonmetals. Fluorine is the most reactive and forms the strongest bonds. Iodine is the least reactive and forms the weakest bonds.

Major reactions include:

(a) most metals + halogens → metal halides

$$Cu(s) + X_2 \rightarrow CuX_2(s) \quad (X = F, Cl, Br, I)$$

(b) hydrogen + halogens → hydrogen halides

$$H_2(g) + X_2 \rightarrow 2HX(g) \quad (X = F, Cl, Br, I)$$

(c) water + Cl_2, Br_2, I_2 → weakly acidic solutions

$$H_2O(l) + 2I_2(g) \rightleftharpoons H^+(aq) + I^-(aq) + HOI(aq) \text{ (weak acid)}$$

Fluorine displaces oxygen and other nonmetals from compounds:

$$2H_2O(l) + 2F_2(g) \rightarrow 4HF(g) + O_2(g)$$

(d) halide ions + halogens

Each halogen except fluorine displaces lower members of the group from aqueous solutions of their halides.

$$Cl_2(aq) + 2X^-(aq) \rightarrow 2Cl^-(aq) + X_2(aq) \quad X = Br, I$$

(e) halogens + halogens → interhalogen compounds (XY, XY_3 XY_5, XY_7)

$$Cl_2(g) + F_2(g) \rightarrow 2ClF(g)$$

$$KI(s) + 4F_2(g) \rightarrow IF_7(g) + KF(s)$$

Commercial Preparation

F_2: $CaF_2(s) + H_2SO_4(18\ M) \xrightarrow{\text{heat}} 2HF(g) + CaSO_4(s)$

$2HF\ (\text{dissolved in KF}) \xrightarrow{\text{electrolysis}} H_2(g) + F_2(g)$

Cl_2: $2NaCl(aq) + 2H_2O(l) \xrightarrow{\text{electrolysis}} 2NaOH(aq) + Cl_2(g) + H_2(g)$

Br_2: $2Br^-(aq) + Cl_2(g) \rightarrow Br_2(l) + 2Cl^-(aq)$
(from seawater)

I_2: $2I^-(aq) + Cl_2(g) \rightarrow I_2(s) + 2Cl^-(aq)$
(from brine wells and seaweed)

$2IO_3^-(aq) + 5HSO_3^-(aq) \rightarrow I_2(s) + 5SO_4^{2-}(aq) + 3H^+(aq) + H_2O(l)$
(naturally occurring iodates)

Laboratory Preparation

React acidified halides with strong oxidizing agents like $KMnO_4$, $K_2Cr_2O_7$, and MnO_2:

$6Br^-(aq) + Cr_2O_7^{2-}(aq) + 14H^+(aq) \rightarrow 3Br_2(l) + 2Cr^{3+}(aq) + 7H_2O(l)$

Uses

- Aqueous chlorine is an industrial and household bleaching agent and disinfectant.
- Chlorine, I_2, and BrCl are used to disinfect water.
- Cl_2 and Br_2 are used in synthesis of numerous industrial chemicals such as CCl_4 (a solvent)
- $CHCl_3$ (an anesthetic)
- $Cl_2C = CHCl$ and $Cl_2C = CCl_2$ (dry-cleaning solvents)
- Teflon is a fluorocarbon polymer.
- Freons are chlorofluorocarbons that are used as refrigerants and spray can propellants. (Their production and use is being limited, however, because they are thought to deplete the protective ozone layer around the earth.)

Hydrogen Halides and Their Salts

Preparation

Hydrogen halides (HX) are prepared by reacting halide salts with concentrated H_2SO_4 or H_3PO_4.

$CaF_2(s) + H_2SO_4(18M) \rightarrow CaSO_4(s) + 2HF(g)$

Properties

- HF is a weak acid in aqueous solution.
- HCl, HBr, and HI are strong acids.
- HF converts many oxygen-containing compounds to fluorides:

  $SiO_2(s) + 4HF(g) \rightarrow SiF_4(g) + 2H_2O(g)$

  (The above reaction is used to etch glass.)

Uses of the Salts

- NaI helps to prevent goiter.
- NaBr and KBr are mild sedatives.

Oxides, Oxo Acids, and Oxo Anions

Fluorine does not form stable oxo compounds. The other halogens form oxo compounds in which their oxidation states range from +1 to +7.

Oxides

There are many halogen oxides, many of which are unstable. OF_2, Cl_2O, Br_3O_8, and I_2O_7 are examples of halogen oxides. ClO_2 is used to disinfect water and bleach flour.

Oxo acids and oxo anions

The acids and their anions are named according to their oxidation state:

	Acid	Oxidation State	Anion
HOCl	hypochlorous	+1	OCl^- hypochlorite
$HClO_2$	chlorous	+3	ClO_2^- chlorite
$HClO_3$	chloric	+5	ClO_3^- chlorate
$HClO_4$	perchloric	+7	ClO_4^- perchlorate

Practice Drill on Naming Oxo Acids and their Salts: Give the name and formula for each of the blanks.

	X = Br	X = I	X = Cl
HOX	HOBr (hypobromous acid)	?	?
HXO_3	?	?	$HClO_3$ (chloric acid)
XO_4^-	?	IO_4^- (periodate ion)	?

Answers: HOI (hypoiodous acid); HOCl (hypochlorous acid); $HBrO_3$ (bromic acid); HIO_3 (iodic acid); BrO_4^- (perbromate ion); ClO_4^- (perchlorate ion)

Properties

- The acid strength of oxo acids increases with increasing oxygen content for each halogen; hypohalous and halous acids are weak acids while halic and perhalic acids are strong.
- The oxo acids and their ions are oxidizing agents and many are unstable. Their oxidizing strength tends to decrease with increasing oxidation number.

Practice: From each set of compounds select the strongest acid and the strongest oxidizing agent.
(a) HOCl, $HClO_2$, $HClO_3$, $HClO_4$
(b) HOBr, BrO_3^-, $HBrO_4$

Answers: (a) strongest acid is $HClO_4$, strongest oxidizing agent is HOCl (b) $HBrO_4$ strongest acid, HOBr strongest oxidizing agent

Preparation

- Hypohalous acids and their salts are prepared by dissolving halogens in water and in aqueous base respectively:

$$Br_2(g) + H_2O(l) \xrightleftharpoons{\text{cool}} H^+(aq) + Br^-(aq) + HOBr(aq)$$

$$Br_2(g) + 2NaOH(aq) \rightarrow NaBr(aq) + NaOBr(aq) + H_2O(l)$$

- NaOCl is prepared commercially by the electrolysis of salt water and allowing the products to mix:

$$NaCl(aq) + H_2O(l) \xrightarrow{\text{electrolysis, cold}} NaOCl(aq) + H_2(g)$$

- Chlorates are prepared by the electrolysis of hot concentrated NaCl or KCl in which products also mix:

$$KCl(aq) + 3H_2O(l) \xrightarrow{\text{electrolysis, hot}} KClO_3(aq) + 3H_2(g)$$

- Prolonged electrolysis oxidizes ClO_3^- to ClO_4^- (perchlorate)

$$ClO_3^-(aq) + H_2O(l) \rightarrow ClO_4^-(aq) + 2H^+(aq) + 2e^-$$

Uses

- Hypochlorite is a laundry bleach.
- Chlorates and perchlorates are used in matches, fireworks, and explosives.

SELF-TEST

1. Small doses of fluoride ion in drinking water reduce the number of dental caries while high doses can be toxic. How many milligrams of F^- ion are consumed by a person drinking 1 L of water in which the fluoride ion concentration is 1.0 ppm? The density of the drinking water is 1.00 g/mL.

2. Write the equation for the main reaction of the following compounds with water:
   (a) Na_2O_2
   (b) SO_2
   (c) NO_2
   (d) Mg_3N_2
   (e) F_2

3. Match the common name in column A with its chemical name in column B.

A	B
(1) aqua regia	(a) KNO_3
(2) saltpeter	(b) polytetrafluoroethylene
(3) freons	(c) N_2O
(4) teflon	(d) $HNO_3(aq)$ and $HCl(aq)$
(5) laughing gas	(e) chlorofluorocarbons

4. Which property of each substance accounts for its use below.
   (a) Ammonia is a household cleaner.
   (b) Chlorine water is a bleaching agent.
   (c) HF is used to etch designs on glass.
   (d) Hydrazine and methylhydrazines are rocket fuels.

5. Indicate which of the following statements are false. Change the underlined word(s) to make the statement true.
   (a) Nitrogen gas and oxygen gas are allotropes.
   (b) Soluble metal oxides generally react with water to form hydroxides.
   (c) Peroxides, superoxides, nitric acid, and hypochlorites are strong reducing agents with commercial applications.
   (d) The Haber process, which commercially fixes nitrogen, requires a high temperature and pressure and a catalyst.
   (e) N_2 and NO_2 are free radicals that contribute to smog and acid rain.
   (f) HF is a weak acid in aqueous solution.
   (g) The halogens occur in nature as diatomic molecules, X_2.

6. Write the equation for the laboratory preparation of each of the following:
   (a) chlorate anions from Cl_2
   (b) Br_2 from Br^-
   (c) HCl from NaCl

7. Write net ionic equations for the reactions of the following with an acid:
   (a) KNO_2
   (b) NaCN
   (c) NaO_2
   (d) BaO_2
   (e) MgO

8. Complete the following equations
   (a) $2Br^-(aq) + Cl_2(g) \rightarrow$
   (b) $Zn(s) + Cl_2(g) \rightarrow$
   (c) $CaF_2(s) + H_2SO_4(18M) \rightarrow$
   (d) $2HgO(s) \xrightarrow{\text{heat}}$
   (e) $3O_2(g) \xrightarrow{\text{electrical discharge}}$
   (f) $P_4O_{10}(s) + 6H_2O(l) \rightarrow$
   (g) $NaNO_3(l) + C(s) \xrightarrow{\text{heat}}$
   Hint: C(s) is a reducing agent
   (h) $3NO_2(g) + H_2O(l) \rightarrow$

ANSWERS TO SELF-TEST QUESTIONS

1. $1000 \text{ mL} \times 1.00 \text{ g}/1 \text{ mL} \times 1 \text{ g F}^-/10^6 \text{ g} \times 10^3 \text{ mg}/1 \text{ g} = \underline{1.0 \text{ mg F}^-}$

2. (a) $2O_2^{2-}(aq) + 2H_2O(l) \rightarrow O_2(g) + 4OH^-(aq)$
   (b) $SO_2(g) + H_2O(l) \rightarrow H_2SO_3(aq)$
   (c) $3NO_2(g) + H_2O(l) \rightarrow 2HNO_3(aq) + NO(g)$
   (d) $N^{3-}(aq) + 3H_2O(l) \rightarrow NH_3(g) + 3OH^-(aq)$
   (e) $2F_2(g) + 2H_2O(l) \rightarrow 4HF(g) + O_2(g)$

3. (1) d (2) a (3) e (4) b (5) c

4. (a) The OH^- attacks grease films.
   (b) Halogens are strong oxidizing agents.
   (c) HF reacts with the silicates in glass forming the corresponding fluorides.
   (d) They are oxidized readily by liquid oxygen and hydrogen peroxide.

5. (a) O_3 (b) true (c) oxidizing (d) true (e) NO (f) true (g) halide ions, X^-

6. (a) $3Cl_2(g) + 6KOH(g) \xrightarrow{\text{heat}} 5KCl(aq) + KClO_3(aq) + 3H_2O(l)$
   (b) $6Br^-(aq) + Cr_2O_7^{2-} + 14H^+(aq) \rightarrow 3Br_2(l) + 2Cr^{3+}(aq) + 7H_2O(l)$ (MnO_4^- could also be used)
   (c) $NaCl(s) + H_2SO_4(18M) \rightarrow NaHSO_4(s) + HCl(g)$

7. (a) $NO_2^-(aq) + H^+(aq) \rightarrow HNO_2(aq)$
(b) $CN^-(aq) + H^+(aq) \rightarrow HCN(g)$
(c) $4O_2^-(aq) + 4H^+(aq) \rightarrow 2H_2O(l) + 3O_2(g)$
(d) $2O_2^{2-}(aq) + 4H^+(aq) \rightarrow 2H_2O(l) + O_2(g)$
(e) $MgO(s) + 2H^+(aq) \rightarrow Mg^{2+}(aq) + H_2O(l)$

8. (a) $2Br^-(aq) + Cl_2(g) \rightarrow Br_2(l) + 2Cl^-(aq)$
(b) $Zn(s) + Cl_2(g) \rightarrow ZnCl_2(s)$
(c) $CaF_2(s) + H_2SO_4(18\ M) \rightarrow CaSO_4(s) + 2HF(g)$
(d) $2HgO(s) \xrightarrow{\text{heat}} 2Hg(l) + O_2(g)$
(e) $3O_2(g) \xrightarrow{\text{electrical discharge}} 2O_3(g)$
(f) $P_4O_{10}(s) + 6H_2O(l) \rightarrow 4H_3PO_4(aq)$
(g) $NaNO_3(l) + C(s) \xrightarrow{\text{heat}} NaNO_2(l) + CO(g)$
(h) $3NO_2(g) + H_2O(l) \rightarrow 2HNO_3(aq) + NO(g)$

CHAPTER 12
LIQUIDS, SOLIDS, AND INTERMOLECULAR FORCES

SELF-ASSESSMENT

Phase Changes (Section 12.1)

1. Which of the following are gases?
   (a) 5 mL of H_2O that takes the shape of the open test tube that contains it
   (b) H_2O at 102°C and 1 atm
   (c) H_2O on the top of Mt. Everest at 100°C
   (d) H_2O at 25°C and 0.01 torr in a vacuum chamber

2. Fifteen grams of ice at 0°C is added to 250 g of coffee at 95°C in a styrofoam cup. Calculate the final temperature after all of the ice has melted (assume no loss of heat to the surroundings from the cup). The specific heat of coffee is 4.18 J/g°C. The heat of fusion of ice is 3.35×10^2 J/g.

Phase Diagrams (Section 12.2)

3. Complete the following statements as they apply to the phase diagram below for a substance that can exist as solid, liquid, or gas.

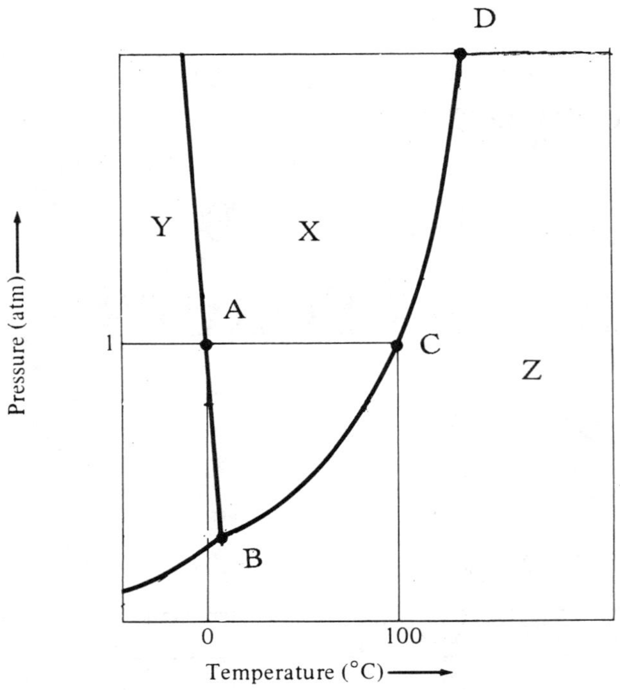

(a) The solid phase is in region ________.
(b) At temperatures higher than point D, this substance can only exist as a _______.
(c) The normal melting point of this substance is ________ and the normal boiling point is _______.
(d) The vapor pressure of the liquid at 100°C is _______.
(e) The triple point is at point _______.
(f) The slope of the fusion curve indicates that the density of the solid phase is _______ than the density of the liquid phase at any one temperature.
(g) The formula for this familiar substance is _______.

Intermolecular Forces (Section 12.3)

4. Rank the following forces in order of increasing strength: dipole-dipole, H-bonding, covalent bonding, London dispersion.

5. Which substance in each pair should have the higher boiling point? Base your selection on intermolecular forces.
(a) N_2, NO
(b) CH_3CH_3, CH_3OH
(c) H_2S, H_2O
(d) Cl_2, I_2

The Liquid State (Section 12.4)

Underline the correct word(s).

6. The liquid state exhibits short-range (order, disorder) and long-range (order, disorder).

7. The viscosity and surface tension of a liquid (increase, decrease, remain the same) with increasing temperature.

The Solid State (Section 12.5)

8. Match the substance in Column A to its classification in Column B.

Column A	Column B
(a) SiO_2 (quartz)	1. amorphous solid
(b) Cu	2. ionic crystal
(c) K_2SO_4	3. network covalent solid
(d) tar (cold)	4. molecular solid
(e) ice	5. metallic crystal

Crystal Structure (Section 12.6)

9. The density of gold is 19.27 g/cm^3, the edge length of the unit cell is 407.9 pm, and the molar mass is 196.97 g.
   (a) How many atoms occupy the unit cell?
   (b) Identify the lattice type.

10. Which packing arrangement is <u>least</u> efficient: body-centered, simple cubic, hexagonal close packing, or cubic close packing?

ANSWERS TO SELF-ASSESSMENT PROBLEMS

If you missed an answer, <u>be sure</u> to study the relevant section in the textbook and study guide.

1. (b), (c), (d)

2. $15 \text{ g} \times 3.35 \times 10^2 \text{ J/g} + 15 \text{ g} \times 4.18 \text{ J/g°C} \times (T_f - 0°C) = 250 \text{ g} \times 4.18 \text{ J/g°C} \times (95 - T_f)$; $T_f = 85°C$

3. (a) Y (b) gas (c) 0°C, 100°C (d) 1 atm (e) B (f) less (g) H_2O

4. London dispersion < dipole-dipole < H-bonding < covalent bonds.

5. (a) NO (b) CH_3OH (c) H_2O (d) I_2

6. short-range order, long-range disorder

7. decrease

8. (a) 3 (b) 5 (c) 2 (d) 1 (e) 4

9. (a) 4 (b) fcc

10. simple cubic

REVIEW

Introduction

The state of a substance–solid, liquid, or gas–represents a compromise between disorder arising from random molecular motion and order imposed by intermolecular attractive forces. An increase in temperature favors the formation of disordered states, while an increase in pressure favors the formation of ordered states. Solids, with fixed volumes and shapes, have strong intermolecular attractions and the most orderly arrangement of molecules. Liquids are less orderly; they have fixed volumes but take the shapes of their containers. Gases, which expand to fill their containers, have weak intermolecular attractions and exhibit the greatest disorder.

Phase Changes (Section 12.1)

Learning Objectives

1. Compare the properties of the three states of matter and relate these properties to molecular motion and attractive forces.
2. Give names for commonly encountered phase changes.
3. Describe the molecular process by which a liquid or solid achieves its equilibrium vapor pressure.
4. State how the vapor pressure of a liquid varies with temperature and with intermolecular attractive forces.
5. Describe the relation between a liquid's vapor pressure and its boiling point.
6. Distinguish between the boiling point of a liquid and its normal boiling point, and describe how the boiling point varies with external pressure.
7. Describe in molecular terms the equilibrium that exists at the melting point.
8. Distinguish between the melting point of a solid and its normal melting point.
9. Given the appropriate data, sketch heating and cooling curves for a substance.
10. Calculate the amount of heat absorbed or evolved during a phase change.
11.* Given the appropriate data, use the Clausius-Clapeyron equation to estimate vapor pressures at various temperatures, boiling points at various pressures, and heats of vaporization.

Review

Phase changes include

melting (solid to liquid) vaporization or evaporation (liquid to gas) sublimation (solid to gas)	endothermic (energy absorbed)
freezing (liquid to solid) condensation (gas to liquid) deposition (gas to solid)	exothermic (energy evolved)

Note: The term vapor is often used for the gaseous form of a substance.

When two opposing phase changes (such as vaporization and condensation) are going on at the same rate, a dynamic equilibrium exists between the two phases. The pressure of the vapor in equilibrium with a liquid or a solid is the equilibrium vapor pressure, often simply called the vapor pressure. Substances with stronger intermolecular forces have lower vapor pressures. All vapor pressures increase with temperature.

When the vapor pressure of a liquid equals the external pressure, the liquid boils (bubbles form inside the liquid). The temperature at which this occurs is the boiling point. The boiling point at one atmosphere external pressure is the normal boiling point.

The melting point (which is the same as the freezing point) is the temperature at which the solid and liquid phase are in equilibrium. The normal melting point is the melting point under one atmosphere external pressure.

A heating curve (see Figure 12.7, p. 487, of your textbook) is a plot of temperature versus heat absorbed; a cooling curve (see Figure 12.8, p. 488, of your textbook) is a plot of temperature versus heat withdrawn. The horizontal portions of such curves represent phase changes; the slanted portions represent phase warming or cooling. Cooling curves often show evidence of supercooling; the substance exists as a liquid below its freezing point.

The molar heat of fusion or enthalpy of fusion, ΔH°_{fus} is the heat required to melt one mole of solid under constant pressure.

The molar heat of vaporization or enthalpy of vaporization, ΔH°_{vap} is the heat required to evaporate one mole of liquid under constant pressure.

Example 12.1: How many kilojoules of heat are needed to vaporize 92.0 g of ethanol (C_2H_5OH) at its normal boiling point?

The ΔH°_{vap} for ethanol is 38.6 kJ/mol.

Answer: $92.0 \text{ g ethanol} \times \frac{1 \text{ mol}}{46 \text{ g}} \times \frac{38.6 \text{ kJ}}{1 \text{ mol}} = \underline{77.2 \text{ kJ}}$

Large heats of fusion and vaporization indicate strong intermolecular attractions. Examine the enthalpy of phase changes for various substances listed in Table 12.3 (p. 485) of your textbook.

Example 12.2: Which requires more heat, the vaporization of 40.0 g of ethanol at its normal boiling point or 40.0 g of water at its normal boiling point?

Answer: 40.0 g of ethanol is 0.91 mol; 40.0 g water is 2.2 mol. Table 12.3 in the text gives ΔH°_{vap} values at the normal boiling points as 38.6 kJ/mol for ethanol and 40.67 kJ/mol for water. Since there are more moles of water as well as more heat required per mole, the water will absorb more heat on evaporation.

Phase Diagrams (Section 12.2)

Learning Objectives

1. Sketch the phase diagram for water, and label the various regions and curves.
2. Locate the normal melting point, normal boiling point, triple point, and critical point on a phase diagram.
3. Use a phase diagram to identify the most stable form of a substance at a given temperature and pressure, and describe what happens to a sample of the substance as its temperature or pressure changes.
4. Use the phase diagram to predict which of two states has the greater density.
5. Use fusion curves to determine the variation of melting point with pressure.
6. State how critical temperatures vary with increasing intermolecular attractions.

Review

A phase diagram is a graph that shows the most stable state of a substance at any given temperature and pressure. The vaporization, fusion, and sublimation curves are lines on the phase diagram that show the temperatures and pressures under which two phases can be at equilibrium.

Fusion curve: solid in equilibrium with liquid so that all points on the curve are melting points.

Vaporization curve: liquid in equilibrium with vapor; all points on this curve are boiling points.

Sublimation curve: solid in equilibrium with vapor; all points on curve are sublimation points.

The triple point is the temperature and pressure under which three phases can coexist in equilibrium.

Increasing pressure favors formation of the state with the greater density. For most substances the solid is more dense than the liquid so that at temperatures near the melting point an increase in pressure tends to favor the solid state. Water is an exception in that the liquid is more dense; an increase in pressure therefore favors the liquid state. Carefully study the phase diagram for water (Fig. 12.10 in your textbook).

If the solid is more dense than the liquid, the slope of the fusion curve is positive (upward to the right). If the liquid is more dense than the solid , the slope of the fusion curve is negative (upward to the left).

The critical temperature is the temperature above which a substance can exist only in the gaseous state; it is at the end of the vaporization curve. A high critical temperature is associated with strong intermolecular attractions.

Intermolecular Forces (Section 12.3)

Learning Objectives

1. Distinguish between intermolecular attractive forces and chemical bonding forces.
2. Describe the origin of dipole-dipole forces, and predict the effect of such forces on melting points, boiling points, and other thermal properties.
3. Describe the hydrogen bond, and identify molecules in which hydrogen bonding might occur.
4. State some of the anomalous properties of water attributable to hydrogen bonding.
5. Describe the origin of London dispersion forces, and predict the effect of such forces on melting points, boiling points, and other thermal properties.

Review

The forces that draw otherwise independent atoms or molecules toward each other are called intermolecular attractive forces. Such forces (often called Van der Waals forces) are usually much weaker than covalent or ionic bonds.

Dipole-dipole forces exist between polar molecules and cause oppositely charged ends of molecules to be nearer to each other than the similarly charged ends.

Hydrogen bonding occurs when a hydrogen atom that is bonded to a very electronegative atom in one molecule is also attracted to a very electronegative atom in an adjacent molecule. Very electronegative atoms are N, O, and F. Strong hydrogen bonds are responsible for the unusually high boiling point of water and the abnormally low density of ice. The hydrogen bond is the strongest intermolecular attractive force.

London dispersion forces exist between nonpolar atoms and molecules. They result from fluctuating charge distributions which polarize neighboring atoms or molecules. London forces are present in all matter, but are greater in substances consisting of large, easily polarized atoms and molecules. London forces account for the fact that the melting and boiling points of nonpolar substances generally increase with increasing molecular mass.

Example 12.3: Predict the principal intermolecular forces in each liquid:

(a) CH_3CH_2OH (ethanol)

(b) liquid helium

(c) $(CH_3)_2CO$ (acetone)

Answer: (a) H-bonding and London forces; (b) London dispersion forces; (c) dipole-dipole and London forces.

The Liquid State (Section 12.4)

Learning Objectives

1. Explain what is meant by the statement that liquids exhibit short-range order.
2. State how the viscosity and surface tension of liquids vary with increasing intermolecular attractions and increasing temperature.
3. Explain why liquids tend to minimize their surface area.

Review

Liquids exhibit short-range order, but disorder prevails over the liquid as a whole.

The viscosity of a liquid is a measure of its resistance to flow. Surface tension is the energy required to stretch the surface of a liquid by some unit amount. It is a measure of the "tightness" of the liquid surface, a property that causes liquids to maintain as small a surface area as possible. Surfactants lower surface tension and therefore increase the wetting ability of a liquid. Both viscosity and surface tension increase with increasing intermolecular attractive forces and decrease with increasing temperature.

The Solid State (Section 12.5)

Learning Objectives

1. Describe the bonding in ionic, molecular, network covalent, and metallic solids, and list the physical properties associated with each type of solid.
2. Give examples of each type of crystalline solid.
3. Distinguish between crystalline and amorphous solids in terms of physical properties and in terms of molecular arrangement.
4. Give examples of commonly encountered amorphous solids.

Review

A crystalline solid consists of atoms, ions, or molecules arranged in a repeating pattern called a crystal lattice.

The properties of a solid depend to a large extent on whether it is crystalline or amorphous.

Crystalline Solids	Amorphous Solids
The particles (atoms, molecules, or ions) are arranged in a repeating pattern called a crystal lattice.	The particle arrangement is irregular.
Have sharp melting points.	Gradually soften over a range of temperature.
Have planar faces that meet at definite angles.	Shapes are irregular.

Other properties of crystalline solids depend to a large extent on the type of bonding, as follows (See also Table 12.7 in the textbook):

Ionic solids are hard, high melting, and brittle.

Molecular solids contain molecules held in place by weak intermolecular forces. They have low melting points, high vapor pressures, and little mechanical strength.

Network covalent solids contain atoms bonded in a continuous network. Like ionic solids, they are hard, high melting, and brittle.

Metallic solids have a luster, are malleable and ductile, and unlike the other types of solids, are good conductors of heat and electricity.

Crystal Structure (Section 12.6)

Learning Objectives

1. Draw simple cubic, body-centered cubic, and face-centered cubic unit cells.
2. State the number of atoms per unit cell and the coordination number for each of the three cubic lattices.
3. Use unit cell data to estimate atomic radii.
4. Calculate the missing quantity given four of the following: type of lattice, edge length, crystal density, molar mass, and Avogadro's number.
5. Calculate the percentage of occupied space in each of the three cubic lattices.
6. Draw the common atomic packing arrangements for crystalline elements.
7. Draw the rock salt structure and its unit cell.
8. Describe some lattice imperfections and state how they affect the properties of crystals.
9.* Derive the Bragg equation and use it to calculate interatomic spacings from measured diffraction angles.

Review

The basic component of a crystal lattice is the unit cell; a three-dimensional arrangement of atoms that repeats itself throughout the lattice. Three different cubic lattices will be considered here: the simple cubic, body centered cubic; and face-centered cubic:

Simple cubic = eight spheres (atoms) at corners of a cube (see Figure 12.31 in your textbook.)
Body-centered cubic = eight spheres (atoms) at corners of a cube and a sphere in the center of the cube (see Figure 12.32 in your textbook.)
Face-centered cubic = eight spheres (atoms) at corners of a cube with a sphere in the center of each face (see Figure 12.33 in your textbook.)

Finding the Number of Atoms in a Unit Cell

A corner sphere contributes 1/8 of itself to the unit cell; a body-centered sphere contributes all of itself to the unit cell; a face-centered sphere contributes 1/2 of itself to the unit cell. With this in mind we can determine the number of atoms in each type of unit cell:

Simple cubic = 8 corners × 1/8 = 1 atom per cell
Face-centered cubic = 8 corners × 1/8 + 6 face atoms × 1/2 = 4 atoms per cell
Body-centered cubic = 8 corners × 1/8 + 1 body atom = 2 atoms per cell

The coordination number of an atom or ion is the number of neighbors in contact with it. The coordination numbers of the atoms in simple, body-centered, and face-centered cubic lattices are 6, 8, and 12, respectively.

Crystal Structure Calculations

Crystal structure calculations involve type of lattice, edge length, density, molar mass, and Avogadro's number. If four of these quantities are given the fifth can be calculated. Atomic or ionic radii can also be calculated. (See examples below.)

Calculating Atomic Radii

The following relate atomic radius, r, to unit cell dimensions:

edge length of a simple cubic unit cell = 2r

edge length of a face-centered = $2\sqrt{2}\ r$

edge length of a body-centered = $4r/\sqrt{3}$

Example 12.4: Nickel crystallizes in a face-centered cubic lattice with a unit cell edge of 352 pm. Estimate the radius of the nickel atom.

Answer: edge length = $2\sqrt{2} \times r$ so that $r = \dfrac{\text{edge length}}{2\sqrt{2}} = \dfrac{352\ \text{pm}}{2 \times \sqrt{2}} = \underline{124\ \text{pm}}$

Calculating Avogadro's Number

Example 12.5: Iron crystallizes in a body-centered cubic lattice with a unit cell edge of 287 pm. The density of iron is 7.86 g/cm^3 and its molar mass is 55.8 g/mol. Determine Avogadro's number.

Given: unit cell = body-centered cubic
edge length = 287 pm
density = 7.86 g/cm^3
molar mass = 55.8 g/mol

Unknown: Avogadro's number

Plan: From the given information find three quantities (1), (2), and (3). Then combine the three as shown:

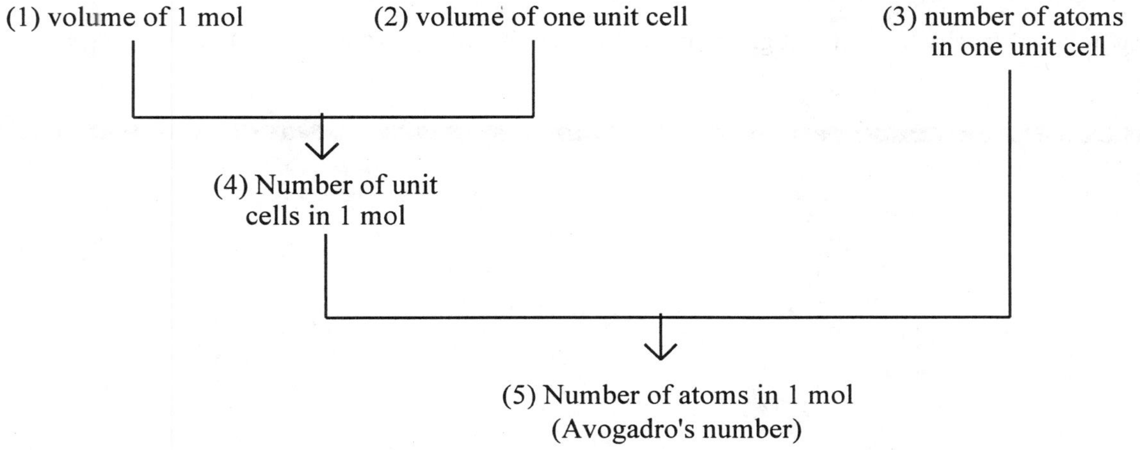

(1) Volume of 1 mol: $\frac{55.8\text{ g}}{1\text{ mol}} \times \frac{1\text{ cm}^3}{7.86\text{ g}} = 7.10\text{ cm}^3/\text{mol}$

(2) Volume of one unit cell: $(\text{edge})^3 = (287 \times 10^{-10}\text{ cm})^3 = 2.36 \times 10^{-23}\text{ cm}^3/\text{unit cell}$

(3) Number of atoms in one unit cell: For a body-centered cubic unit cell;
8 corners × 1/8 + 1 body-centered = 2 atoms/unit cell

(4) Number of unit cells in 1 mol $= \frac{\text{volume of 1 mol (1)}}{\text{volume of one unit cell (2)}}$

$$= \frac{7.10\text{ cm}^3/\text{mol}}{2.36 \times 10^{-23}\text{ cm}^3/\text{unit cell}} = 3.01 \times 10^{23}\text{ unit cells/mol}$$

(5) Avogadro's number $= \frac{3.01 \times 10^{20}\text{ unit cells}}{1\text{ mol}} \times \frac{2\text{ atoms}}{1\text{ unit cell}} = \underline{6.02 \times 10^{23}\text{ atoms/mol}}$

PROBLEM-SOLVING TIP

Crystal Structure Calculations

Crystal structure calculations usually involve type of lattice, edge length, density, molar mass, and Avogadro's number (N_A). If you are given four of these quantities, you can calculate the fifth. To simplify your solutions to crystal structure problems, you can use one formula which relates these five quantities. The formula can be derived by calculating the density of a unit cell:

$$d = \frac{\text{mass of unit cell}}{\text{volume of unit cell}} = \frac{\text{number of atoms/ unit cell} \times \text{molar mass}/N_A}{(\text{edge length})^3}$$

Remember that the number of atoms/unit cell depends on the type of lattice structure: single cubic = 1 atom; body-centered cubic = 2 atoms; face-centered cubic = 4 atoms.

Also be sure to use matching units; if density has units of grams per cubic centimeter then your edge length must have units of centimeters.

Practice Drill on Crystal Structure Calculations: The five problems in this drill all involve the same lithium lattice, but in each problem a different piece of information is missing. Calculate the missing value from the other four.

	Molar Mass	Edge Length	Density	Type of Lattice	Avogadro's Number
(1)	6.94 g	350.8 pm	0.534 g/cm^3	body-centered cubic	?
(2)	?	350.8 pm	0.534 g/cm^3	body-centered cubic lattice	6.02×10^{23}
(3)	6.94 g	?	0.534 g/cm^3	body-centered cubic lattice	6.02×10^{23}
(4)	6.94 g	350.8 pm	?	body-centered cubic lattice	6.02×10^{23}
(5)	6.94 g	350.8 pm	0.534 g/cm^3	?	6.02×10^{23}

Atomic Packing Models

The most efficient atomic packing arrangements for particles of uniform size are hexagonal close packing and cubic close packing (face-centered cubic), each with a coordination number of twelve. Most of the metals and all of the noble gases crystallize in one of the close-packed lattices; most of the remaining metals crystallize in the body-centered cubic lattice.

An ionic compound contains at least two species of ions and its lattice will depend somewhat on the relative number and sizes of the ions. The rock salt structure is made up of interlocking face-centered cubic unit cells of each ion.

Every real lattice has lattice defects, which often tend to weaken the crystal. These include missing or misplaced atoms, impurities, and fragments of extra layers.

*X-ray diffraction patterns are obtained when a beam of x-rays is directed at a mounted crystal, a rotating crystal, or a powdered sample. The Bragg equation can be used to calculate interatomic spacings from x-ray wavelengths and measured scattering angles.

SELF-TEST

1. On the basis of boiling point data, rank the following substances in order of expected
   (a) increasing intermolecular attractions
   (b) vapor pressure at a given temperature
   (c) heat of vaporization (ΔH°_{vap})
   The normal boiling points are: helium = -269°C; diethyl ether = 34.54°C; ethanol = 78.5°C; gold = 2660°C; and water = 100.0°C.

2. As the temperature of a liquid increases the vapor pressure of the liquid (increases, decreases). As the pressure (increases, decreases) the boiling point of the liquid increases.

3. The triple point of CO_2 occurs at -57°C and 5.2 atm. At -57°C and at pressures greater than 5.2 atm the stable state of CO_2 is the solid. Which phase has the greater density, the solid or liquid?

4. Given the heating curve for water, match Column A to Column B.

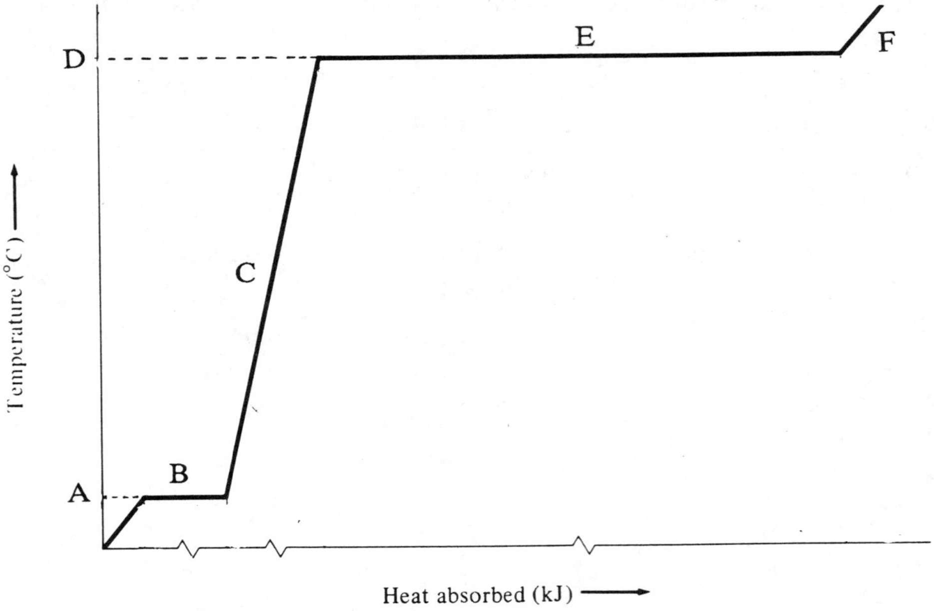

A		B
A	1.	warming of vapor
B	2.	melting point
C	3.	boiling point
D	4.	solid changing to liquid
E	5.	warming of liquid
F	6.	liquid changing to vapor

5. A small pond contains 1.0×10^7 kg of ice. How many kilojoules of heat are needed to melt the ice at 0°C? The molar heat of fusion of water is 6.01 kJ/mol.

6. Use the phase diagram for carbon dioxide given below to determine whether the following statements are true or false. If a statement is false change the underlined word(s) so that it becomes true.

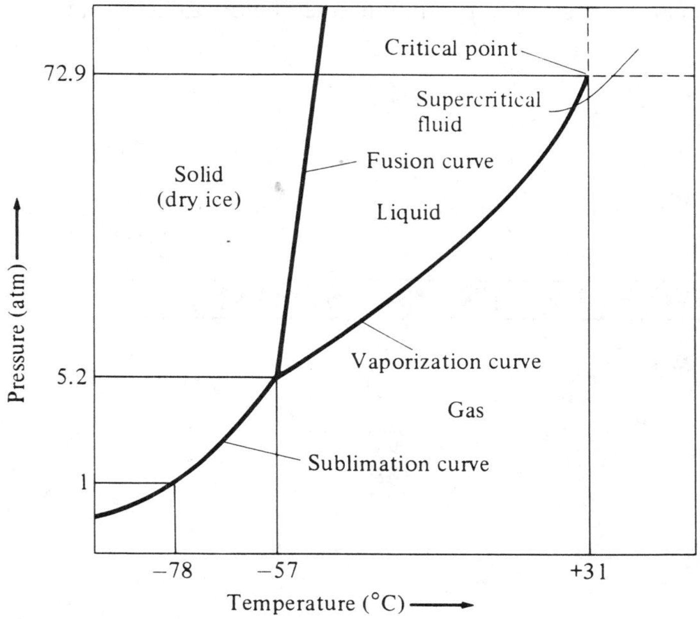

(a) The normal <u>boiling</u> point is -78°C.
(b) Carbon dioxide exists as a <u>gas</u> above 31°C.
(c) A point on the <u>fusion curve</u> is -57°C and 5.2 atm.
(d) If the pressure of carbon dioxide at the triple point is increased <u>solid</u> carbon dioxide will form.
(e) The vapor pressure of carbon dioxide at -70°C is <u>less</u> than 1 atm.

7. Identify the kinds of intermolecular forces that exist between the following species;
(a) $(CH_3)CO$ and $CHCl_3$
(b) Cl_2 and Br_2
(c) CH_3CH_2OH and H_2O

8. On the basis of molecular structure, rank the following liquids in expected order of increasing viscosity:

$CHCl_3$, glycerol ($C_3H_8O_3$), and benzene (C_6H_6).

The structures of glycerol and benzene are:

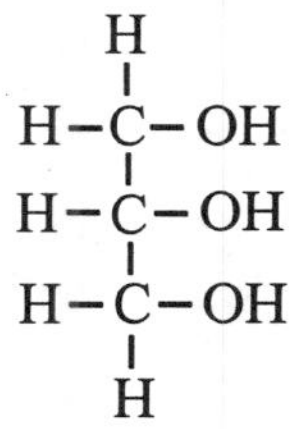

glycerol

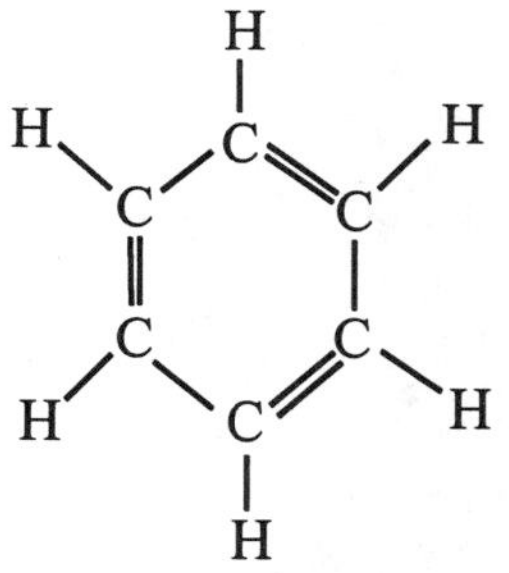

benzene

9. On the basis of bonding, one expects the melting point of NaCl to be (higher, lower) than that of ice. The boiling point of H_2S is expected to be (higher, lower) than that of H_2O.

10. Copper crystallizes in a face-centered cubic lattice. Use the information given below to determine the density of copper.

    unit cell edge = 361.50 pm

    molar mass = 63.546 g/mol

    Avogadro's number = 6.0225×10^{23} atom/mol

ANSWERS TO SELF-TEST QUESTIONS

1. (a) and (c) helium < ethyl ether < ethanol < water < gold

   (b) gold < water < ethanol < ethyl ether < helium

2. increases, increases

3. solid

4. A-2; B-4; C-5; D-3; E-6; F-1

5. $1.0 \times 10^7 \text{ kg} \times \dfrac{10^3 \text{ g}}{1 \text{ kg}} \times \dfrac{1 \text{ mol}}{18.0 \text{ g}} \times \dfrac{6.01 \text{ kJ}}{1 \text{ mol}} = \underline{3.3 \times 10^9 \text{ kJ}}$

6. (a) T (b) T (c) T (d) T (e) greater

7. (a) dipole-dipole and London forces (b) London dispersion forces (c) H-bonding and London forces

8. $C_6H_6 < CHCl_3 <$ glycerol

9. higher, higher

10. 8.9340 g/cm^3

CHAPTER 13
SOLUTIONS

SELF-ASSESSMENT

Solubility (Section 13.1)

Underline the correct word(s) in questions 1–3.

1. The step in the solution process in which solute particles separate from each other is (exothermic, endothermic).
2. Substances dissolve more readily in each other when the intermolecular forces in the solution are (similar to, different from) those in the pure components.
3. As the temperature increases the solubility of sugar in water (increases, decreases) while the solubility of oxygen gas (increases, decreases).

Ideal Solutions and Real Solutions (Section 13.2)

4. Which of the following would be true of an ideal solution?
   (a) Heat is released when components are mixed.
   (b) The volume of the solution is slightly less than the sum of the component volumes.
   (c) The vapor pressure of one component is greater than that predicted by Raoult's law.
   (d) The vapor pressure of each component is proportional to its mole fraction.

5. A solution is prepared by mixing equal numbers of moles of heptanol and benzyl acetate at 25°C. The vapor pressures at 25°C of pure heptanol and benzyl acetate are 10.7 torr and 8.2 torr, respectively. Calculate the partial pressure of each component and the total pressure above the solution assuming ideal behavior.

6. Does a negative Raoult's law deviation indicate an exothermic or an endothermic mixing process? Explain your answer.

Colligative Properties (Section 13.3) and **Molar Masses from Colligative Effects** (Section 13.4)

7. 25 g of NaCl are dissolved in 500 mL of solution. The density of the solution is 1.03 g/mL. Find the molality of the solution.

8. An aqueous solution contains 1.50 g of horse myoglobin (a muscle protein) per liter. The osmotic pressure of the solution is 2.21 torr at 25°C and its density is 1.0 g/mL. Find
   (a) the <u>molar mass</u> of horse myoglobin

(b) the freezing point of the solution
(c) the boiling point elevation (K_f = 1.86°C kg/mol; K_f = 0.512°C kg/mol; R = 0.0821 L atm/mol K)

9. Rank the following aqueous solutions in order of decreasing freezing point:

0.10 *M* K_2SO_4
0.10 *M* NaCl
0.10 *M* glucose
0.05 *M* glucose
0.20 *M* K_2SO_4

Colligative Properties of Ionic Solutions (Section 13.5)

10. (a) Calculate the freezing point depression of a 1.00 m aqueous NaCl solution, assuming complete dissociation (K_f = 1.86°C kg/mol).
(b) The observed FPD is 3.37°C. Why is the observed FPD different from the value calculated in (a)?
(c)* Calculate the van't Hoff factor, i, for NaCl using the information in part (b).

Colloids (Section 13.6)

Underline the correct word(s).

11. A gas dispersed in a liquid is a (foam, gel) while a liquid dispersed in a gas is an (aerosol, emulsion).

12. (Solutions, Colloids) give a Tyndall effect.

ANSWERS TO SELF-ASSESSMENT PROBLEMS

If you missed an answer, be sure to read the relevant section in the textbook and study guide.

1. endothermic

2. similar to

3. increases, decreases

4. (d)

5. $P_{heptane} = 0.5 \times 10.7$ torr = $\underline{5.35 \text{ torr}}$; $P_{benzyl\ acetate} = 0.5 \times 8.2 = \underline{4.1 \text{ torr}}$; $P_T = 5.35 + 4.1 = \underline{9.4 \text{ torr}}$

6. exothermic; evolution of heat and lowering of vapor pressure both result from increased intermolecular attractions.

7. 25 g × 1 mol/58.5 g = 0.43 mol NaCl
500 mL × 1.03 g /mL × 1 kg/1000 g = 0.515 kg of solution
0.515 kg – 0.025 kg = 0.490 kg of solvent
molality = 0.43 mol/0.490 kg = $\underline{0.88 \text{ mol/kg}}$

8. (a) $n = \pi V_{solution}/RT = \dfrac{(2.21/760)\text{atm} \times 1\text{ L}}{0.0821\text{ L atm/mol K} \times 298\text{K}} = 1.19 \times 10^{-4}$ mol

molar mass = 1.50 g/1.19 × 10^{-4} mol = $\underline{1.26 \times 10^4 \text{ g/mol}}$

(b) $\text{molality} = \dfrac{1.19 \times 10^{-4}\ \text{mol}}{1\ \text{kg}} = 1.19 \times 10^{-4}$

$\text{FPD} = 1.86°\text{C kg/mol} \times 1.19 \times 10^{-4}\ \text{mol/kg} = 2.21 \times 10^{-4}\ °\text{C}$

$\text{FP} = 0 - 2.21 \times 10^{-4}°\text{C} = \underline{-2.21 \times 10^{-4}\ °\text{C}}$

(c) $\text{BPE} = 0.512°\text{C kg/mol} \times 1.19 \times 10^{-4}\ \text{mol/kg} = \underline{6.09 \times 10^{-5}\ °\text{C}}$

9. 0.05 *M* glucose > 0.10 *M* glucose > 0.10 *M* NaCl > 0.10 *M* K_2SO_4 > 0.20 K_2SO_4

10. (a) $\text{FPD} = 2 \times 1.86°\text{C kg/mol} \times 1.00\ \text{mol/kg} = \underline{3.72°\text{C}}$

(c) $i = \dfrac{\text{observed colligative effect}}{\text{calculated effect assuming no dissociation}} = 3.37°\text{C}/1.86°\text{C} = \underline{1.81}$

11. foam, aerosol

12. Colloids

REVIEW

Solubility (Section 13.1)

Learning Objectives

1. Describe the effect of solute-solute and solute-solvent attractions on solution formation, and explain why "like dissolves like."
2. Describe how polar water molecules dissolve hydrogen-bonded substances such as C_2H_5OH, ionic solids such as NaCl, and gases such as HCl.
3.* Explain the salting-in and salting-out effects.
4. List each step of the solution process, and state whether the step is endothermic or exothermic.
5. Explain why the heat of solution of a gas in water is usually negative.
6. Explain the effect of temperature on the solubilities of solids and gases.
7. State and explain the effect of increasing pressure on gas solubilities.
8. Use Henry's law to calculate the variation of gas solubility with pressure.

Review

The solubility of one substance in another is a compromise between intermolecular attractions and the natural tendency of substances to mix. A useful rule is "like dissolves like": nonpolar solvents dissolve nonpolar solutes while polar solvents dissolve polar or ionic solutes. Water, the principal solvent on our planet, dissolves

(1) compounds that form hydrogen bonds with it such as ethanol and glucose
(2) many ionic compounds such as NaCl
(3) compounds that react with water to form ions such as HCl

The existence of an ion in aqueous solution is stabilized by a hydration layer that surrounds it. The water in the hydration layer is attracted to the ion by ion-dipole attractive forces.

*The solubility of a sparingly soluble ionic compound in water is increased by the presence of other salts that do not have an ion in common with the compound. This effect is called salting-in. An opposite effect, salting-out, occurs with solutes of low polarity. The atmospheric gases, oxygen and nitrogen, for example, are less soluble in salt water than in pure water.

The heat of solution ($\Delta H^{\circ}_{solution}$) is the enthalpy change that accompanies the dissolving of one mole of solute in a given solvent. It is the sum of the enthalpies of three processes–

(1) separation of solute particles (endothermic)
(2) separation of solvent particles (endothermic; relatively small)
(3) solvation (exothermic)

The dissolving of solids and liquids may be endothermic or exothermic depending on the relative magnitudes of the three enthalpy terms. The dissolving of a gas is always exothermic because the solute particles are already separated.

Most solids become more soluble with increasing temperature. Gases always become less soluble. The effect of temperature on solubility is determined by the heat of solution at the saturation point:

dissolving endothermic at saturation → solubility increases with T

dissolving exothermic at saturation → solubility decreases with T

Gas Solubility and Partial Pressure

Pressure has little effect on the solubilities of solids and liquids. Pressure increases the solubility of a gas. The solubility of a sparingly soluble gas obeys Henry's law, that is, at constant temperature the solubility is directly proportional to the partial pressure of the gas above the solution.

Example 13.1: The solubility of oxygen in water is 1.38×10^{-3} mol/L at 20°C under 1 atm pressure of oxygen gas. Find the solubility of oxygen in water at 20°C under 1 atm pressure of air. (Hint: air is 21 mol% oxygen.)

Answer: Applying Henry's law:

$$\text{solubility} = 1.38 \times 10^{-3}\ \text{mol/L} \times \frac{0.21\ \text{atm}}{1.0\ \text{atm}} = \underline{2.9 \times 10^{-4}\ \text{mol/L}}$$

Ideal Solutions and Real Solutions (Section 13.2)

Learning Objectives

1. List four properties of ideal solutions and explain why an ideal solution would have these properties.
2. Calculate the partial pressure of each component (Raoult's law) and the total vapor pressure above an ideal solution (Dalton's law).
3. Explain why some solutions are almost ideal while other solutions deviate substantially from ideality.
4. Explain the causes of positive and negative deviations from Raoult's law.

Review

An ideal solution is an imaginary solution in which the intermolecular forces are identical to those in the pure components. The following would be true of an ideal solution:

1. Volume is equal to the sum of the component volumes.
2. Heat of solution is zero.

3. Vapor pressure of each component obeys <u>Raoult's law</u>; vapor pressure is proportional to mole fraction in the solution:

   $P_A = X_A P_A°$

   P_A = vapor pressure of component A
   X_A = mole fraction of component A in the solution
   $P°$ = vapor pressure of pure component A

4. Total vapor pressure above the solution obeys <u>Dalton's law of partial pressures</u>; total vapor pressure is the sum of the vapor pressures of the individual components:

   $P_{total} = P_A + P_B + P_C + \ldots$

   $P_A, P_B, P_C + \ldots$ = vapor pressures of components A, B, C, and any others present

<u>Example 13.2</u>: A nearly ideal solution is formed by mixing 0.500 mol of benzene with 2.00 mol of toluene at 20°C . The vapor pressure of pure benzene and toluene at 20°C are 74.7 torr and 22.3 torr, respectively. Estimate the partial pressure of each component and the total vapor pressure above the solution.

Answer: $X_{benzene} = 0.500/2.50 = 0.200$
$X_{toluene} = 2.00/2.50 = 0.800$

$P_{benzene} = 0.200 \times 74.7 \text{ torr} = \underline{14.9 \text{ torr}}$
$P_{toluene} = 0.800 \times 22.3 \text{ torr} = \underline{17.8 \text{ torr}}$
$P_{total} = 14.9 \text{ torr} + 17.8 \text{ torr} = \underline{32.7 \text{ torr}}$

A real solution usually exhibits a Raoult's law deviation, a heat of solution, and a volume change upon mixing of its components. The three possibilities are:

No Deviation from Raoult's Law (see Figure 13.8, p. 542, in textbook)

- Ideal solution; 1.–4. applies (see above)

Negative Deviation from Raoult's Law (see Figure 13.10 in textbook)

- Vapor pressures lower than calculated from Raoult's law
- Intermolecular forces in the solution are greater than in pure components
- Generally accompanied by decrease of volume on mixing
- Generally accompanied by exothermic mixing of components

Positive Deviation from Raoult's Law (see Figure 3.11 in text)

- Vapor pressures higher than calculated from Raoult's law
- Intermolecular forces in the solution are lower than in pure components
- Generally accompanied by increase of volume on mixing
- Generally accompanied by endothermic mixing of components

Colligative Properties (Section 13.3)

Learning Objectives

1. Describe the following colligative effects and explain why they occur: vapor pressure lowering, boiling point elevation, freezing point depression, and osmotic pressure.
2. Given two of the three quantities, calculate vapor pressure lowering, mole fraction of solute, or vapor pressure of solvent.
3. Calculate the molality of a solution from the masses of solute and solvent.
4. Given two of the three quantities, calculate the boiling point of a solution, molality of solute, or K_b for the solvent.

5. Given two of the three quantities, calculate the freezing point of a solution, molality of solute, or K_f for the solvent.
6. Calculate the osmotic pressure of a solution from its molarity or vice versa.

Review

The colligative properties of a solution depend on the concentration of solute particles and not on their nature. The principal colligative properties are

1. Vapor pressure lowering (VPL)
2. Boiling point elevation (BPE)
3. Freezing point depression (FPD)
4. Osmotic pressure(π)

We will treat only solutions in which the solute is nonvolatile. The vapor pressure of a nonvolatile solute (e.g., sugar) is zero.

1. Vapor Pressure Lowering (VPL)

The vapor pressure above a solution of nonvolatile solute is less than the vapor pressure of the pure solvent. The difference is the vapor pressure lowering (VPL).

$VPL = P^{\circ}_{solvent} - P_{solvent}$

$P^{\circ}_{solvent}$ = vapor pressure of the pure solvent
$P_{solution}$ = vapor pressure of the solvent above the solution

The VPL is proportional to the mole fraction of the nonvolatile solute.

$VPL = X_{solute}P^{\circ}_{solvent}$

X_{solute} = mole fraction of solute

Example 13.3: Calculate the vapor pressure of a solution made by dissolving 114 g of sucrose ($C_{12}H_{22}O_{11}$) in 950 g of water at 25°C. The vapor pressure of pure water at 25°C is 23.8 torr.

Answer:

$$X_{C_{12}H_{22}O_{11}} = \frac{\text{moles } C_{12}H_{22}O_{11}}{\text{moles } C_{12}H_{22}O_{11} + \text{moles } H_2O} = \frac{114 \text{ g} \times 1 \text{ mol}/342 \text{ g}}{114 \text{ g} \times 1 \text{ mol}/342 \text{ g} + 950 \text{ g} \times 1 \text{ mol}/18.0 \text{ g}} = \frac{0.333}{0.333 + 52.8} = 0.00630$$

$$VPL = X_{C_{12}H_{22}O_{11}} \times P^{\circ}_{H_2O} = 0.00627 \times 23.8 \text{ torr} = 0.149 \text{ torr}$$

$$P_{solution} = 23.8 \text{ torr} - 0.149 \text{ torr} = \underline{23.7 \text{ torr}}$$

Molality

We will need to use a new concentration unit, molality (m), when we work with freezing points and boiling points.

Molality is the number of moles of solute per kilogram of solvent.

$$\text{molality} = \frac{\text{number of moles of solute}}{\text{number of kilograms of solvent}}$$

Example 13.4: Find the molality of urea ($(NH_2)_2CO$) in a solution containing 3.00 g of urea dissolved in 80.0 g of water.

Answer: moles of urea = 3.00 g × 1 mol/60.0 g = 0.0500 mol

$$\text{molality} = \frac{\text{number of moles of urea}}{\text{number of kilograms of water}} = \frac{0.0500 \text{ mol}}{0.0800 \text{ kg}} = \underline{0.625 \text{ m}}$$

Caution: Be careful not to confuse molality (m) with molarity (*M*); molality has kilograms of solvent in the denominator; molarity has liters of solution in the denominator.

Practice Drill on Molality: Complete the table for the following aqueous solutions of glucose. (Molar mass is 180.2 g):

	Grams of Glucose	Moles of glucose	Kilograms of water	Molality
(a)	45.0 g	?	2.50	?
(b)	?	?	10.0	0.100
(c)	?	0.025	?	0.10
(d)	90.1 g	?	?	0.100

Answers: (a) 0.250 mol; 0.100 m (b) 180 g; 1.00 mol (c) 4.5 g; 0.25 kg (d) 0.500 mol; 5.00 kg

2. Boiling Point Elevation (BPE)

A solution containing a nonvolatile solute boils at a higher temperature than the pure solvent. The difference is the boiling point elevation (BPE).

$BPE = T_b - T_b^\circ$

BPE = boiling point elevation
T_b = boiling point of solution
T_b° = boiling point of pure solvent

For a dilute solution the boiling point elevation is proportional to the molality of the solute:

$BPE = K_b m$

K_b = boiling point elevation constant
m = molality

The value of K_b depends on the solvent. K_b for water is 0.512°C kg/mol.

Example 13.5: Estimate the boiling point of a solution made by mixing 46.0 g of ethylene glycol with 250 g of water. The normal boiling point of water is 100.00°C. (K_b = 0.512°C kg/mol; molar mass of ethylene glycol = 62.1 g)

Answer:

$$\text{molality} = \frac{\text{mol ethylene glycol}}{\text{kg water}} = \frac{46.0\ \text{g} \times 1\ \text{mol} / 62.1\ \text{g}}{250\ \text{g} \times 1\ \text{kg} / 1000\ \text{g}} = 2.96\ \text{m}$$

$$BPE = K_b m = \frac{0.512^\circ\text{C kg}}{1\ \text{mol}} \times \frac{2.96\ \text{mol}}{1\ \text{kg}} = 1.52^\circ\text{C}$$

$BPE = T_b - T_b^\circ$ so that $T_b = T_b^\circ + BPE$ = 100.00°C + 1.52°C = 101.52°C

3. Freezing Point Depression (FPD)

A solution containing a nonvolatile solute freezes at a lower temperature than the pure solvent. The difference is the freezing point depression (FPD). For dilute solutions the freezing point depression is proportional to the molality:

$FPD = T_f° - T_f = K_f m$

$T_f°$ = freezing point of pure solvent
T_f = freezing point of solution
K_f = freezing point depression constant
m = molality of solution

K_f depends on the solvent. K_f for water is 1.86°C kg/mol.

Example 13.6: Estimate the freezing point of the aqueous solution of ethylene glycol described in the above example. (K_f = 1.86°C kg/mol).

Answer:

$$FPD = K_f m = \frac{1.86°\text{C kg}}{1 \text{ mol}} \times \frac{2.96 \text{ mol}}{1 \text{ kg}} = 5.51°\text{C}$$

$T_f = T_f° - FPD = 0.00°C - 5.51°C =$ <u>-5.51°C</u>

Practice Drill on BPE and FPD: Ethylene glycol is used in automobile antifreeze to lower the freezing point and raise the boiling point of water. The molar mass of ethylene glycol is 62.1 g. (K_f = 1.86 kg/mol; K_b = 0.512°C kg/mol) Fill in the table below for the given aqueous solutions of ethylene glycol at 1 atm.

	Grams of Ethylene Glycol	Moles of Ethylene Glycol	Kilograms of Water	Molality	BPE (°C)	T_b (°C)	FPD (°C)	T_f (°C)
(a)	6.21	?	0.100	?	?	?	?	?
(b)	?	?	0.200	?	?	101.024	?	?
(c)	3.1 g	?	?	?	?	?	?	-0.614

Answers: (a) 0.100 mol, 1.00 m, 0.512°C, 100.512°C, 1.86°C, -1.86°C (b) 24.8 g, 0.400 mol, 2.00 m, 1.024°C, 3.72°C, -3.72°C (c) 0.050 mol, 0.15 kg, 0.33 m, 0.17°C, 100.17°C, 0.614°C

4. Osmotic Pressure

Osmosis is the net flow of solvent through a semipermeable membrane from a dilute solution to a more concentrated solution. (A semipermeable membrane allows only selected materials to pass through.) The osmotic pressure (π) of a given solution is the pressure needed to prevent osmosis. The osmotic pressure of a dilute solution is given by

$\pi V_{solution} = n_{solute}RT$

π = osmotic pressure of solution (atm)
n = moles of solute
V = volume of solution (L)
T = temperature (K)

$$R(\text{gas constant}) = 0.0821 \frac{\text{L atm}}{\text{mol K}}$$

or

$\pi = MRT$

$$M = \text{molarity} = \frac{n_{solute}}{V_{solution}} = \text{mol / L}$$

Example 13.7: Estimate the osmotic pressure in torr of an aqueous solution containing 1.50 g of bovine insulin (protein) per liter at 25°C. The molar mass of bovine insulin is 5733 g/mol.

Answer:

Given: 1.50 g of solute in one liter of solution, T = 298K.

Unknown: π (torr)

Plan: Use $\pi V = nRT$, or $\pi = \frac{nRT}{V}$; we must find n_{solute} before we can calculate π.

Calculation: $n = 1.50\ g \times \frac{1\ mol}{5733\ g} = 2.62 \times 10^{-4}\ mol$

$$\pi = \frac{2.62 \times 10^{-4}\ mol \times 0.0821\ L\ atm/mol\ K \times 298K}{1\ L} = 6.40 \times 10^{-3}\ atm = \underline{4.86\ torr}$$

Practice Drill on Osmotic Pressure: Complete the table for the following aqueous solutions of glucose (molar mass = 180.2 g):

	Grams of Glucose	Moles of Glucose	Volume of Solution	T	π
(a)	18.02 g	?	100 mL	298K	? atm
(b)	?	?	0.500 L	37°C	7.7 atm
(c)	?	0.025 mol	1.50 L	? K	252 torr

Answers: (a) 0.100 mol, 24.5 atm; (b) 27.3 g, 0.151 mol; (c) 4.50 g, 242K

Molar Masses From Colligative Properties (Section 13.4)

Learning Objectives

1. Calculate molar masses from colligative effects and concentration data.

Review

Colligative effects are often used to determine the molar masses of very large molecules in the following way:

Laboratory procedure:

1. Dissolve a known mass (g) of the substance in a known quantity of solvent.
2. Measure a colligative effect (FPE, BPE, or π).

Calculations:

1. Find number of moles (n) from the appropriate colligative formula. [Note: the FPD and BPE formulas give the molality (m); m = n/kg solvent.]
2. Divide mass (g) by moles (n) to obtain molar mass in grams per mol.

Example 13.8: A solution contains 1.6 g of cellulose nitrate (gun cotton) per liter of acetone, the solvent. The osmotic pressure of the solution at 20°C is 4.6×10^{-2} torr. Calculate the molar mass of cellulose nitrate.

Answer:

Given: 1.6 g cellulose nitrate per liter
$T = 20°C$
$\pi = 4.6 \times 10^{-2}$ torr

Unknown: molar mass

Plan: Follow the two steps given above.

Calculations: 1. $\pi V = nRT$, or $n = \dfrac{\pi V}{RT}$

$$n = \frac{(4.6 \times 10^{-2} / 760)\text{atm} \times 1\text{ L}}{0.0821\text{ L atm/mol K} \times 293\text{K}} = 2.5 \times 10^{-6}\text{ mol}$$

2. $$\text{molar mass} = \frac{1.6\text{ g}}{2.5 \times 10^{-6}\text{ mol}} = \underline{6.4 \times 10^{5}\text{ g/mol}}$$

Practice Drill on Finding Molar Masses from Colligative Effects: Fill in the table for the following aqueous solutions: ($K_f = 1.86$°C kg/mol; $K_b = 0.512$°C kg/mol; $R = 0.0821$ L atm/mol K)

	Grams of Protein (Solute)	Moles of Protein (Solute)	Kilograms of Water	Liters of Solution	Colligative Effect	Molar Mass
(a)	20.0 g	?		1.00	$\pi = 7.4 \times 10^{-3}$ atm at 25°C	?
(b)	200 g	?	0.900		FPD = 6.66°C	?
(c)	6.0 g	?	0.100		BPE = 0.512°C	?

Answers: (a) 3.0×10^{-4} mol, 6.7×10^{4} g/mol; (b) 3.22 mol, 62.1 g/mol; (c) 0.10 mol, 60 g/mol

Colligative Properties of Ionic Solutions (Section 13.5)

Learning Objectives

1. Calculate the colligative effects of an ionic solute assuming complete dissociation.
2. Explain why the observed colligative effects of an ionic solute are less than the calculated effects.
3. State how the observed colligative effects of an ionic solute vary with increasing ionic charge and increasing concentration of ions.
4.* Calculate van't Hoff factors from observed colligative effects and vice versa.

Review

Ionization produces more than one mole of particles (ions) from a mole of solute; therefore, the colligative effects produced by an ionic solute are greater than that produced by a nonionic solute at the same concentration.

Example 13.9: Calculate the FPD of a 0.200 m KCl solution and compare with the FPD of a 0.200 m aqueous sucrose solution.

Answer: For sucrose, a nonionizing solute,

$$\text{FPD} = K_f m = \frac{1.86°\text{C kg}}{1 \text{ mol}} \times \frac{0.200 \text{ mol}}{1 \text{ kg}} = \underline{0.372°\text{C}}$$

KCl is composed of ions: $KCl = K^+(aq) + Cl^-(aq)$

Hence, dissolving 0.200 mol of KCl will produce 2 × 0.200 mol or 0.400 mol of ions:

$$\text{FPD} = \frac{1.86°\text{C kg}}{1 \text{ mol}} \times \frac{0.400 \text{ mol}}{1 \text{ kg}} = \underline{0.744°\text{C}}$$

The observed effects, however, are less than the effects calculated on the basis of complete dissociation. In a weak electrolyte the difference arises from incomplete ionization. In a strong electrolyte the difference is due to the "ionic atmosphere" of oppositely charged particles in the vicinity of each ion. The colligative effects approach their calculated values with increasing dilution.

*The effective number of moles of ions produced by one mole of solute is given by the van't Hoff factor, i:

$$i = \frac{\text{observed colligative effect}}{\text{calculated effect assuming no dissociation}}$$

Colloids (Section 13.6)

Learning Objectives

1. Compare the properties of solutions, suspensions, and colloidal dispersions.
2. Distinguish between and give examples of sols, gels, emulsions, aerosols, and foams.
3. Distinguish between and give examples of hydrophilic and hydrophobic colloids.
4. Describe how hydrophilic colloids are stabilized by the formation of hydrogen bonds with water.
5. Describe how hydrophobic colloids are stabilized by the adsorption of ions and by the adsorption of emulsifying agents.

Review

Colloids contain particles intermediate in size between those of true solutions and suspensions. Colloidal particles do not settle out on their own, and they pass through ordinary filters. They can, however, be retained by special membranes. Unlike solution particles, colloidal particles scatter light (Tyndall effect). Many commonly encountered materials are colloidal and may occur as *sols*, *gels*, *emulsions*, *aerosols*, or *foams*.

Hydrophilic colloids consist of molecules with polar groups. They are stabilized in aqueous medium by forming hydrogen bonds with surrounding water molecules. Hydrophobic colloids are stabilized by adsorbing ions of the same charge on their surfaces. Mutual repulsion keeps them from coagulating. Hydrophobic colloids are also stabilized by the adsorption of emulsifying agents. The nonpolar end of the emulsifying agent is adsorbed by the colloid; the polar end forms hydrogen bonds with water. Soaps and detergents owe their cleansing ability to this property.

SELF-TEST

1. On the basis of intermolecular forces, list the following compounds in order of increasing expected solubility in H_2O: KNO_3, C_3H_7OH, CH_4, CH_3OH, C_3H_7Cl.

2. The solubility of $O_2(g)$ in water at 25°C is 1.28×10^{-3} mol/L when the partial pressure of O_2 is 1.00 atm.
   (a) Estimate the solubility of O_2 in lake water at 25°C. (Assume the solubility in lake water is the same as in pure water. Remember air is 21 mol % oxygen.)
   (b) Would oxygen be more or less soluble in water at body temperature, 37°C?

3. At 20°C the vapor pressures of methyl alcohol (CH_3OH) and ethyl alcohol (C_2H_5OH) are 94 torr and 44 torr, respectively. A solution is prepared by mixing 64 g of CH_3OH with 69 g of C_2H_5OH. Determine
   (a) the partial pressure of each component
   (b) the total pressure of the vapor
   (c) the mole fraction of each component in the vapor

4. Would you expect the observed values for the solution in question 3 to deviate substantially from your calculated values? Why or why not?

5. You want to prepare an aqueous solution of glycerol (a nonvolatile liquid) that will freeze at -0.25°C. How many milliliters of glycerol ($C_3H_8O_3$) must you add to exactly one kilogram of water to prepare such a solution? (K_f = 1.86°C kg/mol; density of glycerol = 1.26 g/mL)

6. Fluids to be transfused into the bloodstream in large amounts are isotonic (same osmotic pressure) with blood. The osmotic pressure of blood is 7.7 atm at 37°C. How many grams of glucose ($C_6H_{12}O_6$) must be added to 1.5 L of water to prepare a solution that is isotonic with blood at 37°C (body temperature)?

7. The boiling point of ether is 34.51°C and its molal boiling point constant is 2.11°C kg/mol. When 5.30 g of a certain nonvolatile substance is dissolved in 370 g of ether, the boiling point is raised to 34.79°C. Find the molar mass of the substance.

8. A 0.50 m aqueous solution of $CaCl_2$ is left outdoors when the temperature is -1.0°C.
   (a) Will the solution freeze? Assume complete dissociation of $CaCl_2$. (K_f = 1.86°C kg/mol)
   (b) Why will the actual freezing point differ from the calculated freezing point?

9. A solution of acetic acid in benzene freezes at 3.38°C. Predict the boiling point of the solution. The normal freezing and boiling points of benzene are 5.51°C and 80.1°C, respectively. (K_f = 3.90°C kg/mol; K_b = 3.07°C kg/mol)

10. Match each term in Column A to an example in Column B

A	B
(a) colloid	(1) jello
(b) aerosol	(2) milk
(c) foam	(3) fog
(d) gel	(4) salt dissolved in water
(e) solution	(5) paint
(f) suspension	(6) mud
(g) sol	(7) shaving cream

ANSWERS TO SELF-TEST QUESTIONS

1. $CH_4 < C_3H_7Cl < C_3H_7OH < CH_3OH < KNO_3$

2. (a) solubility in lake water $= 1.28 \times 10^{-3}$ mol/L $\times \dfrac{0.21 \text{ atm}}{1 \text{ atm}} = \underline{2.7 \times 10^{-4} \text{ mol/L}}$

   (b) Less; gas solubility decreases with increasing temperature.

3. (a) $X_{CH_3OH} = \dfrac{64/32}{64/32 + 69/46} = 0.57$ $\qquad X_{C_2H_5OH} = 0.43$

   $P_{CH_3OH} = 0.57 \times 94 \text{ torr} = \underline{54 \text{ torr}}$ $\qquad P_{C_2H_5OH} = \underline{19 \text{ torr}}$

   (b) 73 torr

   (c) $X_{CH_3OH} = 54 \text{ torr}/73 \text{ torr} = \underline{0.74}$ $\qquad X_{C_2H_5OH} = \underline{0.26}$

4. No; CH_3OH and C_2H_5OH are similar molecules, so the solution should be almost ideal.

5. m = 0.25°C/1.86°C kg/mol = 0.13 mol/kg

   milliliters of glycerol = 0.13 mol × 92.1 g/mol × 1 mL/1.26 g = <u>9.5 mL</u>

6. n = 7.7 atm × 1.5 L/0.0821 L atm/mol K × 310K = 0.45 mol

   grams of glucose $= 0.45 \text{ mol} \times \dfrac{180 \text{ g}}{1 \text{ mol}} = \underline{82 \text{ g}}$

7. BPE = 0.28°C; m = 0.28°C/2.11°C kg/mol = 0.133 mol/kg

   number of moles = 0.133 mol/kg × 0.370 kg = 0.0492 mol

   molar mass = 5.30 g/0.0492 mol = <u>108 g/mol</u>

8. (a) FPD = 3 × 0.50 mol/kg × 1.86°C kg/mol = 2.79°C; no

   (b) Ions are surrounded by other ions of opposite charge, so that the effective number of moles of ions is less than 3.

9. FPD = 5.51°C – 3.38°C = 2.13°C

   molality = 2.13°C/3.90°C kg/mol = 0.546 m

   BPE = 3.07°C kg/mol × 0.546 mol/kg = 1.68°C

   T_b = 80.1°C + 1.68°C = <u>81.8°C</u>

10. (a) 2 (b) 3 (c) 7 (d) 1 (e) 4 (f) 6 (g) 5

CHAPTER 14
CHEMICAL KINETICS

SELF-ASSESSMENT

Reaction Rates and the Factors that Influence Them (Section 14.1)

1. There are four factors that influence the rate of a reaction. Which factor accounts for each of the following?
   (a) An explosion occurs when wood dust is thrown into a furnace.
   (b) Nitrogen oxides are produced from the reaction between N_2 and O_2 in hot air.
   (c) Reactions that occur slowly outside the body may occur rapidly within the body where enzymes are present.
   (d) An acidified solution of sucrose reacts with water to form glucose and fructose. The concentration of glucose produced per unit time can be increased by adding more sucrose.
   (e) The reaction between $Ag^+(aq)$ and $Cl^+(aq)$ is more rapid than the reaction between CO_2 and H_2O.

Measuring Reaction Rates (Section 14.2)

2. At elevated temperatures NO_2 decomposes into NO and O_2:

   $2NO_2(g) \rightarrow 2NO(g) + O_2(g)$

   The rate of appearance of O_2 is measured as 0.30 mol/L s. Find (a) $-\Delta[NO_2]/\Delta t$ (b) $\Delta[NO]/\Delta t$

Rate Laws (Section 14.3)

3. The following initial rates were obtained for $CO(g) + NO_2(g) \rightarrow CO_2(g) + NO(g)$:

Run	$[CO]$, mol / L	$[NO_2]$, mol / L	$-\frac{\Delta[NO_2]}{\Delta t}$, mol / L s
1	0.10	0.30	0.015
2	0.10	0.60	0.060
3	0.20	0.30	0.015

   (a) Write the experimental rate law and state the order of the reaction.
   (b) Calculate k.
   (c) Determine the rate of disappearance of NO_2 in mol/L s when [CO] and $[NO_2]$ are each 0.20 *M*.

4. The units of the rate constant for a particular reaction are $\frac{L^2}{mol^2 s}$. What is the order of the reaction?

First- and Second-Order Reactions (Section 14.4)

5. The following data were obtained for the reaction $A \rightarrow C + D$

t (hours)	0	3.15	6.30	9.45	12.60	15.75
[A]	2.20	1.10	0.550	0.225	0.1175	0.05875

(a) Show by means of a graph that the above reaction follows first-order kinetics.
(b) What is the half-life of reactant A?

6. The decomposition reaction $2N_2O_5(g) \rightarrow 4NO_2(g) + O_2(g)$ is first order. 14% of N_2O_5 decomposes in 5.0 minutes at 45°C.
(a) Calculate k for the reaction at 45°C.
(b) How much of a 1.5 mol/L sample of N_2O_5 will remain after 23 minutes?
(c) Determine the half-life, $t_{1/2}$, for the decomposition of N_2O_5 at 45°C.

7. The rate constant for the first-order decomposition of ethylene oxide at 670K is $4.5 \times 10^{-5}\ s^{-1}$. How many hours would it take for the concentration of ethylene oxide to drop from 0.500 *M* to 0.125 *M*?

Reaction Mechanisms (Section 14.5)

8. The rate law for the decomposition of NO_2Cl

$$2NO_2Cl(g) \rightarrow 2NO_2(g) + Cl_2(g)$$

is rate = $k[NO_2Cl]$. Devise a possible mechanism.

***Steady-State Approximation** (Section 14.5)

9. Nitric oxide (NO) from car exhaust reacts with O_2 to form the brown gaseous pollutant NO_2. The proposed mechanism is

$$2NO(g) \xrightarrow{k_1} N_2O_2(g) \quad \text{fast}$$

$$N_2O_2(g) + O_2(g) \xrightarrow{k_2} 2NO_2(g) \quad \text{slow}$$

Use the steady-state approximation to derive the rate law.

Free Radical Chain Reactions (Section 14.6)

10. Label each of the reactions given below as a chain initiation, propagation, or termination step.
(a) $Br_2 \rightarrow 2Br\cdot$
(b) $Br\cdot + CH_4 \rightarrow CH_3\cdot + HBr$
(c) $CH_3\cdot + Br\cdot \rightarrow CH_3Br$

Activation Energy (Section 14.7)

11. For the reaction $H_2(g) + I_2(g) \rightarrow 2HI(g)$ the energy of activation, E_a, is 167.4 kJ/mol. By what factor will the rate of reaction increase if the temperature is increased from 200°C to 400°C?

More About Catalysts (Section 14.8)

12. A catalyst increases the rate of a reaction by
    (a) increasing the concentration of the reactants
    (b) providing a new mechanism with a lower activation energy
    (c) increasing the temperature of the reaction
    (d) all of the above

ANSWERS TO SELF-ASSESSMENT PROBLEMS

If you missed an answer, be sure to study the relevant section in the textbook and study guide.

1. (a) state of subdivision
   (b) temperature
   (c) presence of a catalyst
   (d) concentration
   (e) nature of reactants; ionic compounds react more rapidly than covalent compounds

2. (a) 0.60 mol/L s
   (b) 0.60 mol/L s

3. (a) rate = $k[NO_2]^2$ order = 2
   (b) k = 0.17 L/mol s
   (c) 0.0068 mol/L s

4. third order

5. (b) The time required to reduce 2.20 mol/L of A to 1.10 mol/L is the half-life, 3.15 h.

6. (a) $\ln 0.86 = -k \times 5.0\text{ min}$; $\underline{k = 0.030\text{ min}^{-1}}$

   (b) $\ln \frac{[N_2O_5]}{1.5} = -0.030\text{ min}^{-1} \times 23\text{ min}$; $[N_2O_5] = \underline{0.75\text{ mol/L}}$

   (c) $t_{1/2} = 0.693/0.030\text{ min}^{-1} = \underline{23\text{ min}}$ (See part (b))

7. $\ln \frac{0.125}{0.500} = -4.5 \times 10^{-5}\text{ s}^{-1} \times t$; $t = 3.1 \times 10^4\text{ s} = \underline{8.6\text{ h}}$

8. $NO_2Cl(g) \rightarrow NO_2(g) + Cl(g)$ slow

   $NO_2Cl(g) + Cl(g) \rightarrow NO_2(g) + Cl_2(g)$ fast

9. rate = $k_2[N_2O_2][O_2]$

   $k_1[NO]^2 = k_2[N_2O_2][O_2]$

   $$[N_2O_2] = \frac{k_1}{k_2}\frac{[NO]^2}{[O_2]}$$

   $$\text{rate} = \frac{k_2k_1}{k_2}[NO]^2 = k[NO]^2$$

10. (a) initiation
    (b) propagation
    (c) termination

11. $\ln \frac{k_1}{k_2} = \frac{167.4 \times 10^3 \text{ J/mol}}{8.314 \text{ J/mol K}} \left(\frac{1}{473} - \frac{1}{673} \right)$ $\quad \frac{k_1}{k_2} = \underline{12.7}$

12. b

REVIEW

Reaction Rates and the Factors That Influence Them (Section 14.1)

Learning Objectives

1. Give several examples of the units used to describe reaction rates.
2. List the factors that influence reaction rates.
3. Distinguish between homogeneous and heterogeneous reactions.
4. Use the photographers' rule of thumb to estimate the effect of a temperature change on a reaction rate.
5. Distinguish between homogeneous and heterogeneous catalysis.

Review

Chemical kinetics is the study of reaction rates and mechanisms. A chemical reaction rate can be expressed as the change in the amount or concentration of a reactant or of a product per unit time. Commonly used rate units are moles per second (mol/s), liters per second (L/s), and moles per liter second (mol/L s).

Reaction rates are influenced by the following factors:

(a) **Nature of the reactants:** ions tend to react quickly while covalent molecules react slowly. Molecules with strong bonds react more slowly than molecules with weak bonds.

(b) **Concentration and state of subdivision:** increasing a reactant concentration usually (not always) increases the reaction rate.

In a homogeneous reaction all substances are dissolved in the same solution. In a heterogeneous reaction the reactants are in different states. Increasing the surface area in heterogeneous reactions also increases the reaction rate.

(c) **Temperature:** the rate of a reaction increases with increasing temperature. Many rates approximately double for each ten degree rise in Celsius temperature.

(d) **Catalysts:** a catalyst increases the rate of a reaction without being consumed in the reaction. In homogeneous catalysis, the catalyst is in solution with the reactants; in heterogeneous catalysis, a contact catalyst provides a surface on which the reaction takes place. Enzymes are catalysts for biological reactions.

Measuring Reaction Rates (Section 14.2)

Learning Objectives

1. Use the balanced equation for a reaction to relate rates of change of products and reactants.
2.* Obtain average reaction rates and instantaneous reaction rates from concentration versus time data.

Review

A reaction rate can be expressed in terms of any one of the changing reactants or products:

$$\text{rate} = \frac{\text{change in concentration of reactant or product}}{\text{time interval}} = \frac{\text{final concentration} - \text{initial concentration}}{\text{final time} - \text{initial time}}$$

Reaction rates are always positive; therefore, a positive sign must precede a rate of formation of a product while a negative sign must precede a rate of disappearance of a reactant.

Example 14.1: Ozone and nitric oxide (NO) react in polluted air:

$NO(g) + O_3(g) \rightarrow NO_2(g) + O_2(g)$

Express the rate of the reaction in terms of (a) disappearance of NO and (b) formation of NO_2.

Answer:

(a) $\text{Rate} = \dfrac{\text{final concentration of NO} - \text{initial concentration of NO}}{\text{final time} - \text{initial time}} = \dfrac{-\Delta[NO]}{\Delta t}$ (The Greek delta, Δ, means "change in")

(b) $\text{Rate} = \dfrac{\Delta[NO_2]}{\Delta t}$

The different expressions for the rate of a given reaction are related by the coefficients in the balanced equation.

Example 14.2: Given the equation $H_2(g) + Cl_2(g) \rightarrow 2HCl(g)$ show the relationship between the rate of reaction of H_2 and Cl_2 and the rate of formation of HCl.

Answer: The coefficients show that, for each mole of H_2 and for each mole of Cl_2 consumed, 2 mol of HCl are formed.

The rate of formation of HCl is twice the rate of disappearance of each reactant:

$$\frac{\Delta[HCl]}{\Delta t} = \frac{-2\Delta[Cl_2]}{\Delta t} = \frac{-2\Delta[H_2]}{\Delta t}$$

Practice Drill on Rates of Reaction: Fill in the missing rates for the reaction

$2N_2O_5(g) \rightarrow 4NO_2(g) + O_2(g)$

$\frac{-\Delta[N_2O_5]}{\Delta t}$	$\frac{\Delta[NO_2]}{\Delta t}$	$\frac{\Delta[O_2]}{\Delta t}$
(a) ?	?	5.0×10^{-6} mol/L s
(b) ?	2.0×10^{-6} mol/L s	?
(c) 1.5×10^{-6} mol/L s	?	?

Answers: (a) 1.0×10^{-5} mol/L s; 2.0×10^{-5} mol/L s; (b) 1.0×10^{-6} mol/L s; 5.0×10^{-7} mol/L s; (c) 3.0×10^{-6} mol/L s; 7.5×10^{-7} mol/L s

*Reaction rates usually vary with time and the average reaction rate is seldom equal to the instantaneous reaction rate. Both average and instantaneous rates can be obtained from the slopes of concentration versus time curves.

Rate Laws (Section 14.3)

Learning Objectives

1. Use the method of initial reaction rates to find the rate law and rate constant for a reaction.
2. Determine the reaction order from the rate law.
3. Use the rate law to calculate the reaction rate for any given set of concentrations.

Review

A rate law is an experimentally derived expression relating the rate of a chemical reaction to the concentration of one or more species in the chemical reaction.

The rate law often has the general form

$\text{rate} = k[A]^m[B]^n$

The order of the reaction is the sum of the exponents in the rate law. For the reaction above

order of reaction = m + n

The exponents m and n are the orders with respect to reactant A and B where A and B are substances that participate in the reaction.

The exponents can be positive or negative integers or even fractions. They usually differ from the coefficients in the equation and can be obtained experimentally from initial rate versus initial concentration data (method of initial reaction rates).

The constant k in the rate law is called a rate constant; its value depends on the particular reaction and the temperature. Reactions with large rate constants are faster than reactions with small rate constants. The rate constant for any particular reaction will increase with increasing temperature. The units of k depend on the order of the reaction.

Example 14.3: Give the rate law for the reaction $A + B \rightarrow C + D$ if the reaction is
(a) second order in A and first order in B.
(b) third order in A and the reaction order is five.

Answer:
(a) rate = $k[A]^2[B]$
(b) order of reaction = 5 = 3 + ? Reaction is therefore second order in B; rate = $k[A]^3[B]^2$

Example 14.4: NH_3 in waste water reacts with the disinfectant hypochlorous acid (HOCl) to form NH_2Cl (chloramine) and H_2O:

$NH_3(aq) + HOCl(aq) \rightarrow NH_2Cl(aq) + H_2O(l)$

The reaction is first order with respect to NH_3 and to HOCl.

The rate of production of NH_2Cl is 10.2 mol/L s when $[NH_3]$ is 2.0×10^{-3} *M* and [HOCl] is 1.0×10^{-3} *M*.
(a) Determine k.
(b) Find the reaction rate when $[NH_3]$ and [HOCl] are each 0.010 *M*.

Answer:
(a) rate = $k[NH_3][HOCl]$

Solving for k and substituting in given values:

$$k = \frac{\text{rate}}{[NH_3][HOCl]} = \frac{10.2 \text{ mol / L s}}{2.0 \times 10^{-3} \text{ mol / L} \times 1.0 \times 10^{-3} \text{ mol / L}} = \underline{5.1 \times 10^6 \text{ L / mol s}}$$

(b) rate = 5.1×10^6 L/mol s × 0.010 mol/L × 0.010 mol/L = $\underline{5.1 \times 10^2 \text{ mol / L s}}$

Practice Drill on Rate Laws: The rate law for the reaction between NO and Br_2 is rate = $k[NO]^2[Br_2]$. Use the rate law to fill in the following table:

	Initial Rate (mol/L s)	Initial Concentration of NO (mol/L)	Initial Concentration of Br_2 (mol / L)	k $L^2 / mol^2\ s$
(a)	12	0.10	0.10	?
(b)	24	0.10	?	
(c)	?	0.20	0.10	

Answers: (a) 1.2×10^4 L^2/mol^2 s (b) $[Br_2]$ = 0.20 mol/L (c) rate = 48 mol/L s

Determining Rate Laws

A rate law can be determined only from experiments. The order with respect to each specific substance is found by running two reactions using different starting concentrations of that substance, and the same initial concentrations of all other substances. The initial rates of the two runs are measured and compared.

Example 14.5: Given the following data for the reaction $2NO(g) + Cl_2(g) \rightarrow 2NOCl(g)$ write the rate law and state the order of the reaction.

Run	Initial NO Concentration (mol/L)	Initial Cl_2 Concentration (mol/L)	Initial Rate (mol/L s)
1	0.10	0.15	3.80×10^{-6}
2	0.10	0.30	7.60×10^{-6}
3	0.20	0.15	15.2×10^{-6}

Answer: To find the order with respect to Cl_2 compare run 1 with run 2. [NO] is the same in both runs. $[Cl_2]$ in run 2 is twice $[Cl_2]$ in run 1. The rate of run 2 is also twice the rate of run 1. Because the rate changes by the same factor (2) as the concentration of Cl_2, the reaction must be first order in Cl_2.

To find the order with respect to NO, compare runs 1 and 3. The concentration of NO increased by a factor of 2 and the rate increased by a factor of 4 or 2^2. Hence, the reaction must be second order in NO. The rate law is

rate = $k[Cl_2][NO]^2$ and the order of the reaction is $\underset{\text{order of } Cl_2}{\underset{\uparrow}{1}} + \underset{\text{order of NO}}{\underset{\uparrow}{2}} = 3$.

Example 14.6: The following initial rates were obtained for the reaction

$F_2(g) + 2ClO_2(g) \rightarrow 2FClO_2(g)$

Run	$[F_2]$ (mol / L)	$[ClO_2]$ (mol / L)	$\frac{-\Delta[F_2]}{\Delta t}$ (mol / L s)
1	0.10	0.010	1.2×10^{-3}
2	0.10	0.040	4.8×10^{-3}
3	0.20	0.010	2.4×10^{-3}

(a) Write the experimental rate law.

(b) Calculate k.

(c) Determine $\frac{-\Delta[F_2]}{\Delta t}$ when $[F_2] = 0.010\ M$ and $[ClO_2] = 0.020\ M$.

Answer:

(a) To find the order with respect to ClO_2, compare runs 1 and 2; the concentration of ClO_2 increases by a factor of 4 and the rate also increases by a factor of 4 ($4.8 \times 10^{-3}/1.2 \times 10^{-3}$). The reaction is first order in ClO_2.

To find the order with respect to F_2, compare runs 1 and 3; the concentration of F_2 doubles and the rate also doubles. The reaction is first order in F_2. The rate law is rate = $k[F_2][ClO_2]$.

(b) Choose one of the runs and substitute the initial concentration and rate into the rate law. For run 1

1.2×10^{-3} mol/L s = k × 0.10 mol/L × 0.010 mol/L

Solve for k; $k = \dfrac{1.2 \times 10^{-3} \text{ mol / L s}}{0.10 \text{ mol / L} \times 0.010 \text{ mol / L}} = \underline{1.2 \text{ L/mol s}}$

(c) Substitute given concentrations and the value of k determined in part (b) into the rate law:

rate = 1.2 L/mol s × 0.010 mol/L × 0.020 mol/L = $\underline{2.4 \times 10^{-4} \text{ mol / L}}$

First- and Second-Order Reactions (Section 14.4)

Learning Objectives

1. Use concentration versus time data to identify first-order reactions and to calculate their rate constants.
2. Given three of the four quantities, [A], $[A]_o$, k, and t, for a first-order reaction, calculate the missing quantity.
3. Calculate the half-life of a first-order reaction from its rate constant and vice-versa.
4. Use concentration versus time data to identify second-order reactions and to calculate their rate constants.
5. Given three of the four quantities, [A], $[A]_o$, k, and t, for a second-order reaction, calculate the missing quantity.
6. Calculate half-life periods for second-order reactions.
7.* Derive the integrated rate laws for first- and second-order reactions.

Review

The rate of a first-order reaction is proportional to the first power of a reactant concentration:

rate = k[A]

[A] = the concentration (moles/liter of reactant A
k = rate constant (s^{-1})

In first-order reactions, the concentration of the reactant varies with time according to the equation:

$$\ln\frac{[A]}{[A]_o} = -kt$$

In = natural logarithm
[A] = concentration of reactant A after passage of time, t
$[A]_o$ = initial concentration of A at t = 0
k = rate constant
t = time

Note that $\dfrac{[A]}{[A]_o}$ is the fraction of A remaining at time, t.

Rearranging gives us another form of the equation:

$\ln [A] = -kt + \ln [A]_o$

which describes a straight line (y = mx + b). The slope is -k and the y intercept is $\ln[A]_o$.

A graph of ln [A] versus time (t) will be a straight line for a first-order reaction. The value of k can be found from the slope.

Using the First-Order Rate Equation

The rate equation $\ln \frac{[A]}{[A]_o} = -kt$ contains four quantities, [A], $[A]_o$, k, and t; therefore, given three of these quantities the fourth can be calculated. (In some cases it may be convenient to view the fraction of reactant remaining, $[A]/[A]_o$, as one quantity so that, given two of the three quantities $[A]/[A]_o$, t, or k, the third can be determined.)

Example 14.7: The decomposition of hydrogen peroxide $H_2O_2(aq) \rightarrow H_2O(l) + O_2(g)$ in the presence of a manganese dioxide catalyst follows first-order kinetics.

In one experiment the initial H_2O_2 concentration was 0.032 *M* and decreased to 0.013 *M* after 30 min. Calculate the rate constant.

Answer:

Given: $[A]_o = [H_2O_2]_o = 0.032\ M$; $[A] = [H_2O_2] = 0.013\ M$; $t = 30$ min

Unknown: k

Plan: Use $\ln \frac{[A]}{[A]_o} = -kt$, so that $\ln = \frac{[H_2O_2]}{[H_2O_2]_o} = -kt$

We know $[H_2O_2]$, $[H_2O_2]_o$ and t so we can solve for k.

Calculations: $\ln \frac{0.013}{0.032} = -k \times 30$ min; Rearrange and solve for k:

$-k = \frac{\ln(0.013/0.032)}{30\text{ min}} = -0.030\text{ min}^{-1}$; $k = \underline{0.030\text{ min}^{-1}}$

Example 14.8: The hydrolysis of sucrose to form glucose and fructose is a first-order reaction with $k = 2.9 \times 10^{-3}\text{ min}^{-1}$.
(a) How many hours would it take to hydrolyze 25% of the initial concentration of sucrose?
(b) When the initial concentration of sucrose is 2.5 *M*, what concentration will remain after one hour?

Answer:

(a) Given: $k = 2.9 \times 10^{-3}\text{ min}^{-1}$ 25% hydrolyzed so that 75% remains;

$\frac{[A]}{[A_o]} = \frac{[\text{sucrose}]}{[\text{sucrose}]_o} = 0.75$

Unknown: t in hours

Plan: Use $\ln \frac{[A]}{[A_o]} = -kt$, so that $\ln \frac{[\text{sucrose}]}{[\text{sucrose}]_o} = -kt$.

We know $\frac{[\text{sucrose}]}{[\text{sucrose}]_o}$ and k so we can solve for t.

Calculations: Substitute given values

$\ln 0.75 = -2.9 \times 10^{-3}\text{min}^{-1} \times t$

Rearrange and solve for t

$$t = \frac{\ln 0.75}{(-2.9 \times 10^{-3}\ \text{min}^{-1})} = \frac{(0.29)}{(-2.9 \times 10^{-3}\ \text{min}^{-1})} = 99\ \text{min}$$

$$t \text{ in hours} = 99\ \text{min} \times \frac{1\ \text{h}}{60\ \text{min}} = \underline{1.7\ \text{h}}$$

(b) Given: $[A]_o = [\text{sucrose}]_o = 2.5\ M$; $k = 2.9 \times 10^{-3}\ \text{min}^{-1}$; $t = 1$ h or 60 min

Unknown: [sucrose]

Plan: Use $\ln \frac{[A]}{[A_o]} = -kt$, so that $\ln \frac{[\text{sucrose}]}{[\text{sucrose}]_o} = -kt$.

Because we know $[\text{sucrose}]_o$, k, and t we can find [sucrose].

Calculations: Substitute given values

$$\ln \frac{[\text{sucrose}]}{2.5} = -2.9 \times 10^{-3}\ \text{min}^{-1} \times 60\ \text{min} = -0.174$$

Rearrange and solve for ln [sucrose]. Remember the ln of a fraction is the difference between the ln of the numerator and ln of denominator:

$\ln [\text{sucrose}] - \ln 2.5 = -0.174$
$\ln [\text{sucrose}] - 0.916 = -0.174$
$\ln [\text{sucrose}] = 0.742$

Now take the antiln of 0.742 using your calculator: [sucrose] = <u>2.1 mol/L</u>

Practice Drill on the First-Order Rate Equation. Complete the following table:

	$[A]$	$[A_2]$	$[A]/[A_2]$	k	t
(a)	0.090	0.400		?	40 min
(b)	?	0.400		$0.0277\ \text{min}^{-1}$	20.0 min
(c)			0.35	$0.0277\ \text{min}^{-1}$	?

<u>Answers</u>: (a) $k = 0.037\ \text{min}^{-1}$ (b) $[A] = 0.230$ mol/L (c) $t = 37.9$ min

Half-Life

The <u>half-life</u>, $t_{1/2}$, of a reactant is the time required for its concentration to decrease to one-half of its initial value. The half-life in a <u>first-order</u> reaction depends only on the rate constant. The relationship is

$k \times t_{1/2} = 0.693$ k = rate constant
$t_{1/2}$ = half-life

Therefore, if $t_{1/2}$ is known then k can be calculated and vice versa.

Example 14.9: The half-life of a first-order reaction is 2.7×10^5 s. How many hours will it take for the reaction to be 85% complete?

Answer:

Given: $t_{1/2} = 2.7 \times 10^5$ s; the reaction is 85% complete so that 0.15% of A remains;

$$\frac{[A]}{[A_o]} = 0.15$$

Unknown: time in hours

Plan:

1. Use $k \times t_{1/2} = 0.693$ to find k.
2. Use $\ln \frac{[A]}{[A_o]} = -kt$ to find t.

Calculations:

1. $k = \frac{0.693}{t_{1/2}} = \frac{0.693}{2.7 \times 10^5 \text{ s}} = 2.6 \times 10^{-6} \text{ s}^{-1}$
2. $\ln \frac{[A]}{[A_o]} = -kt$ Substitute in known values and solve for t:

$$\ln 0.15 = -2.6 \times 10^{-6} \text{ s}^{-1} \times t$$

$$t = \frac{-\ln 0.15}{2.6 \times 10^{-6} \text{ s}^{-1}} = \frac{-(-1.9)}{2.6 \times 10^{-6} \text{ s}^{-1}} = 7.3 \times 10^5 \text{ s}$$

$$\text{time in hours} = 7.3 \times 10^5 \text{ s} \times \frac{1 \text{ h}}{3600 \text{ s}} = \underline{203 \text{ h}}$$

Practice Drill on the First-Order Rate Equation and Half-Life. Fill in the blanks for the decomposition of ethylene oxide ($(CH_2)_2O$) a first-order reaction:

$$2(CH_2)_2O(g) \rightarrow 2CH_4(g) + O_2(g)$$

	$\frac{[(CH_2)_2O]}{[(CH_2)_2O]_o}$	$[(CH_2)_2O]$	$[(CH_2)_2O]_o$	k	t	$t_{1/2}$
(a)	?			?	25.2 min	56.3 min
(b)		0.80 *M*	0.100 *M*		?	56.3 min

Answers: (a) $k = 0.0123 \text{ min}^{-1}$; $[(CH_2)_2O]/[(CH_2)_2O]_o = \underline{0.733}$

(b) $t = \frac{\ln 0.80}{(-0.0123 \text{ min}^{-1})} = \frac{-(-0.22)}{0.0123 \text{ min}^{-1}} = \underline{18 \text{ min}}$

Second-Order Reactions

The simplest second-order reaction is one in which the rate is proportional to the square of a single reactant concentration (rate = $k[A]^2$). For such a reaction the half-life varies with concentration as well as with k:

$$\frac{1}{[A]} = kt + \frac{1}{[A]_o} \quad \text{and} \quad t_{1/2} = \frac{1}{k[A]_o}$$

$[A]_o$ = initial concentration of reactant A (t = 0)
[A] = concentration of reactant t after passage of time, t
k = rate constant
$t_{1/2}$ = half-life

Reaction Mechanisms (Section 14.5)

Learning Objectives

1. State the molecularity and write the rate law for any given elementary step.
2. State the requirements for a reaction mechanism.
3. Propose mechanisms for reactions in which the first step is rate-determining.
4.* Use the steady-state approximation to find rate laws for reactions in which steps other than the first are rate-determining.

Review

The mechanism of a reaction is the series of consecutive elementary steps leading from reactants to products. The number of reactant particles in an elementary step is called its molecularity.

Most elementary steps are bimolecular, that is, two particles collide and react. An example is the bimolecular, elementary reaction in which ozone is decomposed by nitric oxide in the atmosphere:

$O_3(g) + NO(g) \rightarrow O_2(g) + NO_2(g)$

The exponents in the rate law for an elementary step are the same as the coefficients in the equation for the step.

The rate law for the above reaction is therefore: rate = $k[O_3][NO]$

Practice. Give the rate laws and the molecularities for the following elementary steps:

(a) $Cl_2(g) \rightarrow 2Cl(g)$
(b) $2NO(g) + 2Cl(g) \rightarrow N_2(g) + 2ClO(g)$
(c) $2ClO(g) \rightarrow Cl_2(g) + O_2(g)$

Answers: (a) rate = $k[Cl_2]$; 1 (b) rate = $k[NO]^2[Cl]^2$; 4 (c) rate = $k[ClO]^2$; 2

Devising Reaction Mechanisms

A proposed reaction mechanism must account for the products of the reaction and be consistent with its rate law. The rate law for the reaction is determined by the rate law for the slowest or rate-determining step in the mechanism.

Example 14.10: Nitrogen dioxide (NO_2) reacts with F_2 to form NO_2F:

$2NO_2(g) + F_2(g) \rightarrow 2NO_2F(g)$

The experimental rate law is rate = $k[NO_2][F_2]$. Devise a two-step mechanism with a slow first step.

Answer: A slow first step of the form $NO_2(g) + F_2(g) \rightarrow$ products would account for the observed rate law. Because two NO_2F molecules are formed in the reaction; one of these is the most likely product in step 1, with F as the other product;

Step 1: $NO_2(g) + F_2(g) \rightarrow NO_2F(g) + F(g)$

Two NO_2 molecules react in the balanced equation so another NO_2 molecule must react in step 2. Because F(g) does not appear in the equation for the total reaction it must be a reactant in step 2. (F is an intermediate, a substance produced in one step and used up in another.) NO_2F must be the product in step 2 because the overall equation shows two NO_2F molecules being formed.

Step 2: $NO_2(g) + F(g) \rightarrow NO_2F(g)$

The mechanism is therefore:

(rate-determining) $NO_2(g) + F_2(g) \rightarrow NO_2F(g) + F(g)$ slow

$NO_2(g) + F(g) \rightarrow NO_2F(g)$ fast

Practice: The experimental rate law for the reaction

$2NO(g) + 2H_2(g) \rightarrow N_2(g) + 2H_2O(g)$

is rate $= k[NO]^2[H_2]$. Devise a two-step mechanism with a slow first step.

Answer: $2NO(g) + H_2(g) \rightarrow N_2(g) + H_2O_2(g)$ slow

$H_2O_2(g) + H_2(g) \rightarrow 2H_2O(g)$ fast

*One method for checking a proposed reaction mechanism uses the steady-state approximation, which assumes that the rate of formation of an intermediate in a reacting mixture is equal to its rate of consumption.

Free Radical Chain Reactions (Section 14.6)

Learning Objectives

1. Identify the initiation, propagation, and termination steps in a chain mechanism.
2. Describe the mechanism by which CFCs deplete the ozone layer.

Review

A chain reaction is one in which an initial step leads to a succession of repeating steps that continues indefinitely. The chain initiation step is generally an endothermic dissociation that produces free radicals. (A free radical has an unpaired electron often indicated by a dot.) The repeating or chain propagation steps transform one free radical into another. A chain termination step occurs when two free radicals combine. Explosive reactions often include chain branching steps in which one free radical produces two or more free radicals. The depletion of the protective ozone layer in the stratosphere over various parts of the world is thought to result from chain reactions.

Example 14.11: Water in the atmosphere can absorb radiation from the sun and dissociate:

$H_2O \rightarrow H\cdot + HO\cdot$ (initiation) followed by

$H\cdot + HO\cdot \rightarrow H_2 + \cdot O\cdot$ (propagation) and

$2\cdot O\cdot \rightarrow O_2$ (termination).

Practice: Label the following reactions as initiation, propagation or termination:

(a) $CCl_2F_2 \rightarrow \cdot CClF_2 + Cl\cdot$

(b) $Cl\cdot + O_2 \rightarrow \cdot ClO + \cdot O\cdot$

(c) $\cdot ClO + \cdot O\cdot \rightarrow Cl\cdot + O_2$

(d) $Cl\cdot + Cl\cdot \rightarrow Cl_2$

Answers: (a) initiation (b), (c) propagation (d) termination

Activation Energy (Section 14.7)

Learning Objectives

1. Use values of E_a and ΔH° to sketch the potential energy profile for an elementary step.
2. Given two of the three quantities, E_a (forward), E_a (reverse), and ΔH°, calculate the missing quantity.
3. Use activation energies to identify the slowest step in a mechanism.
4. Use rate constant versus temperature data to calculate activation energies.
5. Use the activation energy and rate constant measured at one temperature to calculate the rate constant at another temperature.

6.* State how the collision frequency varies with concentration and with temperature.

7.* State how the fraction of reactive collisions varies with temperature and with activation energy.

8.* Use the activation energy to calculate the effect of changing temperature on the rate of a reaction.

9.* State how the orientation factor varies with increasing molecular complexity.

Review

The activation energy E_a is the minimum energy required for a reaction to occur. This energy is needed to break bonds within the colliding molecules; therefore, its magnitude depends on the nature of the reactants. The activation energy is the difference between the potential energy of the activated complex and the initial potential energy of the reactants. The potential energy profile for the reaction $H_2 + Br \rightarrow HBr + H$ is given below:

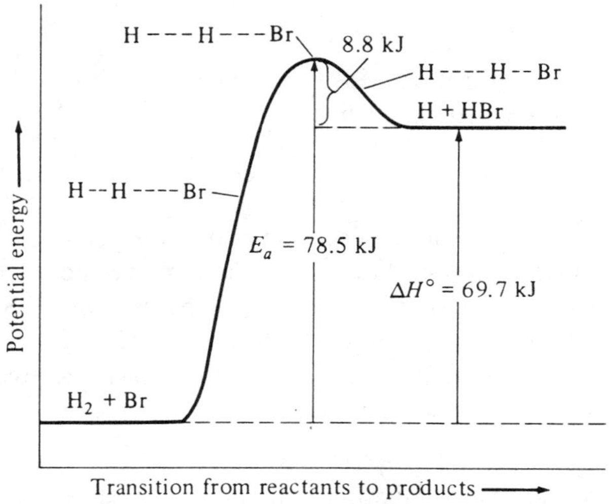

ΔH°, the enthalpy change for the reaction shown above, is equal to the forward activation energy (78.5 kJ) minus the reverse activation energy (8.8 kJ). Note that the reaction shown is endothermic; ΔH° is positive (69.7 kJ). For reactions involving several steps, the step with the greater activation energy is generally the slow or rate-determining step.

Practice: Draw the potential energy profile for an exothermic reaction; $\Delta H°$ is negative.

Determining Activation Energies

For most reactions the change of k with temperature can be described by the Arrhenius equation:

$k = Ae^{-E_a/RT}$

k = rate constant
R = gas constant (J/mol K)
T = Kelvin temperature
E_a = activation energy (J/mol)
A = constant for a given reaction

If k values are known at several temperatures, E_a can be determined from a graph of the above equation.

A plot of ln k versus 1/T is linear and the activation energy is found from the slope; slope = $-E_a/R$.

The following form of the Arrhenius equation can be used for just two values of k and T.

$$\ln \frac{k_1}{k_2} = \frac{E_a}{R}\left(\frac{1}{T_2} - \frac{1}{T_1}\right)$$

k_1 and k_2 = rate constants for reactions occurring at T_1 and T_2 respectively
E_a = activation energy of reaction
R = gas constant

This equation can be used to find E_a from k and T values, or it can be used to calculate a rate constant at any temperature if E_a and the rate constant at some other temperature are known:

Example 14.12: The activation energy, E_a, for $2NO_2(g) \rightarrow 2NO(g) + O_2(g)$ is 27,200 cal. The rate constant, k, at 700K is 20 L/mol s. Find k at 800K.

Given: k_1 = 20 L/mol s, T_1 = 700K, T_2 = 800K, E_a = 27,200 cal

Unknown: k_2

Plan: Use $\ln \frac{k_1}{k_2} = \frac{E_a}{R}\left(\frac{1}{T_2} - \frac{1}{T_1}\right)$

Calculations: $\ln \frac{20}{k_2} = \frac{27{,}200}{1.99}\left(\frac{1}{800} - \frac{1}{700}\right)$

$\ln 20 - \ln k_2 = -2.46$. Rearranging and taking the antiln gives

$\ln k_2 = 2.46 + 3.00 = 5.46$; k_2 = <u>235 mol/L s</u>

*The collision theory of reaction rates assumes that the rate of a bimolecular elementary step is equal to $Z \times f_a \times p$. The collision frequency, Z, is the total number of bimolecular collisions per second. For gaseous reactions, Z is proportional to the concentration of each of the colliding molecules and to the square root of the Kelvin temperature. The activation energy factor, $f_a = e^{-E_a/RT}$, is the fraction of collisions having energies greater than or equal to E_a. This fraction, which increases with increasing temperature and decreases with increasing E_a, is usually more important than Z in determining the effect of temperature on reaction rates. The orientation factor, p, is the fraction of collisions that have the geometry necessary for a reaction to take place. This fraction generally decreases with increasing molecular size.

More About Catalysts (Section 14.8)

Learning Objectives

1. Explain why a reaction goes faster in the presence of a catalyst.
2. Describe the mechanism by which platinum catalyzes the reactions of hydrogen and certain other gases.
3. Explain why many heterogeneous catalysis reactions are zero-order.

Review

A catalyst speeds up a reaction by providing a new mechanism with a lower activation energy. In heterogeneous catalysis, the reacting molecules are adsorbed onto the surface of a contact catalyst. Weak chemical bonds form between the surface atoms and the adsorbed molecules, a phenomenon called chemisorption. The original bonds are stretched so that the adsorbed material can more easily react with molecules from the liquid or gas. The H–H bond for example is stretched when $H_2(g)$ is adsorbed onto Pt so that H atoms are available to other reactants for much less energy. A contact catalyst is inactivated or poisoned by adsorbed impurities.

In many heterogeneous catalysis reactions the gas pressures of reactants are sufficiently high to occupy all of the surface sites available. Consequently, the reaction rate will not increase with the addition of more reactant gases making these reactions zero-order.

SELF-TEST

1. List three ways of increasing the rate of a homogeneous reaction.

2. For the reaction between ozone and nitric oxide in photochemical smog

   $O_3(g) + NO(g) \rightarrow O_2(g) + NO_2(g)$ the following data were obtained:

Run	$[O_3]$ mol / L	$[NO]$ mol / L	$\frac{\Delta[NO_2]}{\Delta t}$ mol / L s
1	2.5×10^{-7}	2.5×10^{-7}	7.5×10^{-7}
2	5.0×10^{-7}	2.5×10^{-7}	1.5×10^{-6}
3	2.5×10^{-7}	2.5×10^{-6}	7.5×10^{-6}

   (a) Write the experimental rate law and give the order with respect to $[O_3]$.
   (b) Calculate k.
   (c) When the concentration of NO in polluted air is 4×10^{-8} mol/L, what must be the concentration of O_3 if 9.6×10^{-9} mol/L of NO_2 are formed per second?

3. Use the data given below for the reaction

   $BrO_3^-(aq) + 5Br^-(aq) + 6H^+(aq) \rightarrow 3Br_2(aq) + 3H_2O(l)$

Run	$[BrO_3^-]$ mol / L	$[Br^-]$ mol / L	$[H^+]$ mol / L	$\frac{-\Delta[BrO_3^-]}{\Delta t}$ mol / L s
1	0.150	0.150	2.00×10^{-4}	5.27
2	0.300	0.150	2.00×10^{-4}	10.54
3	0.150	0.300	2.00×10^{-4}	10.54
4	0.300	0.150	6.00×10^{-4}	94.86

to determine
(a) the rate law
(b) the order of the reaction
(c) rate of reaction when $[BrO_3^-] = [Br^-] = 0.0350\ M$ and $[H^+] = 3.00 \times 10^{-4}\ M$
(d) the rate of production of Br_2 when Br^- is disappearing at a rate of 0.020 mol/L s

4. The rate of elimination of drugs from blood plasma often follows first-order kinetics. For practical purposes, a drug is usually considered to be eliminated in about 3.3 half-lives. What percentage of the drug remains in blood plasma after the passage of 3.3 half-lives?

5. The decomposition of $SO_2Cl_2(g)$ to form $SO_2(g)$ and $Cl_2(g)$ is a first-order reaction.
(a) When the initial concentration of SO_2Cl_2 is 0.010 mol/L, the rate of formation of $SO_2(g)$ is 0.2×10^{-7} mol/L s. Calculate the half-life, $t_{1/2}$, of SO_2Cl_2.
(b) How many hours will it take for 75% of the SO_2Cl_2 to be used up?

6. The experimental rate law for

$NO_2(g) + CO(g) \rightarrow CO_2(g) + NO(g)$ is rate = k $[NO_2]^2$. Propose a mechanism.

7. The rate constant, k, for $2NO_2(g) \rightarrow 2NO(g) + O_2(g)$ at 600K is 0.75 L/mol s. At 700K the rate constant is 20 L/mol s. Find the activation energy, E_a, in kilojoules.

8. The action of photons (hν) on water produces the hydroxyl radicals, •OH, that lead to the removal of CO from the atmosphere.

$H_2O + h\nu \rightarrow HO\cdot + H\cdot$

$HO\cdot + CO \rightarrow CO_2 + H\cdot$

$H\cdot + O_2 \rightarrow HO_2\cdot$

$HO_2\cdot + CO \rightarrow HO\cdot + CO_2$

(a) Which of the above reactions is the chain initiation step? Which are chain propagation steps?
(b) Write three possible chain termination reactions.

9. A catalyst
(a) increases the concentrations of reactants
(b) increases the temperature
(c) decreases the activation energy
(d) increases the activation energy

ANSWERS TO SELF-TEST QUESTIONS

1. (1) increase the concentration of reactants (3) add a catalyst
   (2) increase the temperature

2. (a) rate = k $[O_3][NO]$; first order with respect to $[O_3]$.
   (b) 1.2×10^7 L/mol s
   (c) 2.0×10^{-8} mol/L

3. (a) rate = $k[Br^-][BrO_3^-][H^+]^2$
   (b) 4
   (c) rate = 5.86×10^9 L^3/mol^3 s $\times 0.0350 \times 0.0350 \times (3.00 \times 10^{-4})^2 = \underline{6.46 \times 10^{-1}\text{ mol/L s}}$
   (d) 0.020 mol/L s × 3/5 = $\underline{0.012\text{ mol/L s}}$

4. 10%

5. (a) $k = \dfrac{2.2 \times 10^{-7}\text{ mol/L s}}{0.010\text{ mol/L}}$; $t_{1/2} = \dfrac{0.693}{2.2 \times 10^{-5}} = \underline{3.2 \times 10^4\text{ s}}$

   (b) two half-lives must pass; 3.2×10^4 s × 2 = 6.4×10^4 s = $\underline{18\text{ h}}$

6. $NO_2(g) + NO_2(g) \rightarrow NO_3(g) + NO(g)$ slow

   $NO_3(g) + CO(g) \rightarrow NO_2(g) + CO_2(g)$ fast

7. $E_a = 11.5$ kJ

8. (a) $H_2O + h\nu \rightarrow HO\bullet + H\bullet$ initiation; others are chain propagation steps

   (b) $HO_2\bullet + HO_2\bullet \rightarrow H_2O_2 + O_2$
   $H\bullet + H\bullet \rightarrow H_2$
   $HO_2\bullet + H\bullet \rightarrow H_2O_2$

9. (c)

CHAPTER 15
CHEMICAL EQUILIBRIUM

SELF-ASSESSMENT

Reversible Reactions and the Equilibrium State (Section 15.1)

1. Which of the following statements is incorrect?:
   (a) The forward and reverse reactions occur simultaneously in a reversible reaction.
   (b) All reversible reactions tend toward a state of chemical equilibrium.
   (c) At equilibrium all reactants and products have equal concentrations.
   (d) The concentration of each substance remains constant at equilibrium.
   (e) At equilibrium the rate of the reverse reaction is the same as the rate of the forward reaction.

Reaction Quotients and Equilibrium Constants (Section 15.2)

2. Two moles of N_2O_4 are placed in a 20-L reaction flask and allowed to reach equilibrium at 100°C:

   $N_2O_4(g) \rightleftharpoons 2NO_2(g)$

   If $[NO_2]$ is 0.120 mol/L at equilibrium, find
   (a) the equilibrium concentration of N_2O_4
   (b) K_c
   (c) the fraction of N_2O_4 dissociated

3. $K_c = 1.4 \times 10^{-3}$ at 25°C for the equilibrium $CH_2ClCOOH(aq) \rightleftharpoons H^+(aq) + CH_2ClCOO^-(aq)$
   In which direction will the reaction proceed when
   (a) $[CH_2ClCOOH] = 0.10\ M$; $[H^+] = [CH_2ClCOO^-] = 0.050\ M$?
   (b) $[CH_2ClCOOH] = 1.0\ M$; $[H^+] = [CH_2ClCOO^-] = 0.030\ M$?
   (c) $[CH_2ClCOOH] = 0.050\ M$; $[H^+] = 1.4 \times 10^{-3}\ M$; $[CH_2ClCOO^-] = 0.050\ M$?

Calculations Using K_c (Section 15.3)

4. For the equilibrium reaction described in Exercise 2, determine the equilibrium concentration of NO_2 and N_2O_4 when initial concentrations are
   (a) $[N_2O_4] = 1.00\ M$ and $[NO_2] = 0$
   (b) $[N_2O_4] = 1.00\ M$ and $[NO_2] = 1.50\ M$

5. For the equilibrium $N_2(g) + O_2(g) \rightleftharpoons 2NO(g)$ K_c is 1.0×10^{-5} at 1500K. Initially $[N_2] = 0.80\ M$ and $[O_2] = 0.20\ M$. Find the equilibrium concentration of NO. (This reaction contributes to air pollution when gasoline burns in air in automobile engines.)

Equilibria Involving Solvents (Section 15.4)

6. Ammonia (NH_3), reacts with water according to the reaction $NH_3(aq) + H_2O(l) \rightleftharpoons NH_4^+(aq) + OH^-(aq)$ Find the equilibrium concentration of OH^- when initial concentrations are $[NH_3] = [NH_4^+] = 0.15$ *M* and $K_c = 1.8 \times 10^{-5}$.

Gas Phase Equilibrium: Q_p and K_p (Section 15.5)

7. Refer to Exercise 15.2. Calculate
   (a) the equilibrium partial pressure of N_2O_4 and NO_2 in the flask
   (b) K_p for the reaction

8. K_p for the reaction $CO_2(g) + H_2(g) \rightleftharpoons CO(g) + H_2O(g)$ at 773K is 0.170. If 1.00 atm of CO_2 and 1.00 atm of H_2 are initially introduced into a reaction flask, what is the equilibrium pressure of each gas?

9. Which of the following relationships between K_p and K_c is correct?:
   (a) $K_c = K_p (RT)^{\Delta n}$
   (b) $K_p = K_c (RT)^{\Delta n}$
   (c) $K_c = RT (K_p)^{\Delta n}$
   (d) $K_p = K_c (RT)^{-\Delta n}$

Form of the Reaction Quotient (Section 15.6)

10. The following equilibrium constants were determined at 773K:

    $CO(g) + H_2O(g) \rightleftharpoons CO_2(g) + H_2(g)$ $K_p = 5.88$
    $CH_4(g) + 2H_2O(g) \rightleftharpoons CO_2(g) + 4H_2(g)$ $K_p = 0.633$

    Find K_p's for the reactions

    (a) $1/2CO_2(g) + 2H_2(g) \rightleftharpoons 1/2CH_4(g) + H_2O(g)$
    (b) $CO(g) + 3H_2(g) \rightleftharpoons CH_4(g) + H_2O(g)$

Heterogeneous Equilibria (Section 15.7)

11. K_p is 1.2×10^{-12} at 900K for $UF_6(g) \rightleftharpoons UF_4(s) + F_2(g)$. Find the equilibrium partial pressure of F_2 when UF_6 at an initial partial pressure of 20.0 atm is allowed to decompose.

12. Limestone decomposes with heating according to the equation $CaCO_3(s) \rightleftharpoons CaO(s) + CO_2(g)$. K_p is 10.5 at 1000°C. What equilibrium pressure will develop if $CaCO_3$ is heated to 1000°C in a reaction flask?

Le Chatelier's Principle (Section 15.8)

13. For the reaction $N_2(g) + O_2(g) \rightleftharpoons 2NO(g)$ K_c is 1×10^{-30} at 25°C and 0.10 at 2000°C. Predict the effect of each of the following changes on the equilibrium concentration of NO:
    (a) decreasing temperature
    (b) increasing partial pressure of O_2
    (c) addition of argon gas

ANSWERS TO SELF-ASSESSMENT PROBLEMS

If you missed an answer, be sure to study the relevant section in the textbook and study guide.

1. (c)

2. (a) 0.040 mol/L
   (b) $K_c = [NO_2]^2/[N_2O_4] = (0.120)^2/0.040 = \underline{0.36}$
   (c) $0.0600/0.100 = \underline{0.600}$

3. (a) $Q_c = (0.050)^2/0.10 = 2.5 \times 10^{-2} > K_c$; to the left (c) $Q_c = K_c$; at equilibrium.
   (b) $Q_c = (0.030)^2/1.0 = 9.0 \times 10^{-4} < K_c$; to the right

4. (a) $Q_c < K_c$; $K_c = (2x)^2/(1.00 - x)$
   $[NO_2] = \underline{0.52 \text{ mol/L}}$; $[N_2O_4] = \underline{0.74 \text{ mol/L}}$

   (b) $Q_c > K_c$; $K_c = (1.50 - 2x)^2/(1.00 + x) = 0.36$; $x = 0.396$
   $[NO_2] = \underline{0.71 \text{ mol/L}}$; $[N_2O_4] = \underline{1.40 \text{ mol/L}}$

5. $1.0 \times 10^{-5} = (2x)^2/(0.80 \times 0.20)$; $x = 6.3 \times 10^{-4}$; $[NO] = \underline{1.3 \times 10^{-3}\ M}$

6. Using approximations;

   $$K_c = \frac{(0.15)(x)}{(0.15)}\text{; } [NH_3] = [NH_4^+] = 0.15\ M\text{; } [OH^-] = \underline{1.8 \times 10^{-5}\ M}$$

7. (a) $P_{NO_2} = 3.67$ atm; $P_{N_2O_4} = 1.22$ atm (b) 11.0

8. $K_p = x^2/(1.00 - x)^2$; $P_{CO} = P_{H_2O} = 0.292$ atm
   $P_{CO_2} = P_{H_2} = 0.708$ atm

9. (b)

10. (a) 1.26 (b) $5.88 \times 1/0.633 = \underline{9.29}$

11. $K_p = x/20.0$; $x = \underline{2.4 \times 10^{-11} \text{ atm}}$

12. 10.5 atm

13. (a) decreases (b) increases (c) no effect

REVIEW

Reversible Reactions and the Equilibrium State (Section 15.1)

Learning Objectives

1. State the characteristics common to all chemical equilibria.
2. Explain why the concentration of each substance in an equilibrium mixture does not change with time.

Review

A <u>reversible</u> reaction proceeds in both the forward and backward directions; reactants form products and products form reactants simultaneously. All reversible reactions tend toward an equilibrium state. At <u>equilibrium</u> the rates of the two opposing reactions are the same so that the concentration of each substance remains constant.

Reaction Quotients and Equilibrium Constants (Section 15.2)

Learning Objectives

1. Write reaction quotients for reversible reactions.
2. Use equilibrium concentrations to calculate the value of an equilibrium constant.
3. State the law of chemical equilibrium.
4. Use Q_c and K_c to predict whether a reaction is moving in the forward or reverse direction.
5.* Derive the law of chemical equilibrium from the kinetic relations that exist in the equilibrium state.

Review

The reaction quotient, Q_c , for a reversible reaction is obtained by multiplying the product concentrations and then dividing by the reactant concentrations. Each concentration is raised to a power equal to its coefficient in the balanced equation.

Example 15.1: Write Q_c for the reaction $2NO(g) \rightleftharpoons N_2(g) + O_2(g)$

Answer: $Q_c = \dfrac{[N_2][O_2]}{[NO]^2}$

Practice Drill on Reaction Quotients: Write reaction quotients for the following reversible reactions:

(a) $NO(g) + Br_2(g) \rightleftharpoons 2NOBr(g)$ (b) $N_2(g) + 3H_2(g) \rightleftharpoons 2NH_3(g)$

Answers: (a) $Q_c = [NOBr]^2/[NO][Br_2]$ (b) $Q_c = [NH_3]^2/[N_2][H_2]^3$

As the reaction mixture moves toward equilibrium, the concentrations of reactants and products change, causing the reaction quotient to change as well. Once equilibrium is reached the concentrations remain constant, giving Q_c a constant value. This value is the equilibrium constant, K_c , which can be found by substituting equilibrium concentrations into the Q_c expression.

Example 15.2: Find K_c at 300°C for the reaction $N_2(g) + 3H_2(g) \rightleftharpoons 2NH_3(g)$ given the equilibrium concentrations:

$[N_2] = 0.10$ mol/L; $[H_2] = 0.15$ mol/L; $[NH_3] = 5.8 \times 10^{-3}$ mol/L

Answer: $Q_c = \dfrac{[NH_3]^2}{[N_2][H_2]^3}$ At equilibrium $K_c = Q_c$ so that $K_c = \dfrac{(5.8 \times 10^{-3})^2}{(0.10)(0.15)^3} = 0.10$

(Note that the units are generally omitted for reaction quotients and equilibrium constants.)

Practice Drill on Equilibrium Constants: Fill in the table for the reaction

$N_2(g) + O_2(g) \rightleftharpoons 2NO(g)$ at 2000°C

	$[N_2]$	$[O_2]$	$[NO]$	K_c
(a)	0.20 mol/L	0.20 mol/L	0.064 mol/L	?
(b)	?	0.50 mol/L	0.50 mol/L	see (a)
(c)	0.10 mol/L	0.20 mol/L	?	see (a)

Answers: (a) $K_c = 0.10$ (b) $[N_2] = 5.0$ mol/L; $K_c = 0.10$ (c) $[NO] = 0.045$ mol/L; $K_c = 0.10$

Finding Equilibrium Concentrations from Initial Concentrations: Constructing Concentration Summaries

Many problems ask you to calculate the concentration of each species at equilibrium. To do this it is a good idea to construct a concentration summary around the balanced equation by following the three steps listed below.

Step (1) Find the initial concentration of each species.
Step (2) Find the change in each concentration using the balanced equation.
Step (3) Add (1) and (2) to get equilibrium concentrations.

Example 15.3: H_2 and CO_2 react according to the equation

$H_2(g) + CO_2(g) \rightleftharpoons CO(g) + H_2O(g)$

When 5.0 mol of H_2 and 5.0 mol of CO_2 were placed in a 10.0-L flask and allowed to reach equilibrium, the final concentration of CO was 0.14 mol/L. Calculate K_c.

Given: Initial amount of each reactant and volume; 5.0 mol of H_2 in 10.0 L; 5.0 mol of CO_2 in 10.0 L. Equilibrium concentration of CO = 0.14 mol/L

Unknown: K_c

Plan: We will construct a concentration summary by following the three steps listed above and then substitute equilibrium concentrations into the expression for K_c.

Steps	$H_2(g)$	+	$CO_2(g)$	$\rightleftharpoons$	$CO(g)$	+	$H_2O(g)$
(1) Initial Concentration (mol/L)	$\frac{5.0}{10.0} = 0.50$		$\frac{5.0}{10.0} = 0.50$		0		0
(2) Changes in Concentration	?		?		?		?
(3) Equilibrium Concentrations	?		?		0.14 mol		?

Calculations:

(2) Find the changes in step 2. The CO changed from 0 to 0.14 mol/L, or +0.14 mol/L. The H_2O change is the same as for CO (same coefficients in the equation). The H_2 and CO_2 concentrations must decrease by this same amount (again, same coefficients); each of these changes is –0.14 mol/L. The line for step 2 in the summary becomes

-0.14 mol/L   -0.14 mol/L   -0.14 mol/L   -0.14 mol/L

(3) Add the values in steps (1) and (2) to obtain the final concentrations in step (3). The completed summary is

	$H_2(g)$	+	$CO_2(g)$	$\rightleftharpoons$	$CO(g)$	+	$H_2O(g)$
Step (1)	0.50 mol/L		0.50 mol/L		0		0
Step (2)	-0.14 mol/L		-0.14 mol/L		+0.14 mol/L		+0.14 mol/L
Step (3)	0.36 mol/L		0.36 mol/L		0.14 mol/L		0.14 mol/L

K_c can now be obtained by substituting the equilibrium concentrations into the reaction quotient expression.

$$K_c = \frac{[CO][H_2O]}{[H_2][CO_2]} = (0.14)^2/(0.36)^2 = \underline{0.15}$$

Practice: Complete the following concentration summary when 2.0 mol/L of H_2 are mixed with 2.0 mol/L of CO_2:

	$H_2(g)$	+	$CO_2(g)$	$\rightleftharpoons$	$CO(g)$	+	$H_2O(g)$
Initial (mol/L)	2.0		2.0		0		0
Change (mol/L)	?		?		?		?
Equilibrium (mol/L)	?		1.44		?		?

Answers:	$H_2(g)$	+	$CO_2(g)$	$\rightleftharpoons$	$CO(g)$	+	$H_2O(g)$
Initial	2.0		2.0		0		0
Change	-0.56		-0.56		+0.56		+0.56
Equilibrium	1.44		1.44		0.56		0.56

The Law of Chemical Equilibrium

The law of chemical equilibrium states that the equilibrium constant, K_c, is independent of initial concentrations and varies only with temperature. A reaction moves in the forward direction when Q_c is less than K_c; in the reverse direction when Q_c is greater than K_c; and no change occurs when Q_c is equal to K_c.

Example 15.4: The equation and K_c for the decomposition of chlorine gas at 2000°C are

$Cl_2(g) \rightleftharpoons 2Cl(g)$ $\quad K_c = 3.6 \times 10^{-2}$

In which direction will the reaction move when the concentration of Cl_2 and Cl are both 0.050 mol/L?

Answer: Substituting 0.050 mol/L for $[Cl_2]$ and [Cl] into Q_c gives

$$Q_c = \frac{[Cl]^2}{[Cl_2]} = \frac{(0.050)^2}{(0.050)} = 0.050$$

Because Q_c (0.050) > K_c (0.036) the reaction will move in the reverse direction increasing $[Cl_2]$ and decreasing [Cl] until $Q_c = K_c$.

Practice Drill: Complete the following table for the reaction at 25°C:

$3O_2(g) \rightleftharpoons 2O_3(g)$ $K_c = 1.6 \times 10^{-56}$

	Initial Concentrations $[O_2]$	$[O_3]$	Q_c	Compare Q_c with K_c	Direction of Reaction
(a)	1×10^{-3}	1×10^{-9}	?	?	?
(b)	2×10^{-3}	2×10^{-35}	?	?	?
(c)	2×10^{-3}	0	?	?	?

Answers: (a) $Q_c = 1 \times 10^{-9}$; $Q_c > K_c$; reverse (b) $Q_c = 5 \times 10^{-62}$; $Q_c < K_c$; forward
(c) $Q_c = 0$; $Q_c < K_c$; forward

Calculations Using K_c (Section 15.3)

Learning Objectives

1. Use K_c and the initial concentrations of a mixture to calculate its equilibrium composition.
2. Use simplifying approximations to solve equilibrium problems and check to see if the approximations are valid.

*3. Use the method of successive approximations to solve equilibrium problems.

Review

The equilibrium constant can be used to calculate equilibrium concentrations from initial concentrations.

Example 15.5: K_c for the reaction $I_2(g) + Cl_2(g) \rightarrow 2ICl(g)$ is 9.1. The concentrations in a particular reaction mixture are initially 0.0200 mol/L ICl and 0.100 mol/L each for I_2 and Cl_2. Calculate equilibrium concentrations.

Given: Initial concentrations; $[ICl] = 0.0200$ mol/L; $[I_2] = [Cl_2] = 0.100$ mol/L; $K_c = 9.1$

Unknown: Equilibrium concentrations

Plan:
(1) Calculate Q_c and determine the direction of the reaction.
(2) Use the 3-step summary to establish equilibrium concentrations in terms of *x*.
(3) Use the value of K_c to solve for *x*.

Calculations:

$$Q_c = \frac{[ICl]^2}{[I_2][Cl_2]} = \frac{(0.0200)^2}{(0.100)^2} = 0.0400$$

Because Q_c (0.0400) < K_c (9.1) the reaction will move in the forward direction.
We can now apply the three steps to find equilibrium concentrations:

		$I_2(g)$	+	$Cl_2(g)$	$\rightleftharpoons$	$2ICl(g)$
Step 1:	Initial (mol/L)	0.100		0.100		0.0200
Step 2:	Change (mol/L)	?		?		?
Step 3:	Equilibrium (mol/L)	?		?		?

Step 1: Initial concentrations are given.

Step 2: The balanced equation tells us that I_2 and Cl_2 decrease by the same number of moles as the reaction moves forward. Because we do not know exactly how many moles of I_2 and Cl_2 react, we will let x be the decrease in the concentrations of I_2 and Cl_2. The balanced equation tells us that the number of moles of ICl that form is twice the number of moles of I_2 and Cl_2 that react. Therefore, the concentration of ICl will increase by $2x$.

		$I_2(g)$	+	$Cl_2(g)$	$\rightleftharpoons$	$2ICl(g)$
Step 1:	Initial (mol/L)	0.100		0.100		0.0200
Step 2:	Change (mol/L)	$-x$		$-x$		$+2x$
Step 3:	Equilibrium (mol/L)	$0.100 - x$		$0.100 - x$		$0.0200 + 2x$

Substitute equilibrium concentrations into K_c:

$$K_c = \frac{[ICl]^2}{[I_2][Cl_2]} \qquad 9.1 = \frac{(0.0200 + 2x)^2}{(0.100 - x)(0.100 - x)}$$

Take the square root of both sides:

$$3.0 = \frac{0.0200 + 2x}{0.100 - x}$$; now solve for x to get $x = 0.056$ mol/L

The equilibrium concentrations are

$[I_2] = [Cl_2] = 0.100 - x = 0.100 - 0.056 =$ <u>0.044 mol/L</u>

$[ICl] = 0.0200 + 2x = 0.0200 + 2 \times 0.056 =$ <u>0.132 mol/L</u>

Practice Drill on Equilibrium Calculations: Complete the concentration summary given below for a reaction mixture in which the initial concentrations are $[H_2] = [N_2] = 0.200$ mol/L and $[NH_3] = 0.0100$ mol/L. The equation is:

	$3H_2(g)$	+	$N_2(g)$	$\rightleftharpoons$	$2NH_3(g)$
Initial (mol/L)	0.200		0.200		0.0100
Change (mol/L)	?		$-x$		?
Equilibrium (mol/L)	?		?		?

<u>Answers</u>: Change: $-3x$, $-x$, $+2x$; Equilibrium: $0.200 - 3x$, $0.200 - x$, $0.0100 + 2x$

Using the Quadratic Formula in Equilibrium Calculations

<u>Example 15.6</u>: K_c is 0.10 for the reaction

$N_2(g) + O_2(g) \rightleftharpoons 2NO(g)$

Calculate the equilibrium concentration of each gas when initial concentrations are $[N_2] = 0.900$ mol/L, $[O_2] = 0.700$ mol/L, and $[NO] = 0.200$ mol/L.

Given: Initial concentrations: $[N_2] = 0.900$ mol/L, $[O_2] = 0.700$ mol/L, $[NO] = 0.200$ mol/L $K_c = 0.10$

Unknown: Equilibrium concentrations of N_2, O_2, and NO.

Plan:
(1) Calculate Q_c and determine the direction of the reaction.
(2) Follow the three steps to determine expressions for equilibrium concentrations.
(3) Substitute these equilibrium expressions into K_c.

Calculations:

$$Q_c = \frac{[NO]^2}{[N_2][O_2]} = \frac{(0.200)^2}{(0.900)(0.700)} = 0.0635$$

$Q_c(0.0635) < K_c\ (0.100)$ The reaction proceeds in the forward direction.

		$N_2(g)$	+	$O_2(g)$	$\rightleftharpoons$	$2NO(g)$
Step 1:	Initial (mol/L)	0.900		0.700		0.200
Step 2:	Change (mol/L)	?		?		?
Step 3:	Equilibrium (mol/L)	?		?		?

Step 1: Initial concentrations are given.

Step 2: Let x be the moles per liter of N_2 that reacts by the time equilibrium is reached. Therefore x will also be the moles per liter of O_2 that reacts and $2x$ the moles per liter of NO that forms.

		$N_2(g)$	+	$O_2(g)$	$\rightleftharpoons$	$2NO(g)$
Step 1:	Initial (mol/L)	0.900		0.700		0.200
Step 2:	Change (mol/L)	$-x$		$-x$		$+2x$
Step 3:	Equilibrium (mol/L)	$0.900 - x$		$0.700 - x$		$0.200 + 2x$

$$K_c = \frac{[NO]^2}{[N_2][O_2]}$$

Substituting the equilibrium concentrations and the value for K_c gives

$$0.100 = \frac{(0.200 + 2x)^2}{(0.900 - x) \times (0.700 - x)}$$

To solve for x we must rearrange the equation so it is in the form of a quadratic equation, $ax^2 + bx + c = 0$, and then apply the quadratic formula,

$$x = \frac{-b \pm \sqrt{b^2 - 4ac}}{2a}$$

To put the equilibrium equation into the quadratic form, we will start by multiplying both sides of the equation by the denominator:

$$0.100 \times (0.900 - x) \times (0.700 - x) = (0.200 + 2x)^2$$
$$0.100 \times (0.630 - 1.600x + x^2) = 0.0400 + 0.800x + 4x^2$$
$$0.0630 - 0.160x + 0.100x^2 = 0.0400 + 0.800x + 4x^2$$

Collecting all terms and bringing them to the same side of the equation gives the quadratic equation

$3.900x^2 + 0.960x - 0.0230 = 0$

where a = 3.900, b = 0.960 and c = -0.0230. We can now substitute these values into the quadratic formula:

$$x = \frac{-0.960 \pm \sqrt{(0.960)^2 - 4 \times 3.900 \times (-0.0230)}}{2 \times 3.900} = 0.0220$$

The calculated value for x was obtained by using only the positive sign before the square root. The negative sign would generate a negative x; this means that the reaction is moving in the opposite direction. We know the reaction moved forward, so the negative root was rejected.

Equilibrium concentrations:

$[N_2] = 0.900 - x = 0.900 - 0.0220 = \underline{0.878\ \text{mol/L}}$

$[O_2] = 0.700 - x = 0.700 - 0.0220 = \underline{0.678\ \text{mol/L}}$

$[NO] = 0.200 + 2x = 0.200 + 2 \times 0.022 = \underline{0.244\ \text{mol/L}}$

Using Approximations to Simplify Calculations

We can sometimes avoid using the quadratic formula by making the approximation that certain initial concentrations are the same as equilibrium concentrations. This approximation can only be made when K_c is very large or very small.

When K_c is Very Small

Example 15.7: K_c for the reaction $N_2(g) + O_2(g) \rightleftharpoons 2NO(g)$ is 4.8×10^{-31} at room temperature. Calculate the concentration of NO that would develop if 0.100 mol of N_2 and 0.200 mol of O_2 come to equilibrium in a 1.00-L flask.

Given: Initial concentrations; $[N_2] = 0.100$ mol/L, $[O_2] = 0.200$ mol/L, $[NO] = 0$. $K_c = 4.8 \times 10^{-31}$

Unknown: Equilibrium concentration of NO

Plan: Since the initial concentration of NO is zero, ($Q_c = 0$), the reaction proceeds to the right. We will again follow the three steps to get expressions for the equilibrium concentrations and then substitute these expressions into K_c.

Calculations:

Step 1: The quantities were put into a 1.00-L flask. Therefore the initial concentrations are 0.100 mol/L for N_2 and 0.200 mol/L for O_2.

Step 2: Let x be the decrease in concentration of N_2 and O_2. The balanced equation tells us that 2 mol of NO are produced for each mole of N_2 used up. Therefore, the increase in concentration of NO is equal to $2x$.

		$N_2(g)$	+	$O_2(g)$	$\rightleftharpoons$	$2NO(g)$
Step 1:	Initial (mol/L)	0.100		0.200		0
Step 2:	Change (mol/L)	$\underline{-x}$		$\underline{-x}$		$\underline{+2x}$
Step 3:	Equilibrium (mol/L)	$0.100 - x$		$0.200 - x$		$2x$
	Approximate (mol/L)	0.100		0.200		$2x$

Because K_c is small, $2x$ will be very small, so that the concentrations of N_2 and O_2 will not change very much from their initial values. Hence, the equilibrium concentrations are approximately $[N_2] = 0.100\ M$ and $[O_2] = 0.200\ M$.

(When x is small compared to the initial reactant concentrations it can be ignored in the expressions for the equilibrium concentrations.)

Substituting the approximate equilibrium concentrations into K_c gives

$$K_c = \frac{[NO]^2}{[N_2][O_2]} \qquad 4.8 \times 10^{-31} = \frac{(2x)^2}{(0.100)(0.200)}$$

Cross multiplying and taking the square root gives $2x = \underline{9.8 \times 10^{-17}\ \text{mol/L}} = [NO]$

Practice Drill: Complete the concentration summaries for reactions (a), (b), and (c). Make approximations where valid for the equilibrium concentrations.

(a)	$2H_2O(g)$	$\rightleftharpoons$	$2H_2(g)$	+	$O_2(g)$	$K_c = 7.3 \times 10^{-18}$ at
Initial (mol/L)	0.150		0		0	1000°C
Change (mol/L)	______		______		$+x$	
Equilibrium (mol/L)	______		______		______	

(b)	$CH_4(g)$	+	$H_2O(g)$	$\rightleftharpoons$	$CO(g)$	+	$3H_2(g)$	$K_c = 5.67$ at 1500°C
Initial (mol/L)	1.00		1.00		0		0.700	
Change (mol/L)	______		______		$+x$		______	
Equilibrium (mol/L)	______		______		______		______	

(c)	$H_2(g)$	+	$Br_2(g)$	$\rightleftharpoons$	$2HBr(g)$	$K_c = 2.0 \times 10^9$ at
Initial (mol/L)	0		0		0.200	25°C
Change (mol/L)	______		$+x$		______	
Equilibrium (mol/L)	______		______		______	

Answers:

(a)	$2H_2O(g)$	$\rightleftharpoons$	$2H_2(g)$	+	$O_2(g)$
Initial	0.150		0		0
Change	$-2x$		$+2x$		$+x$
Equilibrium	0.150		$2x$		x

(b)	$CH_4(g)$	+	$H_2O(g)$	$\rightleftharpoons$	$CO(g)$	+	$3H_2(g)$
Initial	1.00		1.00		0		0.700
Change	$-x$		$-x$		$+x$		$+3x$
Equilibrium	$1.00 - x$		$1.00 - x$		x		$0.700 + 3x$

(c)	$H_2(g)$	+	$Br_2(g)$	$\rightleftharpoons$	$2HBr(g)$
Initial	0		0		0.200
Change	$+x$		$+x$		$-2x$
Equilibrium	x		x		0.200

***The Method of Successive Approximations**

The rule of thumb for neglecting x is that it is less than 5% of the initial concentration of the reactant. When this is not the case or when a more accurate answer is desired the method of successive approximation can be used.

Equilibria Involving Solvents (Section 15.4)

Learning Objectives

1. Write Q_c and K_c expressions and solve equilibrium problems for reactions involving a solvent.

Review

When the solvent is also a reactant or product its concentration is omitted from Q_c and K_c. This is valid because the concentration of solvent remains essentially constant during the reaction.

Example 15.8: Write the K_c expression for the following reaction in aqueous solution.

$CH_3OH(aq) + CH_3COOH(aq) \rightarrow CH_3COOCH_3(aq) + H_2O(l)$

Answer: $K_c = \dfrac{[CH_3COOCH_3]}{[CH_3OH][CH_3COOH]}$

Example 15.9: Acetic acid (CH_3COOH) reacts with water according to the reaction

$CH_3COOH(aq) + H_2O(l) \rightleftharpoons CH_3COO^-(aq) + H_3O^+(aq)$

Find the equilibrium concentration of H_3O^+ when 0.100 mol of acetic acid is initially present in one liter of aqueous solution. $K_c = 1.79 \times 10^{-5}$.

Given: Initial concentrations; $[CH_3COOH] = 0.100$ mol/L, $[CH_3COO^-] = [H_3O^+] = 0$; $K_c = 1.79 \times 10^{-5}$

Unknown: Equilibrium concentration of H_3O^+

Plan: The reaction proceeds in the forward direction since only reactants are initially present. We can set up the concentration summary following the three steps and then substitute the equilibrium expression into K_c.

Calculations:

Step 1: Initial concentrations are given.

Step 2: Let x be the concentration of CH_3COOH that is consumed as the reaction proceeds toward equilibrium; the decrease in concentration of CH_3COOH is $-x$, and $[CH_3COO^-]$ and $[H_3O^+]$ will increase by $+x$.

Concentration summary:		$CH_3COOH(aq) + H_2O(l)$	$\rightleftharpoons$	$CH_3COO^-(aq)$	+ $H_3O^+(aq)$
Step 1:	Initial (mol/L)	0.100		0	0
Step 2:	Change (mol/L)	$-x$		$+x$	$+x$
Step 3:	Equilibrium mol/L	$0.100 - x$		x	x

(The H_2O concentration is omitted from the concentration summary because it is a solvent and therefore not included in the expression for K_c.)

Substituting the above equilibrium concentrations into the K_c expression gives

$$K_c = \frac{[CH_3COO^-][H_3O^+]}{[CH_3COOH]} \qquad 1.79 \times 10^{-5} = \frac{(x)(x)}{(0.100 - x)}$$

Because K_c is small we can make the approximation that the initial concentration of CH_3COOH is equal to its equilibrium concentration; $0.100 - x = 0.100$. Thus,

$1.79 \times 10^{-5} = \dfrac{x^2}{0.100}$; Cross multiplying and taking the square root of both sides gives

$\underline{x = 1.34 \times 10^{-3}\ \text{mol/L}} = [H_3O^+]$

PROBLEM-SOLVING TIP

To find equilibrium concentrations, given K_c and initial concentrations: A summary

1. Find initial concentrations in moles per liter of reactants and products. (The solvent concentration, if any, is omitted.)

2. Determine the direction of the reaction.
   (a) If any substance is not initially present (concentration = 0), the reaction will move to produce it.
   (b) Otherwise calculate the value of Q_c. If $Q_c < K_c$, the the reaction moves forward (to the right). If $Q_c > K_c$, the reaction moves in reverse (to the left).

3. Define the concentration changes in terms of x. Let x be the change in some substance whose coefficient is 1 in the equation. (This choice helps keep things simple.) The change will be $+x$ if the chosen substance is increasing, and $-x$ if the substance is decreasing.

   The other concentration changes are proportional to their coefficients. For example, a substance with a coefficient of 2 will have a change of $2x$, $+2x$ if the substance is increasing, and $-2x$ if the substance is decreasing.

4. Obtain the equilibrium concentrations in terms of x.
   (a) Add the initial concentration values and the concentration changes (1 and 3 above) to obtain the equilibrium concentrations.
   (b) Make simplifying approximations where possible, that is, if x will be small, ignore any x or $2x$, etc. that is added to or subtracted from a much larger concentration. Use the value of K_c to help you decide whether x should be neglected. (The neglected quantity must be less than 5% of the initial concentration.)

5. Substitute the final equilibrium concentrations into the expression for K_c.

Multiply out, collect terms, and solve for x. If you have a quadratic equation, $ax^2 + bx + c = 0$, use the quadratic formula

$$x = \frac{-b \pm \sqrt{b^2 - 4ac}}{2a}.$$

This will give two answers because of the ±, so pick the answer that makes sense. (For example, concentrations cannot be negative.)

Gas Phase Equilibrium: Q_p and K_p (Section 15.5)

Learning Objectives

1. Write Q_p and K_p expressions.
2. Use partial pressures in gas phase equilibrium calculations.
3. Convert K_p to K_c and vice versa.

Review

For gas phase reactions partial pressures are often used instead of concentrations in moles per liter. An equilibrium constant composed of partial pressures is called K_p. Because the partial pressure of a gas is proportional to its concentration in moles per liter equilibrium calculations using partial pressures and K_p are analogous to those using moles per liter and K_c.

Example 15.10: The equilibrium $N_2(g) + O_2(g) \rightleftharpoons 2NO(g)$ has a K_p value of 1.0×10^{-30} at 25°C. Find the equilibrium partial pressure of NO when 0.79 atm of N_2 and 0.21 atm of O_2 are mixed in a closed container at 25°C.

Given: Initial partial pressures, $P_{N_2} = 0.79$ atm, $P_{O_2} = 0.21$ atm. $K_p = 1.0 \times 10^{-30}$.

Unknown: Equilibrium partial pressures of N_2, O_2, and NO.

Plan: Analogous to our previous equilibrium calculations using concentrations, we will follow the same three steps to construct a pressure summary. We will then substitute expressions for equilibrium partial pressures into K_p and solve for x.

Calculations: Pressure Summary

		$N_2(g)$	+	$O_2(g)$	$\rightleftharpoons$	$2NO(g)$
Step 1:	Initial (atm)	0.79		0.21		0
Step 2:	Change (atm)	$-x$		$-x$		$+2x$
Step 3:	Equilibrium (atm)	$0.79 - x$		$0.21 - x$		$2x$
	Approximate (atm)	0.79		0.21		$2x$

(Because K_p is so small very little of the N_2 and O_2 will react; so that $0.79 - x = 0.79$ and $0.21 - x = 0.21$.)

$K_p = \frac{P_{NO}{}^2}{P_{N_2}P_{O_2}}$ $\qquad$ $1.0 \times 10^{-30} = \frac{(2x)^2}{(0.79)(0.21)}$

Cross multiplying and taking the square root of both sides gives:

$2x = 4.1 \times 10^{-16}$

At equilibrium $P_{NO} = 2x = \underline{4.1 \times 10^{-16} \text{ atm}}$

The Relationship Between K_p and K_c

K_p can be derived from K_c and vice versa using the formula

$K_p = K_c\,(RT)^{\Delta n}$

K_p = equilibrium constant using partial pressures in atm
K_c = equilibrium constant using concentrations in mol/L
R = 0.0821 L atm/mol K
T = temperature in Kelvin degrees
Δn = number of moles of gas products in balanced equation
– number of moles of gas reactants in balanced equation

Forms of the Reaction Quotient (Section 15.6)

Learning Objectives

1. Calculate the new equilibrium constant from the original constant when the equation for a reaction is reversed or multiplied by some factor.
2. Calculate the equilibrium constant for the sum of two or more reactions whose individual equilibrium constants are known.

Review

There are three rules you should know about relating the equilibrium constant to the form of the equation.

(1) When you reverse an equation the equilibrium constant becomes its reciprocal.
(2) When you multiply an equation by a factor the equilibrium constant is raised to a power equal to that factor.
(3) The equilibrium constant for the sum of two or more equations is the product of two or more equilibrium constants.

Example 15.11: Given the following reactions and their equilibrium constants

(i) $2SO_3(g) \rightleftharpoons 2SO_2(g) + O_2(g)$ $\qquad K_c = 2.5 \times 10^{-3}$

(ii) $NO_2(g) \rightleftharpoons NO(g) + 1/2O_2(g)$ $\qquad K_c = 0.45$

Calculate equilibrium constants for

(a) $SO_2(g) + 1/2O_2(g) \rightleftharpoons SO_3(g)$ and (b) $SO_2(g) + NO_2(g) \rightleftharpoons SO_3(g) + NO(g)$

Answer:

(a) Equation (a) is obtained by reversing equation (i) and multipling all coefficients by 1/2. Therefore K_c for equation (a) is the reciprocal of K_c for equation (i) raised to the 1/2 power (the square root).

$$K_c = \left(\frac{1}{2.5 \times 10^{-3}}\right)^{1/2} = \underline{20}$$

(b) Equation (b) is the sum of equation (a) and equation (ii). Therefore the K_c for equation (b) is the product of K_c for equation (a) and K_c for equation (ii).

$K_c = 20 \times 0.45 = \underline{9.0}$

Practice Drill on the Form of the Equilibrium Constant: Given the equation

$2C(s) + O_2(g) \rightleftharpoons 2CO(g)$ and $K_p = 1.28 \times 10^{48}$

find K_c when (a) the equation is reversed (b) the coefficients are multiplied by a factor of 1/2 and (c) the equation

$2CO_2(g) \rightleftharpoons 2C(s) + 2O_2(g)$ $\qquad K_p = 6.01 \times 10^{-139}$

is added to it.

Answers: (a) 7.81×10^{-49} (b) 1.13×10^{24} (c) 7.69×10^{-91}

Heterogeneous Equilibria (Section 15.7)

Learning Objectives

1. Write reaction quotient and equilibrium constant expressions for heterogeneous equilibria.
2. Perform equilibrium calculations involving heterogeneous equilibria.

Review

Heterogeneous equilibria involve opposing reactions that occur at the surface of a solid or liquid. *Pure solids and pure liquids are omitted from K and Q expressions for heterogeneous reactions.*

Example 15.12: Write K_c and K_p expressions for the reaction $4CuO(s) \rightleftharpoons 2Cu_2O(s) + O_2(g)$

Answer: $K_c = [O_2]$; $\quad K_p = P_{O_2}$

Practice Drill on Heterogeneous Equilibria: Write K_c and K_p expressions for:

(a) $C(s) + S_2(g) \rightleftharpoons CS_2(g)$
(b) $SnO_2(s) + 2H_2(g) \rightleftharpoons Sn(s) + 2H_2O(g)$

Answers: (a) $K_c = [CS_2]/[S_2]$ $\ K_p = P_{CS_2}/P_{S_2}$ (b) $K_c = [H_2O]^2/[H_2]^2$ $\ K_p = P^2_{H_2O}/P^2_{H_2}$

Note: Concentration summaries on heterogeneous equilibria should omit pure solids and liquids.

Le Châtelier's Principle (Section 15.8)

Learning Objectives

1. State Le Châtelier's principle and use it to predict the effect of changes in concentration, volume, pressure, and temperature on an equilibrium system.
2. Use the sign of $\Delta H°$ to predict the effect of temperature changes on the magnitude of K.
3. Explain why catalysts decrease the time required to reach equilibrium.
4.* Explain why K values are not constant at high concentrations and pressures.
5.* Calculate the value of K at one temperature from the measured value at another temperature.

Review

A change in conditions or concentrations may push a system out of equilibrium. The system will then move to a new equilibrium state, which can be predicted by Le Châtelier's principle: When a system in equilibrium is disturbed, the system adjusts to a new equilibrium in a way that partially counteracts the disturbance. We will consider the effect of changes in concentration, pressure, temperature, and the addition of nonreacting substances and catalysts.

A concentration change favors the reaction that partially reverses the concentration change: If the concentration of a reactant is increased, the system will move to the right to decrease reactant concentrations. At the same time product concentrations increase.

If the concentration of a reactant is decreased, the system will move to the left causing an increase in reactant concentrations and a decrease in product concentrations.

Practice: Predict the effects of (a) decreasing and (b) increasing the concentration of products on an equilibrium system.

A pressure change favors the reaction that reverses the pressure change: The compression of an equilibrium mixture favors formation of substances that take up less volume, and thus relieve the pressure; expansion of the mixture favors formation of substances that take up more volume and thus restore some of the pressure.

A temperature increase favors the reaction that absorbs heat; a temperature decrease favors the reaction that produces heat. In other words, increasing the temperature of an equilibrium mixture favors the endothermic reaction (ΔH is positive); decreasing the temperature of an equilibrium mixture favors the exothermic reaction (ΔH is negative).

Nonreacting substances do not affect the equilibrium. Catalysts also have no effect on equilibrium because they increase the forward and reverse reaction rates by the same factor.

Example 15.13: Given the system $2CO_2(g) \rightleftharpoons 2CO(g) + O_2(g)$, $\Delta H° = 566$ kJ, predict the effect of each of the following changes on the concentration of CO_2.
(a) increase in P_{CO}
(b) compression of the equilibrium mixture
(c) decrease in temperature
(d) addition of argon, an inert gas, to the equilibrium mixture

Answer:

(a) An increase in P_{CO} causes an increase in [CO]. As a result the reaction will proceed in the reverse direction to consume some of the added CO. A new equilibrium mixture containing more CO_2 and less O_2 will form.

(b) The reaction will move to relieve the pressure. Gas pressures are proportional to the number of moles; 2 mol of CO_2 exert less pressure than 2 mol of CO and 1 mol of O_2. The reaction will therefore move in the reverse direction. More CO_2 will therefore form.

(c) The forward reaction is endothermic, therefore the reverse reaction is exothermic. Decreasing the temperature favors the exothermic reaction. The equilibrium mixture at the new temperature will contain more CO_2 and less CO and O_2.

(d) Argon does not react with the substances in the equilibrium mixture so the partial pressures and concentrations of CO_2, CO, and O_2 will not change.

Practice Drill on Le Châtelier's Principle: Predict the effect of each of the following changes on the equilibrium concentration of NOCl in the system

$2NO(g) + Cl_2(g) \rightleftharpoons 2NOCl(g) \qquad \Delta H° = -18.42\ kcal$

(a) removal of Cl_2
(b) increase in temperature
(c) increase in volume
(d) addition of NO
(e) addition of a metal catalyst

Answers: (d) increases [NOCl] (a), (b), (c) decreases [NOCl] (e) no change

SELF-TEST

Answer Questions 1 and 2 as true or false.

1. At equilibrium the concentrations of reactants and products are constant.

2. At equilibrium the rates of the forward and reverse reactions are zero.

3. Ammonia gas is a valuable fertilizer that is sometimes pumped directly into the soil. It is synthesized by the reaction $N_2(g) + 3H_2(g) \rightleftharpoons 2NH_3(g)$. When 1.97 mol of H_2 and 1.36 mol of N_2 are placed in a 1.00-L flask at 500°C, the equilibrium concentration of NH_3 is 0.412 mol/L. Calculate
   (a) the equilibrium concentrations of N_2 and H_2
   (b) K_c

4. Consider the equilibrium $H_2(g) + I_2(g) \rightleftharpoons 2HI(g)$. K_c is 49.5 at 440°C. Find the equilibrium concentrations of the three gases when the initial quantities in a 20.0-L flask were 0.400 mol of H_2 and 0.400 mol of I_2.

5. Ascorbic acid ($H_2C_6H_6O_6$) reacts with H_2O according to the equation

   $H_2C_6H_6O_6(aq) + H_2O(l) \rightleftharpoons HC_6H_6O_6^-(aq) + H_3O^+(aq)$

   K_c is 7.9×10^{-5} at 25°C. Calculate the equilibrium concentration of H_3O^+ in a solution made by dissolving 17.6 g of ascorbic acid in 1.00 L of solution.

6. K_p is 2.9×10^{61} at 25°C for $2HCl(g) + F_2(g) \rightleftharpoons 2HF(g) + Cl_2(g)$. Compute the equilibrium concentrations of all four gases in a mixture that initially contained HF at 0.020 atm and Cl_2 at 0.010 atm.

7. K_p at 27°C is 1×10^{-11} for $SnO_2(s) + 2H_2(g) \rightleftharpoons Sn(s) + 2H_2O(g)$
   (a) Calculate the equilibrium concentrations of H_2O and H_2 when H_2 at 1.00 atm is mixed with $SnO_2(s)$ in a reaction flask and allowed to come to equilibrium.
   (b) Find K_c.

8. The production of nitric oxide (NO) in the stratosphere by supersonic jets is thought to contribute to the destruction of the ozone layer:

   $NO(g) + O_3(g) \rightleftharpoons NO_2(g) + O_2(g)$

   The reaction is exothermic. Predict the effect of each of the following on the concentration of O_3:
   (a) More supersonic jets are flying.
   (b) Sunlight causes NO_2 to decompose.
   (c) The temperature of the stratosphere decreases at night.
   (d) Some nonreacting gaseous pollutants are added to the stratosphere.

9. Inhaled carbon monoxide reduces the blood's ability to transport oxygen by forming a complex with hemoglobin.

   $CO(g) + Hem \cdot O_2(aq) \rightleftharpoons O_2(g) + Hem \cdot CO(aq) \quad K_c = 210$

   Find:
   (a) the concentration of CO when the concentration of Hem • CO is 2.0% of the concentration of Hem • O_2. Use $[O_2] = 8.2 \times 10^{-3}$ mol/L (the concentration of O_2 in air). (Mental activity is significantly impaired at this concentration of CO. In many urban areas CO concentrations are significantly higher.)
   (b) the concentration of CO in cigarette smoke that causes the concentration of Hem • CO to be 5.0% of the concentration of Hem • O_2. Why is it unwise to smoke when doing chemistry problems?

ANSWERS TO SELF-TEST QUESTIONS

1. T

2. F

3. (a) $[N_2] = 1.36 - 0.206 = \underline{1.15 \text{ mol/L}}$; $[H_2] = 1.97 - 3 \times 0.206 = \underline{1.35 \text{ mol/L}}$
   (b) $K_c = (0.412)^2/(1.15)(1.35)^3 = \underline{6.00 \times 10^{-2}}$

4. $K_c = (2x)^2/(0.0200 - x)^2$; $x = 0.0156$

   $[HI] = \underline{0.0312 \text{ mol/L}}$; $[H_2] = [I_2] = \underline{0.0044 \text{ mol/L}}$

5. Using approximations $K_c = x^2/0.100$; $x = \underline{2.8 \times 10^{-3} \text{ mol/L}} = [H_3O^+]$

6. Using approximations K_p for the reverse reaction is $3.4 \times 10^{-62} = (2x)^2(x)/(0.020)^2(0.010)$;
   $x = 2.6 \times 10^{-23}$; $[F_2] = \underline{3.2 \times 10^{-23} \text{ atm}}$; $[HCl] = \underline{5.2 \times 10^{-23} \text{ atm}}$; $[HF] = \underline{0.020 \text{ atm}}$; $[Cl_2] = \underline{0.010 \text{ atm}}$

7. (a) $1 \times 10^{-11} = x^2/(1.00 - x)^2$; Using the approximation $1.00 - x = 1.00$ $x = 3 \times 10^{-6}$
$[H_2O] = \underline{3 \times 10^{-6}\text{ atm}}$; $[H_2] = \underline{1.00\text{ atm}}$
(b) $K_c = K_p = \underline{1 \times 10^{-11}}$

8. (a) decreases
(b) decreases
(c) decreases
(d) no effect

9. (a) $210 = \dfrac{[O_2][\text{Hem}\cdot CO]}{[CO][\text{Hem}\cdot O_2]} = \dfrac{(8.2 \times 10^{-3})0.020}{[CO]}$ $[CO] = \underline{7.8 \times 10^{-7}\text{ mol/L}}$

(b) 2.0×10^{-6} mol/L

Aside from the health consequences of cigarette smoking, the carbon monoxide may reduce your performance.

CHAPTER 16
ACIDS AND BASES

SELF-ASSESSMENT

Brønsted-Lowry Acids and Bases (Section 16.1)

1. Which of the following species are Brønsted acids? Which are Brønsted bases?:

   H_3O^+, H_2SO_4, OH^-, HCN, CH_3COOH, CH_3NH_2.

2. Write net ionic equations for the reactions of $H_2PO_4^-$ (an amphiprotic ion) with an aqueous solution of (a) HCl and (b) KOH

Conjugate Acid-Base Pairs (Section 16.2)

3. Give the conjugate base for each acid:
   (a) HI
   (b) HSO_4^-
   (c) H_2CO_3
   (d) $Zn(H_2O)_4^{2+}$

4. Give the conjugate acid for each base:
   (a) NH_3
   (b) O^{2-}
   (c) H_2O
   (d) PO_4^{3-}

5. Identify the conjugate acid-base pairs in each of the following reactions:
   (a) $HCl(aq) + HPO_4^{2-}(aq) \rightarrow H_2PO_4^-(aq) + Cl^-(aq)$
   (b) $CH_3O^-(aq) + H_2O(l) \rightarrow CH_3OH(aq) + OH^-(aq)$
   (c) $HNO_3(aq) + CH_3OH(aq) \rightarrow CH_3OH_2^+(aq) + NO_3^-(aq)$

6. The following list ranks acids in order of increasing acid strength. Write their conjugate bases in order of decreasing base strength.

   $HClO_4 > H_3O^+ > HNO_2 > NH_4^+$

Other Proton Transfer Reactions (Section 16.3)

7. Complete the following equations. Use single or double arrows as appropriate:
   (a) $HClO_4(aq)$ + $H_2O(l) \rightarrow$
   strong acid
   (b) $CH_3NH_2(aq)$ + $H_2O(l) \rightarrow$
   weak base
   (c) $HNO_3(aq)$ + $CH_3NH_2(aq) \rightarrow$
   strong acid weak base
   (d) $Ca(OH)_2(aq)$ + $2HCOOH(aq) \rightarrow$
   strong base weak acid
   (e) $NH_3(l) + NH_3(l) \rightarrow$
   (f) $H_2C_6H_6O_6(aq)$ + $H_2O(l) \rightarrow$
   ascorbic acid
   weak acid

Acidic, Basic, and Neutral Solutions: pH (Section 16.4)

8. The normal pH range of blood is 7.35 to 7.45. The pH of a blood sample from a patient suffering from acidosis is 7.20. Find $[H^+]$, pOH, and $[OH^-]$ of this blood sample.

9. Rank the following solutions in order of increasing acidity:
   (a) Sample of acid rain; pH = 4.0
   (b) Sample of human saliva; $[H^+] = 1.5 \times 10^{-5}$ mol/L
   (c) Blood sample; $[OH^-] = 2.5 \times 10^{-7}$ mol/L
   (d) Sample of stomach acid (HCl); pOH = 11.20

10. Fill in the chart for solutions (a), (b), and (c).

	$[H^+]$	$[OH^-]$	pH	pOH
(a)	1.5×10^{-5}	______	______	___
(b)	______	2.5×10^{-7}	______	___
(c)	______	______	______	11.20

11. 50 mL of a 0.10 *M* HCl solution is added to 50 mL of a 0.20 *M* HNO_3 solution. (a) Find the pH of each solution before they are mixed. (b) Find the pH of the final solution. Assume volumes are additive.

Molecular Structure and Acid Strength (Section 16.5)

12. Predict based on structure which is the stronger acid from each of the following pairs:
   (a) $H_2PO_4^-$ or HPO_4^{2-}
   (b) H_2S or H_2Se
   (c) H_2SO_4 or H_2SO_3
   (d) NH_3 or OH^-
   (e) $Pb(H_2O)_6^{2+}$ or $Fe(H_2O)_6^{3+}$

Lewis Acids and Bases (Section 16.6)

13. Indicate the Lewis acid, Lewis base, electrophile, and nucleophile in each of the following reactions:
   (a) $Ag^+(aq) + 2NH_3(aq) \rightarrow Ag(NH_3)_2^+(aq)$
   (b) $H_3O^+(aq) + OH^-(aq) \rightarrow 2H_2O(l)$
   (c) $Pb^{2+}(aq) + 2Cl^-(aq) \rightarrow PbCl_2(s)$
   (d) $CH_3O^-(aq) + CH_3COOH(aq) \rightarrow CH_3OH(aq) + CH_3COO^-(aq)$

14. Which of the reactions above are also Brønsted-Lowry proton transfers?

ANSWERS TO SELF-ASSESSMENT PROBLEMS

If you missed an answer, be sure to study the relevant section in the textbook and study guide.

1. Brønsted Acids: H_3O^+, H_2SO_4, HCN, and CH_3COOH
   Brønsted Bases: CH_3NH_2, OH^-

2. (a) $H_2PO_4^-(aq) + H^+(aq) \rightarrow H_3PO_4(aq)$ or $H_2PO_4^-(aq) + H_3O^+(aq) \rightarrow H_3PO_4(aq) + H_2O(l)$
   (b) $H_2PO_4^-(aq) + OH^-(aq) \rightarrow HPO_4^{2-}(aq) + H_2O(l)$

3. (a) I^- (b) SO_4^{2-} (c) HCO_3^- (d) $Zn(H_2O)_3(OH)^+$

4. (a) NH_4^+ (b) OH^- (c) H_3O^+ (d) HPO_4^{2-}

5. (a) acid: HCl; base: Cl^-; acid: $H_2PO_4^-$; base: HPO_4^{2-}
   (b) acid: H_2O; base: OH^-; acid: CH_3OH; base: CH_3O^-
   (c) acid: HNO_3; base: NO_3^-; acid: $CH_3OH_2^+$; base: CH_3OH

6. $NH_3 > NO_2^- > H_2O > ClO_4^-$

7. (a) $HClO_4(aq) + H_2O(l) \rightarrow H_3O^+(aq) + ClO_4^-(aq)$
   (b) $CH_3NH_2(aq) + H_2O(l) \rightleftharpoons OH^-(aq) + CH_3NH_3^+(aq)$
   (c) $HNO_3(aq) + CH_3NH_2(aq) \rightarrow CH_3NH_3^+(aq) + NO_3^-(aq)$
   (d) $Ca(OH)_2(aq) + 2HCOOH(aq) \rightarrow 2H_2O(l) + Ca(HCOO)_2(aq)$
   (e) $NH_3(l) + NH_3(l) \rightleftharpoons NH_2^-(l) + NH_4^+(l)$
   (f) $H_2C_6H_6O_6(aq) + H_2O(l) \rightleftharpoons HC_6H_6O_6^-(aq) + H_3O^+(aq)$

8. $7.20 = -\log[H^+]$; $[H^+] = \text{antilog}(-7.20) = 6.3 \times 10^{-8}$
   $pOH = 14 - 7.20 = \underline{6.80}$; $[OH^-] = \text{antilog}(-6.80) = \underline{1.6 \times 10^{-7}}$

9. (d) < (c) < (b) < (a)

10. (a) $[OH^-] = 6.7 \times 10^{-10}$; pH = 4.82; pOH = 9.18
    (b) $[H^+] = 4.0 \times 10^{-8}$; pH = 7.40; pOH = 6.60
    (c) $[H^+] = 1.6 \times 10^{-3}$; $[OH^-] = 6.3 \times 10^{-12}$; pH = 2.80

11. (a) For HCl: pH = -log(0.10) = 1.00 For HNO_3: pH = -log(0.20) = 0.70
    (b) (0.050 L × 0.20 mol/L) + (0.050 L × 0.10 mol/L) = 0.015 mol H^+
    $[H^+]$ = 0.015 mol/0.100 L = 0.15 *M* pH = -log(0.15) = <u>0.82</u>

12. (a) $H_2PO_4^-$ (b) H_2Se (c) H_2SO_4 (d) NH_3 (e) $Fe(H_2O)_6^{3+}$

13. (a) Lewis acid (electrophile): Ag^+ Lewis base (nucleophile): NH_3
    (b) Lewis acid (electrophile): H_3O^+ Lewis base (nucleophile): OH^-
    (c) Lewis acid (electrophile): Pb^{2+} Lewis base (nucleophile): Cl^-
    (d) Lewis acid (electrophile): CH_3COOH Lewis base (nucleophile): CH_3O^-

14. (b) and (d)

REVIEW

Brønsted-Lowry Acids and Bases (Section 16.1)

Learning Objectives

1. Distinguish between the Arrhenius and Brønsted-Lowry definitions of acid, base, and neutralization.
2. Give examples of Brønsted-Lowry acids and bases.
3. Use hydronium ion notation in chemical equations.
4. Give examples of amphiprotic molecules and ions, and write equations for their reactions with acids and bases.
5. Write equations for the self-ionization of amphiprotic liquids.

Review

Arrhenius defined an acid to be any substance that produces $H^+(aq)$ ions in water, a base to be any substance that produces $OH^-(aq)$ ions, and neutralization to be the reaction of $H^+(aq)$ with $OH^-(aq)$ to form water.

Brønsted and Lowry defined an acid as a proton donor and a base as a proton acceptor. Every Brønsted base has one or more lone electron pairs. Some examples of Brønsted acids (with ionizable hydrogens underlined) are $\underline{H}Br$, $\underline{H}CN$, $CH_3COO\underline{H}$, and \underline{H}_2SO_4.

Examples of Brønsted bases are $:\ddot{O}-H^-$, $H-\overset{\displaystyle H}{\underset{\displaystyle H}{N}:}$, and $:C\equiv N:$

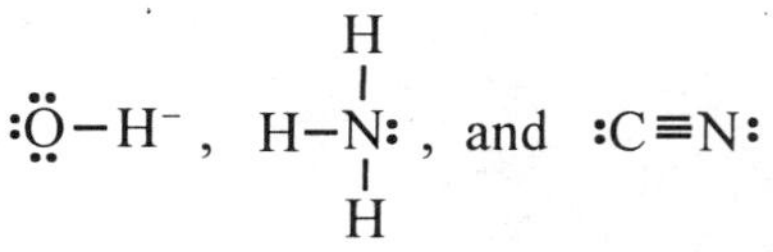

Neutralization is the transfer of a proton from a Brønsted acid to a Brønsted base:

$$\textcircled{H}-Br(aq) + :C\equiv N^- \rightarrow HCN(aq) + Br^-(aq)$$
acid base

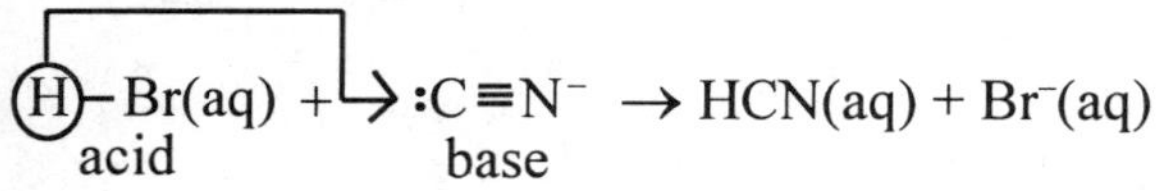

An amphiprotic molecule or ion can donate protons to a base and accept protons from an acid. Dihydrogen phosphate $(H_2PO_4^-)$, is an example:

$$H_2PO_4^-(aq) + OH^-(aq) \rightarrow HPO_4^{2-}(aq) + H_2O(l)$$
acid base

$$H_2PO_4^-(aq) + HCl(aq) \rightarrow H_3PO_4(aq) + Cl^-(aq)$$
base

Amphiprotic liquids undergo autoionization reactions. Water, methanol (CH_3OH), and ammonia are examples. The autoionization reaction for water is

$$H-\ddot{O}: \quad + \quad H-\ddot{O}: \quad \rightleftharpoons \quad H_3O^+(aq) + OH^-(aq)$$

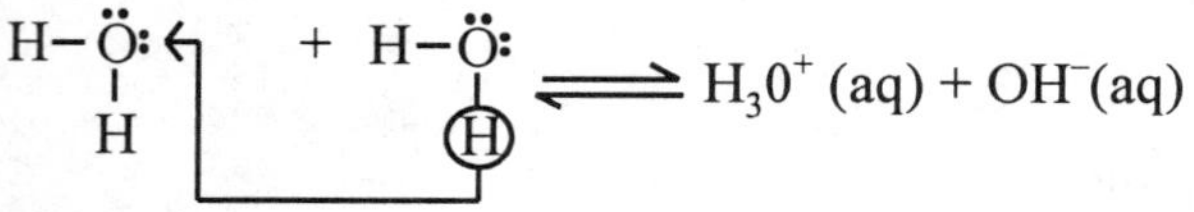

Practice: Write an equation for the autoionization reaction of CH_3OH.

Conjugate Acid-Base Pairs (Section 16.2)

Learning Objectives

1. Given the formula of an acid or a base, write the formula of its conjugate acid or base.
2. Identify the conjugate acid-base pairs in a proton transfer reaction.
3. Explain how acid strengths are compared.
4. Relate the strength of an acid to the strength of its conjugate base.
5. Identify the strongest acid and base that can exist in aqueous solution.
6.* Identify the strongest acid and base that can exist in a given amphiprotic solvent.
7.* Explain how strong acids can be differentiated from each other.

Review

A conjugate acid-base pair consists of an acid and a base whose formulas differ by a single proton.

Some examples of conjugate pairs include

Acid	Base
HCN	CN^-
H_3PO_4	$H_2PO_4^-$
HCO_3^-	CO_3^{2-}

To find the conjugate acid of a given base just add a proton to the formula; to find the conjugate base of a given acid remove an ionizable proton from the formula. Keep in mind that a proton carries a positive charge.

Practice on Conjugate Acid-Base Pairs: Fill in the blanks:

Acid	Base
H_2O	OH^-
NH_3	NH_2^-
HBr	Br^-
H_3O^+	H_2O

Answers: H_2O, NH_2^-, Br^-, H_3O^+

Numbers 1 and 2 are used below to identify the two conjugate pairs in the proton transfer reaction.

$HClO_4(aq)$	+	$H_2O(l)$	→	$ClO_4^-(aq)$	+	$H_3O(aq)$
acid 1		base 2		base 1		acid 2

$HClO_4$ donates a proton to form its conjugate base, ClO_4^-. H_2O accepts a proton to form its conjugate acid, H_3O^+.

Acid and Base Strengths

The strength of an acid is its ability to relinquish protons. The strength of a base is its ability to accept protons. The relative strengths of acids are compared by testing them with the same base. This base is usually water. Stronger acids produce more hydronium ions per mole than weaker acids. Strong acids have weak conjugate bases and vice versa.

Example 16.1: The following are listed in order of increasing acid strength: $NH_3 < H_2O < HCOOH < HCl$

Rank their conjugate bases in order of increasing base strength.

Answer: The base strength order is the reverse of the acid strength order: $Cl^- < HCOO^- < OH^- < NH_2^-$

Acids that are stronger than H_3O^+ are considered to be strong acids; bases stronger than OH^- are strong bases. Strong acids and bases are completely ionized in aqueous solution. Therefore, all strong acids appear to be equally strong in water; water is said to have a leveling effect on them. Water also exerts a leveling effect on bases stronger than OH^-.

Commit the following list to memory:

Strong acids

Perchloric	$HClO_4$	Hydrochloric	HCl
Nitric	HNO_3	Sulfuric (first proton only)	H_2SO_4

Strong bases

The soluble metal hydroxides; NaOH, KOH, $Ca(OH)_2$, etc.

A leveling effect can be seen in other amphiprotic solvents; the strongest acid that exists in a given solvent is the conjugate acid of the solvent and the strongest base is the conjugate base of the solvent.

A solvent that shows up differences in acid and base strength is called a differentiating solvent. Water is a differentiating solvent for acids weaker than H_3O^+ and bases weaker than OH^- (aq).

Other Proton Transfer Reactions (Section 16.3)

Learning Objectives

1. Write equations for proton transfer reactions.
2. Explain why the enthalpy of neutralization is the same for all strong acid-strong base neutralizations and why it varies for neutralizations involving weak acids and bases.
3. Write the equation for each step in the ionization of a polyprotic acid.

Review

The different types of proton transfer reactions covered in Section 16.3 of the text are

Weak acids and weak bases in water: Acids that are weaker than H_3O^+ and bases that are weaker than OH^- are only partially ionized in aqueous solution:

weak acid: $CH_3COOH(aq) + H_2O(l) \rightleftharpoons CH_3COO^-(aq) + H_3O^+(aq)$

weak base: $CH_3NH_2(aq) + H_2O(l) \rightleftharpoons CH_3NH_3^+(aq) + OH^-(aq)$

Strong acids with strong bases in water: Strong acids and strong bases are completely converted to H_3O^+ and OH^-, respectively. Thus the neutralization reaction is

$H_3O^+(aq) + OH^-(aq) \rightarrow 2H_2O(l)$

[The enthalpy of neutralization for the reaction between a strong acid and base is always -55.8 kJ/mol of $H_3O^+(aq)$ or $OH^-(aq)$.]

Strong acid and weak base; strong base and weak acid: A strong acid will donate its proton to a weak base; a strong base will remove a proton from a weak acid. The reactions go virtually to completion.

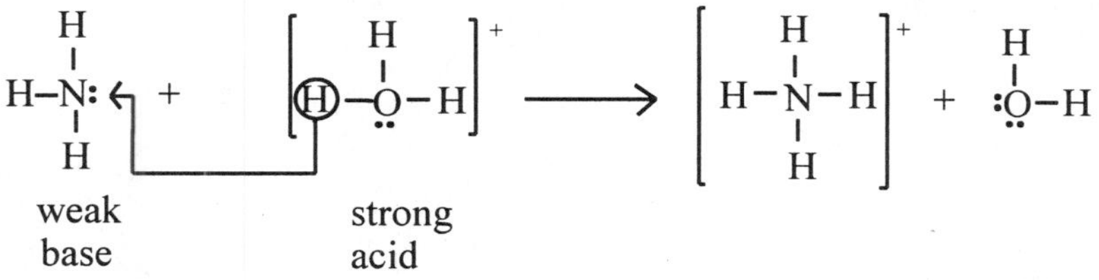

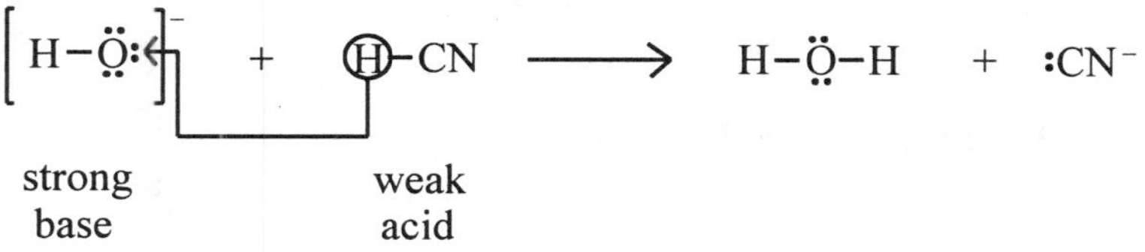

(The enthalpies of neutralizations of weak acids and weak bases are all different.)

Hydrated cations with water: Hydrated cations are acids. Some are strong enough to transfer protons to water:

$Al(H_2O)_6^{3+}(aq) + H_2O(l) \rightleftharpoons Al(H_2O)_5(OH)^{2+}(aq) + H_3O^+(aq)$

Proton transfer reactions: The gas phase reaction between $NH_3(g)$ and $HCl(g)$ is an example:

$NH_3(g) + HCl(g) \rightarrow NH_4Cl(s)$

Polyprotic acids ionization: A polyprotic acid has more than one acidic hydrogen atom. Examples are H_2SO_4, H_2CO_3, H_3PO_4, and hydrated cations such as $Fe(H_2O)_6^{3+}$. The ionization of a polyprotic acid occurs in steps. The equations for the ionization of H_3PO_4 are:

$H_3PO_4(aq) + H_2O(l) \rightleftharpoons H_2PO_4^-(aq) + H_3O^+(aq)$
$H_2PO_4^-(aq) + H_2O(l) \rightleftharpoons HPO_4^{2-}(aq) + H_3O^+(aq)$
$HPO_4^{2-}(aq) + H_2O(l) \rightleftharpoons PO_4^{3-}(aq) + H_3O^+(aq)$

Acidic, Basic, and Neutral Solutions (Section 16.4)

Learning Objectives

1. Given $[H^+]$, $[OH^-]$, pH, or pOH, calculate the other three quantities.

Review

The autoionization of water may be written as $H_2O(l) \rightleftharpoons H^+(aq) + OH^-(aq)$ and the value of its equilibrium constant K_w is $K_w = [H^+][OH^-] = 1.00 \times 10^{-14}$ at 25°C.

The K_w value leads to the following definitions of acidic, basic, and neutral solutions:

Neutral solution (including pure water): $[H^+] = [OH^-] = 1.00 \times 10^{-7}$

Acidic solution: $[H^+] > 1.00 \times 10^{-7}$; $[OH^-] < 1.00 \times 10^{-7}$

Basic solution: $[H^+] < 1.00 \times 10^{-7}$; $[OH^-] > 1.00 \times 10^{-7}$

Example 16.2: Find $[H^+]$ and $[OH^-]$ at 25°C in 0.50 *M* HNO_3.

Answer: HNO_3, a strong acid, ionizes completely so that $[H^+] = \underline{0.50\ M}$.

The only source of OH^- is the autoionization of water. Hence,

$$[OH^-] = \frac{K_w}{[H^+]} = \frac{1.00 \times 10^{-14}}{0.50} = \underline{2.0 \times 10^{-14}\ M}$$

pH: The definition of pH is

$pH = \log[H^+]$

The lower the pH, the more acidic the solution; the higher the pH, the more basic the solution.

pOH: The definition of pOH is

$pOH = -\log[OH^-]$

The lower the pOH, the more basic the solution; the higher the pOH, the more acidic the solution.
In summary:

	$[H^+]$	$[OH^-]$	pH	pOH
Acidic	$> 1.00 \times 10^{-7}$	$< 1.00 \times 10^{-7}$	< 7	> 7
Basic	$< 1.00 \times 10^{-7}$	$> 1.00 \times 10^{-7}$	> 7	< 7
Neutral	1.00×10^{-7}	1.00×10^{-7}	7	7

All solutions: $[H^+][OH^-] = 1.00 \times 10^{-14}$; $pH + pOH = 14.00$

Example 16.3: Calculate the pH of (a) 0.50 *M* HNO_3 and (b) 0.50 *M* KOH.

(a)

Given: $[H^+] = 0.50\ M$

Unknown: pH

Plan: $pH = -\log[H^+] = -\log(0.50)$. Do this operation on an electronic calculator. Enter 0.50 and then press the log key. Multiply the result by -1.
pH = 0.30

(b)

Given: $[OH^-] = 0.50\ M$

Unknown: pH

Plan: Find $[H^+]$ and then take -log to find pH.

Calculation: $$[H^+] = \frac{K_w}{[OH^-]} = \frac{1.00 \times 10^{-14}}{5.0 \times 10^{-1}} = 2.0 \times 10^{-14}\ M$$

Hence, $pH = -\log[H^+] = -\log(2.0 \times 10^{-14})$

To find the log of a product of two numbers find the log of each number and then add:

$\log(2.0 \times 10^{-14}) = \log 2.0 + \log(10^{-14}) = 0.30 + (-14.00) = -13.70$

pH = -(-13.70) = <u>13.70</u>

Alternate procedure: Find pOH and then subtract pOH from 14.00 to get pH;
$[OH^-] = 0.50\ M$; pOH = -log(0.50) = 0.30 pH = 14.00 - pOH = 14.00 - 0.30 = <u>13.70</u>

Logarithms, Significant Figures, and Calculators

Only the digits after the decimal place are significant in a logarithm. The pH is a logarithm. Therefore the number of <u>decimal</u> places in the pH value must equal the number of significant figures in the $[H^+]$ value. If, for example,

$[H^+] = 5.0 \times 10^{-2}$ M then

$\log(5.0 \times 10^{-2})$	=	-1.30 and	pH=	1.30
↑		↑		↑
2 significant figures		2 decimal places		2 decimal places

<u>Example 16.4</u>: The normal pH of stomach acid ranges from 1.00 to 3.00. The pH of a specific sample of stomach acid is measured as 1.80. Find $[H^+]$.

Given: pH = 1.80

Unknown: $[H^+]$

Plan: $pH = -\log[H^+]$
$[H^+]$ = antilog(-pH) = antilog(-1.80)
Take the antilog of -1.80

Calculations: Use your electronic calculator to find the antilog. Look for an inverse key (INV), Y^x key, or a 10^x key.

To use an inverse key:
Enter -1.80
Press INV
Press LOG

If you have a Y^x key:
Enter 10
Press Y^x
Enter -1.80
Press =

If you have a 10^x key:
Enter -1.8
Press 10^x
Press =

$[H^+] = 0.0158\ M$; Our answer can only have two significant figures (two decimal places in log) so that $[H^+] = \underline{1.6 \times 10^{-2}\ M}$

PROBLEM-SOLVING TIP

pH Calculations

Given $[H^+]$, $[OH^-]$, pH, or pOH, you can find the other three quantities by using one or more of the following relationships:

$[H^+][OH^-] = 10^{-14}$
$pH = -\log[H^+]$
$pOH = -\log\ [OH^-]$
$pH + pOH = 14.00$

Practice Drill on pH Calculations: Fill in the blanks for four different aqueous solutions.

	$[H^+]$	$[OH^-]$	pH	pOH
(a)	0.010 M	1×10^{-12}.	______	______
(b)	______	______	10.0	______
(c)	______	1.0 *M*	______	______
(d)	______	______	______	1.5

Answers: (a) $[OH^-] = 1.0 \times 10^{-12}\ M$; pH = 2.00; pOH = 12.00
(b) $[H^+] = 1 \times 10^{-10}\ M$; $[OH] = 1 \times 10^{-4}\ M$; pOH = 4.0
(c) $[H^+] = 1.0 \times 10^{-14}\ M$; pH = 14.00; pOH = 0.00
(d) $[H^+] = 3 \times 10^{-13}\ M$; pH = 12.5; $[OH^-] = 3 \times 10^{-2}\ M$

Rank the above solutions in order of increasing acidity. Answers: (c) < (d) < (b) < (a)

Molecular Structure and Acid Strength (Section 16.5)

Learning Objectives

1. Predict the effect of charge on acid and base strengths.
2. Predict which of two oxo acids is the stronger acid.
3. Predict which of two oxo anions is the stronger base.
4. Predict which of two binary acids is the stronger acid.
5. Predict which of two hydrated cations is the stronger acid.

Review

A Lewis structure often helps us to estimate the strength of an acid or base. The following generalizations are useful.

(1) For species that differ in the number of protons, acid strength increases with increasing positive charge; base strength increases with increasing negative charge.

For example, H_3O^+ is a stronger acid than H_2O; HPO_4^{2-} is a stronger base than $H_2PO_4^-$.

(2) The strength of an oxo acid increases with the polarity of the O–H bond. The polarity, in turn, increases with
 (a) the electronegativity of the central atom. (HOCl is a stronger acid than HOI.)
 (b) the number of oxygen atoms bonded to the central atom. (HNO_3 is a stronger acid than HNO_2.)
 (c) the presence of other electronegative atoms in the molecule. ($CH_2ClCOOH$ is a stronger acid than CH_3COOH.)

(3) A binary acid contains hydrogen and only one other element. The strength of binary acids increases from left to right across a period and from top to bottom in a periodic group. The increase across a period is due to increasing bond polarity. The increase down a group is caused by increasing anion size. (H_2O is a stronger acid than NH_3; H_2S is a stronger acid than H_2O.)

(4) Hydrated cations such as $Al(H_2O)_6^{3+}$ and $Zn(H_2O)_4^{2+}$ are acidic because the waters of hydration are polarized by the cation. The acidity is most pronounced when the metal ion is small and highly charged. ($Al(H_2O)_6^{3+}$ is a stronger acid than $Ni(H_2O)_6^{2+}$.)

Example 16.5: Indicate the stronger acid for each pair:

(a) NH_4^+, NH_3
(b) $HClO_4$, HClO
(c) ClCOOH, HCOOH
(d) $HClO_3$, $HBrO_3$
(e) $Fe(H_2O)_6^{3+}$, $Fe(H_2O)_6^{2+}$

Answers:
(a) NH_4^+ (See (1) above)
(b) $HClO_4$ (See (2) above)
(c) ClCOOH (See (2) above)
(d) $HClO_3$ (See (2) above)
(e) $Fe(H_2O)_6^{3+}$ (See (4) above)

Lewis Acids and Bases (Section 16.6)

Learning Objectives

1. Distinguish between the Lewis and Brønsted-Lowry definitions of acid, base, and neutralization.
2. Use Lewis structures to identify the acid and base in a Lewis acid-base neutralization.

Review

A <u>Lewis acid</u> is a substance that accepts electron pairs; a <u>Lewis base</u> donates electron pairs. Lewis acids are <u>electrophiles</u>, seekers of electrons; Lewis bases are <u>nucleophiles</u>, seekers of a positive nucleus. <u>Neutralization</u> occurs when the base shares its electron pair with the acid:

$$Cu^{2+} \quad + \quad 4 :NH_3 \quad \longrightarrow \quad \left[\begin{array}{c} NH_3 \\ H_3N:Cu:NH_3 \\ NH_3 \end{array} \right]^{2+}$$

Lewis Acid Lewis Base
(Electrophile) (Nucleophile)

(Lewis acid-base reactions include the proton transfer reactions of Brønsted theory and many other reactions which do not involve protons. Lewis theory is the most comprehensive of the three acid-base theories.)

SELF-TEST

1. State which reactant is the Brønsted acid and which the base:
 (a) $C_6H_5NH_2(aq) + HClO_4(aq) \rightarrow C_6H_5NH_3^+(aq) + ClO_4^-(aq)$
 (b) $HS^-(aq) + HNO_3(aq) \rightarrow H_2S(g) + NO_3^-(aq)$
 (c) $HS^-(aq) + O^{2-}(aq) \rightarrow S^{2-}(aq) + OH^-(aq)$

2. Complete the following equations:
 (a) $OH^-(aq) + HS^-(aq) \rightarrow ? + ?$
 (b) $HCl(aq) + HS^-(aq) \rightarrow ? + ?$
 (c) $CH_3CH_2OH + CH_3CH_2OH \rightleftharpoons ? + ?$

3. Match the acid in column A with its conjugate base in column B:

Acid		Base	
(a)	$H_2PO_4^-$	(1)	SO_3^{2-}
(b)	HCO_3^-	(2)	CO_3^{2-}
(c)	H_3PO_4	(3)	HPO_4^{2-}
(d)	HPO_4^{2-}	(4)	$H_2PO_4^-$
(e)	HSO_3^-	(5)	HSO_3^-
(f)	H_2SO_3	(6)	PO_4^{3-}

4. Complete the following equations and identify the conjugate acid-base pairs:
 (a) $HCOOH(aq) + OH^-(aq) \rightarrow ? + H_2O(l)$
 (b) $H_3O^+(aq) + NH_3(aq) \rightarrow ? + H_2O(l)$
 (c) $HCl(aq) + HSO_3^-(aq) \rightarrow ? + Cl^-(aq)$
 (d) $H^-(aq) + H_2O(l) \rightarrow ? + OH^-(aq)$

5. The following are listed in order of decreasing basicity. Rank their conjugate acids in order of increasing acidity:
 $S^{2-} > HCO_3^{2-} > H_2O > HSO_4^-$

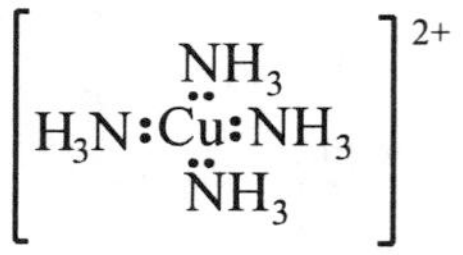

6. Write equations for the
 (a) ionization of aniline ($C_6H_5NH_2$) (a weak base) in water
 (b) autoionization in anhydrous H_2SO_4 (without water)
 (c) stepwise ionization of H_2S in water
 (d) reaction between HBr, a strong acid, and NaCN in water
 (e) reaction between KOH and H_2SO_3 in water

7. Calculate the pH, pOH, and [OH⁻] for
 (a) a solution of ascorbic acid that has ionized to give 2.0×10^{-6} mol of H^+ per liter
 (b) a 0.10 M solution of nitric acid
 (c) a 0.10 M solution of $Ca(OH)_2$

8. Calculate the pH of a solution made by dissolving 800 mg of $Al(OH)_3$ in one liter of 0.100 M HNO_3 solution.

9. Indicate the stronger base in each pair:
 (a) HS^- or S^{2-} (d) CCl_3COO^- or CH_3COO^-
 (b) SO_3^{2-} or SO_4^{2-} (e) HS^- or HSe^-
 (c) $Fe(H_2O)_5(OH)^+$ or $Fe(H_2O)_5(OH)^{2+}$

10. State which reactant is the Lewis acid and which the Lewis base in each equation:

 (a)
 $$F-\overset{\overset{\textstyle F}{|}}{\underset{\underset{\textstyle F}{|}}{B}} \quad + \quad :NH_3 \quad \longrightarrow \quad F-\overset{\overset{\textstyle F}{|}}{\underset{\underset{\textstyle NH_3}{|}}{B}}-F$$

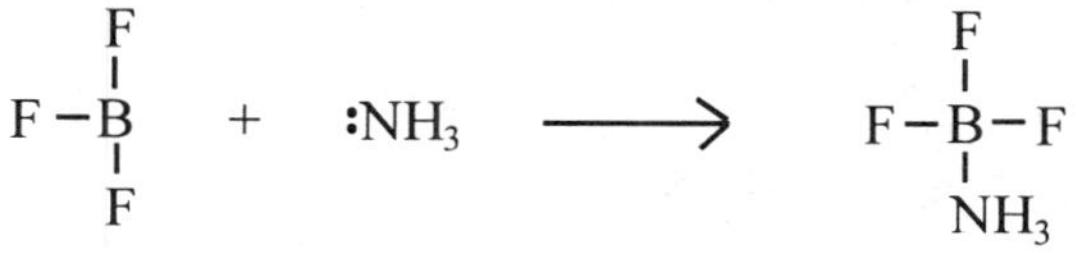

 (b) $HCl(aq) + HSO_3^-(aq) \rightarrow H_2SO_3(aq) + Cl^-(aq)$

 (c) $C_6H_5NH_2(aq) + HNO_3(aq) \rightarrow C_6H_5NH_3^+(aq) + NO_3^-(aq)$

11. Identify the Brønsted-Lowry acids and Brønsted-Lowry bases above.

ANSWERS TO SELF-TEST QUESTIONS

1. (a) acid: $HClO_4$; base: $C_6H_5NH_2$ (c) acid: HS^-; base: O^{2-}
 (b) acid: HNO_3; base: HS^-

2. (a) $S^{2-}(aq) + H_2O(l)$ (c) $CH_3CH_2OH_2^+ + CH_3CH_2O^-$
 (b) $H_2S(g) + Cl^-(aq)$

3. (a) 3 (b) 2 (c) 4 (d) 6 (e) 1 (f) 5

4. (a) $HCOO^-(aq)$: acid: HCOOH; base: $HCOO^-$ (c) $H_2SO_3(aq)$: acid: HCl; base: Cl^-
 acid: H_2O; base: OH^- acid: H_2SO_3; base: HSO_3^-
 (b) $NH_4^+(aq)$: acid: H_3O^+; base: H_2O (d) $H_2(g)$: acid: H_2O; base: OH^-
 acid: NH_4^+; base: NH_3 acid: H_2; base: H^-

5. $H_2S < H_2CO_3 < H_3O^+ < H_2SO_4$

6. (a) $C_6H_5NH_2(aq) + H_2O(l) \rightleftharpoons C_6H_5NH_3^+(aq) + OH^-(aq)$
(b) $H_2SO_4(l) + H_2SO_4(l) \rightleftharpoons HSO_4^-(l) + H_3SO_4^+(l)$
(c) $H_2S(aq) + H_2O(l) \rightleftharpoons HS^-(aq) + H_3O^+(aq)$
$HS^-(aq) + H_2O(l) \rightleftharpoons S^{2-}(aq) + H_3O^+(aq)$
(d) $HBr(aq) + NaCN(aq) \rightarrow HCN(g) + NaBr(aq)$
(e) $2KOH(aq) + H_2SO_3(aq) \rightarrow 2H_2O(l) + K_2SO_3(aq)$

7. (a) $pH = -\log(2.0 \times 10^{-6}) = 5.70$; $pOH = 8.30$; $[OH^-] = 5.0 \times 10^{-9}$ mol/L
(b) $pH = -\log(0.10) = 1.00$; $pOH = 13.00$; $[OH^-] = 1.0 \times 10^{-13}$ mol/L
(c) $[OH^-] = 0.20$ mol/L; $pOH = -\log(0.20) = 0.70$; $pH = 13.30$

8. $$[OH^-] = \frac{0.800 \text{ g Al(OH)}_3}{1 \text{ L}} \times \frac{1 \text{ mol Al(OH)}_3}{78.0 \text{ g Al(OH)}_3} \times \frac{3 \text{ mol OH}^-}{1 \text{ mol Al(OH)}_3} = \underline{0.0308 \text{ mol/L}}$$

$[H^+] = (0.100 - 0.0308)$ mol/L $= 0.069$ mol/L $pH = \underline{1.16}$

9. (a) S^{2-} (b) SO_3^{2-} (c) $Fe(H_2O)_5(OH)^+$ (d) CH_3COO^- (e) HS^-

10. (a) Lewis acid: BF_3; Lewis base: NH_3
(b) Lewis acid: HCl; Lewis base: HSO_3^-
(c) Lewis acid: HNO_3; Lewis base: $C_6H_5NH_2$

11. acids: HCl, H_2SO_3, HNO_3, $C_6H_5NH_3^+$ bases: HSO_3^-, Cl^-, $C_6H_5NH_2$, NO_3^-

CHAPTER 17

ACID - BASE EQUILIBRIA IN AQUEOUS SOLUTION

SELF-ASSESSMENT

Acid and Base Ionization Constants (Sections 17.1 and 17.2)

1. The pK_a of nicotinic acid at 25°C is 4.85. Find its K_a.

2. The pH of a 0.10 *M* dichloroacetic acid solution is 1.37. Find
   (a) the H^+ concentration
   (b) the K_a of dichloroacetic acid

3. Using your answers for questions 1 and 2, determine which acid, nicotinic or dichloroacetic, is the stronger acid.

4. Calculate
   (a) the OH^- concentration and
   (b) the pH of a 0.10 *M* trimethylamine ($(CH_3)_3N$) solution ($K_b = 6.3 \times 10^{-5}$)

Percent Ionization (Section 17.3)

5. The percent ionization of a 0.10 *M* HOCl solution is 0.053%. Find its K_a.

Polyprotic Acid Equilibria (Section 17.4)

6. Given a 0.15 *M* H_2S solution, calculate
   (a) the pH
   (b) the S^{2-} ion concentration ($K_{a_1} = 1.02 \times 10^{-7}$; $K_{a_2} = 1 \times 10^{-13}$)

Acidic and Basic Salts (Section 17.5)

7. Which of the following salts produce acidic solutions in water?:
   (a) $LiNO_3$
   (b) $AlCl_3$
   (c) $NaHPO_4$
   (d) $(CH_3)_2NH_2^+Cl^-$

8. Determine the pH of a 0.15 *M* solution of sodium oxalate ($Na_2C_2O_4$). K_{a_2} for oxalic acid is 6.40×10^{-5}.

The Common Ion Effect (Section 17.6)

9. How many moles of Na_2HPO_4 are added to 1.00 L of a 0.20 *M* NaH_2PO_4 solution to make a solution whose pH is 6.92? ($K_{a_2} = 6.23 \times 10^{-8}$)

Buffer Solutions (Section 17.7)

10. A buffer solution is prepared by combining 50 mL of 0.100 *M* formic acid with 100 mL of 0.050 *M* sodium formate (HCOONa).
    (a) Determine the pH of the solution. ($pK_a = 3.72$)
    (b) What will the pH be after 1.00 mL of 1.0 *M* NaOH is added to the buffer of part (a)?

11. How many milliliters of 10 *M* HCl must be added to one liter of a 0.10 *M* solution of Na_2HPO_4 to produce a buffer solution of pH = 7.40? ($pK_{a_2} = 7.21$)

A Closer Look at Acid-Base Titrations (Section 17.8)

12. Sketch the titration curve for 50.0 mL of 0.200 *M* barbituric acid (a monoprotic acid) with 0.100 *M* NaOH. K_a for barbituric acid is 9.8×10^{-5}. Include in your sketch the following pH values:
    (a) initial pH
    (b) pH at half-neutralization
    (c) equivalence point
    (d) pH 1.0 mL past the equivalence point

ANSWERS TO SELF-ASSESSMENT PROBLEMS

If you missed an answer, be sure to study the relevant section in the textbook and study guide.

1. $K_a = \text{antilog}(-4.85) = \underline{1.4 \times 10^{-5}}$

2. (a) $[H^+] = \text{antilog}(-1.37) = \underline{0.043\ M}$ (b) $K_a = (0.043)^2/(0.10 - 0.043) = \underline{3.2 \times 10^{-2}}$

3. dichloroacetic acid; K_a is larger

4. (a) $6.3 \times 10^{-5} = x^2/0.10$; $x = \underline{2.5 \times 10^{-3}} = [OH^-]$
   (b) $[H^+] = 10^{-14}/2.5 \times 10^{-3} = 4.0 \times 10^{-12}$; $pH = -\log(4.0 \times 10^{-12}) = \underline{11.40}$

5. $K_a = (0.10 \times 0.00053)^2/0.10 = \underline{2.8 \times 10^{-8}}$

6. (a) $1.02 \times 10^{-7} = x^2/0.15$; $x = [H^+] = 1.2 \times 10^{-4}$; $pH = \underline{3.91}$
   (b) $[S^{2-}] = K_{a_2} = \underline{1 \times 10^{-13}}$

7. (b) $Al(H_2O)_6^{3+}(aq) + H_2O(l) \rightleftharpoons Al(H_2O)_5OH^{2+}(aq) + H_3O^+(aq)$
   (d) $(CH_3)_2NH_2^+(aq) + H_2O(l) \rightleftharpoons (CH_3)_2NH(aq) + H_3O^+(aq)$

8. $C_2O_4^{2-}(aq) + H_2O(l) \rightleftharpoons HC_2O_4^-(aq) + OH^-(aq)$
   $K_b = 1.00 \times 10^{-14}/6.40 \times 10^{-5} = x^2/0.15$; $x = 4.8 \times 10^{-6}$
   pOH = 5.32; pH = $\underline{8.68}$

9. $6.23 \times 10^{-8} = (x)(10^{-6.92})/0.20$; $x = \underline{0.10\ \text{mol}}$

10. (a) $pH = 3.72 + \log \dfrac{(0.100\ L \times 0.050\ M)}{(0.050\ L \times 0.100\ M)} = \underline{3.72}$

(b) $pH = 3.72 + \log \dfrac{(0.100\ L \times 0.050\ M + 0.00100\ L \times 1.0\ M)}{(0.050\ L \times 0.100\ M - 0.00100\ L \times 1.0\ M)} = \underline{3.90}$

11. $pH = pK_{a_2} + \log[HPO_4^{2-}]/[H_2PO_4^-] = 7.40 = 7.21 + \log \dfrac{(0.10 - x)}{x}$

$\text{antilog}(0.19) = 1.55 = \dfrac{0.10 - x}{x}$; $x = 0.0392\ M$

$\text{milliliters of HCl} = \dfrac{1000\ ml}{10\ mol} \times 0.0392\ mol = \underline{3.9\ mL}$

12. See Figure 17.4 in your textbook for the shape of the curve and then locate the following pH values:
(a) $9.8 \times 10^{-5} = x^2/0.200$; $x = 4.4 \times 10^{-3}$; $pH = \underline{2.36}$
(b) $pH = pK_a = \underline{4.01}$
(c) $1.00 \times 10^{-14}/9.8 \times 10^{-5} = x^2/(0.050 \times 0.200/0.150)$; $x = [OH^-] = 2.6 \times 10^{-6}$; $pH = \underline{8.42}$
(d) $[OH^-] = \dfrac{0.0010\ L \times 0.100\ mol/L}{0.151} = 6.6 \times 10^{-4}\ M$; $pH = \underline{10.82}$

REVIEW

Acid Ionization Constants (Section 17.1)

Learning Objectives

1. Write the K_a expression for a weak acid.
2. Use K_a and pK_a values to compare acid strengths.
3. Solve problems relating K_a, equilibrium concentrations, pH, and the initial molarity of a weak acid.

Review

The <u>acid ionization constant</u>, K_a, is the equilibrium constant for the ionization of a weak acid in water. For example, nitrous acid (HNO_2) ionizes in water according to

$HNO_2(aq) \rightleftharpoons H^+(aq) + NO_2^-(aq)$

Its ionization constant at 25°C is $K_a = \dfrac{[H^+][NO_2^-]}{[HNO_2]} = 6.0 \times 10^{-4}$

<u>The larger the ionization constant the stronger the acid</u>. Relative acidities can also be compared by examining <u>pK_a</u> values of acids:

$pK_a = -\log K_a$

The pK_a of nitrous acid at 25°C is

$pK_a = -\log(6.0 \times 10^{-4}) = 3.22$

<u>The smaller the pK_a value the stronger the acid</u>.

Example 17.1: List the following acids in order of decreasing acidity:

Acetic acid: $pK_a = 4.75$
Nitrous acid: $K_a = 6.0 \times 10^{-4}$
Nicotinic acid: $pK_a = 4.85$
Lactic acid: $K_a = 1.37 \times 10^{-4}$

Answer: First we convert the K_a's of nitrous and lactic acid to pK_a's. The values are 3.22 and 3.863, respectively. Listing the acids in order of increasing pK_a values (decreasing acidity) gives: nitrous acid > lactic acid > acetic acid > nicotinic acid.

Solving Problems Involving K_a

These problems relate the K_a and initial concentration of a weak acid to the equilibrium concentrations and pH. It is helpful to use a concentration summary similar to those used in the equilibrium calculations of Chapter 15.

Finding the value of a K_a

Example 17.2: The pH of a 0.2152 *M* formic acid (HCOOH) solution is 2.20 at 25°C. Find K_a for formic acid. The equation for the ionization of formic acid is

$HCOOH(aq) \rightleftharpoons H^+(aq) + HCOO^-(aq)$

Given: pH = 2.20; initial concentration of weak acid (HCOOH) is 0.2152 *M*.

Unknown: K_a

Plan: We will find equilibrium concentrations by following the three steps outlined in Chapter 15 of the study guide and text. We will then substitute these values into the K_a expression to find the value of K_a.

Calculations:		HCOOH(aq)	$\rightleftharpoons$	$H^+(aq)$	+	$HCOO^-(aq)$
Step 1:	Initial (mol/L)	0.2152		0		0

(H^+ from the self-ionization of water is ignored because it is less than 10^{-7} mol/L. See Section 16.4)

Step 2 and Step 3: The equilibrium concentration of $H^+(aq)$ is calculated from the given pH:
$pH = 2.20 = -\log[H^+]$

$[H^+] = \text{antilog}(-2.20) = 10^{-2.20} = 6.3 \times 10^{-3}\ M$

According to the ionization equation, 6.3×10^{-3} *M* of $HCOO^-$ also forms and HCOOH decreases by 6.3×10^{-3} *M*.

The concentration summary is therefore

		HCOOH(aq)	$\rightleftharpoons$	$H^+(aq)$	+	$HCOO^-(aq)$
Step 1:	Initial (mol/L)	0.2152		0		0
Step 2:	Change (mol/L)	−0.0063		+0.0063		+0.0063
Step 3:	Equilibrium (mol/L)	0.2089		0.0063		0.0063

Substituting the equilibrium concentrations into the K_a expression gives

$$K_a = \frac{[H^+][HCOO^-]}{[HCOOH]} = \frac{(6.3 \times 10^{-3})^2}{0.2089} = \underline{1.9 \times 10^{-4}}$$

Finding pH from the K_a and the Initial Concentration of the Weak Acid

Example 17.3: The K_a for the ionization of hydrocyanic acid (HCN) is 4.93×10^{-10}. Find the pH of 0.20 *M* hydrocyanic acid.

Given: $K_a = 4.93 \times 10^{-10}$; initial concentration of weak acid (HCN) is 0.20 *M*.

Unknown: pH

Plan: To calculate the pH we must find the equilibrium concentration of $H^+(aq)$. Let this be x. Then the equilibrium concentrations of $CN^-(aq)$ and HCN(aq) are x and $0.20 - x$, respectively.

Calculations: Concentration Summary

	HCN(aq)	⇌	$H^+(aq)$	+	$CN^-(aq)$
Initial (mol/L)	0.20		0		0
Change (mol/L)	$-x$		$+x$		$+x$
Equilibrium (mol/L)	$0.20 - x$		x		x

Substituting the equilibrium concentrations into K_a gives

$$K_a = \frac{[H^+][CN^-]}{[HCN]} = \frac{x^2}{0.20 - x}$$

The quadratic formula will give us an exact solution for x. However, we can avoid using the quadratic formula because K_a is small and very little HCN will ionize. Hence, x is negligible relative to the initial concentration (0.20 *M*) of HCN. We can use the approximation $0.20 - x = 0.20$:

$$4.93 \times 10^{-10} = \frac{x^2}{0.20}\ ;\ x = 9.9 \times 10^{-6} = [H^+]$$

Note: The above approximation is valid only when the neglected quantity (9.9×10^{-6}) is less than 5% of the retained quantity (0.20).

The pH is

$pH = -\log[H^+] = -\log(9.9 \times 10^{-6}) = \underline{5.00}$

Practice Drill on Weak Acids: Fill in the blanks for the three solutions of acetic acid (CH_3COOH) at 25°C given below. The equation for the ionization of acetic acid is

$$CH_3COOH(aq) \rightleftharpoons H^+(aq) + CH_3COO^-(aq)$$

Solution	pH	Initial Concentration of CH_3COOH, mol/L	Equilibrium Concentration of CH_3COO^-, mol/L	K_a
A	?	0.0500	9.3×10^{-4}	?
B	?	0.200	?	same as above
C	4.00	?	?	same as above

Answers: A: pH = 3.03, $K_a = 1.8 \times 10^{-5}$; B: pH = 2.72, $[CH_3COO^-] = 1.9 \times 10^{-3}$; C: $[CH_3COOH] = 5.6 \times 10^{-4}$; $[CH_3COO^-] = 1.0 \times 10^{-4}$

Using the Quadratic Formula

The quadratic formula may be needed when the type of approximation made in Example 17.3 would not be valid. These cases may arise when K_a is unusually large (Example 17.4) or the initial concentrations unusually small.

Example 17.4: Find the pH of a 0.10 *M* solution of chloroacetic acid ($CH_2ClCOOH$) given that its K_a is 1.5×10^{-3}.

Answer: Concentration Summary

	$CH_2ClCOOH(aq)$	$\rightleftharpoons$	$H^+(aq)$	+	$CH_2ClCOO^-(aq)$
Initial (mol/L)	0.10		0		0
Change (mol/L)	$-x$		$+x$		$+x$
Equilibrium (mol/L)	$0.10 - x$		x		x

Substituting the equilibrium concentrations into the K_a expression gives

$$1.5 \times 10^{-3} = \frac{x^2}{0.10 - x}$$

If we solve for x using the same approximation as in Example 17.3 ($0.10 - x = 0.10$), we get $x = 0.012$. Because this value for x is more than 5% of the initial concentration (0.10 M), the approximation is not valid. To find the correct value for x, we put the substituted K_a expression into the standard quadratic form

$(ax^2 + bx + c = 0)$

and use the quadratic formula

$$x = \frac{-b \pm \sqrt{b^2 - 4ac}}{2a}$$

Applying this procedure gives $x^2 + 1.5 \times 10^{-3}x - 1.5 \times 10^{-4} = 0$ ($a = 1$, $b = 1.5 \times 10^{-3}$, $c = -1.5 \times 10^{-4}$)

Therefore, $x = \dfrac{-1.5 \times 10^{-3} \pm \sqrt{(1.5 \times 10^{-3})^2 - 4 \times 1 \times (-1.5 \times 10^{-4})}}{2 \times 1}$; $x = 1.15 \times 10^{-2}$

(The negative root is rejected because it would give a negative H^+ concentration.)

$pH = -\log[H^+] = -\log(1.15 \times 10^{-2}) = \underline{1.94}$

Practice Drill on Making Approximations: Given the K_a and the initial concentration of weak acid, predict whether the approximation would be valid.

	K_a	Initial Concentration of Weak Acid
(a)	1.79×10^{-5}	0.18 *M*
(b)	3.3×10^{-2}	0.010 *M*
(c)	1.79×10^{-5}	1.0×10^{-4} *M*

Answers: (a) yes (b) no (K_a is large) (c) no (initial concentration low)

Base Ionization Constants (Section 17.2)

Learning Objectives

1. Write the K_b expression for a weak base.
2. Use K_b and pK_b values to compare base strengths.
3. Solve problems relating K_b, equilibrium concentrations, pH, and the initial molarity of a weak base.
4. Given K_a, calculate K_b for the conjugate base and vice versa.

Review

A base ionization constant (K_b) is the equilibrium constant for the ionization of a weak base in water. Methyl amine (CH_3NH_2) is a weak base.

$$CH_3NH_2(aq) + H_2O(l) \rightleftharpoons CH_3NH_3^+(aq) + OH^-(aq)$$

$$K_b = \frac{[CH_3NH_3^+][OH^-]}{[CH_3NH_2]} = 3.70 \times 10^{-4} \text{ and } pK_b = -\log(3.70 \times 10^{-4}) = 3.432$$

The larger the K_b the stronger the base; the smaller the pK_b the stronger the base.

Calculations Involving Weak Bases

Calculations involving weak bases are similar to those for weak acids.

Example 17.5: Find the OH^- concentration and pH of a 0.020 *M* aniline solution:

$$C_6H_5NH_2(aq) + H_2O(l) \rightleftharpoons C_6H_5NH_3^+(aq) + OH^-(aq) \quad K_b = 4.3 \times 10^{-10}$$

Given: Initial concentration of $C_6H_5NH_2$ is 0.020 *M*; $K_b = 4.3 \times 10^{-10}$

Unknown: $[OH^-]$ and pH

Plan: We will calculate the OH^- concentration by constructing a concentration summary and then substituting the equilibrium concentrations into the K_b expression. Once we know $[OH^-]$ we can then find the pH.

Calculations: Defining x as the concentration of OH^- at equilibrium gives

Concentration Summary	$C_6H_5NH_2(aq) + H_2O(l)$	$\rightleftharpoons$	$C_6H_5NH_3^+(aq)$	+	$OH^-(aq)$
Initial (mol/L)	0.020		0		0
Change (mol/L)	$-x$		$+x$		$+x$
Equilibrium (mol/L)	$0.020 - x$		x		x

$$K_b = \frac{x^2}{0.020 - x} = 4.3 \times 10^{-10}$$

Because K_b is so small we can make the usual approximation that $0.020 - x = 0.020$:

$$4.3 \times 10^{-10} = \frac{x^2}{0.020}\ ;\ x = [OH^-] = \underline{2.9 \times 10^{-6}\ M}$$

$$pOH = -\log[OH^-] = -\log(2.9 \times 10^{-6}) = 5.54$$

$$pH = 14.00 - pOH = 14.00 - 5.54 = \underline{8.46}$$

Relationship Between K_a and K_b of a Conjugate Acid-Base Pair

Let K_a be the ionization constant for a weak acid HA in water

$$HA \rightleftharpoons H^+ + A^- \qquad K_a = \frac{[H^+][A^-]}{[HA]}$$

and K_b be the ionization constant for its conjugate base A^- in water

$$A^- + H_2O \rightleftharpoons HA + OH^- \qquad K_b = \frac{[HA][OH^-]}{[A^-]}$$

Multiplying $K_a \times K_b$ gives $K_a \times K_b = \frac{[H^+][A^-]}{[HA]} \times \frac{[HA][OH^-]}{[A^-]} = [H^+][OH^-] = K_w$

The above relationship ($K_a \times K_b = K_w$) is true for the K_a and K_b of any conjugate acid-base pair.

Thus if K_a is given then K_b can be calculated and vice versa. The above relationship shows the inverse relationship between the strengths of the acid-base pair: the smaller the K_a (the weaker the acid) the larger the K_b (the stronger its conjugate base) and vice versa.

Example 17.6: K_a for HCN is 4.93×10^{-10} at 25°C.
(a) Find the value of K_b for CN^-.
(b) Write the equation for the base ionization of CN^- and write the K_b expression.

Answer:
(a) $K_a \times K_b = 1.00 \times 10^{-14}$; $K_b = 1.00 \times 10^{-14}/4.93 \times 10^{-10} = \underline{2.03 \times 10^{-5}}$

(b) $CN^-(aq) + H_2O(l) \rightleftharpoons HCN(aq) + OH^-(aq)$; $K_b = \frac{[HCN][OH^-]}{[CN^-]}$

Example 17.7: K_b for NH_3 is 1.8×10^{-5} at 25°C.
(a) Find K_a for NH_4^+.
(b) Write the equation for the acid ionization of NH_4^+ and write the K_a expression.

Answer:
(a) $K_a = 1.00 \times 10^{-14}/1.8 \times 10^{-5} = \underline{5.6 \times 10^{-10}}$
(b) $NH_4^+(aq) + H_2O(l) \rightleftharpoons NH_3(aq) + H_3O^+(aq)$

$$K_a = \frac{[NH_3][H_3O^+]}{[NH_4^+]}$$

Example 17.8: The K_b of $C_6H_5NH_2$ (aniline) is 4.3×10^{-10}.
(a) Calculate K_a for the reaction of $C_6H_5NH_3^+$ (anilinium ion) with water.
(b) Find the pH of a solution that contains 0.10 *M* anilinium chloride ($C_6H_5NH_3^+Cl^-$).

Answer:

(a) $K_a = \dfrac{K_w}{K_b} = \dfrac{1.00 \times 10^{-14}}{4.3 \times 10^{-10}} = \underline{2.3 \times 10^{-5}}$

(b) $C_6H_5NH_3^+ (aq) + H_2O (l) \rightleftharpoons C_6H_5NH_2 (aq) + H_3O^+ (aq)$

Let $x = [H_3O^+]$ so that $2.3 \times 10^{-5} = x^2/0.10 - x$

Using the approximation $0.10 - x = 0.10$ and solving for x gives

$x = 1.5 \times 10^{-3} = [H_3O^+]$

pH = $-\log(1.5 \times 10^{-3}) = \underline{2.82}$

Practice: Rank the following conjugate bases in order of increasing basicities given the ionization constants of their conjugate weak acids.

Conjugate Base	Conjugate Acid	K_a of Conjugate Acid
CH_3COO^-	CH_3COOH	1.79×10^{-5}
CH_2ClCOO^-	$CH_2ClCOOH$	1.40×10^{-3}
$CHCl_2COO^-$	$CHCl_2COOH$	3.3×10^{-2}
CCl_3COO^-	CCl_3COOH	2×10^{-1}

Answer: $CCl_3COO^- < CHCl_2COO^- < CH_2ClCOO^- < CH_3COO^-$

Percent Ionization (Section 17.3)

Learning Objectives

1. Use the ionization constant and initial molarity to calculate the percent ionization of a weak acid or weak base.

Review

The percent ionization of a weak acid or weak base is

$$\%\ \text{ionization} = 100\% \times \text{fraction of molecules ionized} = 100\% \times \frac{\text{moles of ionized acid or base/ liter}}{\text{initial moles of acid or base/ liter}}$$

The percent ionization can be calculated from the ionization constant and the initial molarity.

Example 17.9: Calculate the percent ionization of a
(a) 0.20 *M* benzoic acid solution
(b) 0.020 *M* benzoic acid solution

The ionization equation for benzoic acid is

$C_6H_5COOH(aq) + H_2O(l) \rightleftharpoons C_6H_5COO^-(aq) + H_3O^+(aq)$ and $K_a = 6.46 \times 10^{-5}$

Given: Initial concentration of benzoic acid = 0.20 *M*; $K_a = 6.46 \times 10^{-5}$

Unknown: % ionization

Plan: (a) Use $\% \text{ ionization} = 100\% \times \dfrac{\text{moles of acid ionized / liter}}{\text{initial moles of acid / liter}}$

Because we are given the initial moles per liter of acid (0.20 *M*) all we need to do is determine the moles per liter of acid ionized which is the value of $[H^+]$ and $[C_6H_5COO^-]$. Therefore, we can write a concentration summary similar to the one in Example 17.3 and solve for x:

$$K_a = \frac{[H^+][C_6H_5COO^-]}{[C_6H_5COOH]}$$

$$6.46 \times 10^{-5} = \frac{x^2}{0.20}\ ;\ x = [H^+] = 3.6 \times 10^{-3}\ M$$

$$\% \text{ ionization} = 100\% \times \frac{3.6 \times 10^{-3}\ M}{0.20\ M} = \underline{1.8\%}$$

(b) The initial concentration of benzoic acid is now 0.020 *M*. We can follow the same procedure as in (a):

$$K_a = \frac{[H^+][C_6H_5COO^-]}{[C_6H_5COO]}$$

$$6.46 \times 10^{-5} = \frac{x^2}{0.020}\ ;\ x = [H^+] = 1.14 \times 10^{-3}$$

$$\% \text{ ionization} = 100\% \times \frac{1.14 \times 10^{-3}}{0.020} = \underline{5.7\%}$$

Note: The above example (17.9) shows that the % ionization increases with dilution.

Polyprotic Acid Equilibria (Section 17.4)

Learning Objectives

1. Write the K_a expression for each step in the ionization of a polyprotic acid.
2. Calculate the concentration of each ion in a polyprotic acid solution.

Review

A <u>polyprotic acid</u> has two or more acidic hydrogens, and therefore has two or more ionization steps. Each step has its own K_a. For phosphoric acid the ionization steps and equilibrium expressions are

$$H_3PO_4(aq) \rightleftharpoons H^+(aq) + H_2PO_4^-(aq)$$

$$K_{a_1} = \frac{[H^+][H_2PO_4^-]}{[H_3PO_4]} = 7.52 \times 10^{-3}$$

$$H_2PO_4^-(aq) \rightleftharpoons H^+(aq) + HPO_4^{2-}(aq)$$

$$K_{a_2} = \frac{[H^+][HPO_4^{2-}]}{[H_3PO_4^-]} = 6.23 \times 10^{-8}$$

$$HPO_4^{2-}(aq) \rightleftharpoons H^+(aq) + PO_4^{3-}(aq)$$

$$K_{a_3} = \frac{[H^+][PO_4^{3-}]}{[HPO_4^{2-}]} = 4.5 \times 10^{-13}$$

Note: Usually the successive ionization constants of a polyprotic acid differ by several orders of magnitude: $K_{a_1} > K_{a_2} > K_{a_3}$.

The following rules apply to most solutions that contain only polyprotic acids:

(1) The $[H^+]$ and pH can be calculated from the first $K_a(K_{a_1})$ only. The other ionization steps can be ignored because they are so small.
(2) The concentration of the singly charged anion (e.g., $[H_2PO_4^-]$, $[HS^-]$) equals the H^+ concentration.
(3) The concentration of the doubly charged anion (e.g., $[HPO_4^{2-}]$, $[S^{2-}]$) equals K_{a_2}. (See proof in text.)

<u>Example 17.10</u>: Calculate the concentration of
(a) H^+ and $H_2PO_4^-$
(b) HPO_4^{2-} in a 0.20 M H_3PO_4 solution. Use the ionization constants given above.

Given: $K_{a_1} = 7.52 \times 10^{-3}$
$K_{a_2} = 6.23 \times 10^{-8}$
Initial concentration of $H_3PO_4 = 0.20\ M$

Unknown: Equilibrium concentrations of H^+, $H_2PO_4^-$, and HPO_4^{2-}

Plan: (a) To find $[H^+]$ and $[H_2PO_4^-]$ we will use K_{a_1} and set up the concentration summary ignoring the subsequent small ionizations. We will follow the same procedure described in Section 17.3 to find x, the H^+ concentration.

Calculation:

<u>Concentration Summary</u>	$H_3PO_4(aq)$	$\rightleftharpoons$	$H_2PO_4^-(aq)$	+	$H^+(aq)$
Initial (mol/L)	0.20		0		0
Change (mol/L)	$-x$		$+x$		$+x$
Equilibrium (mol/L)	$0.20 - x$		x		x

$$7.52 \times 10^{-3} = \frac{[H^+][H_2PO_4^-]}{[H_3PO_4]} = \frac{x^2}{0.20 - x}$$

Because K_{a_1} is not small relative to the initial concentration of H_3PO_4, we cannot make the approximation that $0.20 - x = x$. We must use the quadratic formula to find x.

$x^2 + 7.52 \times 10^{-3}x - 1.5 \times 10^{-3} = 0$

$$x = \frac{-7.52 \times 10^{-3} \pm \sqrt{(7.52 \times 10^{-3})^2 - 4 \times 1 \times (-1.5 \times 10^{-3})}}{2 \times 1}$$

$x = \underline{3.5 \times 10^{-2}\ M} = [H^+] = [H_2PO_4^-]$

Plan: (b) Applying Rule (3) given above

$[HPO_4^{2-}] = K_{a_2} = \underline{6.23 \times 10^{-8}\ M}$

Practice: Calculate the concentration of PO_4^{3-} using K_{a_3} and the results of Example 17.8 ($K_{a_3} = 4.5 \times 10^{-13}$). Answer: $[PO_4^{3-}] = 8.0 \times 10^{-19}\ M$

Acidic and Basic Salts (Section 17.5)

Learning Objectives

1. State whether a given ion is acidic, basic, or neutral with respect to water.
2. Predict whether the aqueous solution of a given salt will be acidic, basic, or neutral.
3. Calculate the pH of a salt solution.

Review

Salts are ionic compounds whose ions are neither H^+ nor OH^-. Many salts, however, have ions that react with water to form H^+ or OH^- ions and thus produce acidic or basic solutions. The reaction of an anion or cation with water is often called hydrolysis.

Anions

Every anion is the conjugate base of some acid. However, the anions of strong acids (acids above H_3O^+ in Table 16.2 of your text) are too weakly basic to remove protons from water and therefore have no effect on the pH of an aqueous solution. Examples of such anions are Br^-, Cl^-, and NO_3^-.

Anions of weak acids are strong enough to remove protons from water and form basic solutions. The weaker the acid the more basic its anion.

For example, an aqueous solution of sodium formate is basic due to the reaction between the formate anion ($HCOO^-$) and water:

$HCOO^-(aq) + H_2O(l) \rightleftharpoons HCOOH(aq) + OH^-(aq)$

(Note that an aqueous solution is basic if hydroxide ions are formed.)

Example 17.11: Rank the following aqueous solutions in order of increasing basicity given the acid ionization constants of their conjugate acids:

Salt	K_a
0.10 M sodium acetate (CH_3COONa)	1.76×10^{-5} (CH_3COOH)
0.10 M sodium formate ($HCOONa$)	1.9×10^{-4} ($HCOOH$)
0.10 M sodium phosphate (Na_3PO_4)	4.5×10^{-13} (HPO_4^{2-})

Answer: The larger the K_a, the stronger the acid and the weaker its conjugate base. Therefore, the weakest base is the formate anion and the strongest base is the phosphate anion:

$$\xrightarrow[\text{increasing basicity}]{\text{formate} < \text{acetate} < \text{phosphate}}$$

Cations

Some cations are conjugate acids of weak bases. The ammonium ion (NH_4^+) is the conjugate acid of NH_3. Protonated amines such as methylammonium ion ($CH_3NH_3^+$) and anilinium ion ($C_6H_5NH_3^+$) are conjugate acids of amines – in this case, methylamine (CH_3NH_2) and aniline ($C_6H_5NH_2$).

The reaction of ammonium ion with water is

$$NH_4^+(aq) + H_2O(l) \rightleftharpoons NH_3(aq) + H_3O^+(aq)$$

K for the above reaction is the acid ionization constant (K_a) for NH_4^+ and is equal to K_w/K_b where K_b is the base ionization constant for NH_3. (Recall for a conjugate acid-base pair $K_w = K_a \times K_b$.)

In general, monovalent metal cations and divalent Group 2A cations (except Be) are neutral in water. Most other metal cations are acidic due to the effect of small size and high charge on the water of hydration. An example is

$$Al(H_2O)_6^{3+}(aq) + H_2O(l) \rightleftharpoons Al(H_2O)_5OH^{2+}(aq) + H_3O^+(aq)$$

The following chart summarizes the behavior of ions in water.

Reaction of Ions in Water

	Anions	Examples	Effect on pH
1.	conjugate bases of strong acids	Cl^-, NO_3^-	neutral
2.	conjugate bases of weak acids (acids weaker than H_3O^+)	NO_2^-, CH_3COO^-	basic
	Cations		
1.	small, highly charged, metal ions	Al^{3+}, Cu^{2+}	acidic
2.	monovalent metal cations and divalent Group 2A cations (except Be)	Na^+, Ca^{2+}	neutral
3.	NH_4^+ and protonated amines	NH_4^+, $CH_3NH_3^+$	acidic

The pH of Salt Solutions

The pH of a salt solution depends on the reaction of its ions with water. Each ion must be examined (see the above chart).

Example 17.12: Rank the following solutions in order of increasing pH:
(a) $NaNO_2$ (b) $MgCl_2$ (c) NH_4Cl

Answer:
(a) Na^+ (a monovalent cation) does not affect the pH. NO_2^- is the conjugate base of the weak acid HNO_2. It will therefore accept protons from water producing OH^- and a basic solution.
(b) Mg^{2+} (a Group 2 cation) is neutral with respect to water and so is the chloride anion (conjugate base of a strong acid). A solution of $MgCl_2$ is therefore neutral.
(c) NH_4^+ is acidic. Cl^- (the conjugate base of a strong acid) has no effect on pH. A solution of NH_4Cl is therefore acidic.

$$\xrightarrow[\text{increasing pH}]{NH_4Cl < MgCl_2 < NaNO_2}$$

If both the cation and anion react with water, the net effect depends on which ion is stronger; the stronger ion has the larger ionization constant.

Example 17.13: Would you expect a solution of ammonium cyanide (NH_4CN) to be acidic, basic, or neutral? The K_a for HCN is 4.93×10^{-10} and the K_b for NH_3 is 1.77×10^{-5}.

Answer: Both ions react with water. The two competing ionizations and their ionization constants are:

$$NH_4^+(aq) + H_2O(l) \rightleftharpoons NH_3(aq) + H_3O^+(aq)$$

$$K_a = K_w/K_b = 1.00 \times 10^{-14}/1.77 \times 10^{-5} = 5.65 \times 10^{-10}$$

$$CN^-(aq) + H_2O(l) \rightleftharpoons HCN(aq) + OH^-(aq)$$

$$K_b = K_w/K_a = 1.00 \times 10^{-14}/4.93 \times 10^{-10} = 2.03 \times 10^{-5}$$

Because the K_b of the cyanide anion is larger than the K_a of the ammonium cation the solution is basic.

Practice Drill: Which of the following aqueous solutions are basic?

$Ca(NO_2)_2$, $AlCl_3$, $(NH_4)_3PO_4$

K_b for NH_3 is 1.77×10^{-5} and K_{a_3} for H_3PO_4 is 4.5×10^{-13}.

Answer: $Ca(NO_2)_2$ and $(NH_4)_3PO_4$

Calculating the pH of a Salt Solution

The pH can be calculated from the initial concentration of the salt and the appropriate ionization constant.

Example 17.14: Find the pH of a 0.20 *M* sodium nitrite ($NaNO_2$) solution. K_a for HNO_2 is 6.0×10^{-4}

Given: Initial concentration of salt = 0.20 *M*; K_a for $HNO_2 = 6.0 \times 10^{-4}$

Unknown: pH of solution

Plan: The sodium ion is neutral. The nitrous anion (NO_2^-) is the conjugate base of the weak acid HNO_2 and therefore reacts with water to form a basic solution:

$$NO_2^-(aq) + H_2O(l) \rightleftharpoons HNO_2(aq) + OH^-(aq)$$

$$K_b = \frac{K_w}{K_a}$$

We will construct a concentration summary and then substitute equilibrium concentrations into the K_b expression.

Calculations:

Concentration Summary	$NO_2^-(aq) + H_2O(l)$	$\rightleftharpoons$	$HNO_2(aq)$	+	$OH^-(aq)$
Initial (mol/L)	0.20		0		0
Change (mol/L)	$-x$		$+x$		$+x$
Equilibrium (mol/L)	$0.20 - x$		x		x

$K_b = K_w/K_a = 1.0 \times 10^{-14}/6.0 \times 10^{-4} = 1.7 \times 10^{-11}$

$$1.7 \times 10^{-11} = \frac{[HNO_2][OH^-]}{[NO_2^-]} = \frac{x^2}{0.20}$$

$x = 1.8 \times 10^{-6}\ M = [OH^-]$

The pH can be calculated by subtracting the pOH from 14.00.

$pOH = -\log(1.8 \times 10^{-6}) = 5.74$

$pH = 14 - 5.74 =$ 8.26

Amphiprotic Ions

Amphiprotic ions such as $H_2PO_4^-$ and HSO_3^- can either accept or donate protons; they can act as either bases or acids in water. If the ability of an amphiprotic ion to accept protons is greater than its ability to donate protons, the ion will produce a basic solution; if the ability of the ion to donate protons is greater, it will form an acidic solution. The predominant reaction of the the amphiprotic anion with water is the one with the larger K value.

Finding the pH of a Solution Containing an Amphiprotic Ion

Example 17.15: Will an aqueous solution of $KHPO_4$ be acidic or basic? ($K_{a_2} = 6.23 \times 10^{-8}$; $K_{a_3} = 4.5 \times 10^{-13}$)

Answer: HPO_4^{2-} can act as an acid or base in water:

acid $\quad HPO_4^{2-}(aq) \rightleftharpoons H^+(aq) + PO_4^{3-}(aq)$

$K_a = 4.5 \times 10^{-13}$

base $\quad HPO_4^{2-}(aq) + H_2O(l) \rightleftharpoons H_2PO_4^-(aq) + OH^-(aq)$

$$K_b = \frac{K_w}{K_{a_2}} = \frac{1.00 \times 10^{-14}}{6.23 \times 10^{-8}} = 1.60 \times 10^{-7}$$

The basic reaction predominates because its equilibrium constant (K_b) is about one million times larger than K_a.

The Common Ion Effect (Section 17.6)

Learning Objectives

1. Calculate the pH of a solution containing a weak acid (or a weak base) and its salt.
2. Calculate the pH of (a) a solution containing both a weak and a strong acid and (b) a solution containing both a weak and a strong base.

Review

The common ion effect is a shift in equilibrium caused by the addition of an ion that participates in the equilibrium. The addition of a solute that has an ion in common with a weak electrolyte will shift the equilibrium in the direction that reduces the ionization of the weak electrolyte. For example, the ionization of formic acid (HCOOH) is decreased by the addition of sodium formate:

$$HCOOH(aq) \rightleftharpoons HCOO^-(aq) + H^+(aq)$$

↑
Adding formate anions
shifts equilibrium left.

Note that decreasing the ionization of a weak acid increases the pH; decreasing the ionization of a weak base decreases the pH.

Practice: What effect does the addition of NH_4Cl have on the pH of a solution of ammonia (NH_3)?
Answer: pH decreases

Calculating the pH of a Solution Containing a Weak Acid (or a Weak Base) and Its Salt

The salt of a weak acid contains the conjugate base of the acid; the salt of a weak base contains the conjugate acid of the base. The pH of a solution containing such a conjugate acid-base pair can be found from the initial molarities and the appropriate K value.

Example 17.16: Calculate the pH of 0.025 *M* benzoic acid (C_6H_5COOH) that also contains 0.020 *M* potassium benzoate (C_6H_5COOK). The K_a of benzoic acid is 6.46×10^{-5}.

Given: Initial concentration of benzoic acid = 0.025 *M*. Initial concentration of benzoate (conjugate base) = 0.020 *M*. (Remember the salt is 100% ionized.) $K_a = 6.46 \times 10^{-5}$

Unknown: pH of solution containing benzoic acid and its conjugate base

Plan: Construct a concentration summary following the same procedure as with other equilibrium problems. Note that in this case two (not one) of the initial concentrations are known.

Calculations:

Concentration Summary	$C_6H_5COOH(aq)$	$\rightleftharpoons$	$C_6H_5COO^-(aq)$	$+H^+(aq)$
Initial (mol/L)	0.025		0.020	0
Change (mol/L)	$-x$		$+x$	$+x$
Equilibrium (mol/L)	$0.025 - x$		$0.020 + x$	x

The common ion effect tells us that x will be small since the presence of benzoate represses the ionization of benzoic acid. Therefore, we can make the approximation that $0.025 - x = 0.025$ and $0.020 + x = 0.020$.

Substituting these values into K_a gives

$$K_a = \frac{[C_6H_5COO^-][H^+]}{[C_6H_5COOH]}$$

$$6.46 \times 10^{-5} = \frac{(0.020)(x)}{0.025}$$

$$x = 8.1 \times 10^{-5}\ M = [H^+]$$

$$pH = -\log [H^+] = -\log(8.1 \times 10^{-5}) = \underline{4.09}$$

Practice: Calculate the pH of a solution that is 0.025 *M* benzoic acid and 0.025 *M* benzoate ion.
$K_a = 6.46 \times 10^{-5}$
Answer: $K_a = [H^+]$; $-\log(6.46 \times 10^{-5}) = 4.190$

The ionization of a weak acid is repressed by a strong acid; hence, most of the hydrogen ions in a mixture of a strong and a weak acid come from the strong acid. The pH of such a solution will be essentially that produced by the strong acid.

Similarily, the pH of a mixture of a strong and a weak base is that produced by the strong base.

Buffer Solutions (Section 17.7)

Learning Objectives

1. Explain, with equations, how a buffer solution resists changes in pH.
2. Solve problems that relate pH to the composition of a buffer solution.
3. Calculate the pH changes produced by the addition of small amounts of strong acid or strong base to a buffer solution.
4. Use a table of ionization constants to choose an appropriate buffer system for a given pH.

Review

A buffer solution resists changes in pH. It contains both an acid and a base component so that addition of a small amount of acid will be neutralized by the base component, and addition of a small amount of base will be neutralized by the acid component.

Buffer solutions can be prepared from weak acids (acid component) and their conjugate base (base component) or from weak bases (base component) and their conjugate acid (acid component). Examples of buffer systems are acetic acid and sodium acetate (acetate is the base), ammonia and ammonium chloride (ammonium is the acid), NaH_2PO_4 and Na_2HPO_4. Note that the latter buffer involves two salts; $H_2PO_4^-$ acts as the acid component and HPO_4^{2-} is the base component.

Finding the pH of a Buffer Solution

The pH for a buffer system can be found by simply substituting the initial concentrations of the acid and base into the appropriate K and then solving for x the H_3O^+ concentration. (Remember the common ion effect represses ionization so that the initial concentrations are approximately the same as the equilibrium concentrations.) A more direct method uses a different form of K_a called the Henderson-Hasselbalch equation:

$$pH = pK_a + \log \frac{[base]}{[acid]}$$

$pK_a = -\log K_a$
$[base]$ = equilibrium concentration of base
$[acid]$ = equilibrium concentration of acid

The concentration ratio [base]/[acid] is called the buffer ratio. The equation shows that pH of a buffer depends on the buffer ratio; the pH will increase if the ratio increases, and decrease if the ratio is decreased.

Example 17.17: Find the pH of a solution that is 0.025 *M* benzoic acid and 0.020 *M* sodium benzoate. The K_a of benzoic acid is 6.46×10^{-5}.

Given: Initial concentration of acid = 0.025 *M*; Initial concentration of base = 0.020 *M*
$K_a = 6.46 \times 10^{-5}$

Unknown: pH

Plan: Substitute the given concentrations of acid and base into the Henderson-Hasselbalch equation. (We are making the approximation that initial concentrations = equilibrium concentrations.)

$$pH = pK_a + \log \frac{[base]}{[acid]}$$

$$pk_a = -\log(K_a) = -\log(6.46 \times 10^{-5}) = 4.19$$

$$pH = 4.19 + \log \frac{0.20}{0.25} = 4.19 + (-0.097) = \underline{4.09}$$

Note: We found the same pH for this solution by substituting into the K_a expression in Example 17.16.)

Practice Drill on Buffer Systems: Fill in the blanks for three different buffer solutions containing acetic acid and sodium acetate (base). (K_a of acetic acid 1.76×10^{-5})

	[Sodium Acetate]	[Acetic Acid]	pH
(a)	0.10	0.010	?
(b)	?	0.050	4.75
(c)	0.10	?	3.75

Answers: (a) pH = 5.75 (b) [acetate] = 0.050 *M* (c) [acetic acid] = 1.0 *M*

Adding Small Amounts of Acid or Base to a Buffer Solution

The pH of a buffer system changes slightly when a small amount of acid or base is added.

Example 17.18: A particular buffer solution contains 0.020 M NH_3 and 0.015 M NH_4Cl. K_b of NH_3 is 1.77×10^{-5}.
(a) Find the pH of the buffer solution.
(b) Find the pH of the buffer solution after 0.20 mL (about 4 drops) of 5.0 M HCl is added to one liter of the buffer solution.
(c) Find the pH of the buffer solution after 0.20 mL of 5.0 M NaOH is added to one liter of the buffer solution.

Calculations:

(a) $$pH = pK_a + \log\frac{[NH_3] \text{ base}}{[NH_4^+] \text{ acid}}$$

$$pH = -\log\frac{1.00 \times 10^{-14}}{1.77 \times 10^{-5}} + \log\frac{0.020\ M}{0.015\ M} = 9.248 + 0.12 = \underline{9.37}$$

(b) We expect that adding acid will lower the pH. Let us find by how much. Because HCl is a strong acid it will all react with the base component (NH_3) of the buffer system converting NH_3 into NH_4^+.

$H^+(aq) + NH_3(aq) \rightarrow NH_4^+(aq)$

The number of moles of HCl added to one liter of buffer is

$$0.20\text{ mL} \times \frac{5.0\text{ mol}}{1000\text{ mL}} = 1.0 \times 10^{-3}\text{ mol}$$

This number of moles of NH_3 (1.0×10^{-3} mol) are therefore converted to NH_4^+. The new concentrations are
$[NH_4^+] = 0.015\text{ M} + 0.0010\text{ M} = 0.016\ M$
$[NH_3] = 0.020\text{ M} - 0.0010\text{ M} = 0.019\ M$

$$pH = pKa + \log\frac{[NH_3]}{[NH_4^+]} = -\log\frac{(1.00 \times 10^{-14})}{(1.77 \times 10^{-5})} + \log\frac{0.019}{0.016} = 9.248 + 0.075 = \underline{9.321}$$

(c) We expect base to raise the pH. Hydroxide ion reacts with the acid component of the buffer (NH_4^+) according to

$NH_4^+(aq) + OH^-(aq) \rightarrow NH_3(aq) + H_2O(l)$

Therefore the 1.0×10^{-3} mol OH^- will form 1.0×10^{-3} mol of NH_3 and will use up 1.0×10^{-3} mol of NH_4^+.

The new concentrations are:
$[NH_3] = 0.020\ M + 0.0010\ M = 0.021\ M$;
$[NH_4^+] = 0.015\ M - 0.0010\ M = 0.014\ M$.

The new pH is

$$pH = 9.248 + \log\frac{0.021}{0.014} = 9.248 + 0.18 = \underline{9.43}$$

Preparing Buffer Solutions

The buffer capacity is the number of moles of strong acid or strong base required to change the pH of one liter of buffer by one unit.

The buffer capacity is therefore a measure of the buffer's ability to resist changes in pH; the greater the buffer capacity, the more resistant the buffer to changes in pH.

The buffer capacity increases with increasing concentrations of conjugate acid and base and also varies with the buffer ratio.

The most effective buffers have a buffer ratio of 1.0, so that

$pH = pK_a + \log 1 = pK_a$

$pH = pK_a$

Most buffers have a pH within one unit of the pK_a.

Example 17.19: How many moles of sodium acetate must be added to 500 mL of 0.50 *M* acetic acid to make a buffer solution of pH = 5.00? ($K_a = 1.76 \times 10^{-5}$)

Given: pH = 5.00; 500 mL of 0.50 *M* acetic acid

Unknown: moles of sodium acetate

Plan: Use $pH = pK_a + \log [base]/[acid]$ to find [base]. (Note that this pH is slightly higher than the pK_a, so that the base concentration must exceed the acid concentration.)

Calculations: $$5.00 = -\log(1.76 \times 10^{-5}) + \log \frac{[\text{base}]}{0.50} = 4.75 + \log \frac{[\text{base}]}{0.50}$$

Rearrangement gives

$$\log \frac{[\text{base}]}{0.50} = 0.25\ ; \qquad \frac{[\text{base}]}{0.50} = \text{antilog}(0.25) = 1.8$$

$[\text{base}] = 1.8 \times 0.50\ M = 0.90\ M$

The number of moles of sodium acetate that must be added to 500 mL of the solution is

$$0.500\text{ L} \times \frac{0.90\text{ mol}}{1\text{ L}} = \underline{0.45\text{ mol sodium acetate}}$$

A Closer Look at Acid-Base Titration (Section 17.8)

Learning Objectives

1. Calculate the equivalence point pH for a given titration.
2. Sketch and compare titration curves for (a) a strong acid titrated with a strong base, (b) a weak acid titrated with a strong base, and (c) a weak base titrated with a strong acid.
3. Use a titration curve to choose a suitable indicator for the titration.

Review

In an acid-base titration, one reactant is gradually added to a known amount of the other reactant until a color change or a meter reading indicates the endpoint. The endpoint should be close to the equivalence point, the point at which neither reactant is in excess. The pH at the equivalence point depends on the salt formed during the titration. The acid-base behavior of the salt depends on the strengths of the acid and base used in the titration:

strong acid – strong base	pH = 7.0
weak acid – strong base	pH > 7.0
strong acid – weak base	pH < 7.0

Example 17.20: Calculate the pH at the equivalence point when 50.00 mL of 0.1500 *M* methylamine is titrated with 0.1500 *M* HCl. (K_b for methylamine is 3.70×10^{-4})

Given: 50.00 mL of 0.1500 *M* base; 0.1500 *M* HCl added; $K_b = 3.70 \times 10^{-4}$

Unknown: pH at equivalence point

Plan:

(1) Use the initial moles of base to determine the moles of salt formed. The equation is

$$CH_3NH_2(aq) + HCl(aq) \rightarrow CH_3NH_3^+(aq) + Cl^-(aq)$$

(2) Find the molarity of the salt by dividing the moles (from Step 1) by the final volume.

(3) Determine which ion of the salt will react with water, then construct a concentration summary to find pH.

Calculations:

(1) moles of salt = initial moles of base $= 0.05000\ L \times \dfrac{0.1500\ mol}{1\ L} = 7.500 \times 10^{-3}\ mol\ CH_3NH_3^+$

(2) The added volume of HCl was that which contains 7.500×10^{-3} mol HCl or 50.00 mL. (This was easy since the acid and base concentrations were equal, their volumes had to be equal as well.)

Therefore, final volume = 50.00 mL + 50.00 mL = 100.00 mL

$$\underset{\text{methylamine}}{\underset{\uparrow}{50.00\ mL}} + \underset{\text{HCl}}{\underset{\uparrow}{50.00\ mL}}$$

$$\text{molarity of salt} = \frac{7.500 \times 10^{-3}\ mol}{0.1000\ L} = 0.07500\ M$$

(3) The Cl^- (conjugate base of a strong acid) does not react with water. $CH_3NH_3^+$ (conjugate base of a weak acid) does. The K_a for $CH_3NH_3^+$ will be calculated from K_b for CH_3NH_2.

Concentration Summary	$CH_3NH_3^+(aq) + H_2O(l)$	$\rightleftharpoons$	$CH_3NH_2(aq)$	+ $H_3O^+(l)$
Initial (mol/L)	0.07500		0	0
Change (mol/L)	$-x$		$+x$	$+x$
Equilibrium (mol/L)	$0.07500 - x$		x	x
Making approximations (mol/L)	0.07500		x	x

$$K_a = K_w/K_b = 1.00 \times 10^{-14}/3.70 \times 10^{-4} = 2.70 \times 10^{-11} = \frac{x^2}{0.07500}$$

$x = 1.42 \times 10^{-6} = [H_3O^+]$; pH = -log $[H_3O^+]$ = <u>5.848</u>

Answer Check: pH should be < 7.0 because $CH_3NH_3^+$ is a conjugate acid.

Titration Curves

A <u>titration curve</u> is a plot of pH versus volume of acid or base added during a titration. The shape of the curve is determined by the strengths of the acid and base. You should be familiar with the curves and be able to calculate the pH at any point for three types of titration:

(i) strong acid with strong base (see Figure 17.3, p. 724, of your textbook)

(ii) weak acid with strong base (see Figure 17.4, p. 727, of your textbook)

(iii) weak base with strong acid (see Figure 17.5, p. 728, of your textbook)

PROBLEM-SOLVING TIP

Summary of pH Calculations for Three Types of Titrations

	Strong Acid-Strong Base	(HX) Weak Acid-Strong Base	(B) Weak Base-Strong Acid
1. Before titration begins.	$HA \rightarrow H^+ + A^-$ $[H^+]$ = HA molarity (Note; pH < 7.0)	$HX \rightleftharpoons H^+ + X^-$ Use K_a and initial concentration of HX to find pH (Note: pH < 7.0)	$B + H_2O \rightarrow HB^+ + OH^-$ Use K_b and initial concentration of B to find OH^- then pH. (Note: pH > 7.0)
2. Before equivalence point.	$H^+ + OH^- \rightarrow H_2O$ (1) moles of H^+ = initial moles of HA minus moles of OH^- added. (2) divide (1) by the total volume to get $[H^+]$. (Note: pH < 7.0)	$HX + OH^- \rightarrow X^- + H_2O$ (1) moles of HX = initial moles of HX minus moles of OH^- added. (2) moles of X^- = moles OH^- added. (3) use $pH = pK_a + \log X^-/HX$ (Note: pH < 7.0)	$B + H^+ \rightarrow HB^+$ (1) moles of B = initial moles of B minus moles of H^+ added. (2) moles of HB^+ = moles of H^+ added. (3) use $pH = pK_a + \log B/BH^+$. (Note pH > 7.0)
3. At equivalence point.	pH = 7.0	$X^- + H_2O \rightleftharpoons HX + OH^-$ (1) moles of X^- = initial moles of HX (2) divide (1) by total volume to get $[X^-]$ (3) use K_b to get $[OH^-]$. (Note: pH > 7.0)	$HB^+ + H_2O \rightleftharpoons H_3O^+ + B$ (1) moles of HB^+ = initial moles of B (2) divide (1) by total volume to get $[HB^+]$ (3) use K_a to find $[H^+]$. (Note: pH < 7.0)
4. After equivalence point.	$[OH^-]$ is in excess (1) moles of OH^- in excess = moles of OH^- added minus initial moles of H^+ (2) divide (1) by total volume to get $[OH^-]$. (Note: pH > 7.0)	$[OH^-]$ is in excess (1) moles of OH^- in excess = moles of OH^- added minus initial moles of acid (2) divide (1) by total volume to get $[OH^-]$ (Note: pH > 7.0)	$[H^+]$ is in excess (1) moles of H^+ in excess = moles of H^+ added minus initial moles of B (2) divide (1) by total volume to get $[H^+]$. (Note: pH < 7.0)

PROBLEM-SOLVING TIP

Acid-Base Problems

Many students find acid-base chemistry confusing because there seems to be so many different kinds of problems. It is therefore a good idea to list the different types of problems and understand their solutions. As a guide, some pH problems and their solution plans are listed below. Think of others with different unknowns and add to the list. Be sure to always define x since the meaning of x depends on the particular problem. See below.

	Given	Unknown	Plan
I	Initial concentration of weak acid, K_a General equation: Weak acid + $H_2O \rightleftharpoons$ conjugate base + H_3O^+ *Example:* CH_3COOH (aq) + $H_2O(l) \rightleftharpoons CH_3COO^-(aq) + H_3O^+(aq)$	pH	1. Let $x = [H^+]$ = [conjugate base] and [weak acid] = initial concentration – x. 2.◆ Decide whether the approximation [weak acid] = initial concentration is valid. 3. Substitute concentrations into K_a expression and solve for x. 4. pH = -log x
II	Initial concentration of weak base, K_b General equation: Weak base + $H_2O \rightleftharpoons$ Conjugate acid + OH^- *Example:* $NH_3(aq) + H_2O(l) \rightleftharpoons NH_4^+(aq) + OH^-(aq)$	pH	1. Let $x = [OH^-]$ = [conjugate acid] and [weak base] = initial concentration – x. 2.◆ Decide on whether the approximation [weak base] = initial concentration is valid. 3. Substitute concentrations into K_b expression. 4. pOH = -log x; pH = 14.00 – pOH or $[H^+] = 1.00 \times 10^{-14}/x$; pH = -log$[H^+]$
III	Initial concentration of a salt of a weak acid (conjugate base), K_a General equation: conjugate base + $H_2O \rightleftharpoons$ weak acid + OH^- *Example:* $CH_3COO^-(aq) + H_2O(l) \rightleftharpoons CH_3COOH(aq) + OH^-(aq)$	pH	1. Let $x = [OH^-]$ = [weak acid] and [base] = initial concentration – x. 2.◆ Decide on whether [base] = initial concentration. 3. Substitute concentrations into K_b; $K_b = K_w/K_a$ 4. See 4, in II above
IV	Initial concentration of a salt of a weak base (conjugate acid), K_b General equation: conjugate acid + $H_2O \rightleftharpoons$ weak base + H_3O^+ *Example:* $NH_4^+(aq) + H_2O(l) \rightleftharpoons NH_3(aq) + H_3O^+(aq)$	pH	1. Let $x = [H^+]$ = [weak base]; [acid] = initial concentration – x 2.◆ Same as 2. in III above but for [acid]. 3. Substitute concentrations into K_a; $K_a = K_w/K_b$ 4. pH = -log x

	Given	Unknown	Plan
V	Initial concentration of weak acid, Initial concentration of salt of weak acid [conjugate base], K_a;	pH	1. Let $x = [H^+]$; [weak acid] = initial concentration [base] = initial concentration 2. Substitute concentrations into K_a and solve for x or use $pH = pK_a + \log([base]/[acid])$; $pK_a = -\log K_a$
VI	Initial concentration of weak acid, concentration of added strong base, K_a General equation: Weak acid + OH^- $\rightarrow$ conjugate base + H_2O *Example:* $CH_3COOH(aq) + OH^-(aq) \rightarrow CH_3COO^-(aq) + H_2O(l)$	pH	1. Let $x = [H^+]$; [weak acid] = initial concentration – concentration of strong base added ; [conjugate base] = concentration of strong base added. 2. Same as 2. in V above.
VII	Initial concentration of weak base, Initial concentration of salt of weak base, K_b	pH	?

♦ Approximation is valid when neglected quantity (x) is less than 5% of retained quantity (initial concentration). This is generally true when K is small and initial concentrations are not very small.

SELF-TEST

1. Find the pH of a 0.20 M NH_4Cl solution. K_b for NH_3 is 1.8×10^{-5}.

2. Rank the following aqueous solutions in order of increasing pH:
   (a) 0.10 M HCl
   (b) 0.20 M HCl
   (c) 0.10 M CH_3COOH ($K_a = 1.8 \times 10^{-5}$)
   (d) 0.10 M C_6H_5OH ($K_a = 1.28 \times 10^{-10}$)
   (e) 0.10 M H_3PO_4 ($pK_a = 2.124$)
   (f) 0.10 M NH_3 ($K_b = 1.8 \times 10^{-5}$)
   (g) 0.10 M $NaNO_2$ (K_a for $HNO_2 = 6.0 \times 10^{-4}$)
   (h) 0.10 M NaCl

3. (a) Find the percent ionization of 0.10 M CH_3COOH.. ($K_a = 1.8 \times 10^{-5}$)
   (b) The percent ionization of 0.010 M CH_3COOH is (greater, less than) the percent ionization calculated in (a).

4. Find the S^{2-} concentration in a 0.15 M H_2S solution. ($K_{a_1} = 1.02 \times 10^{-7}$; $K_{a_2} = 1 \times 10^{-13}$)

5. How many moles of NH_4Cl must be dissolved per liter of solution to produce a pH of 5.00? (K_b for NH_3 is 1.77×10^{-5})

6. Some solid NaHS is dissolved in each of the following solutions. Indicate whether the pH will increase, decrease, or remain the same. For H_2S $K_{a_1} = 1.02 \times 10^{-7}$; $K_{a_2} = 1 \times 10^{-13}$
   (a) 0.10 M H_2S
   (b) 0.10 M KHS
   (c) 0.10 M Na_2S

7. A buffer solution is prepared by dissolving 0.030 mol of NaH_2PO_4 and 0.025 mol of Na_2HPO_4 in one liter of water. (pK_{a_2} of H_3PO_4 = 7.206)
   (a) Find the pH of the solution.
   (b) Find the pH after 0.0010 mol of HCl are added to the above buffer solution.

8. How many grams of sodium acetate must be added to one liter of a 0.20 *M* solution of acetic acid to produce a buffer solution if pH = 5.0? ($pK_a = 4.75$)

ANSWERS TO SELF-TEST QUESTIONS

1. $1.00 \times 10^{-14}/1.8 \times 10^{-5} = x^2/0.20$; $x = [H^+] = 1.1 \times 10^{-5}$ M; pH = $-\log(1.1 \times 10^{-5})$ = <u>4.96</u>

2. (b) < (a) < (e) < (c) < (d) < (h) < (g) < (f)

3. (a) $1.8 \times 10^{-5} = x^2/0.10$; $x = 1.3 \times 10^{-3}$; % ionization = $(1.3 \times 10^{-3}/0.10) \times 100$ = <u>1.3%</u>
   (b) greater (% ionization increases with dilution)

4. $[S^{2-}] = K_{a_2} = 1 \times 10^{-13}$ *M*

5. $NH_4^+(aq) + H_2O(l) \rightleftharpoons NH_3(aq) + H_3O^+(aq)$; $K_a = 1.00 \times 10^{-14}/1.77 \times 10^{-5} = (1.0 \times 10^{-5})^2/x$; x = <u>0.18 moles</u>

6. (a) increase (adding the conjugate base of H_2S makes the solution more basic.)
   (b) increase (concentration of HS^-; a basic anion, is increased.)
   (c) decrease (adding conjugate acid of S^{2-} makes the solution more acidic.)

7. (a) $\text{pH} = 7.206 + \log \dfrac{(0.025)}{(0.030)} = \underline{7.13}$

   (b) $\text{pH} = 7.206 + \log \dfrac{(0.025 - 0.0010)}{(0.030 + 0.0010)} = \underline{7.09}$

8. $5.0 = 4.75 + \log(x/0.20)$; $x = 0.36$ *M*; 0.36 mol/L × 82 g/mol = <u>30 g</u>

CHAPTER 18

SOLUBILITY AND COMPLEX ION EQUILIBRIA

SELF-ASSESSMENT

The Solubility Product (Section 18.1)

1. The molar solubility of $BaSO_4$ is 1.05×10^{-5} *M*. Calculate the K_{sp} value for $BaSO_4$.

2. Find the molar solubility of $BaSO_4$ in a 0.10 *M* $BaCl_2$ solution. (Use your answer to problem 1.)

3. Calculate the pH of a solution saturated with $Mg(OH)_2$. ($K_{sp} = 7.1 \times 10^{-12}$)

Precipitation and the Ion Product (Section 18.2)

4. The concentration of Pb^{2+} in a sample of drinking water is 1.0×10^{-6} *M*. Will a precipitate form when the pH of the solution is (a) 7.0 (b) 10.0? The K_{sp} for $Pb(OH)_2$ is 1.2×10^{-15}.

5. The K_{sp} for $CaCrO_4$ is 7.1×10^{-4} and the K_{sp} for $PbCrO_4$ is 2.8×10^{-13}.
   (a) Calculate the K_2CrO_4 concentration that would give the maximum separation of Ca^{2+} and Pb^{2+} in a solution that is 0.025 *M* in each ion.
   (b) Find the concentration of each ion remaining in solution after precipitation.

Sparingly Soluble Hydroxides (Section 18.3)

6. The K_{sp} for $Zn(OH)_2$ is 4.5×10^{-17}.
   (a) Calculate the maximum concentration of Zn^{2+} in a solution of pH 6.0.
   (b) At what pH will 75% of Zn^{2+} in a 0.25 *M* $Zn(NO_3)_2$ solution precipitate out as $Zn(OH)_2$?

7. A solution contains 1.0 *M* $Cu(NO_3)_2$ and 0.25 *M* acetic acid. What is the maximum number of moles of sodium acetate that can be added to one liter and precipitation not occur? The K_{sp} for $Cu(OH)_2$ is 1.6×10^{-19}; K_a for acetic acid is 1.8×10^{-5}.

8. $Al(OH)_3$ is amphoteric while $Ca(OH)_2$ is not. What reagent could be used to distinguish between these two white solids? Write the equation.

Solubility and pH (Section 18.4)

9. Indicate which of the following sparingly soluble salts dissolve in strong acid and write the relevant molecular equations: $PbBr_2$, $SrCO_3$, Ag_2S, and CaF_2.

10. Calculate the molar solubility of CuS in a saturated H_2S solution with pH = 1.0. $K_{spa} = 6 \times 10^{-16}$

11. Determine the pH that will give the maximum separation of Fe^{2+} and Sn^{2+} when the solution is 0.010 *M* in both $Fe(NO_3)_2$ and $Sn(NO_3)_2$ and is also saturated with H_2S (0.10 *M*). The K_{spa} of FeS is 6×10^2; the K_{spa} of SnS is 1×10^{-5}.

Complex Ions (Section 18.5)

12. Estimate the concentration of Ni^{2+} and $Ni(NH_3)_6^{\;2+}$ in a solution made by dissolving 0.10 mol of $Ni(NO_3)_2$ in one liter of 1.0 *M* NH_3. ($K_f = 5.5 \times 10^8$)

13. Will a precipitate form when 0.10 mol of Na_2S is added to one liter of the solution described in problem 12? The K_{sp} for NiS is 4×10^{-20}.

14. Three complexing agents, NH_3, CN^-, and $S_2O_3^{\;2-}$, are each added in excess to 0.10 *M* $AgNO_3$. Which complex ion has the greatest concentration? K_f's: $Ag(NH_3)_2^{\;2+}$, 1.6×10^7; $Ag(CN)_2^{\;-}$, 1×10^{21}; $Ag(S_2O_3)_2^{\;3-}$, 1.7×10^{13}.

ANSWERS TO SELF-ASSESSMENT PROBLEMS

If you missed an answer, be sure to study the relevant section in the textbook and study guide.

1. $K_{sp} = [Ba^{2+}][SO_4^{\;2-}] = (1.05 \times 10^{-5})^2 = \underline{1.10 \times 10^{-10}}$

2. $K_{sp} = [Ba^{2+}][SO_4^{\;2-}] = 1.10 \times 10^{-10} = (0.10)[SO_4^{\;2-}]$; $[SO_4^{\;2-}] = \underline{1.1 \times 10^{-9}\ M}$ = molar solubility

3. $K_{sp} = [Mg^{2+}][OH^-]^2 = (x)(2x)^2 = 7.1 \times 10^{-12}$; $x = 1.2 \times 10^{-4}\ M$; $[OH^-] = 2.4 \times 10^{-4}$ M
$[H^+] = 1.0 \times 10^{-14}/2.4 \times 10^{-4} = 4.2 \times 10^{-11}$; pH = $\underline{10.38}$

4. (a) $Q = [Pb^{2+}][OH^-]^2 = (1.0 \times 10^{-6})(1.0 \times 10^{-7})^2 = 1.0 \times 10^{-20}$; no precipitate
(b) $Q = (1.0 \times 10^{-6})(1.0 \times 10^{-4})^2 = 1.0 \times 10^{-14}$; precipitate forms

5. (a) $7.1 \times 10^{-4} = (0.025)[CrO_4^{\;2-}]$; $[CrO_4^{\;2-}] = \underline{0.028\ M} = [K_2CrO_4]$
(b) $[Ca^{2+}] = 0.025\ M$; $[Pb^{2+}] = 2.8 \times 10^{-13}/(0.028) = \underline{1.0 \times 10^{-11}\ M}$

6. (a) $4.5 \times 10^{-17} = [Zn^{2+}](1.0 \times 10^{-8})^2$; $[Zn^{2+}] = \underline{0.45\ M}$
(b) $[Zn^{2+}] = 0.25\ M \times 0.25 = 0.0625\ M$; $4.5 \times 10^{-17} = (0.0625)[OH^-]^2$; $[OH^-] = 2.7 \times 10^{-8}$
$[H^+] = 1.0 \times 10^{-14}/2.7 \times 10^{-8} = 3.7 \times 10^{-7}$; pH = $\underline{6.43}$

7. $1.6 \times 10^{-19} = 1.0[OH^-]^2$; $[OH^-] = 4.0 \times 10^{-10}\ M$; $[H^+] = 1.0 \times 10^{-14}/4.0 \times 10^{-10} = 2.5 \times 10^{-5}\ M$
K_a = [H^+][acetate]/[acetic acid] = (2.5×10^{-5})[acetate]/(0.25) = 1.8×10^{-5}
[acetate] = $\underline{0.18\ M}$ (0.18 mol to one liter)

8. Add excess of NaOH to dissolve $Al(OH)_3$ according to $Al(OH)_3(s) + OH^-(aq) \rightarrow Al(OH)_4^{\;-}(aq)$

9. $SrCO_3$: $CO_3^{2-}(aq) + 2H^+(aq) \rightarrow H_2CO_3(aq)$
Ag_2S: $S^{2-}(aq) + 2H^+(aq) \rightarrow H_2S(aq)$
CaF_2: $F^-(aq) + H^+(aq) \rightarrow HF(aq)$

10. $$\frac{[Cu^{2+}](0.10)}{(1 \times 10^{-1})^2} = 6 \times 10^{-16}$$

$[Cu^{2+}] = \underline{6 \times 10^{-17}\ M}$

11. $$\frac{(0.01)(0.10)}{[H^+]^2} = 6 \times 10^2$$

$[H^+] = 1 \times 10^{-3}\ M$; pH = $\underline{3.0}$

12. $[NH_3] = 1.6\ M - (6 \times 0.10\ M) = 1.0\ M$

$$K_d = \frac{1}{K_f} = \frac{[Ni^{2+}][NH_3]^6}{[Ni(NH_3)_6{}^{2+}]} = \frac{[Ni^{2+}](1.0)^6}{(0.10)} = \frac{1}{5.5 \times 10^8}$$

$[Ni^{2+}] = \underline{1.8 \times 10^{-10}\ M}$

13. (a) $Q = (1.8 \times 10^{-10})(0.10) = 1.8 \times 10^{-11}$; $Q > K_{sp}$; precipitate forms

14. $Ag(CN)_2^-$; (largest K_f)

REVIEW

The Solubility Product (Section 18.1)

Learning Objectives

1. Write K_{sp} expressions for sparingly soluble solids.
2. Calculate K_{sp} values from solubilities and vice versa.
3. Calculate the solubility of a sparingly soluble solid in solution containing a common ion.

Review

The solubility product constant (K_{sp}) is the equilibrium constant for a solid in equilibrium with its dissolved ions. For example, the K_{sp} expression for the equilibrium

$CuCl(s) \rightleftharpoons Cu^+(aq) + Cl^-(aq)$ is $K_{sp} = [Cu^+][Cl^-] = 1.7 \times 10^{-7}$.

Note: K_{sp}'s, like other equilibrium expressions, do not include solids. The K_{sp} expression can be determined from the formula of the compound alone; the exponents of the concentrations in the expression are the same as the subscripts in the formula.

Example 18.1: Write the equilibrium equations and give the K_{sp} expressions for $Al(OH)_3$ and MgF_2.

Answer: The equilibria are

$Al(OH)_3(s) \rightleftharpoons Al^{3+}(aq) + 3OH^-(aq)$ $\quad K_{sp} = [Al^{3+}][OH^-]^3$
$MgF_2(s) \rightleftharpoons Mg^{2+}(aq) + 2F^-(aq)$ $\quad K_{sp} = [Mg^{2+}][F^-]^2$

Practice Drill on Writing K_{sp} Expressions: Write the equilibrium equations and K_{sp} expressions for the following compounds:

(a) AgBr
(b) $Pb(Br)_2$
(c) $Sr_3(PO_4)_2$

Answers: (a) $AgBr(s) \rightarrow Ag^+(aq) + Br^-(aq)$ $\quad K_{sp} = [Ag^+][Br^-]$
(b) $PbBr_2(s) \rightarrow Pb^{2+}(aq) + 2Br^-(aq)$ $\quad K_{sp} = [Pb^{2+}][Br^-]^2$
(c) $Sr_3(PO_4)_2 \rightarrow 3Sr^{2+}(aq) + 2PO_4^{3-}(aq)$ $\quad K_{sp} = [Sr^{2+}]^3[PO_4^{3-}]^2$

K_{sp}'s are limited to sparingly soluble salts where ionic concentrations are low and interionic attractions can be ignored.

A K_{sp} value can be calculated from the solubility of the salt and vice versa provided that the ions of the salt do not react with water or any other dissolved species.

Calculating K_{sp} Values From Solubility Data

Example 18.2: The solubility of SnF_2 in water is 0.12 g/mL at 20°C. Find its K_{sp}.

Answer: The K_{sp} expression is

$K_{sp} = [Sn^{2+}][F^-]^2$.

Therefore, to determine the value of the K_{sp}, we must first find the concentration of Sn^{2+} and F^- ions in solution. We can get these values from the given solubility of the salt.

The steps are

1. Convert grams per liter to moles per liter for SnF_2.
2. Find molarity of each ion.
3. Substitute molarities into the K_{sp} expression.

1. The molar mass of SnF_2 is 156.7 g. Its molar solubility is

$$\frac{0.12 \text{ g SnF}_2}{1 \text{ L}} \times \frac{1 \text{ mol SnF}_2}{156.7 \text{ g SnF}_2} = 7.7 \times 10^{-4} \text{ mol SnF}_2\text{/L}$$

2. The formula SnF_2 indicates that the number of moles of Sn^{2+} in solution equals the number of moles of SnF_2, and the moles of F^- is twice that number.

$[Sn^{2+}] = 7.7 \times 10^{-4}$ mol/L; $[F^-] = 2 \times 7.7 \times 10^{-4}$ mol/L $= 15.4 \times 10^{-4}$ mol/L

3. Substituting the concentrations into the K_{sp} expression gives

$K_{sp} = [Sn^{2+}][F^-]^2 = (7.7 \times 10^{-4})(15.4 \times 10^{-4})^2 = \underline{1.8 \times 10^{-9}}$

Calculating Solubilities From K_{sp} Values

Treat this type of problem like any other equilibrium problem in which you are given the equilibrium constant and are asked to find a concentration.

Example 18.3: The K_{sp} of calcium fluoride is 3.9×10^{-11} at 25°C. Calculate its solubility in milligrams per liter of solution.

Answer: We will construct a concentration summary in which x is the solubility of CaF_2 in mol/L. The equation shows that x mol of CaF_2 produces x mol of Ca^{2+} and $2x$ mol of F^-.

Concentration Summary:	$CaF_2(s) \rightleftharpoons$	$Ca^{2+}(aq)$	+	$2F^-(aq)$
Initial (mol/L)		0		0
Changes (mol/L)		$+x$		$+2x$
Equilibrium (mol/L)		x		$2x$

Substituting the above equilibrium concentrations into the K_{sp} expression gives

$K_{sp} = [Ca^{2+}][F^-]^2 = 3.9 \times 10^{-11}$;
$(x)(2x)^2 = 3.9 \times 10^{-11}$;
$4x^3 = 3.9 \times 10^{-11}$;
$x^3 = 9.8 \times 10^{-12}$

Taking the cube root of each side gives $x = \underline{2.1 \times 10^{-4}\ M}$

Note: Many calculators have y^x keys that you can use to find cube roots; in the above example, $y = 9.8 \times 10^{-12}$ and $x = 1/3$ or 0.3333.

The molar mass of CaF_2 is 78.08 g; the solubility in milligrams per liter is calculated as follows:

$$\frac{78.08 \text{ g } CaF_2}{1 \text{ mol } CaF_2} \times \frac{2.1 \times 10^{-4} \text{ mol } CaF_2}{1 \text{ L}} \times \frac{1000\text{mg}}{1 \text{ g}} = \underline{16 \text{ mg/L}}$$

Practice Drill on K_{sp} and Solubility: Fill in the following table (Molar masses: AgCl, 143.4 g; AgBr, 187.8 g):

Salt	$[Cl^-]$	Solubility in grams per 100 mL	K_{sp}
AgCl	________	1.92×10^{-4}	________
AgBr	________	________	5.0×10^{-13}

Answers: AgCl: $1.34 \times 10^{-5}\ M$; $K_{sp} = 1.79 \times 10^{-10}$ AgBr: $7.1 \times 10^{-7}\ M$; 1.3×10^{-5} g/100 mL

Common Ion Effect

The solubility of a sparingly soluble salt is reduced in a solution that already contains one of the ions of the salt. This is an example of the common ion effect discussed in Section 17.6 of your text and study guide.

<u>*Example 18.4*</u>: What is the molar solubility of stannous fluoride (SnF_2) in a 0.080 *M* NaF solution? The K_{sp} of SnF_2 is 1.8×10^{-9}.

Given:	Initial concentration of $F^- = 0.080\ M$; $K_{sp} = 1.8 \times 10^{-9}$
Unknown:	Molar solubility of SnF_2
Plan and Calculations:	Construct a concentration summary defining x as the moles per liter of SnF_2 that dissolves. The concentration of Sn^{2+} is x and the concentration of F^- is $0.080\ M + 2x$.

<u>Concentration Summary</u>: $SnF_2(s) \rightleftharpoons$	$Sn^{2+}(aq)$ +	$2F^-(aq)$
Initial (mol/L)	0	0.080 M
Changes (mol/L)	$+x$	$+2x$
Equilibrium (mol/L)	x	$0.080 + 2x$
Approximate (mol/L)	x	0.080

We expect x to be small because SnF_2 is a sparingly soluble salt (small K_{sp}). Hence, we can make the approximation that $0.080 + 2x = 0.080$. Substituting the equilibrium concentrations into the K_{sp} gives

$K_{sp} = [Sn^{2+}][F^-]^2 = 1.8 \times 10^{-9}$; $(x)(0.080)^2 = 1.8 \times 10^{-9}$, $x = \underline{2.8 \times 10^{-7}\ M}$

Note: The solubility of SnF_2 in a sodium fluoride solution is much less than in pure water (see Example 18.1).

Practice Drill on the Common Ion Effect and Solubility: Fill in the blanks for two salts added to a fluoride containing solution:

	<u>Initial F^-</u>	<u>Solubility (mol/L)</u>	<u>K_{sp}</u>
(a) CaF_2 added to NaF	0.10 *M*	________	3.9×10^{-11}
(b) MgF_2 added to KF	________	3.7×10^{-6}	3.7×10^{-8}

<u>Answers</u>: (a) 3.9×10^{-9} mol/L (b) 0.10 *M*

Precipitation and the Ion Product (Section 18.2)

Learning Objectives

1. Perform calculations to predict whether the mixing of two solutions will produce a precipitate.
2. Calculate the concentration of an ion required to maintain a given concentration of some other ion.
3. Calculate the concentration of a precipitating agent that will result in the maximum separation of two ions.
4. Calculate the concentration of ions remaining in solution after the addition of a precipitating agent.

Review

The reaction quotient Q for dissolving an ionic solid is called an ion product. In a saturated solution the ion product equals the K_{sp}. When the ion product is less than the K_{sp}, the solution is unsaturated and more solid can dissolve. When the ion product is greater than the K_{sp}, a precipitate will form. The final solution will be saturated, with the precipitate in equilibrium with the dissolved ions.

To predict whether a precipitate will form when two solutions are mixed,
1. Find the molarity of each ion in the final solution.
2. Substitute the molarities into the ion product expression and calculate Q. Compare Q with the K_{sp} value.

$Q < K_{sp}$	No precipitate (Solution is unsaturated.)
$Q = K_{sp}$	No precipitate (Solution is saturated.)
$Q > K_{sp}$	Precipitate forms. (Final solution is saturated.)

Example 18.5: A solution is made by mixing 100 mL of 0.010 M $Mg(NO_3)_2$ with 200 mL of 0.025 M NaOH. Will $Mg(OH)_2$ precipitate?

Given: 100 mL of 0.010 M $Mg(NO_3)_2$
200 mL of 0.025 M NaOH
$K_{sp} = 7.1 \times 10^{-12}$

Unknown: precipitate?

Plan and Calculations: Follow the two steps given above:

1. The number of moles of each ion is
   Mg^{2+}: 0.100 L × 0.010 mol/L = 0.0010 mol
   OH^-: 0.200 L × 0.025 mol/L = 0.0050 mol

   The volume of the solution is 100 mL + 200 mL = 300 mL. The molarities are
   Mg^{2+}: 0.0010 mol/0.300 L = 0.0033 mol/L
   OH^-: 0.0050 mol/0.300 L = 0.017 mol/L

2. Substituting the above molarities into Q gives
   $Q = [Mg^{2+}][OH^-]^2 = (0.0033)(0.017)^2 = 9.5 \times 10^{-7}$
   $Q > K_{sp}$: A precipitate will form.

Practice: One liter of a 1.0×10^{-3} M $Pb(NO_3)_2$ solution is combined with one liter of a 2.5×10^{-3} M NaCl solution. The K_{sp} of $PbCl_2$ is 1.7×10^{-5}. Will precipitation occur? (Ans: No)

K_{sp} values can also be used to determine the concentration of an ion needed to maintain a given concentration of some other ion.

Example 18.6: The desired concentration of Sn^{2+} in a particular solution is 2.8×10^{-7} mol/L. The K_{sp} of SnF_2 is 1.8×10^{-9}. What concentration of NaF is required?

Given: $[Sn^{2+}] = 2.8 \times 10^{-7}$ M; $K_{sp} = 1.8 \times 10^{-9}$

Unknown: $[NaF] = [F^-]$

Plan: Substitute $[Sn^{2+}]$ into the K_{sp} expression and solve for $[F^-]$.

Calculations: $K_{sp} = [Sn][F^-]^2 = 1.8 \times 10^{-9}$
$(2.8 \times 10^{-7})[F^-]^2 = 1.8 \times 10^{-9}$
$[F^-] = 0.080\ M$

Practice Drill on Precipitation Reactions: Fill in the blanks for these three solutions that contain Pb^{2+} and Cl^- ions. ($K_{sp} = 1.7 \times 10^{-5}$)

	$[Pb^+]$	$[Cl^-]$	Q	Will a precipitate form?
(a)	1.0×10^{-3} *M*	________	1.7×10^{-5}	________
(b)	0.025 *M*	0.10 *M*	________	________
(c)	1.0×10^{-5}	0.020 *M*	________	________

Answers: (a) 0.13 *M*; no (b) 2.5×10^{-4}; yes (c) 4.0×10^{-9}; no

Fractional Precipitation

Ions can be separated by fractional precipitation, a procedure in which the concentration of precipitating agent is adjusted so as to precipitate less soluble salts and leave those that are more soluble in solution.

Example 18.7: A solution contains 0.25 *M* Mg^{2+} and 0.25 *M* Pb^{2+}. What concentration of CO_3^{2-} will precipitate the maximum amount of $PbCO_3$ without precipitating $MgCO_3$? (K_{sp}'s: $PbCO_3$, 7.4×10^{-14}; $MgCO_3$, 1.6×10^{-6})

Answer: The K_{sp} values indicate that $MgCO_3$ is less soluble than $PbCO_3$. The carbonate concentration should be close to saturating the solution with respect to $MgCO_3$. Therefore we will use the K_{sp} expression for $MgCO_3$:

$$K_{sp} = [Mg^{2+}][CO_3^{2-}] = 1.6 \times 10^{-6}$$
$$(0.25)[CO_3^{2-}] = 1.6 \times 10^{-6}$$
$$[CO_3^{2-}] = \underline{6.4 \times 10^{-6}\ M}$$

Note: A carbonate concentration greater than 6.4×10^{-6} *M* will precipitate both $MgCO_3$ and $PbCO_3$; a smaller concentration will precipitate only the less soluble $PbCO_3$.

The efficiency of the separation of the two cations in Example 18.7 can be determined by comparing the Pb^{2+} concentration in solution before and after the separation.

Example 18.8: Refer to Example 18.7 and calculate the concentration of Pb^{2+} in the solution after the maximum separation has occurred.

Answer: In Example 18.7 we found that maximum separation occurs when the concentration of carbonate ion is 6.4×10^{-6} *M*. At this concentration the solution contains some Pb^{2+} in equilibrium with solid $PbCO_3$ and all of the Mg^{2+}. We will therefore use the K_{sp} for $PbCO_3$ to calculate the concentration of the remaining Pb^{2+}.

$$K_{sp} = [Pb^{2+}][CO_3^{2-}] = 7.4 \times 10^{-14}$$
$$= [Pb^{2+}](6.4 \times 10^{-6}) = 7.4 \times 10^{-14}$$
$$[Pb^{2+}] = \underline{1.2 \times 10^{-8}}$$

The Pb^{2+} concentration has decreased from 0.25 to 1.2×10^{-8} *M*, therefore, most of the Pb^{2+} has precipitated out of solution. Only 4.8×10^{-6} % (1.2×10^{-8} *M*/0.25 *M* × 100%) of the initial Pb^{2+} concentration remains in solution.

Practice: A solution contains 0.20 *M* Hg_2^{2+}, 0.20 *M* Pb^{2+} and Br^- ions.

(a) Calculate the concentration of Br^- that will give the best separation of the cations.

(b) Find the concentration of both Hg_2^{+} and Pb^{2+} in the solution remaining after the separation has occurred.

K_{sp}'s: Hg_2Br_2, 1.3×10^{-22}; $PbBr_2$, 2.1×10^{-6}

Answers: (a) 3.2×10^{-3} *M* (b) $[Pb^{2+}] = 0.20$ *M*; $[Hg_2^{2+}] = 1.3 \times 10^{-17}$ *M*

Sparingly Soluble Hydroxides (Section 18.3)

Learning Objectives

1. Calculate the OH^- concentration needed to precipitate a sparingly soluble hydroxide.
2. Calculate the pH below which a sparingly soluble hydroxide will not precipitate.
3. Calculate the composition of a buffer that will prevent a sparingly soluble hydroxide from precipitating.
4. Write equations for the reaction of amphoteric hydroxides and their corresponding oxides with strong acids and strong hydroxide bases.

Review

Raising the pH (increasing the OH^- concentration) increases the ion product of sparingly soluble hydroxides and favors precipitation; lowering the pH (decreasing the OH^- concentration) decreases the ion product, and favors dissolving. Therefore, sparingly soluble hydroxides tend to precipitate in basic solution and dissolve in acidic solution.

Example 18.9: A sample of groundwater contains 5.0×10^{-3} *M* Ca^{2+}. Above what pH will $Ca(OH)_2$ precipitate? The K_{sp} of $Ca(OH)_2$ is 6.5×10^{-6}.

Given: $[Ca^{2+}] = 5.0 \times 10^{-3}$ *M*; $K_{sp} = 6.5 \times 10^{-6}$

Unknown: pH value above which precipitate forms.

Plan:
1. Substitute given $[Ca^{2+}]$ into the K_{sp} expression and solve for OH^-.
2. Calculate pH.

Calculations:

1. $K_{sp} = [Ca^{2+}][OH^-]^2 = 6.5 \times 10^{-6}$
   $(5.0 \times 10^{-3})[OH^-]^2 = 6.5 \times 10^{-6}$
   $[OH^-] = 3.6 \times 10^{-2}$ *M*
2. $K_w = [OH^-][H^+] = 1.0 \times 10^{-14}$
   $(3.6 \times 10^{-2})[H^+] = 1.0 \times 10^{-14}$
   $[H^+] = 2.8 \times 10^{-13}$ *M*
   $pH = -\log [H^+] = -\log (2.8 \times 10^{-13})$
   pH = 12.55

Practice: The same groundwater sample contains 1.0×10^{-3} *M* Mg^{2+}. Below what pH will $Mg(OH)_2$ not precipitate? The K_{sp} for $Mg(OH)_2$ is 7.1×10^{-12}. Answer: 9.92

Use of Buffers

Precipitation of sparingly soluble hydroxides can be prevented by buffering the solution to maintain the OH^- concentration below a certain value, which will depend on the concentration of the metal ion in the solution.

Example 18.10: A solution is 2.0×10^{-3} M Pb^{2+} and 0.10 M NH_3. Calculate the minimum NH_4^+ concentration needed to prevent precipitation of $Pb(OH)_2$. [$K_{sp} = 1.2 \times 10^{-15}$; $K_b(NH_3) = 1.8 \times 10^{-5}$]

Note: NH_3 ionizes to form OH^- ions, which could precipitate $Pb(OH)_2$. Therefore, we need to repress the ionization of NH_3. The NH_4^+ ion is a common ion that will do this.

Given: $[Pb^{2+}] = 2.0 \times 10^{-3}$ M $\quad K_{sp} = 1.2 \times 10^{-5}$
$[NH_3] = 0.10$ M $\quad K_b = 1.8 \times 10^{-5}$

Unknown: $[NH_4^+]$ needed to prevent precipitation.

Plan:
1. Substitute the given $[Pb^{2+}]$ into the K_{sp} expression and solve for $[OH^-]$.
2. Substitute $[OH^-]$ from step 1 and given $[NH_3]$ into K_b expression for NH_3 and solve for $[NH_4^+]$.

Calculations:
1. $K_{sp} = [Pb^{2+}][OH^-]^2 = 1.2 \times 10^{-15}$
$(2.0 \times 10^{-3})[OH^-]^2 = 1.2 \times 10^{-15}$
$[OH^-] = 7.7 \times 10^{-7}$

2. $$K_b = \frac{[NH_4^+][OH^-]}{[NH_3]} = 1.8 \times 10^{-5}; \quad \frac{[NH_4^+](7.7 \times 10^{-7})}{(0.10)} = 1.8 \times 10^{-5}$$

$[NH_4^+] = \underline{2.3\ M}$

Practice Drill A: Preventing precipitation with a NH_3/NH_4^+ buffer. (K_{sp}: $Mg(OH)_2$, 7.1×10^{-12}; $Ca(OH)_2$, 6.5×10^{-6}; K_b for NH_3 is 1.8×10^{-5})

[Metal ion]	$[NH_3]$	$[NH_4^+]$ needed
(a) 0.010 M Mg^{2+}	0.10 M	______
(b) 0.10 M Ca^{2+}	1.0 M	______

Answers: (a) 6.8×10^{-2} M; (b) 2.2×10^{-3} M

Practice Drill B: Predicting hydroxide precipitation in a NH_3/NH_4^+ buffer. (K_{sp}: $Pb(OH)_2$, 1.2×10^{-15})

$[NH_3]$	$[NH_4^+]$	$[OH^-]$
0.10 M	0.10 M	______

Maximum possible concentration of cation:

$[Mg^{2+}]$ ______ $\quad [Pb^{2+}]$ ______

Answers: $[OH^-] = 1.8 \times 10^{-5}$; $[Mg^{2+}] = 0.022$ M; $[Pb^{2+}] = 3.7 \times 10^{-6}$ M

Amphoteric Hydroxides

Amphoteric hydroxides have the ability to neutralize both acids and bases. For example, $Al(OH)_3$ dissolves in acidic and basic solutions according to the equations

$Al(OH)_3(s) + 3H^+(aq) \rightarrow Al^{3+}(aq) + 3H_2O(l)$
$Al(OH)_3(s) + OH^-(aq) \rightarrow Al(OH)_4^-(aq)$

(See Table 18.2 in your text for other amphoteric hydroxides and the ions they form.)

Note: An oxide will undergo the same reactions as the corresponding hydroxide.

Solubility and pH (Section 18.4)

Learning Objectives

1. Predict whether the solubility of a given salt will be affected by pH.
2. Write equations for the reactions that occur when salts of weak acids dissolve in strong acids.
3. Calculate the concentration of metal ion that will remain unprecipitated in an H_2S solution at a given pH.
4. Calculate the pH at which two ions can be separated using H_2S as the precipitating agent.

Review

The solubility of a sparingly soluble salt of a weak acid increases with increasing acidity of the solution. For example, the solubility of $PbCO_3$ increases when nitric acid is added to the solution. The equation for the reaction with a small concentration of nitric acid is

$HNO_3(aq) + CO_3^{2-}(aq) \rightarrow HCO_3^-(aq) + NO_3^-(aq)$

(Greater concentrations of HNO_3 will convert the CO_3^- ion to H_2O and CO_2.)

The carbonate ion (a conjugate base of a weak acid) reacts with the strong acid. The carbonate ion concentration decreases, the ion product is lowered, and more $PbCO_3$ can dissolve.

Note: Generally, any reaction that removes or ties up an ion of the salt also increases the solubility of the salt.

pH Control of Sulfide Precipitations

Metal ions in solution are often identified by their reaction with H_2S to form sulfides. In an aqueous solution of H_2S the following equilibria exist:

$H_2S(aq) \rightleftharpoons H^+(aq) + HS^-(aq) \qquad K_{a_1} = 1.02 \times 10^{-7}$
$HS^-(aq) \rightleftharpoons H^+(aq) + S^{2-}(aq) \qquad K_{a_2} = 1 \times 10^{-19}$

Note that the second equilibrium can be ignored due to the small K_{a_2} value so that the $[S^{2-}]$ is negligible. In acidic solution reverse reactions are favored so that all of the sulfide will be in the form of $H_2S(aq)$. Precipitation of metal sulfides will be determined by equilibria such as

$MnS(s) + 2H^+(aq) \rightleftharpoons Mn^{2+}(aq) + H_2S(aq)$

$$K_{spa} = \frac{[Mn^{2+}][H_2S]}{[H^+]^2} = 3 \times 10^{10}$$

(K_{spa} = solubility product in acid solution)

Saturated solutions of H_2S are often used in precipitation reactions in which $[H_2S] = 0.10$ M at room temperature.

Example 18.11: Calculate the maximum concentration of unprecipitated Sn^{2+} in a 0.15 M HCl solution that is saturated with H_2S.

$K_{spa} = 1 \times 10^{-5}$ for SnS.

Given: $[H^+] = 0.15$ M
$[H_2S] = 0.10$ M (for saturated solution)
$K_{spa} = 1 \times 10^{-5}$

Unknown: $[Sn^{2+}]$

Plan: Substitute the $[H^+]$ and $[H_2S]$ concentrations into the K_{spa} expression for SnS and Solve for Sn^{2+}.

Calculation:

$$K_{spa} = \frac{[Sn^{2+}][H_2S]}{[H^+]^2} = 1 \times 10^{-5}$$

$$\frac{[Sn^{2+}](0.10)}{(0.15)^2} = 1 \times 10^{-5}\ ;\ [Sn^{2+}] = 2.5 \times 10^{-6}\ M$$

A $[Sn^{2+}]$ greater than 2.5×10^{-6} M will shift the equilibrium

$$SnS(s) + 2H^+(aq) \rightleftharpoons Sn^{2+}(aq) + H_2S(aq)$$

to the left causing SnS to precipitate.

Practice Drill on pH Control of Sulfide Precipitation Reactions: Fill in the blanks for the two manganese solutions saturated with H_2S given below. The K_{spa} for MnS is 3×10^{10}. At room temperature $[H_2S] = 0.10$ M in a saturated solution.

	$[Mn^{2+}]$	Minimum $[H^+]$ below which precipitation will occur	pH below which precipitation will not occur
(a)	2.5×10^{-4} *M*	__________	__________
(b)	0.025 *M*	__________	__________

Answers: (a) $[H^+] = 3 \times 10^{-8}$ *M*; pH = 7.5 (b) $[H^+] = 3 \times 10^{-7}$ *M*; pH = 6.5

Most metal sulfides are sparingly soluble so that metal ions can often be separated by the differential precipitation of their insoluble sulfide salts from solutions containing H_2S. The pH of these solutions are adjusted to maximize the separation of the metal ions. See Example 18.12 below.

Example 18.12: A saturated solution of H_2S (0.10 *M*) contains 0.15 *M* $Fe(NO_3)_2$ and 0.15 *M* $Cd(NO_3)_2$. What pH will give the best separation of Fe^{2+} and Cd^{2+}? K_{spa}'s: FeS, 6×10^2; CdS, 8×10^{-7}

Given: $[Fe^{2+}] = 0.15$ *M* $K_{spa} = 6 \times 10^2$ for FeS
$[Cd^{2+}] = 0.15$ *M* $K_{spa} = 8 \times 10^{-7}$ for CdS
$[H_2S] = 0.10$ *M*

Unknown: pH for maximum separation

Plan:

1. Use the K_{spa} for FeS because it is more soluble (greater K_{spa}). Substitute given Fe^{2+} and H_2S concentrations into the K_{spa} expression for FeS and solve for $[H^+]$. (This concentration of H^+ will cause most Cd^{2+} to precipitate out of solution and leave most Fe^{2+}.)
2. Calculate the pH.

Calculations:

1. $$\frac{[Fe^{2+}]^2[H_2S]}{[H^+]^2} = 6 \times 10^2$$

$$\frac{(0.15)(0.10)}{[H^+]^2} = 6 \times 10^2;\ [H^+] = 5 \times 10^{-3}$$

3. pH = <u>2.3</u>

Practice: If you added more $Fe(NO_3)_2$ to the solution in Example 18.12 would the pH for maximum separation increase or decrease? If you doubled the Fe^{2+} concentration what would be the required pH for maximum separation? <u>Answers</u>: decrease; pH = 1.7

Complex Ions (Section 18.5)

Learning Objectives

1. Give examples of complex ions and their uses.
2. Use formation and/or dissociation constants to predict which of two complex ions is more stable.
3. Calculate the concentration of uncomplexed metal ion in the presence of excess complexing agent.
4. Calculate the molar solubility of a salt in excess complexing agent.

Review

A complex ion consists of a central metal ion bonded to one or more surrounding molecules or ions called ligands. The charge on a complex ion is the sum of the charges on the metal ion and the ligands. Examples of complex ions are $Ag(NH_3)^{2+}$, $Fe(SCN)^{2+}$, and $Al(OH)_4^-$.

Complex ion formation is used in the laboratory to identify certain ions and to dissolve selectively certain precipitates.

Formation and Dissociation Constants

The equilibrium constant for the formation of a complex ion from the metal ion and complexing agent (ligand) is called a formation constant (K_f). The formation equation and K_f of $PbCl_3^-$ for example is

$$Pb^{2+}(aq) + 3Cl^-(aq) \rightarrow PbCl_3^-(aq) \qquad K_f = \frac{[PbCl_3^-]}{[Pb^{2+}][Cl^-]^3} = 25$$

Large formation constants indicate stable ions. The reciprocal of the formation constant is the dissociation constant, K_d:

$K_d = 1/K_f$

Stable ions have small dissociation constants.

Example 18.13: Rank the following complex ions in order of increasing stability using K_f values from Table 18.4 of your text: $Ni(NH_3)_6^{2+}$ $Fe(SCN)^{2+}$ $Al(OH)_4^-$.

Answer: The K_f values are: $Ni(NH_3)^{2+}$, 5.5×10^8; $Fe(SCN)^{2+}$, 1.2×10^2; $Al(OH)_4^-$, 7.7×10^{33}. The larger the K_f value, the more stable the complex: $Fe(SCN)^{2+} < Ni(NH_3)_6^{2+} < Al(OH)_4^-$.

Calculations Involving Complex Ions

Calculating the Concentration of Metal Ion: When the complexing agent is in large excess, you can usually assume that most of the metal ion is fully complexed. Therefore, the concentration of complex ion is assumed to be equal to the concentration of metal ion before complexation. Once we know the concentration of complex ion we can proceed to construct a concentration summary, in which x is usually the concentration of complex ion that dissociates.

Example 18.14: Estimate the equilibrium concentration of uncomplexed Cu^+ when 0.015 mol of $CuNO_3$ is dissolved in 1.00 L of 2.0 *M* NaCN. The K_f of $Cu(CN)_2^-$ is 1.0×10^{24}.

Given: $[Cu(CN)_2^-] = 0.015\ M$
$[CN^-] = 2.0\ M$
$K_f = 1.0 \times 10^{24}$

Unknown: Equilibrium concentration of uncomplexed Cu^+

Plan:
1. Construct a concentration summary, defining x as the moles of complex dissociating per liter to give x mol of Cu^+ per liter at equilibrium.
2. Calculate K_d.
3. Substitute expressions for concentrations into the K_d expression and solve for the concentration of Cu^+.

Calculations: 1. Concentration Summary

	$Cu(CN)_2^-(aq)$	$\rightleftharpoons$	Cu^+	+	$2CN^-(aq)$
Initial (mol/L)	0.015		0		2.0
Changes (mol/L)	$-x$		x		$+2x$
Equilibrium (mol/L)	$0.015 - x$		x		$2.0 + 2x$
Approximate (mol/L)	0.015		x		2.0

2. $$K_d = \frac{1}{K_f} = \frac{1}{1.0 \times 10^{24}} = 1.0 \times 10^{-24}$$

3. $$K_d = \frac{[Cu^+][CN^-]^2}{[Cu(CN)_2^-]} = 1.0 \times 10^{-24} = \frac{[Cu^+](2.0)^2}{0.015} = 1.0 \times 10^{-24};\ [Cu^+] = \underline{3.8 \times 10^{-27}\ M}$$

Note: You can get the same answer by substituting the above concentration into the K_f expression and using the given K_f value.

Practice: Complete the concentration summary for the complex ion equilibrium give below.

	$Zn(OH)_4^{2-}$ $\rightleftharpoons$	$Zn^{2+}(aq)$ +	$4OH^-(aq)$
Initial (mol/L)	1.0×10^{-2}	0	1.0 M
Changes (mol/L)	________	________	________
Equilibrium (mol/L)	________	________	________
Approximate (mol/L)	________	________	________

Answers: Approximate: 1.0×10^{-2} *M*; *x*; 1.0 *M*

Calculating the Molar Solubility of a Salt in the Presence of a Complexing Agent

The molar solubility of a salt increases in the presence of an agent that complexes one of its ions. When there is a large excess of complexing agent, a combined equilibrium constant ($K_{sp} \times K_f$) can be used to calculate the molar solubility. This combined equilibrium constant is for the equilibrium:

salt + complexing agent $\rightleftharpoons$ complex ion + anion of salt

$$K_{combined} = K_f \times K_{sp} = \frac{[\text{complex}][\text{anion}]}{[\text{complexing agent}]}$$ (The salt itself is a solid and thus omitted.)

Practice: Write the equation and combined equilibrium expression for the dissolving of Ag_2S in NaCN solution. The complex formed is $Ag(CN)_2^-$.

Example 18.15: Calculate the molar solubility of CuCl in a 1.5 *M* NaCN solution. The K_{sp} of CuCl is 1.7×10^{-7}; the K_f of $Cu(CN)_2^-$ is 1.0×10^{24}.

Given: $[CN^-] = 1.5\ M$
$K_{sp} = 1.7 \times 10^{-7}$
$K_f = 1.0 \times 10^{24}$

Unknown: molar solubility of CuCl

Plan:
1. Construct a concentration summary for the combined equilibrium, defining *x* as the complex ion concentration (or the moles per liter of CuCl dissolving). This is also the molar solubility of the salt.
2. Substitute equilibrium concenration expressions from the concentration summary into the combined equilibrium expression and solve for *x*.

Calculation: 1. Concentration Summary

	$CuCl(s) + 2CN^-(aq)$ $\rightarrow$	$Cu(CN)_2^-(aq)$ +	$Cl^-(aq)$
Initial (mol/L)	1.5 *M*	0	0
Change (mol/L)	$-2x$	x	x
Equilibrium (mol/L)	$1.5 - 2x$	x	x

2. $K_{sp} \times K_f = \dfrac{[Cl^-][Cu(CN)_2^-]}{[CN^-]^2} = (1.7 \times 10^{-7}) \times (1.0 \times 10^{24})$

$$\frac{(x)(x)}{(1.5-2x)^2} = 1.7 \times 10^{17}$$

Taking the square root of both sides gives $\dfrac{x}{1.5-2x} = 4.1 \times 10^8$ and $x = \underline{0.75 \text{ mol/L}}$

Practice: Complete the concentration summary for the equilibrium equation given below:

$$AgCl(s) + 2S_2O_3^{2-}(aq) \rightleftharpoons Ag(S_2O_3)_2^{3-}(aq) + Cl^-(aq)$$

	$S_2O_3^{2-}$	$Ag(S_2O_3)_2^{3-}$	Cl^-
Initial (mol/L)	1.5	0	0
Changes (mol/L)	________	________	________
Equilibrium (mol/L)	________	________	________

SELF-TEST

1. Drinking liquids from unglazed ceramic cups painted with $PbCrO_4$, a yellow pigment, may cause lead poisoning. Calculate the maximum concentration of Pb^{2+} in milligrams per liter in water from such a cup, assuming no reactions occur (other than dissolving). The K_{sp} for $PbCrO_4$ is 1.8×10^{-14}.

2. The molar solubility of CaF_2 in water is 2.32×10^{-4} mol/L. Calculate its molar solubility in 0.80 M NaF.

3. Sodium hydroxide is added to well water containing 8.0×10^{-4} *M* Mg^{2+} and 5.0×10^{-3} *M* Ca^{2+}.
   (a) Above what pH will $Mg(OH)_2$ precipitate?
   (b) What are the concentrations of Ca^{2+} and Mg^{2+} at this pH? (K_{sp}'s: $Mg(OH)_2$, 7.1×10^{-12}; $Ca(OH)_2$, 6.5×10^{-6}.)

4. The maximum allowable concentration of Cd^{2+} in public drinking water is 0.15 mg/L according to federal regulations. What concentration of S^{2-} could maintain Cd^{2+} at this level assuming no other reactions occur? ($K_{sp} = 1 \times 10^{-27}$ for CdS)

5. A solution contains 0.010 *M* Ba^{2+} and 0.020 *M* Ca^{2+}.
   (a) Calculate the PO_4^{3-} concentration that would give a maximum separation of the two ions.
   (b) What concentration of each metal ion remains in solution after precipitation? K_{sp}'s: $Ba_3(PO_4)_2$, 6×10^{-39}; $Ca_3(PO_4)_2$, 1.3×10^{-32}.

6. Acid rain lowers the pH of some soils, increasing the solubility of certain compounds that contain metal ions like Cu^{2+}. The water in contact with a certain soil is found to be 3.0×10^{-4} *M* in Cu^{2+}. If the soil contains solid $Cu(OH)_2$ ($K_{sp} = 1.6 \times 10^{-19}$) what must the pH of this water be? (Many metals in high concentration are toxic to plants.)

7. Calculate the concentration ratio of acetate ion to acetic acid that would prevent the precipitation of $Zn(OH)_2$ from a 1.0 *M* $Zn(NO_3)_2$ solution. ($K_{sp} = 4.5 \times 10^{-17}$; $K_a = 1.8 \times 10^{-5}$)

8. A solid white mixture is known to consist of $Pb(OH)_2$, an amphoteric hydroxide, and $Ca(OH)_2$. How could you separate the metal ions from each other? Write the equation(s). (See Table 18.2 in your text if necessary.)

9. Predict which of the following sparingly soluble lead salts are more soluble in nitric acid solution than in neutral solution: $PbBr_2$, PbC_2O_4, $Pb(OH)_2$, PbI_2. Write a molecular equation for any reaction that occurs.

10. Calculate the molar solubility of CdS in a solution buffered to pH = 3.0 and saturated with H_2S. $K_{spa} = 8 \times 10^{-7}$.

11. (a) Identify the precipitate that forms when 0.10 mol of $Pb(NO_3)_2$ and 0.10 mol of $Zn(NO_3)_2$ are added to 1.0 L of a saturated H_2S solution. K_{spa}'s: PbS, 3×10^{-7}; ZnS, 3×10^{-2}
(b) Calculate the pH that would give the best separation of Pb^{2+} and Zn^{2+} in the above solution.

12. Equal numbers of moles of $Cd(NO_3)_2$ and $Cu(NO_3)_2$ are dissolved in one liter of 2.0 M ammonia solution. Which of the two complexes, $Cd(NH_3)_4^{2+}$ or $Cu(NH_3)_4^{2+}$, will be in greater concentration? K_f's: $Cd(NH_3)_4^{2+}$, 1×10^7; $Cu(NH_3)_4^{2+}$, 1.1×10^{13}

13. Geologists dissolve a small amount of ore in acid and then add ammonia to detect copper in their quest for copper ore deposits. Estimate the final Cu^{2+} concentration when 2.0 mL of 5.0 *M* ammonia is added to 8.0 mL of 0.10 *M* $Cu(NO_3)_2$. The formation constant for $Cu(NH_3)_4^{2+}$ is 1.1×10^{13}.

14. Will $Cu(OH)_2$ precipitate if the pH of the solution in problem 13 is raised to 11.0? ($K_{sp} = 1.6 \times 10^{-19}$)

ANSWERS TO SELF-TEST QUESTIONS

1. $1.8 \times 10^{-14} = [Pb^{2+}][CrO_4^{2-}] = x^2$; $x = 1.34 \times 10^{-7}$ *M*
1.34×10^{-7} mol/L × 207.2 g/L × 1000mg/g = 0.0278 mg/L = <u>0.028 mg/L</u>

2. $K_{sp} = (2.32 \times 10^{-4})(2 \times 2.32 \times 10^{-4})^2 = 4.99 \times 10^{-11}$
$4.99 \times 10^{-11} = [Ca^{2+}](0.80)^2$; $[Ca^{2+}] = \underline{7.8 \times 10^{-11}\ M}$ = molar solubility

3. $7.1 \times 10^{-12} = (8.0 \times 10^{-4})[OH^-]^2$; $[OH^-] = 9.4 \times 10^{-5}$ *M*
$[H^+] = 1.0 \times 10^{-14}/9.4 \times 10^{-5} = 1.1 \times 10^{-10}$
pH = -log (1.1×10^{-10}) = 9.96 <u>above 9.96</u>

4. 1×10^{-27} = (0.15 mg/L × 1 g/1000mg × 1 mol/112.4 g)$[S^{2-}]$; $[S^{2-}] = \underline{7.5 \times 10^{-22}\ M}$

5. (a) $1.3 \times 10^{-32} = [Ca^{2+}]^3[PO_4^{3-}]^2 = (0.020)^3[PO_4^{3-}]^2$; $[PO_4^{3-}] = \underline{4.0 \times 10^{-14}\ M}$
(b) $[Ca^{2+}]$ = <u>0.020 M</u>
$6 \times 10^{-39} = [Ba^{2+}]^3[PO_4^{3-}]^2 = [Ba^{2+}]^3(4.0 \times 10^{-14})^2$; $[Ba^{2+}] = \underline{1.6 \times 10^{-4}\ M}$

6. $1.6 \times 10^{-19} = (3.0 \times 10^{-4})[OH^-]^2$; $[OH^-] = 2.3 \times 10^{-8}$ *M*
$[H^+] = 1.0 \times 10^{-14}/2.3 \times 10^{-8} = 4.3 \times 10^{-7}$; pH = <u>6.37</u>

7. $4.5 \times 10^{-17} = (1.0)[OH^-]^2$; $[OH^-] = 6.7 \times 10^{-9}$ *M*

$$1.8 \times 10^{-5} = \frac{[\text{acetate}](1.0 \times 10^{-14} / 6.7 \times 10^{-9})}{(\text{acetic acid})}; \quad \frac{[\text{acetate}]}{[\text{acetic acid}]} = \underline{12}$$

8. Add excess OH^-: $Pb(OH)_2(s) + 2OH^-(aq) \rightarrow Pb(OH)_4^{2-}(aq)$

9. $PbC_2O_4(aq) + 2HNO_3(aq) \rightarrow H_2C_2O_4(aq) + Pb(NO_3)_2(aq)$
   $Pb(OH)_2(aq) + 2HNO_3(aq) \rightarrow 2H_2O(l) + Pb(NO_3)_2(aq)$

10. $[H^+] = 10^{-3.5} = 3 \times 10^{-4}\ M$

$$\frac{[Cd^{2+}](0.10)}{(3 \times 10^{-4})^2} = 8 \times 10^{-7};\ [Cd^{2+}] = 7 \times 10^{-13}\ M$$

11. (a) PbS; PbS has a much smaller K_{spa} and is therefore less soluble in acidic solution.

    (b) $\dfrac{(0.10)(0.10)}{[H^+]^2} = 3 \times 10^{-2};\ [H^+] = 0.6\ M;\ pH = 0.3$

12. $Cu(NH_3)_4^{2+}$

13. $[NH_3] = 5.0\ mol/L \times 0.0020\ L/0.010\ L = 1.0\ M;\ [Cu(NH_3)_4^{2+}] = 0.10\ mol/L \times 0.0080\ L/0.010\ L = 0.080\ M$

$$\frac{[NH_3]^4[Cu^{2+}]}{[Cu(NH_3)_4^{2+}]} = \frac{1}{1.1 \times 10^{13}};\quad \frac{(0.68)^4[Cu^{2+}]}{(0.080)} = 9.1 \times 10^{-14};\ [Cu^{2+}] = \underline{3.4 \times 10^{-14}\ M}$$

14. $Q = (3.4 \times 10^{-14})(1 \times 10^{-3})^2 = 3 \times 10^{-20};\ Q < K_{sp}$; No precipitate

SURVEY OF THE ELEMENTS 3: THE REMAINING NONMETALS

SELF-ASSESSMENT

Group 6A: Sulfur, Selenium, and Tellurium (Section S3.1)

Fill in the blanks with a sulfur-containing compound or ion.

1. ____________ converts NH_3 to $(NH_4)_2SO_4$, which is used in fertilizers.

2. ____________ is often a product of reactions in which H_2SO_4 acts as an oxidizing agent.

3. ____________, an air pollutant, reacts with carbonates in statues causing crumbling.

4. ____________ ion dissolves unexposed AgBr out of photographic film.

5. ____________ is a toxic gas that is used to precipitate metal ions due to its weak acidic properties.

Group 5A: Phosphorus, Arsenic and Antimony (Section S3.2)

6. All of the following statements are true except
   (a) Phosphorus exists in allotropic forms.
   (b) The two most common oxidation states of phosphorous are +5 and +3.
   (c) Phosphoric acid (H_3PO_4) is a triprotic weak acid.
   (d) Phosphine (PH_3) is a stronger base than NH_3.

7. Complete and balance the following equations:
   (a) $Ca_3(PO_4)_2(s) + H_2SO_4(aq, 98\%) \rightarrow$ superphosphate
   (b) $Ca(s) + P_4(s) \xrightarrow{\text{heat}}$
   (c) $P_4O_{10}(s) + H_2O(l) \rightarrow$

Group 4A: Carbon (Section S3.3)

8. List two properties of graphite, diamonds, and fullerenes and give two uses for each.

9. Write balanced equations for the reaction between
   (a) carbon and nickel oxide, heated
   (b) calcium acetylide and water
   (c) carbon and hot, concentrated sulfuric acid

Group 4A: Silicon and Germanium (Section S3.4)

Answer questions 10–16 as true or false. If a statement is false change the underlined word(s) so that it is true.

10. Quartz is an example of a <u>silicate</u>.

11. Each Si atom in a <u>silicate</u> is surrounded tetrahedrally by four O atoms while in a <u>silica</u> each O atom belongs to two SiO_4 tetrahedra.

12. <u>Silicones</u> are silicon-hydrogen analogues of hydrocarbons.

13. <u>Silanes</u> are hydrolyzed by water.

Group 3A: Boron (Section S3.5)

14. Boron occurs naturally associated with <u>hydrogen</u>.

15. Boric acid is a <u>Brønsted</u> acid.

16. <u>Hydrogen</u> bridges exist between boron atoms in boranes.

ANSWERS TO SELF-ASSESSMENT PROBLEMS

If you missed an answer, <u>be sure</u> to study the relevant section in the textbook and study guide.

1. H_2SO_4
2. SO_2
3. SO_2
4. $S_2O_3^{2-}$
5. H_2S
6. (d)

7. (a) $Ca_3(PO_4)_2(s) + 2H_2SO_4(aq, 98\%) \rightarrow Ca(H_2PO_4)_2(s) + 2CaSO_4(s)$
   (b) $6Ca(s) + P_4(s) \xrightarrow{\text{heat}} 2Ca_3P_2(s)$
   (c) $P_4O_{10}(s) + 6H_2O(g) \rightarrow 4H_3PO_4(l)$

9. (a) $C(s) + NiO(s) \xrightarrow{\text{heat}} Ni(l) + CO(g)$
   (b) $CaC_2(s) + 2H_2O(l) \rightarrow Ca(OH)_2(s) + C_2H_2(g)$
   (c) $C(s) + 2H_2SO_4(aq) \rightarrow CO_2(g) + 2SO_2(g) + 2H_2O(g)$

10. F; silica
11. T
12. F; Silanes
13. T
14. F; oxygen
15. F; Lewis
16. T

REVIEW

Group 6A: Sulfur, Selenium, and Tellurium (Section S3.1)

Learning Objectives

1. Describe the occurrence, preparation, and properties of elemental sulfur.
2. Describe the molecular structure of the various forms of solid, liquid, and gaseous sulfur.
3. Write equations for the reactions of sulfur with oxygen, chlorine, fluorine, and metals.
4. Describe the contact process for the preparation of sulfuric acid; include the appropriate equations.
5. Write equations for reactions in which sulfuric acid is an acid, an oxidizing agent, and a dehydrating agent.
6. State the principal uses of sulfuric acid.
7. Write equations for the preparation of SO_2 and its reactions with $H_2O(l)$, $OH^-(aq)$, and $CaCO_3(s)$.
8. Write equations for the preparation of thiosulfate ion and its reactions with $H^+(aq)$, $MnO_4^-(aq)$, $I_2(s)$, and $AgBr(s)$.
9. Describe the properties and uses of hydrogen sulfide.
10. Write formulas for some polysulfide ions, and write equations for their formation.

Review

Sulfur

Occurrence

- mostly in metal sulfides of the earth's crust such as pyrite (FeS_2)
- elemental sulfur found on top of domes of salt deposits

Physical Properties

- bright yellow solid in the form of powder or broken sticks
- allotropic forms (see Table S3.2 of text for physical properties)
- rhombic (Figure S3.3a of text); stable at room temperature
- monoclinic (Figure S3.3b of text); stable above 95.5°C

Chemical Properties

- undergoes combustion $S(s) + O_2(g) \rightarrow SO_2(g)$
- combines directly with all elements except Au, Pt, and noble gases. (See Table S3.3 in your text for principal reactions of sulfur.)

Important Compounds and Their Reactions

<u>The +6 Oxidation State: Sulfuric Acid and Sulfates</u>

Sulfuric acid is the most widely produced industrial chemical. It can be prepared from sulfur and from SO_2.

<u>Step</u>

1. $2SO_2(g) + O_2(g) \xrightarrow[400°\,C]{V_2O_5\ \text{catalyst}} 2SO_3(g)$
2. $SO_3(g) + H_2SO_4(aq, 98\%) \rightarrow H_2S_2O_7$ pyrosulfuric acid
3. $H_2S_2O_7(l) + H_2O(l) \rightarrow 2H_2SO_4(aq, 98\%)$

Reactions and Uses of Sulfuric Acid

(1) As a strong acid
- converts NH_3 to $(NH_4)_2SO_4$ (for fertilizers) and $Ca_3(PO_4)_2$ to $CaHPO_4$ and $Ca(H_2PO_4)_2$ (for fertilizers)
- dissolves metal oxides from iron and steel surfaces before coating with Zn or Sn
- used to prepare more volatile acids such as HCl and HNO_3 from their salts:

$$H_2SO_4(98\%) + NaNO_3(s) \xrightarrow{\text{heat}} NaHSO_4(s) + HNO_3(g)$$

(2) As an oxidizing agent
- oxidizes most metals and many nonmetals. SO_2 is generally a product.

$$Zn(s) + \underset{\text{concentrated}}{2H_2SO_4(aq)} \rightarrow ZnSO_4(aq) + SO_2(g) + 2H_2O(l)$$

$$C(s) + \underset{\text{hot, concentrated}}{2H_2SO_4(aq)} \rightarrow CO_2(g) + 2SO_2(g) + 2H_2O(g)$$

(3) As a dehydrating agent sulfuric acid has a high affinity for water:
- concentrated H_2SO_4 reacts with water to form hydrates ($H_2SO_4{\cdot}nH_2O$; n = 1 to 8) and releases large quantities of heat
- absorbs water vapor from gaseous mixtures such as air
- removes H and O from carbohydrates:

$$\underset{\text{glucose}}{C_6H_{12}O_6(s)} + 6H_2SO_4(98\%) \rightarrow 6C(s) + 6H_2SO_4 \cdot H_2O$$

The +4 Oxidation State: Sulfur Dioxide and Sulfites

Reactions of Sulfur Dioxide
- dissolves in water:

$$SO_2(g) + H_2O(l) \rightarrow H_2SO_3(aq)$$

Note: H_2SO_3 is unstable and has never been isolated.

- reacts with excess base to form sulfites:

$$H_2SO_3(aq) + OH^-(aq) \rightarrow HSO_3^-(aq) + H_2O(l)$$

$$HSO_3^-(aq) + OH^-(aq) \rightarrow \underset{\text{sulfite ion}}{SO_3^{2-}(aq)} + H_2O(l)$$

- reacts with carbonates to form sulfites, which are then oxidized:

$$CaCO_3(s) + SO_2(g) \rightarrow CaSO_3(s) + CO_2(g)$$

$$2CaSO_3(s) + O_2(g) \rightarrow 2CaSO_4(s)$$

The above reaction accounts for the crumbling of marble and limestone statues. It also explains the ability of carbonates to scrub (remove SO_2) smokestack gases.

Reactions of the Thiosulfate ($S_2O_3^{2-}$) Ion
- acid decomposes thiosulfate:

$$S_2O_3^{2-}(aq) + 2H^+(aq) \rightarrow S(s) + \underset{\text{unstable}}{H_2SO_3(aq)}$$

$$H_2SO_3(aq) \rightarrow SO_2(g) + H_2O(l)$$

- strong oxidizing agents convert thiosulfate to sulfate:

  $8MnO_4^-(aq) + 5S_2O_3^{2-}(aq) + 14H^+(aq) \rightarrow 8Mn^{2+}(aq) + 10SO_4^{2-}(aq) + 7H_2O(l)$

- weaker oxidizing agents convert thiosulfate to the tetrathionate ($S_4O_6^{2-}$) ion:

  $2S_2O_3^{2-}(aq) + I_2(s) \rightarrow S_4O_6^{2-}(aq) + 2I^-(aq)$

- dissolves unexposed AgBr out of photographic film (fixer):

  $AgBr(s) + 2S_2O_3^{2-}(aq) \rightarrow Ag(S_2O_3)_2^{3-}(aq) + Br^-(aq)$

The -2 Oxidation State: Sulfides

- H_2S (smell of rotten eggs) is produced when protein decays. It is used as a precipitating agent for metal ions and it has
  weak acidic properties (see Chapters 17 and 18)
- polysulfide ions can be prepared by treating S(s) with a hot sulfide solution:

  $S^{2-}(aq) + S(s) \rightarrow S_2^{2-}(aq)$

  Larger sulfur chains are also produced in the reaction.

Group 5A: Phosphorus, Arsenic, and Antimony (Section S3.2)

Learning Objectives

1. Describe the preparation of elemental phosphorus; include the appropriate equations.
2. Compare the properties of red and white phosphorus.
3. Write equations for the reactions of phosphorus with oxygen, the halogens, sulfur, and active metals.
4. Sketch the structures of P_4, P_4O_6, and P_4O_{10}.
5. Write equations for the reactions of P_4O_6 and P_4O_{10} with water.
6. Write equations for the preparation of phosphorous acid, orthophosphoric acid (two equations), and pyrophosphoric acid.
7. Write Lewis structures for phosphorous acid, orthophosphoric acid, and pyrophosphoric acid.
8. Write equations for the preparation of superphosphate and triple superphosphate.
9. Write Lewis structures for the pyrophosphate ion, the triphosphate ion, and the metaphosphate ion.
10. Write equations for the preparation of ionic phosphides and their reaction with water.
11. List the three principal oxidation states of the Group 5A elements, and illustrate each state with one phosphorus, one arsenic, and one antimony compound.

Review

Occurrence

- in phosphate minerals of rocks
- elemental phosphorus is obtained from phosphate rock in the form of white phosphorus (P_4)

Properties

- allotropic forms:
  white phosphorus: waxy solid, unstable, reacts with O_2
  red phosphorus: red colored solid, most stable form of phosphorus

- reacts with O_2, S, halogens and many metals (see Table S3.7 in your text) to form +5, +3, and -3 oxidation states (The +5 oxidation state is the most stable oxidation state for phosphorus.)

Important Compounds and their Reactions

Oxides

- P_4O_6: formed when phosphorus is burned in limited O_2
- P_4O_{10}: formed when O_2 is in excess, absorbs water:

$$P_4O_{10}(s) + 6H_2O(l \text{ or } g) \rightarrow 4H_3PO_4(l)$$

Oxo Acids and Anions of Phosphorus

- phosphorus acid (H_3PO_3) is formed by the hydration of P_4O_6:

$$P_4O_6(s) + 6H_2O(l) \rightarrow 4H_3PO_3(aq)$$

- phosphoric acid is prepared from phosphate rock ($Ca_3(PO_4)_2(s)$):

$$Ca_3(PO_4)_2(s) + 3H_2SO_4(aq) \rightarrow 2H_3PO_4(85\%) + 3CaSO_4(s)$$

- superphosphate is a mixture of $Ca(H_2PO_4)_2$ and $CaSO_4$ and is used as a fertilizer

$$Ca_3(PO_4)_2(s) + 2H_2SO_4(aq,\ 98\%) \rightarrow \underset{\text{superphosphate}}{Ca(H_2PO_4)_2(s) + 2CaSO_4(s)}$$

- phosphates such as Na_3PO_4 are used in laundry and cleaning products

The -3 Oxidation State: Phosphides and Phosphine

- Phosphide ion (P^{3-}) can be prepared from the reaction:

$$6Ca(s) + P_4(s) \xrightarrow{\text{heat}} 2Ca_3P_2(s)$$

- P^{3-} reacts with water to form basic solutions:

$$Ca_3P_2(s) + 6H_2O(l) \rightarrow \underset{\text{phosphine}}{2PH_3(g)} + 3Ca(OH)_2(aq)$$

- phosphine is a weaker base than NH_3; addition of a proton produces the unstable phosphonium ion (PH_4^+):

$$PH_4Cl(s) + H_2O(l) \rightarrow PH_3(g) + H_3O^+(aq) + Cl^-(aq)$$

Group 4A: Carbon (Section S3.3)

Learning Objectives

1. Compare the structure and properties of graphite, diamond and buckminsterfullerene.
2. Write equations for the reactions of carbon with oxygen, metal oxides, sulfur and fluorine.
3. Describe the properties and uses of carbon dioxide.
4. Describe and write equations for the preparation of sodium hydrogen carbonate by the Solvay process.
5. Describe the molecular structure, production, and toxicity of carbon monoxide.
6. Distinguish between ionic carbides, interstitial carbides and covalent carbides, and give examples of each.
7. Write equations for the reactions of Be_2C, Al_4C_3, and CaC_2 with water.

Review

Carbon exists in several allotropic forms; diamond, graphite and the fullerenes.

Graphite

Graphite consists of adjacent layers of fused planar, six-membered rings. (See Figure 2.18b in your text.)

The presence of electrons in pi molecular orbitals affords unique properties to graphite: it is the only natural nonmetallic substance that conducts electricity. Conductivity increases with pressure so that graphite is used in microphones and telephone receivers where it responds to the changing pressure of sound waves.

Air and water are absorbed between layers of graphite enabling the layers to slide past each other giving graphite its lubricating ability.

Amorphous carbon refers to soot, ashes, and carbon black in which the atoms do not form a consistent lattice. Many atoms are not completely bonded so it is an effective absorbant (activated charcoal).

Diamonds

Each atom in the diamond lattice has a single bond to form four tetrahedrally placed carbon atoms. (see Figure 2.18a in your text.)

Diamonds can scratch every other known substance due to their extreme hardness. Their high refractive index explains their sparkling appearance which coupled with their rarity makes them valuable.

Diamonds conduct heat, are good electrical insulators, and are transparent to visible, UV, x-rays, and most IR radiation. Consequently, they have many industrial uses.

Fullerenes

Fullerenes consist of discrete molecules that contain 12 pentagons and a number of hexagons which form a carbon cage.

Buckminsterfullerene, C_{60} is the principal fullerene with a soccer-ball structure consisting of 12 pentagons and 20 hexagons.

Fullerenes have been detected in nature as well as being prepared in bulk quantities in the laboratory. Research scientists plan to use fullerenes for catalysts, to transport drugs to specific body sites, for lubricants, and for superconductors. See Chemical Insight: Buckyballs and Other Fullerenes in your text.

Important Compounds of Carbon and Their Important Reactions

Carbon Dioxide and Carbonates

Inorganic carbon compounds consist mainly of carbon dioxide and carbonates

Carbon Dioxide

- colorless, odorless gas at ordinary temperature and pressure
- dissolves in water:

$$CO_2(g) + H_2O(l) \rightleftharpoons \underset{\text{(unstable)}}{H_2CO_3(aq)} \rightleftharpoons H^+(aq) + HCO_3^-(aq)$$

- CO_2 dissolved under pressure is used in beverages
- heavier than air so that it can smother fires
- contributes to the "greenhouse effect" (see Chemical Insight on Greenhouse Effect in your textbook)

Carbonates

- $NaHCO_3$ can be prepared by the Solvay process:

  Steps

  (1) $CaCO_3(s) \xrightarrow{\text{heat}} CaO(s) + CO_2(g)$

  (2) $CO_2(g) + NaCl(aq) + H_2O(l) + NH_3(aq) \rightarrow NaHCO_3(s) + NH_4Cl(aq)$

  Ammonia is recovered by

  (3) $\underset{\text{from (1)}}{\underset{\uparrow}{CaO(s)}} + \underset{\text{from (2)}}{\underset{\uparrow}{2NH_4Cl(aq)}} \rightarrow CaCl_2(aq) + H_2O(l) + 2NH_3(g)$

Carbon Monoxide

- CO is produced whenever carbon compounds are oxidized in a limited supply of oxygen
- CO is produced in oxygen at temperatures above 500°C

  $C(s) + CO_2(g) \rightleftharpoons 2CO(g) \quad K_p = 4.8 \times 10^4$ at 2000°C

- CO forms stable complexes with many transition metals including Fe in hemoglobin (Hb)

  $Hb{\cdot}O_2(aq) + CO(g) \rightleftharpoons Hb{\cdot}CO(aq) + O_2(g) \quad K_c = 200$

  (CO binds hemoglobin so that it can no longer bind O_2.)

Carbides and Cyanides

Ionic carbides: Contain active metals and carbon. They react with water:

$Be_2C(s) + 4H_2O(l) \rightarrow 2Be(OH)_2(s) + CH_4(g)$

$CaC_2(s) + 2H_2O(l) \rightarrow Ca(OH)_2(s) + C_2H_2(g)$

Interstitial carbides: Carbon atoms occupy sites in the crystal lattices of less active metals. (Steel contains small grains of Fe_3C.)

Covalent carbides: Carbon atoms form covalent bonds with other nonmetals. CS_2, a solvent, and HCN, a toxic gas, are examples.

Group 4A: Silicon and Germanium (Section S3.4)

Learning Objectives

1. Describe the occurrence, preparation, and purification of silicon.
2. Describe the structure and properties of silicon.
3. Write the structures of the orthosilicate and pyrosilicate ions.
4. Sketch the structures of silicate anions containing tetrahedra linked in rings, double chains, and sheets.
5. Sketch the structure of quartz.
6. Describe the structure and properties of silicones and silanes.

Review

Silicon

- Si represents 25.7% of the earth's crust, giving structure to minerals and rocks as silicates
- crude Si is obtained from the reduction of SiO_2 with coke:

$$SiO_2(s) + 2C(s) \xrightarrow{3000^\circ C} Si(l) + 2CO(g)$$

- pure silicon needed for semiconductors is prepared from zone refining (See Figure S3.24 in your text.)

Silica and Silicates

The strength of the S–O bond in silicates contributes to the prevalence of silicates in nature. Each Si atom in a silicate is surrounded tetrahedrally by four O atoms.

Orthosilicates: contain isolated SiO_4^{4-} ions (See Figure S3.26a in your text.)

Pyrosilicate ($Si_2O_7^{6-}$): contains two tetrahedra (See Figure S3.26b in your text.)

Other silicate anions contain tetrahedra linked in rings, double chains, and sheets. Asbestos, talc, and mica are examples.

Silica (silicon dioxide) has an uncharged framework with the empirical formula SiO_2. Each O atom belongs to two SiO_4 tetrahedra. Quartz is an example of a silica.

Some Synthetic Silicon Compounds

Silicones contain silicon-oxygen chains with hydrocarbon side groups (see Figure S3.30 in your text). They tend to be flexible, hydrophobic, and heat resistant. They are used in brake fluids, as water repellants, lubricants, and in cosmetic products.

Silanes are silicon-hydrogen compounds that are analogous to hydrocarbons, but much less stable. SiH_4 is the simplest silane. Silanes are hydrolyzed by water and ignite spontaneously in air:

$$SiH_4(g) + 2O_2(g) \rightarrow SiO_2(s) + 2H_2O(l)$$

Group 3A: Boron (Section S3.5)

Learning Objectives

1. Describe the occurrence, preparation, and properties of elemental boron.
2. Describe the structures of boron, the tetraborate ion, and boric acid.
3. Write an equation for the preparation of boric acid from borax.
4. Write an equation for (a) the reaction of boric acid with water and (b) the dehydration of boric acid.
5. Explain why boric acid and the boron halides are Lewis acids.
6. Sketch the diborane molecule, and discuss its bonding.

Review

Boron in minerals like silicon occurs associated with oxygen. Borax ($Na_2B_4O_7 \cdot 10H_2O$) is the commercial source of boron. It gives a mildly basic solution:

$$\underset{\text{tetraborate ion}}{B_4O_7^{2-}(aq)} + 7H_2O(l) \rightleftharpoons \underset{\text{boric acid}}{4H_3BO_3(aq)} + 2OH^-(aq)$$

Practice: What is the effect of adding strong acid to the above equilibrium? (Answer: Equilibrium shifts to the right to produce more boric acid.)

Boric acid dehydrates to boric oxide, B_2O_3:

$$2H_3BO_3(s) \xrightarrow{\text{heat}} B_2O_3(s) + 3H_2O(g)$$

Impure boron is prepared by the reduction of B_2O_3 with Mg at elevated temperatures:

$$B_2O_3(s) + 3Mg(s) \xrightarrow{\text{heat}} 2B(s) + 3MgO(s)$$

Boric acid, often written as $B(OH)_3$ is not a Brønsted acid; it acts as a Lewis acid displacing protons and forming $B(OH)_4^-$:

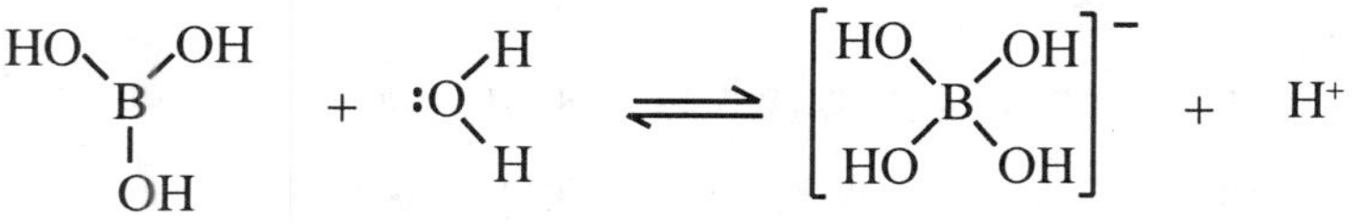

Boranes are boron hydrides of which the simplest is diborane (B_2H_6):

$$2NaBH_4(s) + I_2(s) \rightarrow B_2H_6(g) + 2NaI(s) + H_2(g)$$

sodium borohydride

Many boranes are unstable. All boranes react with water to form boric acid and hydrogen:

$$B_2H_6(g) + 3O_2(g) \rightarrow 2H_3BO_3(s)$$

$$B_2H_6(g) + 6H_2O(l) \rightarrow 2H_3BO_3(s) + 6H_2(g)$$

Boranes have been widely studied due to their unusual multicentered bonding in which hydrogen bridges exist between boron atoms. (See Figure S3.37a in your text.)

Synthetic boron compounds such as the trihalides (BF_3, BCl_3) are strong Lewis acids. BCl_3 is reduced by hydrogen at high temperatures to form very pure boron:

$$2BCl_3(s) + 3H_2(g) \xrightarrow{\text{heat}} 2B(s) + 6HCl(g)$$

SELF-TEST

1. Write Lewis structures for
   (a) thiosulfate ion
   (b) pyrophosphoric acid
   (c) boric acid
   (d) bicarbonate ion

2. Complete and balance the following equations:
   (a) $Zn(s) + H_2SO_4(aq) \rightarrow$ Zn SO_3 + H_2O
   (b) $BaSO_3(s) + 2H^+(aq) \rightarrow$ H_2SO_3 +
   (c) $MgCO_3(s) + SO_2(g) \rightarrow$
   (d) $S_2O_3^{2-}(aq) + I_2(s) \rightarrow$

3. Label the following elements as metals, nonmetals, or semimetals.

S	Non-	Te	metal
B	Non	C	Non
Si	Semi-	P	Non

4. Write equations for the reaction of water with
   (a) Al_4C_3
   (b) pyrosulfuric acid ($H_2S_2O_7$)
   (c) phosphorous pentoxide (P_4O_{10})
   (d) Ca_3P_2

5. Match the element in column A with one of its properties in column B.

A	B
(a) sulfur	(1) semimetal
(b) tellurium	(2) molecule is a puckered ring
(c) phosphorus	(3) burns in a limited supply of oxygen to form the +3 oxidation state
(d) boron	(4) One of its allotropes can scratch every other known substance.
(e) carbon	(5) Zone refining is used to obtain the ultrapure element.
(f) silicon	(6) semimetal and semiconductor

6. Which of the following form basic solutions? $B_4O_7^{2-}$, SO_3, SO_3^{2-}, P_4O_6.

7. Which compound or ion of sulfur does each of the following statements describe?
   (a) This compound is a good oxidizing and dehydrating agent.
   (b) Molecules of this weak diprotic acid are unstable. They are thought to form when sulfur dioxide dissolves in water.
   (c) This air pollutant attacks carbonates of marble and limestone statues.
   (d) This ion is formed when a sulfite solution is heated with sulfur.

8. Write an equation for each of the following reactions.
   (a) copper metal added to concentrated sulfuric acid solution
   (b) iodine added to thiosulfate solution
   (c) calcium phosphate added to sulfuric acid solution to form phosphoric acid as one of the products
   (d) heating lime (CaO) with coke (C)

ANSWERS TO SELF-TEST QUESTIONS

1. (a) 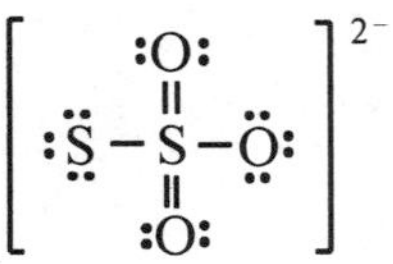

(c) 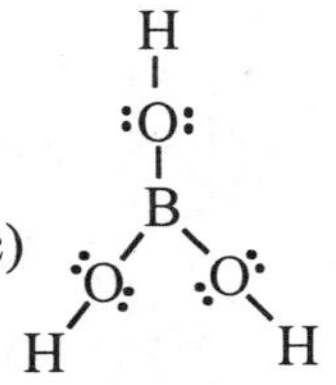

(b) 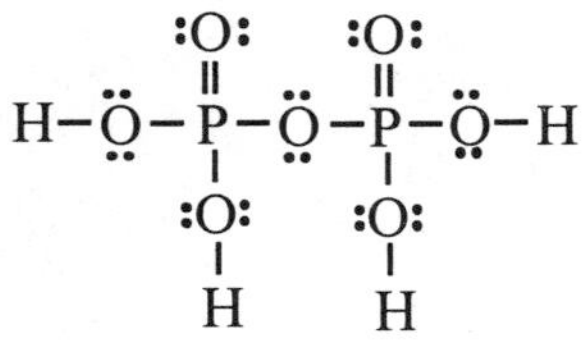

(d) 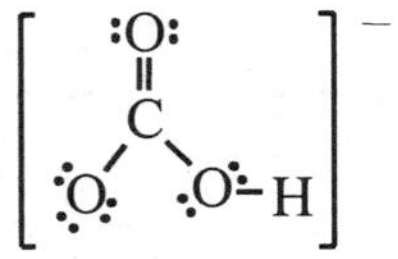

2. (a) $Zn(s) + 2\underset{\text{concentrated}}{H_2SO_4(aq)} \rightarrow ZnSO_4(aq) + SO_2(g) + 2H_2O(l)$
   (b) $BaSO_3(s) + 2H^+(aq) \rightarrow Ba^{2+}(aq) + SO_2(g) + H_2O(l)$
   (c) $MgCO_3(s) + SO_2(g) \rightarrow MgSO_3(s) + CO_2(g)$
   (d) $2S_2O_3^{2-}(aq) + I_2(s) \rightarrow S_4O_6^{2-}(aq) + 2I^-(aq)$

3. Nonmetals: S, C, P; Semimetals: B, Si, Te

4. (a) $Al_4C_3(s) + 12H_2O(l) \rightarrow 4Al(OH)_3(s) + 3CH_4(g)$
   (b) $H_2S_2O_7(l) + H_2O(l) \rightarrow 2H_2SO_4(aq, 98\%)$
   (c) $P_4O_{10}(s) + 6H_2O(l \text{ or } g) \rightarrow 4H_3PO_4(l)$
   (d) $Ca_3P_2(s) + 6H_2O(l) \rightarrow 2PH_3(g) + 3Ca(OH)_2(aq)$

5. (a) 2; (b) 1; (c) 3; (d) 6; (e) 4; (f) 5

6. $B_4O_7^{2-}$, SO_3^{2-}

7. (a) H_2SO_4 (b) H_2SO_3 (c) SO_2 (d) $S_2O_3^{2-}$

8. (a) $Cu(s) + 2H_2SO_4(\text{concentrated}) \rightarrow CuSO_4(aq) + SO_2(g) + 2H_2O(l)$
   (b) $2S_2O_3^{2-}(aq) + I_2(s) \rightarrow S_4O_6^{2-}(aq) + 2I^-(aq)$
   (c) $Ca_3(PO_4)_2(s) + 3H_2SO_4(aq) \rightarrow 2H_3PO_4(85\%) + 3CaSO_4(s)$
   (d) $CaO(s) + 3C(s) \xrightarrow{\text{heat}} CaC_2(s) + CO(g)$

CHAPTER 19

FREE ENERGY, ENTROPY, AND THE SECOND LAW OF THERMODYNAMICS

SELF-ASSESSMENT

Free Energy and Useful Work (Section 19.1)

1. Which of the following changes has the potential for doing useful work?
   (a) Liquid water boiling at 100°C and 1 atm pressure.
   (b) Ice melting at 25°C and 1 atm pressure.
   (c) N_2 and O_2 reacting to form NO at 25°C. (The ΔG for the reaction at 25°C is positive.)
   (d) Solid nickel in a 1.0 *M* $Cu(NO_3)_2$ solution, forming copper metal and Ni^{2+} ions.

2. The mineral calcite ($CaCO_3$) is often used to neutralize acidic water entering homes:

   $CaCO_3(s) + H^+(aq) \rightarrow Ca^{2+}(aq) + HCO_3^-(aq)$

   At 25°C when all concentrations are 1.0 *M*, $\Delta G°$ is -11.3 kJ/mol and $\Delta H°$ is -27 kJ/mol.
   (a) What is the maximum useful work that could be obtained by dissolving 1 mol of $CaCO_3$ in acidic solution under 1 atm pressure?
   (b) How many kilocalories of heat would be evolved if this much work were actually done?

Free Energy Change and Concentration (Section 19.2)

3. In each case indicate whether ΔG is higher than, lower than, or equal to $\Delta G°$:
   (a) Concentrations of all dissolved reactants and products are 1 *M* and the partial pressure of all gaseous reactants and products are 1 atm.
   (b) Concentrations of all products are greater than in their standard states while concentrations of all reactants are less than in their standard states.
   (c) $\Delta G°$ for a reaction is +25 kJ/mol and the system is at equilibrium.
   (d) Concentrations of all reactants are greater than in their standard states while all products are in their standard states.

4. Iron (II) is oxidized by molecular oxygen to form iron(III) hydroxide in aqueous solution:

   $2Fe^{2+}(aq) + 1/2O_2(g) + 5H_2O(l) \rightleftharpoons 2Fe(OH)_3(s) + 4H^+(aq)$

   $\Delta G°$ for the reaction is -32.02 kJ/mol at 25°C. Find K at this temperature. (R = 8.314 J/mol K)

5. Use data from Table 19.2 in your text to determine the maximum useful work that could be obtained from the complete oxidation of 1 mol of benzene (C_6H_6) at 25°C. The equation is

   $C_6H_6(l) + 15/2O_2(g) \rightarrow 6CO_2(g) + 3H_2O(g)$

6. Use the data from Table 19.2 in your text to calculate ΔG for the reaction

   $2NO(g) + O_2(g) \rightarrow 2NO_2(g)$

   at 25°C when the partial pressure of each gas is 0.20 atm. Is the formation of NO_2 from NO and O_2 spontaneous under these conditions?

Entropy and the Second Law of Thermodynamics (Section 19.3)

7. For which of the following changes does the entropy of the system decrease?
   (a) Water vapor condenses at 25°C.
   (b) Octane (C_8H_{18}) combines with oxygen and gives off heat.
   (c) Children line up in a playground.
   (d) Natural gas escapes into a room from a small hole in a gas pipe.
   (e) Organic pollutants in water are trapped on a carbon filter.

Entropy Calculations (Section 19.4)

8. Calculate the entropy change when 150 g of water condenses at its normal boiling point. The heat of vaporization of water is 2.256×10^3 J/g.

9. Calculate the entropy change when 160 g of oxygen gas dissolves in liquid water at 25°C and 1 atm

   [$O_2(g) \rightleftharpoons O_2(aq)$]. The ΔH_f° and ΔG_f° of $O_2(aq)$ are -16 kJ/mol, and 16.4 kJ/mol, respectively.

10. The $\Delta H°$ and $\Delta G°$ values for the reaction

    $2O_3(g) \rightleftharpoons 3O_2(g)$

    at 25°C are -285.4 kJ/mol and -326.4 kJ/mol respectively. The third law entropy (S°) of O_3 is 238.93 J/mol K. Find S° for $O_2(g)$ at 25°C.

The Variation of $\Delta G°$ With Temperature (Section 19.5)

11. At high temperatures $SO_3(g)$ will decompose according to

    $2SO_3(g) \rightarrow 2SO_2(g) + O_2(g)$

    $\Delta H°$ is 197.78 kJ, and $\Delta S°$ is 188.06 J/K. Estimate the temperature above which the decomposition is spontaneous.

ANSWERS TO SELF-ASSESSMENT PROBLEMS

If you missed an answer, be sure to study the relevant section in the textbook and study guide.

1. (b) and (d) are spontaneous reactions and therefore have the potential for doing useful work.

2. (a) maximum useful work = $-\Delta G = -(-11.3 \text{ kJ}) = \underline{11.3 \text{ kJ}}$
(b) $27 \text{ kJ} - 11.3 \text{ kJ} = \underline{16 \text{ kJ}}$

3. (a) $\Delta G = \Delta G°$ (b) $\Delta G > \Delta G°$ (c) $\Delta G = 0$; $\Delta G < \Delta G°$ (d) $\Delta G < \Delta G°$

4. $\Delta G° = -32.02 \text{ kJ/mol} = -RT \ln K$; $\underline{K = 4.1 \times 10^5}$

5. $\Delta G° = 6 \times \Delta G_f^\circ(CO_2) + 3 \times \Delta G_f^\circ(H_2O) - \Delta G_f^\circ(C_6H_6)$; (Because O_2 is an element in its standard state $\Delta G_f^\circ = 0$)

$\Delta G° = 6 \text{ mol} \times (-394.359 \text{ kJ/mol}) + 3 \text{ mol} \times (-228.572 \text{ kJ/mol}) - 124.42 \text{ kJ} = \underline{-3176.29 \text{ kJ}}$

maximum useful work = $-\Delta G° = \underline{3176.29 \text{ kJ}}$

6. $\Delta G° = 2 \times \Delta G_f^\circ(NO_2) - 2 \times \Delta G_f^\circ(NO) = 102.62 \text{ kJ} - 173.10 \text{ kJ} = -70.48 \text{ kJ}$

$$\Delta G = \Delta G° + RT \ln Q_p = -70.48 \text{ kJ/mol} + 8.314 \times 10^{-3} \text{ kJ/mol K} \times 298\text{K} \times \ln \frac{(0.20)^2}{(0.20)^2(0.20)} = \underline{66.49 \text{ kJ/mol}}.$$

The reaction is spontaneous.

7. a, c, e

8. $\Delta S° = \Delta H°/T = (2.256 \times 10^3 \text{ J/g} \times 150 \text{ g})/373\text{K} = \underline{907 \text{ J/K}}$

9. $$\Delta S° = \frac{\Delta H° - \Delta G°}{T} = \frac{-16 \text{ kJ/mol} - 16.4 \text{ kJ/mol}}{298\text{K}} = -0.11 \text{ kJ/mol}; \; 0.11 \text{ kJ/mol} \times 5.00 \text{ mol} = \underline{-0.55 \text{ kJ}}$$

10. $$\Delta S° = \frac{-285.4 \text{ kJ} - (-326.4 \text{ kJ})}{298\text{K}} = +138 \text{ J/K}$$

$+138 \text{ J/K} = 3 \text{ mol} \times S°(O_2) - 2 \text{ mol} \times 238.93 \text{ J/mol K}$; $S°(O_2) = \underline{205 \text{ J/mol K}}$

11. $\Delta G° = \Delta H° - T\Delta S° = 0$

$$T = \frac{\Delta H°}{\Delta T°} = \frac{+197.78 \times 10^3 \text{ kJ}}{+188.06 \text{ J/K}} = \underline{1051.7 \text{ K}}$$

REVIEW

Introduction

All of the chemical reactions we have studied so far, such as acid-base neutralization, combustion, and precipitation reactions to name just a few, occur because the chemical system is not at equilibrium; the concentration of reactants and products are not equilibrium concentrations. To achieve the equilibrium concentrations one of the two opposing reactions occurs at a faster rate than the other. Because it is this faster reaction that we observe, we call it the "spontaneous reaction" and recognize that there is a "net driving force" in the direction of that reaction.

Free Energy and Useful Work (Section 19.1)

Learning Objectives

1. Distinguish between useful work and expansion work.
2. Use the sign of ΔG to predict whether or not a given reaction is spontaneous.
3. Use the magnitude of ΔG to calculate the maximum useful work that can be obtained from a spontaneous reaction.

Review

A spontaneous reaction can do useful work in its drive toward equilibrium. In thermodynamics, useful work is non-PV work. (Remember that PV work is work due to expansion, which pushes back the atmosphere.) Electrical work is one example of useful work that might be done by a spontaneous reaction. A system at equilibrium cannot do useful work.

Chemical systems are said to possess a thermodynamic property, G, the Gibbs free energy, which changes as a reaction proceeds. At constant temperature and pressure, the maximum useful work that can be done is equal to the negative of the free energy change:

maximum useful work = $-\Delta G$

Therefore;
$\Delta G < 0$ for spontaneous reactions; useful work can be obtained from such reactions;
$\Delta G = 0$ at equilibrium; useful work cannot be obtained from reactions at equilibrium;
$\Delta G > 0$ for reactions which are not spontaneous (this means that the reverse reaction would be spontaneous).

Note: The value of ΔG depends only on the free energies of the reactants and products [$\Delta G = G(\text{products}) - G(\text{reactants})$]
ΔG provides no information about the reaction mechanism or the speed of the reaction.

Example 19.1: ΔG for the reaction: $4Al(s) + 3O_2(g, 1\text{ atm}) \rightarrow 2Al_2O_3(s)$ is -3351.4 kJ at 25°C.
(a) Is the reaction spontaneous?
(b) What is the maximum amount of useful work that can be obtained from the oxidation of four moles of Al?

Answer:
(a) $\Delta G < 0$; reaction is spontaneous.
(b) maximum useful work = $-\Delta G = -(-3351.4\text{ kJ}) = 3351.4\text{ kJ}$

Useful work is done at the expense of evolved heat; the more work done, the less heat is evolved.

Example 19.2: $\Delta H°$ for the formation of methane

$C(s) + 2H_2(g) \rightarrow CH_4(g)$

is -74.81 kJ/mol at 25°C. ΔG is -50.72 kJ/mol. When one mole of methane is formed, how many kilojoules of heat would be evolved if
(a) no useful work is done?
(b) the maximum useful work is done?

Answer:
(a) When no useful work is done, all of the energy ($\Delta H°$) will appear as heat; 74.81 kJ.
(b) When the maximum useful work is done ($-\Delta G$), then only 74.81 kJ – 50.72 kJ = 24.09 kJ of heat is evolved.

Practice Drill; Useful Work and Heat: Given: $\Delta G = -569.43$ kJ and $\Delta H° = -601.70$ kJ for the reaction

$Mg(s) + 1/2O_2(g) \rightarrow MgO(s)$

One mole of Mg is oxidized under three different sets of conditions (a, b, c in the table below). Fill in the blanks.

Useful work done	Heat evolved
(a) 0	__________
(b) maximum useful work	__________
(c) ____________________	100.00 kJ

Answers: (a) 601.70 kJ; (b) 32.27 kJ; (c) 501.70 kJ

Free Energy Change and Concentration (Section 19.2)

Learning Objectives

1. Distinguish between ΔG and $\Delta G°$.
2. State how ΔG varies with increasing reactant or product concentrations.
3. Calculate standard free energy changes from K and vice versa.
4. Use tabulated free energies of formation to calculate $\Delta G°$ for a given reaction.
5. Use $\Delta G°$ to calculate ΔG under any set of pressures or concentrations.

Review

ΔG varies with concentration according to the equation:

$\Delta G = \Delta G° + RT \ln Q$

$\Delta G°$ = standard free energy change
R = gas constant (8.314 J/mol K)
T = Kelvin temperature
$\ln Q$ = natural logarithm of the reaction quotient

From the above equation we can see that ΔG becomes more positive with increasing Q and more negative with decreasing Q. The standard free energy change ($\Delta G°$) is the free energy change when each product and each reactant is in its standard state. This means that the gas pressures are 1 atm and all concentrations are 1 *M*. When Q = 1, ln Q = 0 and $\Delta G = \Delta G°$.

Example 19.3: What is the effect of an increase in reactant concentrations on the driving force of a forward reaction?

Answer: Because the form of Q is product concentrations/reactant concentrations, an increase in reactant concentrations will decrease Q. Therefore RT ln Q will also decrease and so will ΔG ($\Delta G = \Delta G° + RT \ln Q$). A lower ΔG means a greater driving force in the forward direction.

Note: "Lower" means either a smaller positive value or a larger negative value.

Practice Drill; The Effect of Q on ΔG: Fill in the blanks with increases, decreases, or remains the same.

	Reactant Concentrations	Product Concentration	Q	ΔG of Forward Reaction	ΔG of Reverse Reaction
(a)	decreases	remains the same	____	__________	__________
(b)	remains the same	decreases	____	__________	__________
(c)	remains the same	increases	____	__________	__________
(d)	decreases	increases	____	__________	__________

(e) In which of the above cases is the driving force of the forward reaction increased?

Answers: (a) increases, increases, decreases; (b) decreases, decreases, increases; (c) increases, increases, decreases; (d) increases, increases, decreases; (e) b (ΔG becomes more negative)

At equilibrium $Q = K$ and $\Delta G = 0$, so that $0 = \Delta G° + RT \ln K$. Solving for $\Delta G°$ gives

$\Delta G° = -RT \ln K$

$\Delta G°$ can therefore be calculated from the equilibrium constant and vice versa.

Example 19.4: Calculate K for the dissociation of $H_2O(l)$ at 25°C, given:

$H_2O(l) \rightleftharpoons H^+(aq) + OH^-(aq) \quad \Delta G° = 79.87$ kJ/mol

Answer: $\Delta G° = -RT \ln K$

$$\ln K = -\frac{\Delta G°}{RT} = \frac{-79.87 \text{ kJ / mol}}{8.314 \times 10^{-3} \text{ kJ / mol K} \times 298\text{K}} = -32.24; \ \underline{K = 1.0 \times 10^{-14}}$$

$\Delta G_f°$ <u>is the standard free energy change for the formation of 1 mol of a substance from its elements in their most stable forms.</u> $\Delta G_f°$ for an element in its most stable form is zero.

The standard free energy change is also equal to the sum of the free energies of formation ($\Delta G_f°$) of the products minus the sum of the free energies of formation of the reactants.

$$\Delta G° = \Sigma n_p \times (\Delta G_f°)_p - \Sigma n_r \times (\Delta G_f°)_r$$

$\Delta G°$ = standard free energy change (J/mol or kJ/mol)

n_p, n_r = number of moles of a product (p) or number of moles of a reactant (r) in the equation

$(\Delta G_f°)_p (\Delta G_f°)_r$ = standard free energy change for the formation of 1 mol of a product (p) or reactant (r) from its elements in their most stable forms. (Values can be obtained from tables.)

Example 19.5: Can $CaCO_3$, an important component of mortar, be prepared from CaO and CO_2 at 1 atm and 25°C? Use $\Delta G_f°$ values from Table 19.2 in your text.

Answer:	CaO(s) ↑	+	$CO_2(g)$ ↑	→	$CaCO_3(s)$ ↑
ΔG_f° (kJ):	-604.03 kJ/mol		-394.359 kJ/mol		-1128.79 kJ/mol

$\Delta G^\circ = -1128.79\ kJ - (-604.03\ kJ - 394.359\ kJ) = \underline{-130.40\ kJ}$

Although the negative ΔG° indicates that the reaction is spontaneous, we do not know whether the reaction occurs fast enough to be observed.

Note: The ΔG_f° for one of the participating substances can be found if ΔG° for the reaction and all of the other ΔG_f° values are known.

Using ΔG_f° Values to Find K

Example 19.6: Find ΔG° and K_p for the reaction $2NO(g) + O_2(g) \rightarrow 2NO_2(g)$

Answer: We can calculate ΔG° by using free energies of formation from Table 19.2 in your text:

	2NO(g)	+	$O_2(g)$	→	$2NO_2(g)$
ΔG_f° (kJ):	2 × 86.55		0		2 × 51.31

$\Delta G^\circ = 2 \times 51.31\ kJ - 2 \times 86.55\ kJ = \underline{-70.48\ kJ}$

To find K_p we will use the relationship $\Delta G^\circ = -RT \ln K_p$

Rearranging gives $\ln K_p = \dfrac{-\Delta G^\circ}{RT} = \dfrac{70.48\ kJ}{8.314 \times 10^{-3}\ kJ / mol\ K \times 298K} = 28.45$

Taking the antiln K_p = antiln (28.45) = $e^{28.45} = \underline{2.3 \times 10^{12}}$

Practice Drill on the Relationship Between ΔG°, ΔG_f° Values and K: Fill in the blanks for reactions a and b at 25°C.

(a) $N_2O_4(g) \rightleftharpoons 2NO_2(g)$
(b) $2O_3(g) \rightleftharpoons 3O_2(g)$

	ΔG_f° (reactant) kJ / mol	ΔG_f° (product) kJ / mol	ΔG° (kJ)	K_p
(a)	97.89	________	________	0.15
(b)	________	________	-326.4	________

Answers: (a) $\Delta G_f^\circ(NO_2)$ = 51.3 kJ/mol; ΔG° = 4.7 kJ; (b) $\Delta G_f^\circ(O_3)$ = 163.2 kJ/mol; $\Delta G_f^\circ(O_2)$ = 0; $K_p = 2 \times 10^{57}$)

Calculating ΔG for Nonstandard Conditions

Use $\Delta G = \Delta G° + RT \ln Q$

ΔG = free energy change (J/mol) under nonstandard conditions
$\Delta G°$ = free energy change (J/mol) when each gas is at 1 atm pressure and each dissolved substance is 1 *M*.
R = 8.314 J/mol K
T = temperature (K)
Q = reaction quotient (Use Q_p for gaseous reactions; use Q_c for reactions in solutions.)

Example 19.7: $\Delta G°$ is +79.87 kJ/mol for the dissociation of water

$H_2O(l) \rightleftharpoons H^+(aq) + OH^-(aq)$ at 25°C.

(a) Calculate ΔG when $[H^+] = 2.0 \times 10^{-2}$ M and $[OH^-] = 5.0 \times 10^{-3}$ M.
(b) Is the dissociation of water spontaneous at the above concentrations?

(a) Given: $\Delta G° = 79.87$ kJ/mol
$T = 25°C = 298K$
$[H^+] = 2.0 \times 10^{-2}$ M; $[OH^-] = 5.0 \times 10^{-3}$ *M*

Unknown: ΔG

Plan: Use $\Delta G = \Delta G° + RT \ln Q_c$; $Q_c = [H^+][OH^-]$

Calculation: $\Delta G = \Delta G° + RT \ln Q_c$
$= 79.87 \text{ kJ/mol} + 8.314 \times 10^{-3} \text{ kJ/mol} \times 298K \times \ln(2.0 \times 10^{-2} \times 5.0 \times 10^{-3}) = \underline{57.05 \text{ kJ/mol}}$

(b) ΔG is positive so that the forward reaction is not spontaneous; the reverse reaction

$H^+(aq) + OH^-(aq) \rightleftharpoons H_2O(l)$

is spontaneous under the given conditions.

Entropy and the Second Law of Thermodynamics (Section 19.3)

Learning Objectives

1. Give examples showing how spontaneous processes are accompanied by an increase in the disorder of the system and/or the surroundings.
2. State the second law of thermodynamics.
3. Predict whether the entropy of the system will increase or decrease during a given physical or chemical change.

Review

During a spontaneous change the system or the surroundings, or both, enter a more disordered state. The thermodynamic property that measures disorder and randomness is called entropy (S).

The second law of thermodynamics states that all spontaneous changes are accompanied by an increase in the entropy of the universe: The entropy change of the universe is the sum of the entropy change of the system plus the entropy change of the surroundings. Hence, for a spontaneous change

$$\Delta S_{universe} = \Delta S_{system} + \Delta S_{surroundings} > 0$$

Note that the second law says that only the entropy of the universe must increase ($\Delta S_{universe} = +$) for a spontaneous change; the entropy of the system can in fact decrease ($\Delta S_{system} = -$) as in freezing of liquids. The entropy change of the surroundings must then be sufficiently positive so that the entropy change of the universe is positive.

The entropy of a system increases with processes such as expansion, warming, melting, evaporation, generation of a gas, and an increase in the number of moles of particles. The reverse processes result in an entropy decrease of the system.

Practice: In which of the following reactions does the entropy of the system increase?
(a) $2KClO_3(s) \rightarrow 2KCl(s) + 3O_2(g)$
(b) $NH_3(g) + HCl(g) \rightarrow NH_4Cl(s)$
(c) explosion of nitroglycerin

Answer: (a) and (c)

Entropy Calculations (Section 19.4)

Learning Objectives

1. Use standard enthalpy and free energy changes to calculate $\Delta S°$ for a phase change or chemical reaction.
2. State the third law of thermodynamics.
3. Use tabulated absolute entropies to calculate $\Delta S°$ of a reaction.
4.* Derive the relation $\Delta G = \Delta H - T\Delta S$.
5.* Show that $\Delta G_{system} = -T\Delta S_{universe}$ for a spontaneous reaction occurring at constant temperature and pressure.

Review

At constant pressure and temperature ΔG, ΔH, and ΔS are related by the equation

$\Delta G = \Delta H - T\Delta S$

ΔG = change in free energy of system (J or kJ)
ΔH = change in enthalpy of system (J or kJ)
T = temperature of system (K)
ΔS = change in entropy of system (J/K or kJ/K)

At equilibrium $\Delta G = 0$, so that $0 = \Delta H - T\Delta S$ and $\Delta S = \dfrac{\Delta H}{T}$

The equation $\Delta S = \Delta H/T$ is a special case of the more general equation $\Delta S = q_{rev}/T$, where q_{rev} is the amount of heat absorbed by a system reacting under reversible conditions.

Example 19.8: H_2S in the atmosphere is converted to SO_2 by oxygen according to

$H_2S(g) + 3/2O_2(g) \rightarrow SO_2(g) + H_2O(g)$

Calculate $\Delta S°$ at 25°C for the above reaction using ΔG_f° and ΔH_f° values from Table 19.2 in your text.

Answer: $\Delta G°$ and $\Delta H°$ for the reaction can be found by using ΔG_f° and ΔH_f° data from the table. Once $\Delta G°$, and $\Delta H°$ are known then $\Delta S°$ can be calculated from $\Delta G° = \Delta H° - T\Delta S°$.

	$H_2S(g)$	+	$3/2O_2(g)$	→	$SO_2(g)$	+	$H_2O(g)$
ΔG°_f (kJ)	-33.56		0		-300.19		-228.572
ΔH°_f (kJ)	-20.63		0		-296.83		-241.818

ΔG° = 1 mol × (-228.572 kJ/mol) + 1 mol × (-300.19 kJ/mol) – 1 mol × (-33.56 kJ/mol) = -495.20 kJ

ΔH° = 1 mol × (-241.818 kJ/mol) + 1 mol × (-296.83 kJ/mol) – 1 mol × (-20.63 kJ/mol) = -518.02 kJ

$\Delta G^\circ = \Delta H^\circ - T\Delta S^\circ$. Rearranging and dividing by T gives

$$\Delta S^\circ = \frac{\Delta H^\circ - \Delta G^\circ}{T}$$

Substituting calculated values for ΔH° and ΔG° gives

$$\Delta S^\circ = \frac{-518.02\text{ kJ} - (-495.20\text{ kJ})}{298\text{K}} = -0.0766\text{ kJ/K} = \underline{-76.6\text{ J/K}}$$

(Note that the entropy change is negative. The decrease in entropy could have been predicted without calculations; the equation shows a decrease in molecules of gas, which indicates a decrease of entropy in the system.)

Practice Drill on $\Delta G = \Delta H - T\Delta S$: Fill in the chart for the following thermodynamic changes:

	ΔG (kJ)	ΔH (kJ)	ΔS (J/K)	T (K)
(a)	-141.74	?	-188	298
(b)	?	118.9	354.8	298
(c)	0	608.6	828	?

Answers: (a) -197.8 kJ (b) 13.2 kJ (c) 735K

Example 19.9: Calculate ΔS° for the vaporization of 100 g of liquid water at 1 atm and 100°C. The heat of vaporization of water under these conditions is 40.7 kJ/mol.

Answer: Vaporization at the boiling temperature is an equilibrium process so the equation $\Delta S = \Delta H/T$ applies. The enthalpy change is

$$\Delta H^\circ = 100\text{ g} \times \frac{1\text{ mol}}{18.02\text{ g}} \times \frac{40.7\text{ kJ}}{\text{mol}} = 226\text{ kJ}$$ Therefore,

$$\Delta S^\circ = \frac{226\text{ kJ}}{373\text{K}} = 606\text{ J/K}$$

Note that the superscript is used because the vaporization occurs at 1 atm pressure (standard conditions).

Absolute Entropies

The third law of thermodynamics states that the entropy of a pure perfectly crystalline element or compound is zero at zero Kelvin. Absolute entropies at other temperatures have been calculated using this law. Table 19.2 in your text lists some standard molar entropies at 25°C.

The standard entropy change of a reaction ($\Delta S°$) can be found by subtracting the tabulated standard molar entropies of the reactants from those of the products.

$$\Delta S = \sum n_p S_p^{\circ} - \sum n_r S_r^{\circ}$$

$\Delta S°$	=	standard entropy change of a reaction
n_p, n_r	=	number of moles of a product (n_p) or reactant (n_r)
S_p°, S_r°	=	absolute entropy of product (p) or reactant (r). (Values are listed in Table 19.2 of your text.)

Example 19.10: Calculate $\Delta S°$ for the oxidation of 1 mol of ethanol, ($C_2H_5OH(l)$), at 25°C.

$C_2H_5OH(l) + 3O_2(g) \rightarrow 2CO_2(g) + 3H_2O(l)$

Use entropy data from Table 19.2 in your text.

Answer:	$C_2H_5OH(l)$	+	$3O_2(g)$	$\rightarrow$	$2CO_2(g)$	+	$3H_2O(l)$
S°(J/K):	160.7		3 × 205.138		2 × 213.74		3 × 69.91

$\Delta S° = 2 \times 213.74$ J/K $+ 3 \times 69.91$ J/K $- 160.7$ J/K $- 3 \times 205.138$ J/K $=$ -138.9 J/K

Note that $\Delta S°$ could also be calculated by finding $\Delta G°$ and $\Delta H°$ from formation data (Table 19.2) and substituting into

$\Delta G° = \Delta H° - T\Delta S°$

*For a spontaneous reaction at constant temperature and pressure, the decrease in free energy of the reacting system is related to the increase in entropy of the universe by the equation

$\Delta G_{system} = -T\Delta S_{universe}$

The Variation of ΔG° With Temperature (Section 19.5)

Learning Objectives

1. Use the values of $\Delta H°$ and $\Delta S°$ at one temperature to estimate $\Delta G°$ at some other temperature.
2. Estimate the temperatures, if any, at which a reaction is spontaneous.

Review

ΔG varies substantially with temperature while ΔH and ΔS do not. Consequently we can use enthalpy and entropy changes measured at one temperature to estimate $\Delta G°$ and K at some other temperature by applying the relationship $\Delta G° = \Delta H° - T\Delta S°$.

Example 19.11: Use data from Table 19.2 in your text to estimate (a) ΔG° and (b) K_p for the oxidation of 1 mol of NO(g) at 0°C to NO_2(g).

Given: NO(g) + 1/2O_2(g) → NO_2(g); T = 0°C; data from Table 19.2

Unknown: (a) ΔG°, (b) K_p.

Plan: (a) Calculate ΔH° and ΔS° for the reaction at 25°C from the ΔH_f° and S° values in Table 19.2. Substitute these calculated values and T = 273K into ΔG° = ΔH° – TΔS° to find ΔG°.
(b) Use ΔG° = -RT ln K_p to calculate K_p.

Calculations:

(a)	NO(g)	+	1/2O_2(g)	→	NO_2(g)
ΔH_f° (kJ)	90.25		0		33.18
S° (J/K)	210.76		1/2 × 205.138		240.06

ΔH° = 33.18 kJ – 90.25 kJ = -57.07 kJ

ΔS° = 240.06 J/K – 1/2 × 205.138 J/K – 210.76 J/K = -73.27 J/K

ΔG° = ΔH° – TΔS° = -57.07 kJ – 273K × (-0.07327 kJ/K) = <u>-37.07 kJ</u>

(b) ΔG° = -RT ln K_p

-37.07 kJ = -8.314 × 10^{-3} kJ/mol K × 273K × ln K_p

Solving for ln K and taking the antiln gives K_p = <u>1 × 10^7</u>.

Note: When ΔH and ΔS have the same sign, there will always be some temperature above (both +) or below (both –) which the reaction is spontaneous. To find this temperature, use ΔG = ΔH – TΔS = 0 and solve for T.

Example 19.12: Nitrogen oxides are air pollutants emitted during the high-temperature combustion of fossil fuels. Estimate the temperature above which N_2(g) and O_2(g) will spontaneously form nitric oxide (NO(g)). Use data from Table 19.2 in your text.

Given: N_2(g) + O_2(g) → 2NO(g) and Table 19.2

Unknown: T above which reaction is spontaneous under standard conditions.

Plan: Use ΔH_f° and S° values to compute ΔH° and ΔS° for the given reaction. Substitute these values into ΔG° = ΔH° – TΔS° = 0 to find the equilibrium temperature. ΔG° will be negative for all temperatures above this equilibrium temperature.

Calculations:	N_2(g)	+	O_2(g)	→	2NO(g)
ΔH_f° (kJ)	0		0		2 × 90.25
S° (J/K)	191.61		205.138		2 × 210.76

ΔH° = 2 mol × 90.25 kJ/mol = 180.50 kJ

ΔS° = 2 × 210.76 J/K – 205.138 J/K – 191.61 J/K = 24.77 J/K

ΔG° = ΔH° – TΔS° = 0; Solving for T and substituting calculated values

T = ΔH°/ΔS° = 180.50 kJ/24.77 × 10^{-3} kJ/K = <u>7287K</u>

PROBLEM-SOLVING TIP

Summary of Important Thermodynamic Relationships

Relationship	Meaning of Variables and Their Common Units	Meaning and Application of Relationships
$\Delta G = -$ maximum useful work	ΔG = change in Gibbs free energy (J or kJ) = G (products) – G (reactants)	A negative ΔG means the change is spontaneous and can do useful work (non-PV work). The maximum useful work is equal to the negative of the ΔG value.
$\Delta G = \Delta G° + RT \ln Q$	$\Delta G°$ = change in free energy under standard conditions: (all concentrations are 1 M, all gases at 1 atm) R = 8.314 J/mol K Q = reaction quotient = [products]/[reactants] T = temperature (K)	The equation is used to calculate the free energy change when one or more concentrations or pressures are not standard.
$\Delta G° = -RT \ln K$	$\Delta G°$, R, T (see above) K = equilibrium constant	The equation can be used to calculate K if $\Delta G°$ is known and vice versa for any given temperature. If gases are involved then K_p is generally used: $\Delta G° = -RT \ln K_p$
$\Delta G° = \Sigma n_p(\Delta G_f^\circ)_p - \Sigma n_r(\Delta G_f^\circ)_r$	n_p = number of moles of each product in balanced equation n_r = number of moles of each reactant in balanced equation ΔG_f° = standard free energy change for the formation of compound from its elements (kJ/mol). (See Table 19.2 of text.) ΔG_f° for the most stable form of any element is 0.	Use data in Table 19.2 to find $\Delta G°$ for a reaction from its balanced equation. The tabulated values are for 25°C.
$\Delta S_{universe} = \Delta S_{system} + \Delta S_{surroundings} > 0$	ΔS = entropy change (J/K or kJ/K)	Statement of the Second Law of Thermodynamics: The entropy of the universe increases with any spontaneous change. If ΔS has a negative value the $\Delta S_{surroundings}$ must have a larger positive value if the change is to be spontaneous.

Relationship	Meaning of Variables and Their Common Units	Meaning and Application of Relationships
$\Delta G = \Delta H - T\Delta S$	ΔH = change in enthalpy (kJ/mol)	The free energy change is the sum of two terms: ΔH (heat of reaction at constant pressure) and $-T\Delta S$. A negative ΔH (exothermic reaction) and a positive ΔS (increase in entropy of the system) favor spontaneity. ΔH and ΔS tend not to change with temperature so their value at one temperature can be used to estimate ΔG at a different temperature.
$\Delta S = q_{rev}/T$ $\Delta S = \Delta H/T$	q_{rev} = heat lost or gained during a reversible change (J or kJ) T = temperature (K)	The entropy change for a reversible process is the heat lost or gained divided by the temperature. At constant pressure, $q_{rev} = \Delta H$ and the equation becomes $\Delta S = \Delta H/T$. **Note:** This equation can be applied to phase changes at their transition temperatures. This equation cannot be applied to spontaneous chemical reactions.
$\Delta S° = \Sigma n_p S_p° - \Sigma n_r S_r°$	$\Delta S°$ = entropy change under standard conditions S° = Third Law Entropy value (Table 19.2) Note: S° for an element is not zero.	The entropy change for a reaction under standard conditions can be calculated from the third law entropy values.

SELF-TEST

Underline the correct word(s) for questions 1-4.

1. If useful work is done during an exothermic reaction, then the amount of heat released is (greater, less than) the heat released when no useful work is done.

2. The expansion of gases in a steam engine is an example of (useful, PV) work.

3. During a lightning storm, atmospheric $N_2(g)$ and $O_2(g)$ combine to form NO(g). The value of ΔG for the reaction under these conditions is therefore (negative, zero, positive).

4. As the concentration of reactants increases, ΔG (increases, decreases, remains the same) and $\Delta G°$ (increases, decreases, remains the same) for a reaction.

5. The $\Delta G°$ value for the reaction

$C_2H_4(g) + H_2(g) \rightarrow C_2H_6(g)$

is -101 kJ at 25°C. Find

(a) K_p at 25°C and

(b) ΔG_f° ($C_2H_6(g)$) at 25°C given that ΔG_f° ($C_2H_4(g)$) = 68.15 kJ/mol.

6. Determine ΔG at 25°C for the reaction

$Mg^{2+}(aq) + 2OH^-(aq) \rightleftharpoons Mg(OH)_2(s)$

when $[Mg^{2+}] = 1.0 \times 10^{-8}$ *M* and pH = 9.00. The ΔG_f° values are: $Mg^{2+}(aq)$ = -456.0 kJ/mol. $OH^-(aq)$ = -157.3 kJ/mol; $Mg(OH)_2(s)$ = -933.9 kJ/mol. Will a $Mg(OH)_2$ precipitate form?

7. Determine the sign of the entropy change of the surroundings for each of the following changes.
(a) Liquid water turns to ice in a freezer.
(b) One gram of liquid ethanol evaporates at room temperature and pressure.
(c) A cell synthesizes insulin.
(d) Ice melts at room temperature.

8. Determine the entropy change when one kilogram of liquid nitrogen vaporizes at its normal boiling point of -195.81°C. The heat of vaporization of N_2 is 0.201×10^3 J/g.

9. The ΔG_f° and ΔH_f° values for calcite ($CaCO_3$), are -1128.79 kJ/mol, and -1206.92 kJ/mol respectively at 25°C.
(a) Calculate ΔS_f° for calcite at 25°C.
(b) Given the following third law entropy values S°(graphite) = 5.740 J/mol K; $S°(O_2)$ = 205.138 J/mol K; and S°(Ca) = 41.42 J/mol K, calculate the third law entropy value for $CaCO_3(s)$ at 25°C.

10. Below what temperature will $CaCO_3$ spontaneously form from its elements? (Use information in Problem 9.)

ANSWERS TO SELF-TEST QUESTIONS

1. less than

2. PV

3. negative

4. decreases (becomes more negative), remains the same

5. (a) $\Delta G° = -RT \ln K_p = -101 \text{ kJ} = -8.314 \times 10^{-3} \text{ kJ/mol K} \times 298\text{K} \times \ln K_p$; $K_p = \underline{5 \times 10^{17}}$
(b) $-101 \text{ kJ} = \Delta G_f^\circ (C_2H_6(g)) - \Delta G_f^\circ (C_2H_4(g)) = \Delta G_f^\circ (C_2H_6(g)) - 68.15 \text{ kJ}$; $\Delta G_f^\circ (C_2H_6(g)) = \underline{-33 \text{ kJ/mol}}$.

6. $\Delta G° = -933.9 \text{ kJ} - 2 \times (-157.3 \text{ kJ}) - (-456.0 \text{ kJ}) = -163.3 \text{ kJ/mol}$
$\Delta G = \Delta G° + RT \ln Q$

$$\Delta G = -163.3 \text{ kJ/mol} + 8.314 \text{ J/mol K} \times 298\text{K} \ln \frac{1}{(1.0 \times 10^{-8})(1.0 \times 10^{-5})^2}$$

$\Delta G = -163.3 \text{ kJ/mol} + 103 \text{ kJ/mol} = \underline{-60 \text{ kJ/mol; yes}}$

7. a, c are positive; b, d are negative.

8. $\Delta S° = \Delta H°/T = (0.201 \times 10^3 \text{ J/g} \times 1000 \text{ g})/77.34\text{K} = \underline{2.60 \text{ kJ/K}}$

9. (a) $\Delta S_f^\circ = \dfrac{\Delta H_f^\circ - \Delta G_f^\circ}{T} = \dfrac{-1206.92 \text{ kJ / mol} - (-1128.79 \text{ kJ / mol})}{298\text{K}} = \underline{-262 \text{ J/mol K}}$

   (b) $Ca(s) + 3/2O_2(g) + C(\text{graphite}) \rightarrow CaCO_3(s)$

   $\Delta S_f^\circ = S°(CaCO_3) - S°(\text{graphite}) - 3/2S°(O_2) \times -S°(Ca)$

   $-262 \text{ J/mol K} = S°(CaCO_3) - 5.740 \text{ J/mol K} - (3/2 \text{ mol} \times 205.138 \text{ J/mol K}) - 41.42 \text{ J/mol K}$

   $S°(CaCO_3) = \underline{93 \text{ J/mol K}}$

10. $0 = \Delta H_f^\circ - T\Delta S_f^\circ;\ T = \Delta H_f^\circ/\Delta S_f^\circ = \dfrac{-1206.92 \text{ kJ / mol}}{-0.262 \text{ kJ / mol K}} = \underline{4607\text{K}}$

    Below approximately 4607K

CHAPTER 20
ELECTROCHEMISTRY

SELF-ASSESSMENT

Galvanic Cells (Section 20.1)

True or False. If the statement is false, change the underlined word(s) to make it true.

1. The sign of the cathode in a galvanic cell is <u>positive</u> and the sign of the <u>anode</u> is negative.

2. In a galvanic cell <u>cations</u> migrate toward the anode.

3. A standard galvanic cell is based on the reaction

   $2Hg(l) + 2Cl^-(aq) + 2RuO_4^-(aq) \rightarrow Hg_2Cl_2(s) + 2RuO_4^{2-}(aq)$

   (a) Suggest electrode materials and electrolyte compositions and indicate electron flow for the standard cell.
   (b) Write the cell notation.

Electrical Work and Free Energy (Section 20.2)

4. E° for the reaction $Cu(s) + Cl_2(g) \rightarrow Cu^{2+}(aq) + 2Cl^-(aq)$ is +1.0181 V.
   (a) Calculate ΔG°.
   (b) What is the maximum electrical work that can be obtained when 1 mol of Cu is oxidized according to the above equation?

Standard Electrode Potentials (Section 20.3)

The reduction potentials for $Ag^+(aq)$, $Cl_2(g)$, $Pb^{2+}(aq)$, and AgCl(s) are 0.799 V, 1.360 V, -0.127 V, and 0.222 V.

5. Which of the above species is the best oxidizing agent?

6. Predict which of the following reactions will occur under standard conditions and write their equations.
   (a) oxidation of Pb(s) by $Ag^+(aq)$
   (b) oxidation of Ag(s) by $Cl_2(g)$ to produce AgCl(s)
   (c) reduction of $Ag^+(aq)$ by $Cl^-(aq)$

7. Calculate E° and ΔG° for the reactions described in 6(a) – (c).

EMF and Concentration (Section 20.4)

8. Answer (a) – (d) as increases, decreases, or remains the same for the redox reaction

$2S_2O_3^{2-}(aq) + I_2(aq) \rightarrow 2I^-(aq) + S_4O_6^{2-}(aq)$

(a) If $Na_2S_2O_3$ is added to the reaction mixture the reaction quotient (Q) __________ and the EMF __________.
(b) If KI is added to the reaction mixture, Q __________ and the EMF __________.
(c) As the reaction proceeds the EMF __________ while K __________.
(d) As the reaction proceeds E° (standard EMF) __________.

9. The half-reaction and standard reduction potential for the conversion of sulfate ion to sulfite ion is

$SO_4^{2-}(aq) + H_2O(l) + 2e^- \rightarrow SO_3^{2-}(aq) + 2OH^-(aq) \quad E° = -0.936\ V$

(a) Calculate the reduction potential when the concentrations of SO_4^{2-} and SO_3^{2-} ions are 1.0×10^{-4} *M* and 1.0×10^{-3} *M*, respectively, at pH = 8.00. Will the reduction potential increase or decrease at lower pH?
(b) One mole of $Cu(NO_3)_2$ is added to 1 L of the solution described in (a) above. Write the equation and calculate the EMF for the net reaction. The standard reduction potential for the Cu/Cu^{2+} half-reaction is 0.339 V.

10. Use the standard reduction potential

$O_2(g) + 4H^+(aq) + 4e^- \rightleftharpoons 2H_2O(l) \quad 1.229\ V$

to calculate the equilibrium constant for

$H_2(g) + 1/2O_2(g) \rightleftharpoons H_2O(l)$

(Hint: Use the standard hydrogen half-reaction.)

Batteries and Fuel Cells (Section 20.5)

11. Answer (a) – (d) as true or false. If a statement is false, change the underlined word(s) to make it true.
(a) <u>Zinc</u> is the anode in both the dry cell and the alkaline dry cell.
(b) The <u>dry cell</u> maintains a fairly constant voltage during its lifetime and is used in cardiac pacemakers.
(c) The <u>alkaline dry cell</u> can be recharged.
(d) A <u>fuel</u> cell is a galvanic cell in which reactants are continually fed in and products removed.

Corrosion (Section 20.6)

12. Account for the following facts:
(a) Steel wool is commonly stored under soapy water.
(b) Iron is often coated with zinc if it is to be in constant contact with water.
(c) Radiator coolants contain corrosion inhibitors.
(d) Acid rain is corrosive.

Electrolysis (Section 20.7) and **Electrochemical Stoichiometry** (Section 20.8)

13. Chromium metal can be electroplated from an acidic solution of CrO_3 according to the equation

$CrO_3(aq) + 6H^+(aq) + 6e^- \rightarrow Cr(s) + 3H_2O(l)$

(a) A 0.150-A current is passed through the solution for 3.00 h. Assuming no other reactions, calculate the grams of Cr deposited (at. wt. of Cr = 52.00).
(b) Using the same current calculate how long it would take to plate 1.00 g of Cr.

14. In the electrolysis of a $CuSO_4$ solution, Cu metal is deposited at one electrode and O_2 gas and H^+ ions are produced at the other electrode.
    (a) Write balanced equations for the half-reactions and for the overall reaction.
    (b) How many liters of O_2 are produced at 273K and 1 atm when a 0.500 A current passes through the cell for 2.00 h?

ANSWERS TO SELF-ASSESSMENT PROBLEMS

If you missed an answer, be sure to study the relevant section in your text and study guide.

1. T

2. F; anions

3. (a) Calomel electrode (anode) consists of liquid mercury, solid Hg_2Cl_2 amd 1 *M* KCl. The cathode consists of 1 *M* $KRuO_4$ and 1 *M* K_2RuO_4 using Pt as an inert electrode.
   (b) $Hg(l) / Hg_2Cl_2(s) / KCl(1\ M) // KRuO_4(1\ M) / K_2RuO_4(1\ M) / Pt$

4. (a) $\Delta G° = -nFE° = -2 \times 96{,}485\ C \times 1.0181\ J/1C = \underline{-196.46\ kJ}$
   (b) 196.46 kJ

5. Cl_2

6. (a) $Pb(s) + 2Ag^+(aq) \rightarrow Pb^{2+}(aq) + 2Ag(s)$
   (b) $2Ag(s) + Cl_2(g) \rightarrow 2AgCl(s)$

7. (a) $E° = +0.127\ V + 0.799\ V = 0.926\ V$; $\Delta G° = -nFE° = -2 \times 96{,}485\ C \times 0.926\ J/1C = -179\ kJ$
   (b) $E° = -0.222\ V + 1.360\ V = 1.138\ V$; $\Delta G° = -nFE° = -2 \times 96{,}485\ C \times 1.138\ J/1C = -219.6\ kJ$
   (c) $E° = 0.799\ V - 1.360\ V = -0.561\ V$; $\Delta G° = -nFE° = -2 \times 96{,}485\ C \times (-0.561\ J/1C) = 108\ kJ$

8. (a) decreases, increases
   (b) increases, decreases
   (c) decreases, remains the same
   (d) remains the same

9. (a) $$E = E° - \frac{0.0592}{n}\log Q = -0.936\ V - \frac{0.0592}{2}\log\frac{[OH^-]^2[SO_3^{2-}]}{[SO_4^{2-}]}$$

$$= -0.936\ V - 0.0296\log\frac{(1.0\times10^{-6})^2(1.0\times10^{-3})}{(1.0\times10^{-4})} = \underline{-0.610\ V}\text{; increase}$$

   (b) $Cu^{2+}(aq) + SO_3^{2-}(aq) + 2OH^-(aq) \rightarrow Cu(s) + SO_4^{2-}(aq) + H_2O(l)$
   $E = E_{red} + E_{oxid} = 0.339\ V + 0.610\ V = \underline{0.949\ V}$

10. $H_2(g) \rightarrow 2H^+(aq) + 2e^-$ $\qquad E°_{oxid} = 0.000$
    $O_2(g) + 4H^+(aq) + 4e^- \rightleftharpoons 2H_2O(l)$ $\qquad E°_{red} = 1.229\ V$

$$E° = \frac{0.0542}{2} \log K = 1.229\ V;\ \log K = \frac{1.229\ V \times 2}{0.0592}$$

$K = \text{antilog } 41.5;\ \underline{K = 3 \times 10^{41}}$

11. (a), (d) T; (b) F; mercury cell; (c) F; storage battery

13. (a) 0.150 C/s × 3600 s/h × 3.00 h × 1 mol Cr/96,485 C × 6 × 52.00 g Cr/1 mol Cr = <u>0.146 g</u>

(b) $1.00\ g\ Cr \times \frac{1\ mol\ Cr}{52.00\ g\ Cr} \times \frac{96{,}485\ C \times 6}{1\ mol\ Cr} \times \frac{1\ s}{0.150\ C} \times \frac{1\ h}{3600\ s} = \underline{20.6\ h}$

14. (a) $2(Cu^{2+}(aq) + 2e^- \rightarrow Cu(s))$ cathode

$2H_2O(l) \rightarrow 4e^- + O_2(g) + 4H^+(aq)$ anode

$2Cu^{2+}(aq) + 2H_2O(l) \rightarrow 2Cu(s) + O_2(g) + 4H^+(aq)$

(b) $\frac{0.500\ C}{1\ s} \times \frac{3600\ x}{1\ h} \times 2.00\ h\ \frac{1\ mol\ O_2}{4 \times 96{,}485\ C} \times \frac{22.4\ L}{1\ mol\ O_2} = \underline{0.209\ L}$

REVIEW

<u>Electrochemistry</u> deals with processes that convert chemical energy into electrical energy and vice versa. Two branches of electrochemistry are reviewed in this chapter; galvanic cells and electrolysis.

Galvanic Cells (Section 20.1)

Learning Objectives

1. Identify the anode, the cathode, the direction of electron flow, and the direction of ion flow in a given galvanic cell.
2. Describe galvanic cells using cell notation.
3. Given the cell reaction or the cell notation for a galvanic cell, sketch a diagram of the cell.
4. Distinguish between standard and nonstandard cells and half-cells.

Review

A <u>galvanic cell</u> uses a spontaneous redox reaction to produce current and do electrical work. It consists of two half-cells, each containing an electrode and an electrolyte solution. The electrodes are connected by an external circuit; the electrolyte solutions may be connected by a salt bridge, by a porous partition, or they may be in direct contact. Study Figures 20.1 and 20.3 in your text which show examples of galvanic cells.

Note: A salt bridge is an inverted U-tube filled with an electrolyte solution such as Na_2SO_4 or KCl.

The electrode at which oxidation occurs is the <u>anode</u>; reduction occurs at the <u>cathode</u>. In a galvanic cell the anode is the electron source or negative electrode, and the cathode is the positive terminal. Electrons flow through the external circuit from anode (–) to cathode (+). Within the cell anions migrate toward the anode and cations toward the cathode.

Electrode	Sign	Half-reaction	Electron Flow	Ion Flow
anode	–	oxidation	anode to cathode	anions to anode
cathode	+	reduction		cations to cathode

A galvanic cell can be represented by cell notation which has the following form

anode / anode electrolyte // cathode electrolyte / cathode

The vertical lines separate the phases and the double vertical line indicates a salt bridge or a porous partition. Concentrations and states are given in parentheses.

Note: Cell notation imitates the flow of electrons from anode to cathode; the anode half-cell provides the electrons and is thus written on the left and the cathode half-cell accepts the electrons and is described to the right of the double vertical line.

Example 20.1: Give the cell notation for a galvanic cell using the reduction-oxidation reaction

$Zn(s) + Fe^{2+}(aq) \rightarrow Zn^{2+}(aq) + Fe(s)$

The concentrations of the electrolytes are 0.20 *M*.

Answer: The half reactions are

$Zn(s) \rightarrow Zn^{2+}(aq) + 2e^-$ anode, oxidation

$Fe^{2+}(aq) + 2e^- \rightarrow Fe(s)$ cathode, reduction

The cell notation is

$Zn(s) / Zn^{2+}(0.20\ M) // Fe^{2+}(0.20\ M) / Fe$

Practice Drill on Redox Reactions and Galvanic Cells: Refer to Table 20.1 (p. 854) in your text and write the relevant half-reactions and cell notation for the two reactions given below. All concentrations are 1 *M* and all gases are at 1 atm.

(a) $Cl_2(g) + Ni(s) \rightarrow Ni^{2+}(aq) + 2Cl^-(aq)$

anode reaction:

cathode reaction:

cell notation:

(b) $Ag(s) + Cl^-(aq) + Fe^{3+}(aq) \rightarrow Fe^{2+}(aq) + AgCl(s)$

anode reaction:

cathode reaction:

cell notation:

Answers:

(a) anode reaction: $Ni(s) \rightarrow Ni^{2+}(aq) + 2e^-$
cathode reaction: $Cl_2(g) + 2e^- \rightarrow 2Cl^-(aq)$
cell reaction: $Ni / Ni^{2+}(1\ M) // Cl_2(1\text{ atm}) / Cl^-(1\ M) / Pt$

(b) anode reaction: $Ag(s) + Cl^-(aq) \rightarrow AgCl(s) + e^-$
cathode reaction: $Fe^{3+}(aq) + e^- \rightarrow Fe^{2+}(aq)$
cell notation: $Ag(s) / AgCl(s) / Cl^-(1\ M) // Fe^{3+}\ (1\ M), Fe^{2+}\ (1\ M) / Pt$

A standard cell is one in which each substance is in its standard state: all concentrations are 1 *M* and all gases are at 1 atm pressure. Each half-cell of a standard cell is called a standard half-cell or a standard electrode.

Practice: Write the cell notation for a standard cell based on reaction (a) above.

Answer: See answer to (a) above.

Electrical Work and Free Energy (Section 20.2)

Learning Objectives

1. Use the cell voltage to calculate the maximum electrical work that can be obtained from the transfer of a given amount of charge.
2. Calculate the free energy change of a cell reaction from the cell voltage and vice versa.
3. Explain why the voltage of a galvanic cell must be positive.

Review

The driving force that causes electrons to move around a circuit is called the electromotive force (EMF) and is measured in volts (V). (One volt is equal to one joule per coulomb.) The EMF in a galvanic cell comes from the redox reaction which supplies the electrons at the anode and removes them at the cathode. The flow of electrons in the external circuit can be harnessed to do work.

The maximum electrical work that can be done by a galvanic cell is

maximum electrical work = Q × E   Q = charge transferred in coulombs (C)
E = EMF of the cell in volts (V) or joules/coulomb (J/C)

Note: The actual work will always be less than the maximum because some energy will be lost as heat.

Example 20.2: Calculate the maximum electrical work that can be done by a 1.5 V battery supplying 0.25 C/s for 3.0 min.

Given: E = 1.5 V
time = 3.0 min
current = 0.25 C/s

Unknown: max work

Plan:
1. Find Q: Q = current × time
2. Use the relationship maximum electrical work = E × Q

Calculations:
1. Q = 0.25 C/s × (3.0 min × 60 s/min) = 45 C
2. max work = 1.5 J/C × 45 C = 68 J

ΔG and Cell Voltage

The free energy change that accompanies a cell reaction is given by the equation

$\Delta G = -nFE$ ΔG = free energy change in joules
n = number of moles of electrons flowing through the cell
F = Faraday (F): 1 F = 96,485 C/mol e^-
E = cell voltage

For a reaction in a standard cell the equation becomes

$\Delta G° = -nFE°$ E° = standard cell voltage

The free energy change of a spontaneous redox reaction is always negative and its voltage positive. The greater the voltage, the greater the driving force of the redox reaction and the more negative the ΔG.

Example 20.3: A potentiometer indicates a cell voltage of 0.360 V for a standard cell based on the reaction

$Zn(s) + Cd^{2+}(aq) \rightarrow Zn^{2+}(aq) + Cd(s)$

Calculate ΔG°. (A potentiometer draws little current and can therefore be used to measure the maximum cell voltage.)

Answer: The oxidation of 1 mol of Zn by Cd^{2+} requires the transfer of 2 mol of electrons. Therefore,

$$\Delta G° = -nFE° = -2 \text{ mol e}^- \times \frac{96{,}485 \text{ C}}{1 \text{ mol e}^-} \times \frac{0.360 \text{ J}}{1 \text{ C}} = -6.95 \times 10^4 \text{ J} = \underline{-69.5 \text{ kJ}}$$

Practice: The ΔG° for the reaction

$Ag(NH_3)_2^+(aq) + Cu^+(aq) \rightarrow Cu^{2+}(aq) + Ag(s) + 2NH_3(aq)$

is -20.36 kJ. Find E° for the reaction. (Answer: 0.211 V)

Standard Electrode Potentials (Section 20.3)

Learning Objectives

1. Diagram the standard hydrogen electrode, and explain how it is used to find half-cell potentials.
2. Convert standard reduction potentials into standard oxidation potentials and vice versa.
3. Use standard reduction potentials to compare the strengths of oxidizing and reducing agents and to predict whether a given redox reaction is spontaneous.
4. Use standard reduction potentials to calculate E° and ΔG° for a given redox reaction.

Review

The voltage of a galvanic cell is the sum of the oxidation (E_{oxid}) and reduction potentials (E_{red}):

$E = E_{oxid} + E_{red}$

E_{oxid} is the contribution of the oxidation half-reaction to the cell voltage; E_{red} is the contribution of the reduction half-reaction. As a result E_{oxid} and E_{red} are called half-cell potentials.

Under standard conditions the above relationship becomes

$E° = E°_{oxid} + E°_{red}$

Half-cell potentials are determined by coupling the half-cell of interest to the standard hydrogen electrode whose half-cell potential is taken as zero; the measured cell voltage is therefore the desired half-cell potential.

The standard hydrogen electrode consists of a Pt electrode immersed in 1 *M* HCl. $H_2(g)$ at 1 atm is bubbled through the solution. (See Figure 20.7 in your text.) The half-cell reaction for the standard hydrogen electrode is

$2H^+(1\ M) + 2e^- \rightleftharpoons H_2(g, 1\ atm)$

The cell notation for the standard hydrogen electrode is

$Pt / H_2(g, 1\ atm) / H^+(1\ M)$

Example 20.4: When the standard hydrogen electrode is coupled to a standard silver electrode the observed cell voltage ($E°$) is 0.799 V and the half-reactions are

anode: $H_2(g, 1\ atm) \rightarrow 2H^+(1\ M) + 2e^-$

cathode: $2Ag^+(1\ M) + 2e^- \rightarrow 2Ag(s)$

sum: $H_2(1\ atm) + 2Ag^+(1\ M) \rightarrow 2H^+\ (1\ M) + 2Ag(s)$ $E° = 0.799\ V$

Find the reduction potential for the standard silver electrode.

Answer: $E° = E^{\circ}_{oxid} + E^{\circ}_{red} = 0.799\ V$; the standard oxidation potential of the hydrogen electrode is zero so that $E^{\circ}_{red} = 0.799\ V$

Note: The equation for the reduction potential of the standard silver electrode is

$Ag^+ + e^- \rightarrow Ag$ $E^{\circ}_{red} = 0.799\ V$

and the notation is $Ag^+(1\ M)/Ag$.

The EMF does not change when the coefficient of the equation for the half-cell reaction is multiplied by some number as it is in the above example.

The positive sign of the standard reduction potential of the silver electrode indicates that the standard silver electrode is the cathode and accepts electrons from the hydrogen reference electrode, the anode in this cell.

The standard oxidation potential is the negative of the standard reduction potential. Its value for the standard silver electrode is -0.799 V:

$E_{oxid} = -E_{red}$

Standard reduction potentials are listed in Table 20.1 of your text. Half-cells that have positive reduction potentials will accept electrons from the standard hydrogen electrode; negative reduction potentials indicate that those half-cells will donate electrons to the standard hydrogen electrode.

Using a Table of Reduction Potentials

A high reduction potential indicates a substance that has a great tendency to accept electrons and to be easily reduced. Such a substance is a strong oxidizing agent. F_2 is the strongest oxidizing agent and therefore has the highest reduction potential. A low or very negative reduction potential indicates that the substance has a great tendency to donate electrons and be easily oxidized. Such a substance is a strong reducing agent.

High E_{red}	tends to be reduced;	strong oxidizing agent
Low E_{red}	tends to be oxidized;	strong reducing agent

Example 20.5: Use a table of reduction potentials to determine which member of the following pairs is the stronger oxidizing agent.

(a) H_2 or I_2 (b) Fe^{3+} or Hg_2^{2+}

Answer:

(a) The stronger oxidizing agent has the higher E_{red}. The standard reduction potentials for H_2 and I_2 are 0.00 and +0.535 V respectively. I_2 is the stronger oxidizing agent.

(b) The standard reduction potentials for Fe^{3+} and Hg_2^{2+} are +0.769 V and +0.796 V. Hg_2^{2+} is the better oxidizing agent.

When two half-reactions are coupled, the reaction with the higher or more positive reduction potential will be the reduction half-reaction leaving the other half-reaction as the oxidation half-reaction.

Example 20.6: Using the reduction reactions and potentials given below predict whether gold will dissolve in an acidic solution of $KMnO_4$.

	E°
$MnO_4^-(aq) + 8H^+(aq) + 5e^- \rightleftharpoons Mn^{2+}(aq) + 4H_2O(l)$	1.512 V
$Au^{3+}(aq) + 3e^- \rightleftharpoons Au(s)$	1.498 V

Answer: The MnO_4^- half-reaction has the higher reduction potential indicating that it is the stronger oxidizing agent. Therefore, it will be reduced and Au will be oxidized and dissolve. The reactions that will occur are

$MnO_4^-(aq) + 8H^+(aq) + 5e^- \rightarrow Mn^{2+}(aq) + 4H_2O(l)$

$Au(s) \rightarrow Au^{3+}(aq) + 3e^-$

Practice: Predict whether mercury will dissolve in 1 *M* HCl given

$Hg^{2+}(aq) + 2e^- \rightleftharpoons Hg(l)$; 0.796 V. (Note: H^+ would have to be reduced) Answer: No

Caution: In deciding whether a given reaction will occur make sure you understand which substances are present. For example, in Example 20.6, the only reaction that you can consider is the oxidation of gold by MnO_4^- because only those two substances are present. You cannot, for example, consider the oxidation of Mn^{2+} since Mn^{2+} is not present.

Calculating Standard Cell Voltages and Free Energy Changes

The voltage of a galvanic cell is the sum of the half-cell voltages.

Example 20.7: A galvanic cell uses the redox reaction

$Sn(s) + 2AgCl(s) \rightarrow 2Ag(s) + Sn^{2+}(aq) + 2Cl^-(aq)$

(a) Write the equations for the half-cell reactions.

(b) Calculate E° for the cell.

Answer: (a) The relevant reduction reactions and their potentials are

$AgCl(s) + e^- \rightleftharpoons Ag(s) + Cl^-(aq)$ 0.222 V

$Sn^{2+}(aq) + 2e^- \rightleftharpoons Sn(s)$ -0.14 V

(b) $E = E_{red} + E_{oxid} = 0.222\ V + 0.14\ V = \underline{0.36\ V}$

Practice: A galvanic cell is based on the reaction

$Cr_2O_7^{2-}(aq) + 14H^+(aq) + 6I^-(aq) \rightarrow 3Cr^{3+}(aq) + 3I_2(s) + 7H_2O(l)$

(a) Use Table 20.1 in your text to write the half-cell reactions.
(b) Calculate E°.

Answers:

(a) $Cr_2O_7^{2-}(aq) + 14H^+(aq) + 6e^- \rightarrow 2Cr^{3+}(aq) + 7H_2O(l)$ $E^\circ_{red} = 1.33\ V$

$3(2I^-(aq) \rightarrow I_2(s) + 2e^-)$ $E^\circ_{oxid} = -0.535\ V$

(b) $E^\circ = 0.795\ V$

We have already learned the equation ΔG = -nFE (Section 20.2) so if we know E for a redox reaction we can calculate ΔG. The equation also tells us that ΔG for a redox reaction is negative (reaction is spontaneous) only when E is positive. Therefore, we can use a table of reduction potentials to find E for any redox reaction and then predict if the reaction is spontaneous under standard conditions.

Example 20.8: Use a table of reduction potentials to
(a) predict whether the reaction

$Cl_2(g) + 2I^-(aq) \rightarrow 2Cl^-(aq) + I_2(s)$

is spontaneous under standard conditions and

(b) calculate ΔG°.

Answer:

Given: $Cl_2(g) + 2I^-(aq) \rightarrow 2Cl^-(aq) + I_2(s)$

Unknown:
(a) E° (sign of E° will tell us if the reaction is spontaneous)
(b) ΔG°

Plan:
(a) 1. We can get the half-reactions from the net equation.
2. The E° for each half-reaction comes from the table of reduction potentials. We reverse the sign for the oxidation reaction.
3. Find $E^\circ = E^\circ_{red} + E^\circ_{oxid}$; if E° is positive, the reaction is spontaneous.

Calculations:
1. $Cl_2(g) + 2e^- \rightleftharpoons 2Cl^-(aq)$

$I_2(s) + 2e^- \rightleftharpoons 2I^-(aq)$ (reversed)

2. $E^\circ_{red} = 1.360\ V$

$E^\circ_{oxid} = -E_{red} = -0.535\ V$

3. $E^\circ = 0.825\ V$

E° is positive, so the reaction is spontaneous under standard conditions.

(b) Use ΔG° = -nFE°: Two moles of electrons are transferred for each mole of Cl_2 reduced; n = 2, E° = 0.825 V = 0.825 J/C. Hence,

$$\Delta G^\circ = -nFE^\circ = -2\ \text{mole e}^- \times \frac{96{,}485\ \text{C}}{1\ \text{mol e}^-} \times \frac{0.825\ \text{J}}{1\ \text{C}} = \underline{-159\ \text{kJ}}$$

Note: $\Delta G°$ is negative as predicted from the positive E°.

Practice:

(a) Predict whether the reaction

$Sn^{2+}(aq) + 2Ag(s) + 2Cl^{-}(aq) \rightarrow Sn(s) + 2AgCl(s)$

is spontaneous under standard conditions.

(b) Calculate $\Delta G°$ for the reverse reaction.

Answers: (a) E° = -0.36 V; not spontaneous (b) $\Delta G°$ = -69 kJ

Practice Drill on the Use of the Table of Reduction Potentials: Use the following reduction reactions and their reduction potentials to answer questions (a) – (c).

	E
$O_2(g) + 4H^+(aq) + 4e^- \rightleftharpoons 2H_2O(l)$	1.229 V
$Fe^{3+}(aq) + e^- \rightleftharpoons Fe^{2+}(aq)$	0.769 V

(a) From the following list choose the most easily oxidized and the strongest oxidizing agent: O_2, H_2O, Fe^{3+}, Fe^{2+}

(b) Predict whether the reaction

$O_2(g) + 4H^+(aq) + 4Fe^{2+}(aq) \rightarrow 4Fe^{3+}(aq) + H_2O(aq)$

is spontaneous under standard conditions.

(c) Calculate $\Delta G°$ for the reaction given in (b).

Answers: (a) Fe^{2+}, O_2 (b) E° = 0.460 V (yes) (c) -178 kJ

EMF and Concentration (Section 20.4)

Learning Objectives

1. Calculate cell voltages and half-cell potentials at concentrations other than standard state conditions.
2. Calculate $\Delta G°$, E°, or K, given one of the three quantities.

Review

Cell and half-cell voltages vary with concentration according to the Nernst equation:

$E = E° - (RT/nF) \ln Q$

E = voltage of cell or half-cell
E° = voltage of standard cell or half-cell
R = 8.314 J/mol
T = temperature in kelvins
n = number of moles of electrons transferred
F = 96,485 C/mol e^-
Q = reaction quotient

At room temperature the Nernst equation can be simplified to

$$E = E° - \frac{0.0592}{n} \log Q$$

The Nernst equation tells us that the EMF of a cell or half-cell depends on Q; as Q decreases (more reactants, less products) EMF increases, as Q increases (more products, less reactants) EMF decreases.

Example 20.9: Calculate the potential of a galvanic cell based on the reaction

$Zn(s) + Ni^{2+}(aq) \rightarrow Ni(s) + Zn^{2+}(aq)$

at 25°C. The concentrations of Zn^{2+} and Ni^{2+} ions are 1.00×10^{-4} *M* and 1 *M*, respectively.

Given: $Zn(s) + Ni^{2+}(aq) \rightarrow Zn^{2+}(aq) + Ni(aq)$
$[Zn^{2+}] = 1.00 \times 10^{-4}\ M$
$[Ni^{2+}] = 1\ M$

Unknown: E

Plan:

1. Use the table of reduction potentials to get the half-cell potentials. Substitute these values into the equation

   $E° = E°_{red} + E°_{oxid}$

   to find E°.

2. Substitute E° and given concentrations into the Nernst equation to calculate E.

Calculations:

1. $Ni^{2+}(aq) + 2e^- \rightleftharpoons Ni(s)$ $\qquad E°_{red} = -0.236\ V$

   $Zn^{2+}(aq) + 2e^- \rightleftharpoons Zn(s)$ $\qquad E°_{oxid} = -E°_{red} = 0.762\ V$

   Sum: $E° = 0.526\ V$

2. $E = E° - \frac{0.0592}{2} \log Q$ and $Q = \frac{[Zn^{2+}]}{[Ni^{2+}]}$

   $$E = 0.526\ V - \frac{0.0592\ V}{2} \times \log(1.00 \times 10^{-4})$$

   $= 0.526\ V - (0.0296)(-4)\ V = 0.526\ V + 0.118\ V = \underline{0.644\ V}$

Note: The voltage calculated above is greater than E°. This is expected because the Nerst equation tells us that EMF increases when there is an increase in the concentration of a reactant or a decrease in the concentration of a product ($[Zn^{2+}]$ is less than 1 *M*). Moreover we would expect EMF values to decrease when products increase or reactants decrease. Check your answers to make sure they are in agreement with your common sense predictions.

Practice: Calculate E_{oxid} for $Zn^{2+}(aq) + 2e^- \rightarrow Zn(s)$ when $[Zn^{2+}] = 1.00 \times 10^{-4}$ *M*.
($E^{o}_{red} = -0.762\ V$)

(<u>Answer</u>: 0.644 V)

Note: The EMF for a reaction can also be calculated by first finding the EMF for each half-reaction and then adding: $E = E_{oxid} + E_{red}$.

E and the Equilibrium Constant

At equilibrium when all reactants and products are at equilibrium concentrations E = 0; there is no net driving force to form products or reactants. At equilibrium the Nernst equation becomes

$E° = (RT/nF) \ln K$ K = equilibrium constant at temperature T

At room temperature

$E° = (0.0592/n) \log K$

The above equation can be used to calculate K from E° and vice versa. It also tells us that redox reactions with high E° values also have large K values; this is expected because high E° values imply a great driving force to form products, thus giving rise to a large K value.

Example 20.10: Calculate K for the reaction given in Example 20.9.

Answer: Substituting n = 2 and E° = 0.526 V into the above equation gives

$E° = (0.0592/n) \log K$

$0.526 \text{ V} = (0.0592 \text{ V}/2) \log K$

Rearranging gives

$\log K = 17.8$ and

$K = \text{antilog } 17.8 = 10^{17.8} = \underline{6 \times 10^{17}}$

Equilibrium constants for reactions that are not redox reactions can sometimes be calculated from EMF values. All we need are two half-reactions from a table of reduction potentials that when added, will give us the reaction of interest and its E°.

Example 20.11: Use reduction potentials from Table 20.1 in your text to calculate K_f for the complex equilibrium

$Zn^{2+}(aq) + 4OH^-(aq) \rightleftharpoons Zn(OH)_4^-(aq)$

Answer: From Table 20.1 in your text we can get the half-reactions

$Zn^{2+}(aq) + 2e^- \rightleftharpoons Zn(s)$ $E°_{red} = -0.762 \text{ V}$

$Zn(OH)_4^{2-} + 2e^- \rightleftharpoons Zn(s) + 4OH^-(aq)$ $E°_{oxid} = -E°_{red} = 1.190 \text{ V}$

$E° = 0.428 \text{ V}$

Substituting n = 2 and E° = 0.428 V into $E° = (0.0592/n) \log K$ gives

$$0.428 \text{ V} = \frac{0.0592 \text{ V}}{2} \log K$$

Rearranging $\log K_f = 14.5$ and $K_f = \underline{3 \times 10^{14}}$

Practice Drill on E, E°, and K Values for a Given Reaction: Fill in the blanks for reactions (a) and (b) using the half-reactions and reduction potentials given.

$AgBr(s) + e^- \rightleftharpoons Ag(s) + Br^-(aq)$	0.0732 V
$Cu^{2+}(aq) + 2e^- \rightleftharpoons Cu(s)$	0.339 V
$Ag^+(aq) + e^- \rightleftharpoons Ag(s)$	0.799 V

(a) $Cu(s) + 2Ag^+(aq) \rightarrow Cu^{2+}(aq) + 2Ag(s)$
(b) $AgBr(s) \rightarrow Ag^+(aq) + Br^-(aq)$

	E°	n	Q	E	K
(a)	________	________	0.500	________	________
(b)	________	________	2.00	________	________

Answers: (a) E° = 0.460 V; n = 2; E = 0.469 V; $K = 3 \times 10^{15}$
(b) E° = -0.726 V; n = 1; E = -0.744 V; $K = 5 \times 10^{-13}$

Batteries and Fuel Cells (Section 20.5)

Learning Objectives

1. Describe the construction and operation of zinc batteries, storage batteries, and the H_2/O_2 fuel cell.
2. Describe the difference between a fuel cell and a battery.

Review

A battery is a package of one or more galvanic cells used to produce electrical energy. In this section we will discuss some well-known batteries.

Zinc Batteries

The common flashlight battery is often called a dry cell because the electrolyte is part of a paste. Zinc is the anode in these cells while a graphite rod immersed in a paste of MnO_2, zinc and ammonium chlorides, carbon and starch powders and some water serves as the cathode. The simplified half-cell reactions are

anode: $Zn(s) \rightarrow Zn^{2+}(aq) + 2e^-$

cathode: $2MnO_2(s) + 2NH_4^+(aq) + 2e^- \rightarrow Mn_2O_3(s) + 2NH_3(aq) + H_2O(l)$

The alkaline dry cell contains an alkaline paste that is less corrosive than the ammonium chloride paste of the dry cell described above. This contributes to a longer shelf life and increased reliability. The reactions are

anode: $Zn(s) + 2OH^-(aq) \rightarrow ZnO(s) + H_2O(l) + 2e^-$

cathode: $2MnO_2(s) + H_2O(l) + 2e^- \rightarrow Mn_2O_3(s) + 2OH^-(aq)$

The mercury cell delivers about 1.35 V for most of its lifetime. Consequently it is used in cardiac pacemakers, hearing aids, etc., where reliability is required. The half-cell reactions are

anode: $Zn(s) + 2OH^-(aq) \rightarrow ZnO + H_2O(l) + 2e^-$

cathode: $HgO(s) + H_2O(l) + 2e^- \rightarrow Hg(l) + 2OH^-(aq)$

Storage Batteries

Storage batteries can be recharged using an outside power source to reverse the electrode reactions. The lead storage battery or car battery is a common example. Each cell consists of alternating sheets of lead and lead dioxide with a 30% solution of sulfuric acid (battery acid) acting as the electrolyte. The discharging half-reactions are

anode: $Pb(s) + HSO_4^-(aq) \rightarrow PbSO_4(s) + H^+(aq) + 2e^-$

cathode: $PbO_2(s) + HSO_4^-(aq) + 3H^+(aq) + 2e^- \rightarrow PbSO_4(s) + 2H_2O(l)$

Recharging reverses the above reactions; the actual voltage needed to recharge a battery exceeds the battery voltage by an amount called the overvoltage.

Fuel Cells

A fuel cell is a galvanic cell in which reactants are continuously fed in and products are continuously removed. Fuel cells differ from batteries in that they do not store electrical energy.

Corrosion (Section 20.6)

Learning Objectives

1. Describe the galvanic mechanism by which iron forms rust.
2. List three ways of preventing corrosion and explain how they work.

Review

Corrosion is a galvanic process during which metals become pitted and eaten away by oxidation. The regions of metal that dissolve are anodic; the protected regions are cathodic.

Iron corrodes when it is covered by water containing dissolved impurities and becomes anodic:

$Fe(s) \rightarrow Fe^{2+}(aq) + 2e^-$

If the water is acidic then atmospheric oxygen is reduced to water:

$4H^+(aq) + O_2(g) + 4e^- \rightarrow 2H_2O(l)$

In a more basic medium a less vigorous reduction occurs:

$O_2(g) + 2H_2O(l) + 4e^- \rightarrow 4OH^-(aq)$

Consequently the corrosion of iron is promoted by acid and inhibited by base.

Corrosion can be retarded with protective coatings, with chemical corrosion inhibitors, and by cathodic protection. These techniques are summarized below:

Protective coatings work by simply covering the metal. Examples of materials used to cover the metals are oil, paint, and a less active metal.

Corrosion inhibitors are added to solutions in contact with the metal. These chemicals are adsorbed onto the metal surface slowing the rate of electron transfer. Red lead (Pb_3O_4) and zinc chromate for example are added to some protective paints.

Cathodic protection tends to be more effective in preventing corrosion. It involves putting a more active metal such as Zn (galvanized iron) or Mg into electrical contact with the iron. The active metal will become anodic and corrode; the iron will become cathodic and will not corrode.

Electrolysis (Section 20.7)

Learning Objectives

1. Diagram and write electrode reactions for the electrolytic production of sodium and aluminum.
2. Diagram and write electrode reactions for the electrolysis of water.
3. Diagram and write electrode reactions for the electrolysis of brine in (a) the diaphragm cell and (b) the membrane cell.
4. Diagram an electroplating cell and write equations for the electrodeposition of a metal from a solution of its salt.
5. Use ΔG to calculate the minimum voltage required for an electrolysis.

Review

Electrolysis is the use of an electric current to bring about an oxidation-reduction reaction. Electrolysis is carried out in an electrolytic cell consisting of two electrodes immersed in an electrolyte. The electrodes are connected to the terminals of a voltage source such as a battery. The electrode connected to the negative terminal receives electrons from the voltage source and becomes negative while the electrode connected to the positive terminal loses electrons and becomes positive.

The anode in an electrolytic cell is positive; the cathode negative. Reduction occurs at the cathode where there is an excess of electrons; oxidation at the anode which is deficient in electrons.

Note: The signs of the electrodes in the electrolytic cell are the reverse of those in the galvanic cell.

Listed below are the overall and electrode reactions of some important electrolytic cells along with the figure number in your text which shows a diagram of the cell.

Electrolytic Production of sodium	Electrode Reactions
(see Figs. 20.19 and 20.20)	cathode: $Na^+(l) + e^- \rightarrow Na(l)$
$2NaCl(l) \rightarrow 2Na(l) + Cl_2(g)$	anode: $2Cl^-(l) \rightarrow Cl_2(g) + 2e^-$
aluminum (see Fig. 20.21: Hall-Heroult Cell)	cathode: $Al^{3+}(l) + 3e^- \rightarrow Al(l)$
	anode: $2O^{2-}(l) \rightarrow O_2(g) + 4e^-$
Electrolysis of water (see Fig. 20.22)	
$2H_2O(l) \rightarrow 2H_2(g) + O_2(g)$	cathode: $2H_2O(l) + 2e^- \rightarrow H_2(g) + 2OH^-(aq)$
	anode: $2H_2O(l) \rightarrow 4H^+(aq) + O_2(g) + 4e^-$

Electrolytic Production of sodium

brine (concentrated salt water)
See Fig. 20.24a, (diaphragm cell) and 20.24 (membrane cell)
$2NaCl(aq) + 2H_2O(l) \rightarrow$
$Cl_2(g) + H_2(g) + 2NaOH(aq)$

Electrode Reactions

cathode: $2H_2O(l) + 2e^- \rightarrow H_2(g) + 2OH^-(aq)$

anode: $2Cl^-(aq) \rightarrow Cl_2(g) + 2e^-$

The minimum voltage needed to bring about the electrode reactions listed above can be ideally calculated by rearranging $\Delta G = -nFE$ to give

$$E = \frac{-\Delta G}{nF}$$

Usually much more voltage is required to make these reactions occur than the calculated minimum amount.

Electroplating

Electroplating is the electrolytic deposition of a metal from a solution of its salt on the surface of another metal. The anode is usually the plating metal and the cathode is the object to be plated.

Electrochemical Stoichiometry (Section 20.8)

Learning Objectives

1. Solve problems relating current, time, and quantity of substance consumed or produced in an electrolytic cell.

Review

The equation for an electrode reaction relates the number of moles of substance consumed or produced during a cell reaction to the quantity of charge passing through the cell. For example the half-reaction

$Hg_2Cl_2(s) + 2e^- \rightarrow 2Hg(l) + 2Cl^-(aq)$

tells us that each mole of Hg_2Cl_2 that reacts picks up 2 mol of electrons (and produces 2 mol of Hg). Because each mole of electrons carries 96,485 C of charge (or one Faraday), each mole of Hg_2Cl_2 gains $2 \times 96{,}485$ C of charge. Hence, if we know how much charge has passed through the cell then we can calculate the amount of reactant used up or product formed. Or, knowing the amount of reactant consumed or product formed, we can calculate the charge that passed through the cell.

Example 20.12: How many moles of nickel metal will be produced when 4000 C pass through a solution of $Ni(NO_3)_2$?

Answer: The equation for the reduction of Ni^{2+} is

$Ni^{2+}(aq) + 2e^- \rightarrow Ni(s)$

The equation indicates that 1 mol of nickel metal is deposited by 2 mol of electrons or $2 \times 96{,}485$ C of charge. Therefore,

$$4000\ C \times \frac{1\ mol\ Ni}{2 \times 96{,}485\ C} = \underline{0.0207\ mol\ Ni}$$

In some cases we know the duration and magnitude of the current. The total number of coulombs that pass through a cell is given by

$Q = I \times t$

Q = charge in coulombs
I = current in amperes (1 ampere = 1 C/s)
t = time in seconds

Example 20.13: How many grams of Ag will be deposited when a 0.50 A current passes through a $Ag(CN)_2^-$ solution for 2.00 h. The reduction half-reaction is

$Ag(CN)_2^-(aq) + e^- \rightarrow Ag(s) + 2CN^-(aq)$

Given: $I = 0.50\ A = 0.50\ C/s$
$t = 2.00\ h$
reduction equation

Unknown: grams of Ag produced

Plan:
1. Find coulombs of charge. Use $Q = I \times t$
2. Use 1 mol of silver metal produced for 1 × 96,485 C of charge passing through the circuit to get moles of silver produced.
3. Convert moles to grams of silver metal.

Calculation:

1. $\frac{0.50\ C}{1\ s} \times \frac{3600\ s}{1\ h} \times 2\ h = 3600\ C$

2. $3600\ C \times \frac{1\ mol\ Ag}{1 \times 96{,}485\ C} = 0.037\ mol\ Ag$

3. $\frac{107.9\ g\ Ag}{1\ mol\ Ag} \times 0.037\ mol\ Ag = \underline{4.0\ g\ Ag}$

Practice: How many hours would it take to produce 8.0 g of silver in the above cell using a current of 1.00 A? Answer: 2.0 h

Note: You can treat electrolytic stoichiometric problems as conversion problems. The conversion scheme for Example 20.13 would be

$(Q \times t) \rightarrow mol\ Ag \rightarrow g\ Ag$

Practice Drill on Electrolytic Stoichiometry Problems: Fill in the blanks for the electroytic cell using the half-reaction

(a) $Zn^{2+}(s) + 2e^- \rightarrow Zn(s)$

	I	t	Number of moles of ion reacting	Number of moles of metal deposited
(a)	0.500 A	___	________	0.500 mol
(b)	1.00 A	2.00 h	________	________

Answers: (a) t = 53.6 h; 0.500 mol Zn^{2+} (b) 0.0373 mol Zn^{2+}; 0.0373 mol Zn

SELF-TEST

For Questions 1–3 fill in the blanks with anode or cathode.

1. Reduction occurs at the __________; oxidation at the __________.

2. Anions migrate toward the __________; electrons flow toward the __________.

3. According to the cell notation Sn(s) / $SnCl_2$(1 *M*) // AgCl(s) / KCl(1 *M*) / Ag(s), $SnCl_2$ is the electrolyte for the ____.

4. The ΔG° for the reaction

   $Sn(s) + 2AgCl(s) \rightarrow 2Ag(s) + Sn^{2+}(aq) + 2Cl^{-}(aq)$

   is -69.2 kJ. (a) Calculate E° for the reaction. (b) Find the maximum work that could be obtained from the oxidation of 5.0 g of Sn according to the equation given above (at. wt. of Sn = 118.7).

5. Your table of Standard Reduction Potentials (Table 20.1 in your text) gives the following

$2Hg^{2+}(aq) + 2e^{-} \rightleftharpoons Hg_2^{2+}$	0.908	$Ag^{+}(aq) + e^{-} \rightleftharpoons Ag(s)$	0.799 V
$Hg^{2+}(aq) + 2e^{-} \rightleftharpoons Hg(l)$	0.852	$Hg_2^{2+}(aq) + 2e^{-} \rightleftharpoons 2Hg(l)$	0.796 V

   (a) Which half-reaction could theoretically be used to reduce Ag^{+}? Write the overall equation.
   (b) Calculate the E° and ΔG° for the reaction.
   (c) Which of the following is the most readily oxidized? the most easily reduced?: Ag^{+}, Hg^{2+}, Hg_2^{2+}, Hg, Ag.

6. A galvanic cell is composed of two H_2O/H_2O_2 half-cells one with an electrolyte pH = 3.00 and the other with a pH = 1.00. The concentration of H_2O_2 is 1 *M* in both half-cells. Given the reduction reaction and the reduction potential

   $H_2O_2(aq) + 2H^{+}(aq) + 2e^{-} \rightarrow 2H_2O(l)$ E° = 1.763 V

   calculate E for the cell.

7. Calculate K_{sp} for AgI using

   $AgI + e^{-} \rightleftharpoons Ag(s) + I^{-}(aq)$ E° = -0.152 V

   $Ag^{+}(aq) + e^{-} \rightleftharpoons Ag(s)$ E° = 0.799 V

8. List three different kinds of batteries and indicate their advantages and disadvantages.

9. What is corrosion? How can it be prevented?

10. A current is passed through a solution of a platinum salt for 2.00 h. Determine the magnitude of the current if 4.00 g of Pt are deposited. Assume that the only reaction occurring at the cathode is

    $Pt^{2+}(aq) + 2e^{-} \rightarrow Pt(s)$ (at. wt. of Pt = 195.08)

11. A current of 0.500 A is passed through a brine solution for 30.0 min.
    (a) Write the half-equations for the electrode reactions and the equation for the overall reaction.
    (b) How many liters of chlorine gas are produced at the anode at 25°C and 1 atm?

ANSWERS TO SELF-TEST QUESTIONS

1. cathode; anode

2. anode; cathode

3. anode

4. (a) $E° = \Delta G°/nF = -69{,}200/2 \times 96{,}485 = \underline{0.359\ V}$

   (b) $w = 5.0\ g \times \dfrac{1\ mol}{118.7\ g} \times 69.2\ kJ/1\ mol = \underline{2.9\ kJ}$

5. (a) $2Ag^{+}(aq) + 2Hg(l) \rightleftharpoons Ag(s) + Hg_2^{2+}(aq)$

   (b) $E° = E^{\circ}_{red} + E^{\circ}_{oxid} = 0.799\ V + (-0.796\ V) = 0.003\ V$

   $\Delta G° = -nFE° = -2 \times 96{,}485\ C \times 0.003\ J/C = \underline{6 \times 10^{2}\ J}$

   (c) Hg, Hg^{2+}

6. E for cell at pH = 1.00; $E = 1.763\ V - \dfrac{0.0592\ V}{2} \log \dfrac{1}{(1.0 \times 10^{-1})^{2}} = 1.704\ V$

   E for cell at pH = 3.00; $E = 1.763\ V - \dfrac{0.0592\ V}{2} \log \dfrac{1}{(1.0 \times 10^{-3})^{2}} = 1.585\ V$

   $E = 1.704\ V - 1.585\ V = \underline{0.119\ V}$

7. $E° = \dfrac{0.0592}{n} \log K$; $E° = -0.152\ V + (-0.799\ V) = -0.951\ V$

   $K = \text{antilog}(-16.1)$; $\underline{K = 8 \times 10^{-17}}$

10. $4.00\ g \times 1\ mol\ Pt/195.1\ g\ Pt \times \dfrac{2 \times 96{,}485\ C}{1\ mol\ Pt} = 3.96 \times 10^{3}\ C$

    $I = Q/t = 3.96 \times 10^{3}\ C/7200\ s = \underline{0.550\ A}$

11. (a) cathode: $2H_2O(l) + 2e^{-} \rightarrow H_2(g) + 2OH^{-}(aq)$
    anode: $2Cl^{-}(aq) \rightarrow Cl_2(g) + 2e^{-}$
    overall: $2NaCl(aq) + 2H_2O(l) \rightarrow Cl_2(g) + H_2(g) + 2NaOH(aq)$

    (b) $0.500\ C/s \times 3600\ s/h \times 0.500\ h \times \dfrac{1\ mol\ Cl_2}{2 \times 96{,}485\ C} \times \dfrac{(0.0821 \times 298)\ L}{1\ mol\ Cl_2} = \underline{0.114\ L}$

CHAPTER 21
METALS AND COORDINATION CHEMISTRY

SELF-ASSESSMENT

Physical and Chemical Properties of Metals (Sections 21.1 and 21.2)

1. All of the following statements are true about metals except
   (a) They conduct heat and electricity.
   (b) Their electrons tend to be tightly held relative to those of nonmetals.
   (c) Their atoms occupy lattice sites in unit cells.
   (d) A Group 2A metal tends to have a higher melting and boiling point than the Group 1A metal in the same period.
   (e) On the average, metal densities are higher than nonmetal densities.

Underline the correct word(s) for Questions 2-5.

2. The covalent character of the metal-oxygen bond in the oxides of representative metals (increases, decreases) as you move from left to right across a period so that the basicity of the oxide (increases, decreases).

3. The metal-oxygen bond is (more, less) ionic in PbO than in PbO_2.

4. As one moves down Group 1A the enthalpy of sublimation (ΔH°_{sub}) and the enthalpy of ionization (ΔH°_{ion}) (increases, decreases).

5. The energy of hydration tends to be (large, small) for highly charged small ions.

6. All of the following statements are true about transition metals except
   (a) Many of their compounds tend to be paramagnetic.
   (b) Many of them exhibit variable oxidation states.
   (c) Due to the presence of d electrons, their atoms are relatively large.
   (d) Due to the presence of d electrons, their compounds tend to be colored.

Band Theory of Conductors, Semiconductors, and Insulators (Section 21.3)

7. Use band theory to explain the relative conductivities of an insulator, a semiconductor, and a metallic conductor.

8. The conductivity of a metallic conductor (increases, decreases) with increasing temperature while the conductivity of a semiconductor (increases, decreases).

9. Doping is a process that (increases, decreases) the conductivity of a semiconductor.

Metal Complexes (Section 21.4)

10. When 2.30 g of $CrCl_3 \cdot 4\ H_2O$ are dissolved in a $AgNO_3$ solution, 1.43 g of AgCl is formed. Assuming all of the chloride ion is in the precipitate, determine the formula of the complex. The molecular weights of $CrCl_3 \cdot 4H_2O$ and AgCl are 230 and 143, respectively.

11. Give the coordination number and oxidation state of the metal ion in each of the following complexes:
    (a) $K[AgF_4]$
    (b) $[Cr(H_2O)_6]Br_3$
    (c) $Pt(NH_3)_2Cl_4$
    (d) $Na_3[Ag(S_2O_3)_2]$
    (e) $[Cu(en)_2]Cl_2$ (en = ethylenediamine)

12. The cyano ligand forms a square planar complex with Ni^{2+} and an octahedral complex with Ni^{3+}. Draw these two complexes.

Isomerism in Metal Complexes (Section 21.5)

13. The complex $[NiBr_2Cl_2]^{2-}$ has two isomers while $[CdBr_2Cl_2]^{2-}$ has only one form. Draw the three complexes.

14. Which of the following complexes can have optical isomers? Geometric isomers? Ionization isomers?
    (a) $[Au(H_2O)_2Cl_4]^-$
    (b) $[Pt(NH_3)F_2Cl]^-$
    (c) $CrSO_4Br \cdot 5NH_3$
    (d) $[Co(en)_2Cl_2]^+$

15. Which of the following are optically active? Sketch the enantiomers.
    (a) tetrahedral $Ni(NH_3)_2Cl_2$
    (b) square planar $Pt(NH_3)_2Cl_2$
    (c) $[Fe(en)_2(NO_2)_2]^+$
    (d) $[Mn(en)_3]^{3+}$

***Bonding in Metal Complexes** (Section 21.6)

16. $Fe(H_2O)_6{}^{2+}$ is a paramagnetic complex while $Fe(CN)_6{}^{4-}$ is not.
    (a) Diagram the d orbital electron configuration for these two complexes.
    (b) Which of the two ligands, cyano or aquo, exerts the strongest ligand field?

For Questions 17-22 underline the correct word(s).

17. The electrostatic model of metal complexes is best applied to complexes containing (main group, transition) metals.

18. Tetrahedral complexes are always (low, high) spin.

19. For any one transition metal ion, substituting a strong-field ligand for a weak-field ligand (increases, decreases) the frequency of the photon absorbed.

20. The energy of the d_{z^2} and $d_{x^2-y^2}$ orbitals of a transition metal in an octahedral complex is (higher, lower) than the energy of the d orbitals of the free ion.

21. The conversion of $[Mn(H_2O)_6]^{2+}$ to $[Mn(CN)_6]^{4-}$ (increases, decreases) the paramagnetism, and (increases, decreases) the wavelength of the color transmitted by the complex.

22. In a square planar complex the d_{xz}, d_{yz}, and d_{z^2} generally have (higher, lower) energy than the d_{xy} and $d_{x^2-y^2}$ orbitals.

ANSWERS TO SELF-ASSESSMENT PROBLEMS

If you missed an answer, be sure to study the relevant section in the textbook and study guide.

1. (b)
2. increases, decreases
3. more
4. decreases
5. large
6. (c)
8. decreases, increases
9. increases

10. moles Cl^- = moles AgCl = 1.43/143 = 0.0100 mol

moles $CrCl_3 \cdot 4H_2O$ = 2.30/230 = 0.0100 mol. Hence, each mole of $CrCl_3 \cdot 4H_2O$ contains 1 mol of Cl^-.

The formula is $[Cr(H_2O)_4Cl_2]Cl$.

11. (a) 4, 3+ (b) 6, 3+ (c) 6, 4+ (d) 2, 1+ (e) 4, 2+

12.

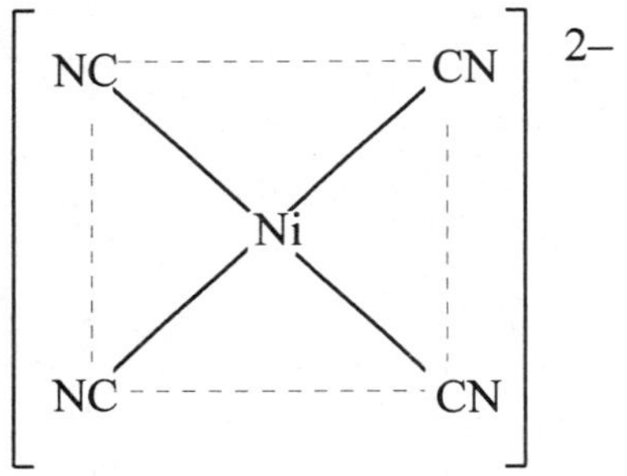

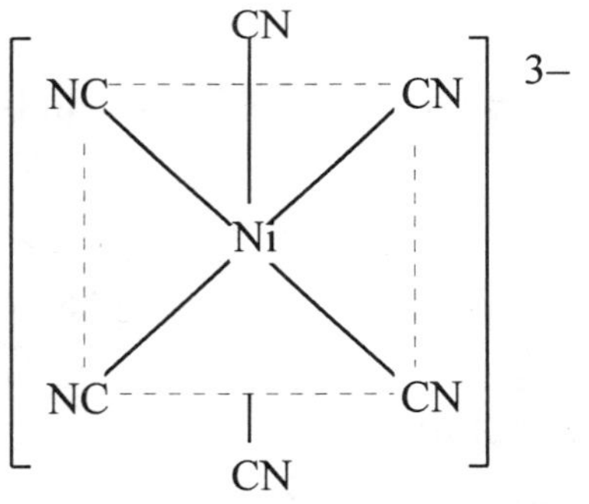

13.

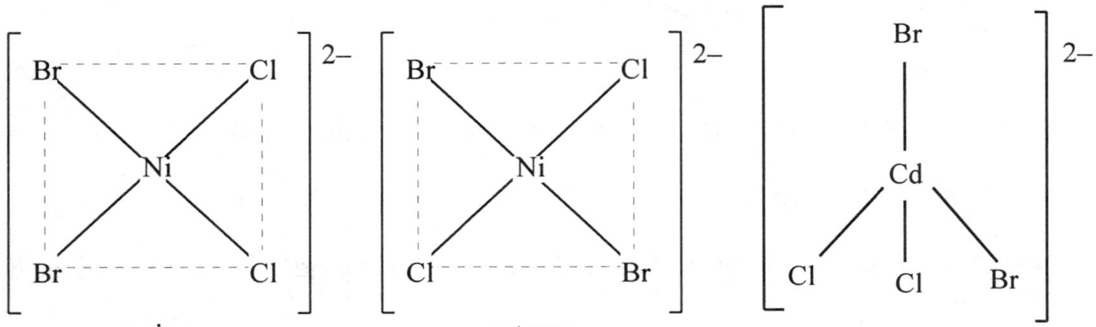

cis trans tetrahedral

square planar

14. (a) geometric (b) geometric (c) ionization and geometric (d) geometric and optical

15. (c)

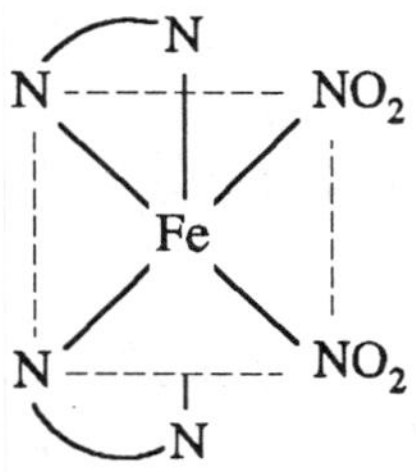

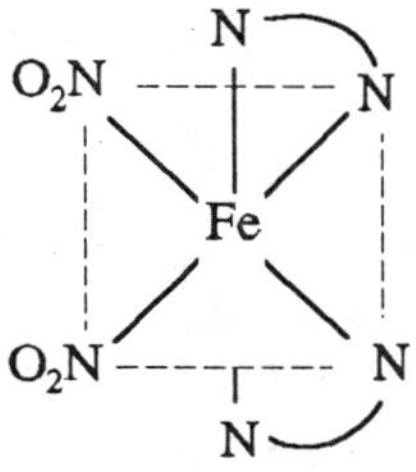

(d)

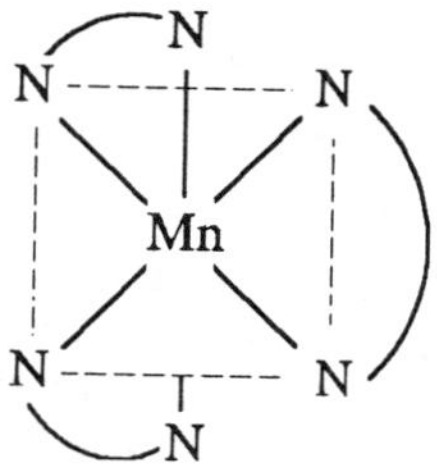

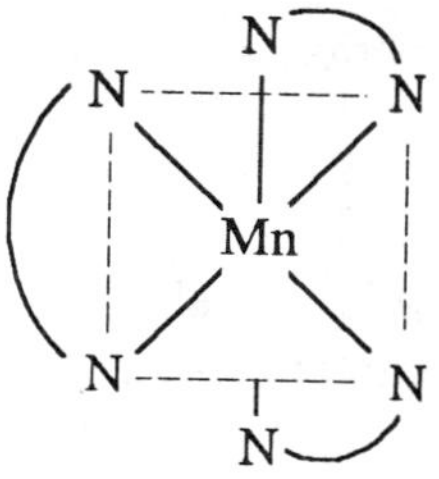

16. (a)

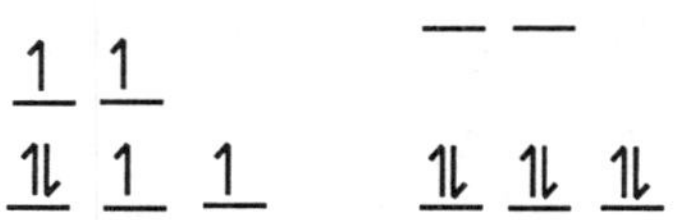

$[Fe(H^2O)^6]^{2+}$ $[Fe(CN)^6]^{4-}$

(b) cyano

17. main group

18. high

19. increases

20. higher

21. decreases, increases

22. lower

REVIEW

Physical Properties of Metals (Section 21.1)

Learning Objectives

1. List the physical properties common to most metals.
2. Distinguish between metals, nonmetals, and semiconductors on the basis of electrical conductivity.
3. Describe how metallic bond strength is measured.

4. List the factors that influence metalic bond strength.
5. State how the hardness, melting points, and boiling points of metals vary across a period and down a group.

Review

Some important physical properties of metals are listed below.

- All metals are lustrous.
- All metals are conductors of heat and electricity. (Loosely held electrons can transport kinetic energy and charge.)
- Metals tend to have large densities due to close packing of their atoms.
- Some metals, such as gold, are highly malleable.
- Metals tend to be hard, and to have high melting and boiling points. (These properties depend on the strength of the metallic bond; the smaller the atomic size and the greater the number of valence electrons the stronger the bond. Melting and boiling points generally increase going across a period and decrease going down a group.)

Chemical Properties of Metals (Section 21.2)

Learning Objectives

1. List the chemical properties associated with metallic character.
2. State how the ionic versus covalent character of representative metal bonds varies with the periodic position and oxidation state of the metal.
3. List each step in the formation of an aqueous cation and describe the periodic trends in energy for each step.
4. Describe how oxidation potentials are calculated from thermodynamic data.
5. State how the acid-base character of a metal oxide varies with the periodic position and oxidation state of the metal.
6. List the special properties of the transition metals.

Review

Metals tend to form positive ions and ionic bonds. Most metals also form basic oxides. As you move across a period, these properties tend to decrease due to an increased tendency for electron sharing and covalent bond character. For a transition or post-transition metal, the higher the oxidation state the more covalent the bonding.

Activity: The Energetics of Cation Formation

An active metal is a strong reducing agent because it readily loses electrons. The oxidation potential ($E_{oxid} = -E_{red}$) is a measure of the activity of a metal in an aqueous environment. Most main group metals are more active than hydrogen. Activity trends are irregular because the formation of a metal ion in aqueous solution from a solid metal is the sum of three steps:

1. Sublimation: $M(s) \rightarrow M(g)$ ΔH°_{sub}
2. Ionization: $M(g) \rightarrow M^{+}(g) + e^{-}$ ΔH°_{ion}
3. Hydration: $M^{+}(g) \rightarrow M^{+}(aq)$ ΔH°_{hyd}

Sum: $M(s) \rightarrow M^{+}(aq) + e^{-}$ ΔH°_{oxid}

General Trends

- Metals with large atoms and few valence electrons have smaller ΔH°_{sub} values.
- Metals with large atoms losing fewer electrons have smaller ΔH°_{ion} values.
- Energies of hydration (ΔH°_{hyd}) are more negative for small ions with greater charge.

The free energy change (ΔG°_{oxid}) can be calculated by subtracting $T\Delta S^\circ_{oxid}$ from ΔH°_{oxid}:

$\Delta G^\circ_{oxid} = \Delta H^\circ_{oxid} - \Delta S^\circ_{oxid}$

E°_{oxid} can now be calculated using

$\Delta G^\circ_{oxid} = -nFE^\circ_{oxid}$

Example 21.1: Table 21.2 in your text lists the calculated activities of some metals. Account for the fact that Al is only slightly less active than Ca although it has a high ionization energy.

Answer: Al^{3+} is a small, highly charged ion so that it releases a large amount of energy upon hydration.

Oxides and Hydroxides

Group 1A and 2A metals (except for Be) form ionic oxides and hydroxides. The basicity of these compounds decreases fom left to right across a period and bottom to top of a group due to an increase in covalent character of the metal-oxygen bond. If a given metal forms more than one oxide the more basic oxide, will contain the metal in the lower oxidation state.

Practice: For each pair choose the more basic compound:
(a) NaO or Al_2O_3
(b) $Mn(OH)_2$ or $Mn(OH)_3$
Answers: NaO, $Mn(OH)_2$

Properties of Transition Metals

Transition metals are in the B groups of the periodic table. Some of their properties are listed below.

- Atomic radii of transition metals are relatively small.
- Most transition metals and transition metal compounds are paramagnetic.
- Transition metals tend to have variable oxidation states. This is due to the d electrons.
- Transition metals tend to form more stable complexes than representative metals.
- Transition metal compounds are often colored due to the excitations of d electrons (see Section 21.6).
- Some transition metals act as catalysts.

Band Theory of Conductors, Semiconductors, and Insulators (Section 21.3)

Learning Objectives

1. Diagram the formation of the valence and conduction bands in an atomic solid.
2. Distinguish between metallic conductors, insulators, and semiconductors in terms of band theory.

Review

Band theory is used to explain many properties of metallic conductors, semiconductors, and insulators.

A band is a collection of molecular orbitals whose energy levels are closely spaced. The valence band contains valence electrons; the conduction band consists of mostly vacant higher energy levels.

In a metallic conductor, the valence band and conduction band are closely spaced, so electrons can easily migrate into empty conduction orbitals and carry current. In an insulator, the energy gap between the full valence band and empty conduction band is large so that electrons cannot easily move into unoccupied orbitals in the conduction band. Hence, current does not flow.

The energy gap in a semiconductor is smaller than that in an insulator so that some very energetic electrons can migrate into the conduction band. Application of a potential difference across a semiconductor will increase the number of electrons in the conduction band, and increase the conductivity.

Example 21.2: Using band theory, predict the effect of (a) temperature and (b) addition of impurities on the conductivity of a conductor and of a semiconductor.

Answers: (a) An increase in temperature and addition of impurities decreases the conductivity of a metallic conductor. In both cases the crystal lattice of the metal is disrupted, thus impeding the flow of electrons.
(b) In semiconductors, an increase in temperature will increase the number of electrons in the conduction band, thus increasing the conductivity. Addition of certain impurities also increases the conductivity by either increasing the number of vacancies in the valence band or by increasing the number of electrons in the conduction band. Addition of these impurities to a semiconductor is called doping.

Metal Complexes (Section 21.4)

Learning Objectives

1. Use conductivity data and precipitation stoichiometry to determine the formula of a complex.
2. Identify the complex in a coordination compound, and determine the oxidation number of its metal atom.
3. Sketch the geometries of two-, four-, and six-coordinate complexes.
4. Give examples of monodentate and polydentate ligands, and identify their coordinating atoms.
5.* Given the formula of a complex, write its IUPAC name, and vice versa.

Review

A metal complex has a central metal atom or ion bonded to one or more groups called ligands. Ligands have one or more lone-pair electrons directed toward the central metal atom or ion. The formula of a complex is written with the metal written first followed by the ligands. For example, $Fe(CO)_6$ is a complex formed from six carbonyl (CO) ligands bonded to an iron atom. If the complex is an ion as in $K_3[Fe(CN)_6]$ and $[Ag(NH_3)_2]Cl$, the complex is written within the brackets.

Given a formula for a compound you can determine the oxidation state of the metal in the complex using the same rules you learned in Section 11.2 about oxidation states.

Practice Drill on Formulas of Complexes: Fill in the blanks.

Compound	Complex Ion	Ligand	Oxidation State of Metal
$K_4[FeCl_6]$			
$Na_3[Co(NO_2)_6]$			
$[Cr(H_2O)_6]Cl_3$			

Answers: $[FeCl_6]^{4-}$, Cl^-, 2+; $[Co(NO_2)_6]^{3-}$, NO_2^-, 3+; $[Cr(H_2O)_6]^{3+}$, H_2O, 3+

A coordinate covalent bond is a bond in which one atom supplies both of the electrons. Because ligands orient their lone-pair electrons toward the metal atom or ion, metal-ligand bonds are often described as coordinate covalent, and their complexes as coordination compounds. Formulas of coordination compounds are often determined from conductivity data and from precipitation reactions. Conductivity measurements can be used to find the number of moles of ions in 1 mol of compound while precipitation reactions show which ions are outside the complex.

Example 21.3: One mole of $CoCl_3{\bullet}4NH_3$ forms 2 mol of ions in solution and reacts with 1 mol of $AgNO_3$ to form a precipitate. Write the formula of the compound.

Answer: The precipitation reaction indicates that there is 1 mol of Cl^- ion per mole of compound. The only formula for the compound consistent with the given observations is $[Co(NH_3)_4Cl_2]Cl$:

$$[Co(NH_3)_4Cl_2]Cl(s) \rightarrow [Co(NH_3)_4Cl_2]^+(aq) + Cl^-(aq)$$

Geometry of Complexes

The number of ligands bonded to the central metal atom or ion is called the coordination number. Common coordination numbers are 2, 4, and 6. Each coordination number is associated with certain preferred geometries:

Coordination number	Geometry	Example
2	linear	$[Ag(CN)_2]^-$
4	tetrahedral square planar (see below)	$Ni(CO)_4$ $[Ni(CN)_4]^{2-}$
6	octahedral (see below)	$[Mn(NH_3)_6]^{2+}$

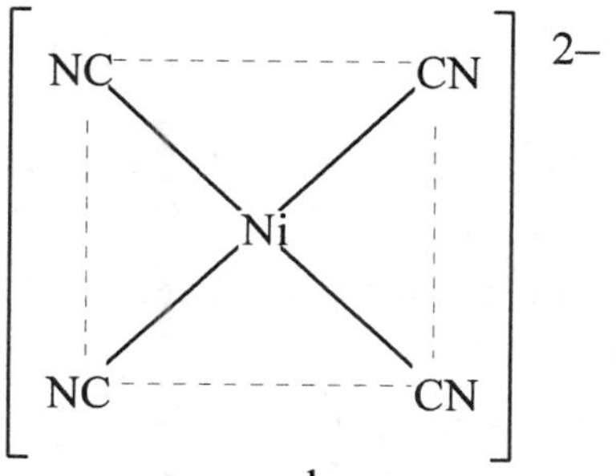

square planar

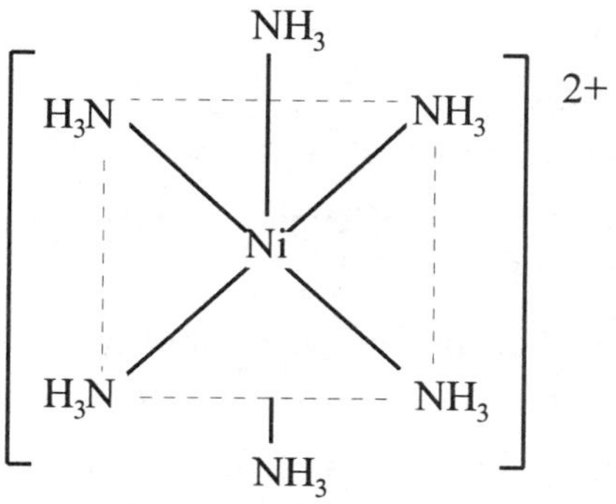

octahedral

See Figure 21.7 in your text for additional sketches of coordination compounds.

Ligands are often described as being <u>monodentate</u>, <u>bidentate</u>, or <u>tridentate</u> and so forth according to whether they have one, two, three, or more coordination atoms (atoms that bond to the metal). A <u>polydentate</u> ligand has two or more coordination sites and is called a <u>chelating agent</u>. Ethylenediamine, a bidentate, is an example:

$$CH_2-CH_2$$

$$H_2\overset{..}{N} \qquad \overset{..}{N}H_2$$

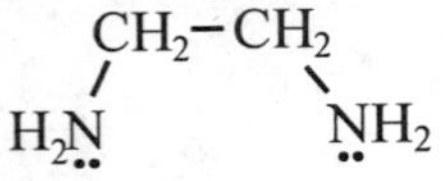

The nitrogen atoms in ethylenediamine donate their electrons to the metal ion.

Practice Drill on Coordination Compounds: Fill in the chart given below.

Compound	Coordination Number of Metal	Oxidation Number of Metal
(a) $Ni(en)_2Cl_2$	_____	_____
(b) $[Fe(en)_3]Br_3$	_____	_____
(c) $Na_3[Ag(S_2O_3)_2]$	_____	_____
(d) $K[PtCl_3(NH_3)]$	_____	_____

<u>Answers:</u> (a) 6, 2+ (b) 6, 3+ (c) 2, 1+ (d) 4, 2+

Isomerism in Metal Complexes (Section 21.5)

Learning Objectives

1. List and give examples of three types of structural isomerism exhibited by coordination compounds.
2. State whether a given complex could have geometric isomers, and sketch the *cis* and *trans* forms.
3. Describe how two enantiomers differ from each other in structure and properties.
4. State whether a given complex could exhibit optical isomerism, and sketch the enantiomers.

Review

<u>Isomers</u> are compounds that have the same formula but different arrangements of atoms. <u>Structural isomers</u> differ in their arrangement of bonds. <u>Stereoisomers</u> have the same bonds but the spatial orientations of the bonds are different. Some examples of isomerism in coordination compounds are given below.

Structural Isomerism

Several types of structural isomers are found in coordination compounds.

<u>Ionization isomers</u>: isomers that have a different ion outside of the complex. For example $[Pt(NH_3)_4Cl_2]Br_2$ and $[Pt(NH_3)_4Br_2]Cl_2$ are ionization isomers.

<u>Coordination isomers</u>: isomers that contain two or more metal atoms with a different distribution of the ligands between the metal atoms.

<u>Linkage isomers</u>: isomers that differ in the atom of a ligand that is bonded to the central metal. For example the NO_2^- ligand can bond to the metal with nitrogen or with oxygen so that isomers containing this ligand are possible.

Stereoisomerism

We will examine two kinds of stereoisomers: geometric isomers and enantiomers.

Geometric isomers or *cis-trans* isomers: isomers in which the orientation of certain bonds differ with respect to a molecular axis or plane. The *cis* isomer has similar substituents (atoms or groups of atoms) on the same side of the molecular axis or plane; the *trans* isomer has similar substituents on opposite sides. For example, the geometric isomers of the square planar $Pt(NH_3)_2Cl_2$ are

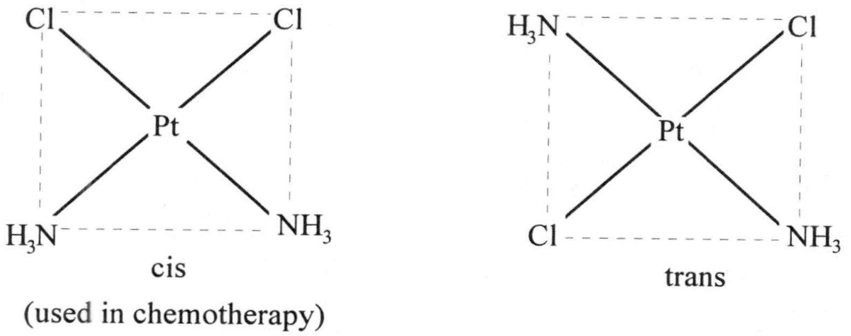

Enantiomers: isomers that are mirror images of each other. (Enantiomers are also called optical isomers.) The enantiomers of $[Cr(H_2O)_2(NH_3)_2Br_2]^+$ are

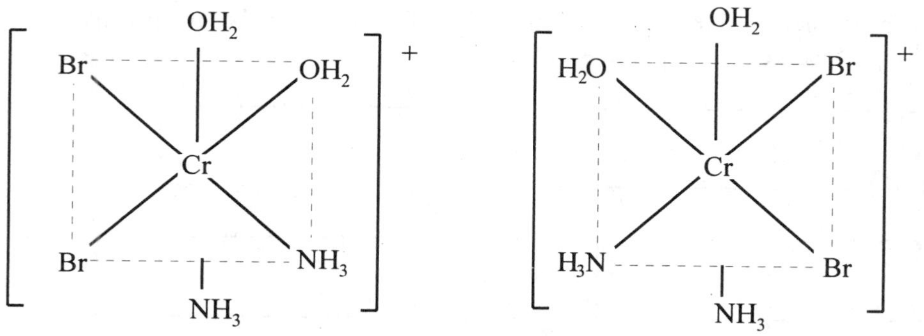

The above compounds are also said to be chiral, meaning that they cannot be superimposed on their mirror images. A chiral substance is optically active; it will rotate the plane of polarized light. Enantiomers rotate the plane of polarized light through equal but opposite angles. The enantiomer that rotates the plane of polarized light to the right is called the dextrorotatory or (+)-isomer, the enantiomer that rotates the plane of polarized light to the left is the levorotatory or (-)-isomer. A racemic mixture is one that contains equal concentrations of both enantiomers; it has no effect on the plane of polarized light.

Achiral molecules are superimposable on their mirror images and have no effect on polarized light. An achiral molecule can be identified by the presence of a plane of symmetry (also called a mirror plane), an imaginary plane which divides the molecule into two halves that are mirror images of each other. H_2S and NH_3 are achiral; they both possess a plane of symmetry, as shown below.

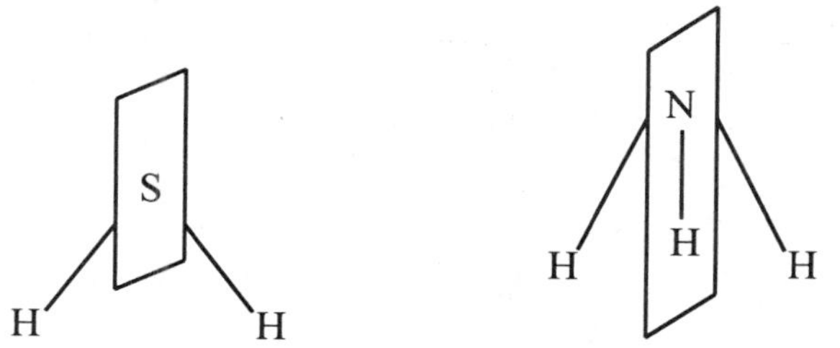

Practice: Which of the following compounds are chiral? Draw the enantiomers.
(a) tetrahedral $[NiCl_2Br_2]^{2-}$
(b) square planar *cis* $Cu(NH_3)_2Cl_2$
(c) octahedral $[Co(en)_3]^{3+}$
Answer: (c)

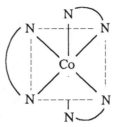

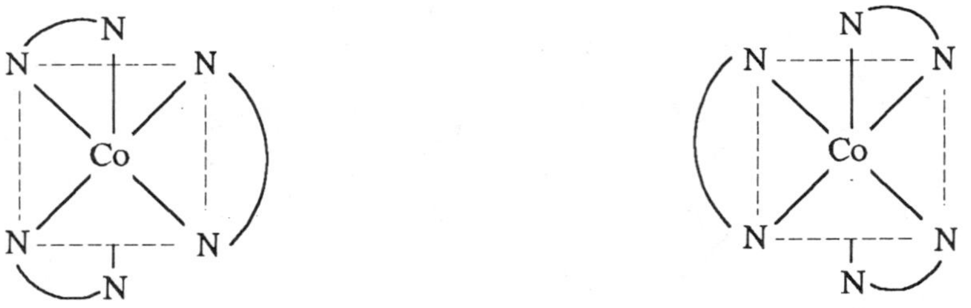

 PROBLEM-SOLVING TIP

Isomers

It is important to understand the different kinds of isomers and their relationships. The diagram given below may be helpful. You should be able to describe each kind of isomer and provide examples.

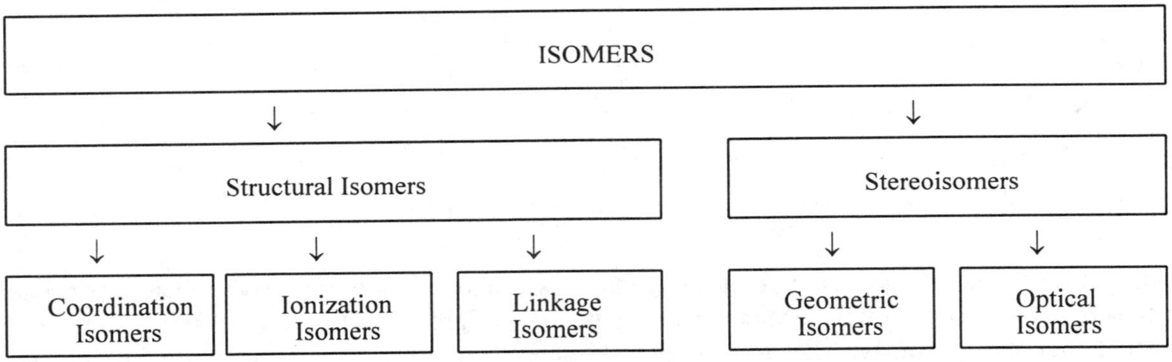

Practice Drill on Isomerism of Coordination Compounds: For the octahedral compounds (a) - (c) write a check (\checkmark) above each line if that kind of isomer could exist.

	Ionization	Linkage	Enantiomers	Geometric
(a) $CrCl_2Br \cdot 4H_2O$	_____	_____	_____	_____
(b) $CoCl_2(NO_2) \cdot 3NH_3$	_____	_____	_____	_____
(c) $CoCl_2 \cdot 4H_2O$	_____	_____	_____	_____

Answers:	(a)	\checkmark	—	—	\checkmark
	(b)	—	\checkmark	\checkmark	\checkmark
	(c)	—	—	—	\checkmark

*Bonding in Metal Complexes (Section 21.6)

Learning Objectives

*1. Describe the electrostatic model of metal-ligand bonding.
*2. List some properties of transition metal complexes that are not explained by the electrostatic model.
*3. Draw an energy level diagram for the orbitals that are occupied by the metal d electrons in an octahedral complex and show the configuratons adopted by one, two, and three d electrons.
*4. Use observed colors to predict relative ligand field strengths.
*5. Use the spectrochemical series to predict color shifts when one ligand is substituted for another in a given metal complex.
*6. Diagram the high- and low-spin configurations for octahedral d^4, d^5, d^6 and d^7 complexes.
*7. Use the observed paramagnetism of a complex to determine whether it has a high-spin or low-spin configuration.

Review

Although an electrostatic model can explain some properties of metal complexes it is clear that covalent bonding plays a major role in many metal ligand interactions. Consequently, both an electrostatic model and molecular orbital model will be discussed below.

The Electrostatic Model

The electrostatic model is particularly useful in explaining properties of main group metal complexes. According to this model, the central metal ion is a uniform sphere of charge and the ligands are viewed as simple spheres or simple dipoles.

The electrostatic model, for example, predicts that the strength of the metal-ligand bond should increase with increasing charge density on the metal. Enthalpies of hydration for main group metals show stronger metal-H_2O bonds as the charge on the metal ion increases and its radius decreases.

The relative stabilities of some transition metal complexes can be roughly predicted from the electrostatic model, but many other properties cannot. A molecular orbital theory explains more properties for transition metal complexes.

The Molecular Orbitals of Octahedral Complexes

In an octahedral complex six ligand orbitals will overlap six metal orbitals that are oriented along the x, y, and z axis. The metal orbitals that significantly participate in bonding are the outer s, p_x, p_y, p_z, $d_{x^2-y^2}$ and d_{z^2} orbitals. The other three d orbitals namely the d_{xy}, d_{xz}, and d_{yz} orbitals of the metal lie between the axes and therefore are considered "nonbonding." As a result the d orbitals once degenerate in the free metal are now split in the complex. The lower energy set, t_{2g}, consists of the three nonbonding d orbitals which are separated from the higher energy set or e_g^* d orbitals by the orbital splitting energy (Δ).

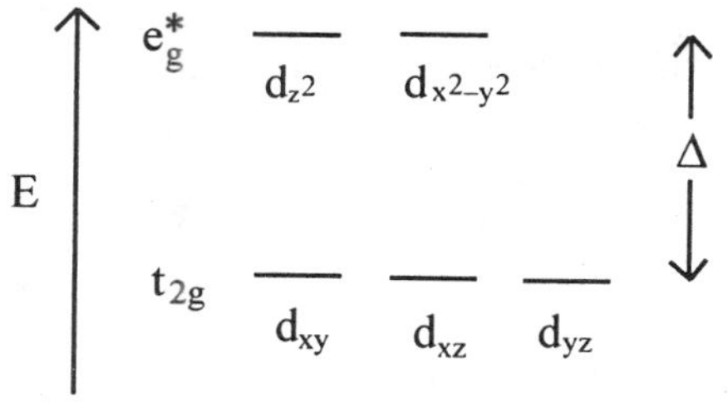

A ligand that causes a large d orbital splitting energy is a strong-field ligand; a ligand that causes a small d orbital splitting energy is a weak-field ligand.

Ligand field strengths can be determined by comparing the colors of complexes of the same metal ion with different ligands. The spectrochemical series lists the ligands in order of increasing field strength (see Table 21.8 in your text).

The magnitude of the orbital energy splitting and the spin pairing energy in a transition metal complex will determine the degree of paramagnetism in the complex. When the central metal atom or ion has more than three electrons the additional electrons can either pair or go into the higher energy d orbitals (e_g^*). High splitting energies and low pairing energies favor pairing of electrons in the lower energy d orbitals (t_{2g}) and low-spin complexes. Low orbital splitting energies and high pairing energies favor the occupation of the higher energy d orbitals and high-spin complexes.

Example 21.4: Cl^- is a weak-field ligand while CN^- is a strong-field ligand. Draw the orbital configurations and predict which of the following is the more paramagnetic complex:

$[Fe(CN)_6]^{4-}$ or $[FeCl_6]^{4-}$

Answer: The CN^- ligand exerts a strong ligand field so that $[Fe(CN)_6]^{4-}$ is a low-spin complex. Its orbital configuration is

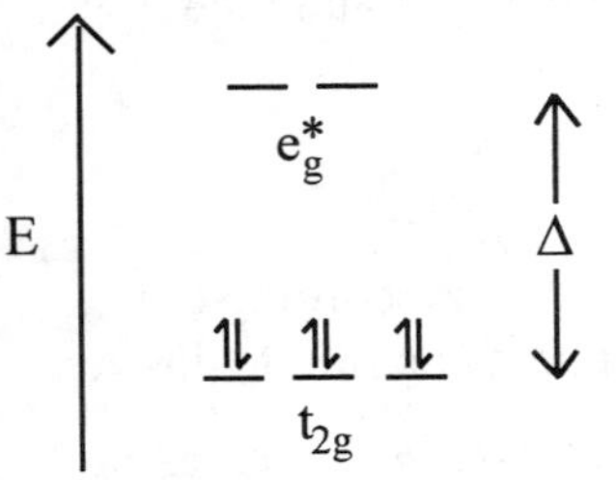

Because Cl^- is a weak-field ligand, it produces high-spin complexes. Its d orbital configuration is

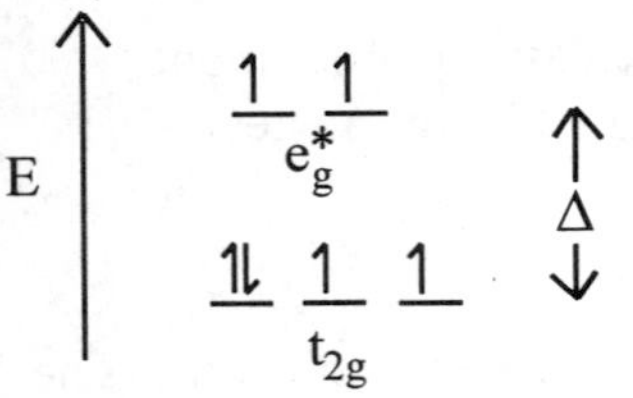

Hence, $[FeCl_6]^{4-}$ is more paramagnetic.

Practice: How many unpaired electrons are in $[Mn(H_2O)_6]^{2+}$? (H_2O is a weak-field ligand.) Answer: 5

Tetrahedral and Square Planar Complexes

The ligands in a tetrahedral complex are between the x, y, and z axes so that the d_{xy}, d_{xz}, and d_{yz} orbitals (t_2^*) have a higher energy than the d_{z^2} and $d_{x^2-y^2}$ orbitals (e).

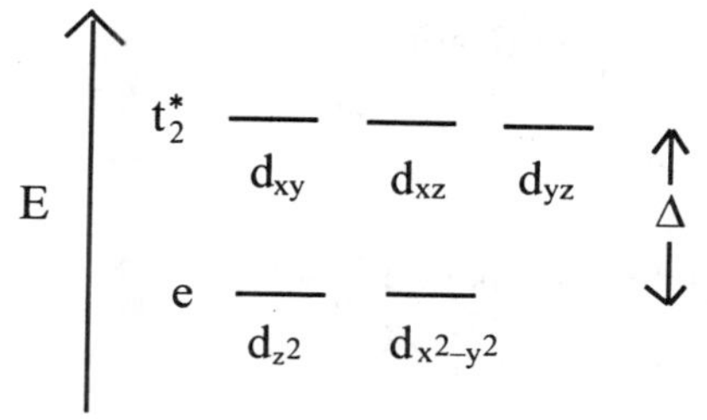

The orbital splitting energy is small because there are only four ligands present so that the tetrahedral complexes are all high-spin.

The ligands of a square planar complex lie along the x and y axes where they raise the energy of the $d_{x^2-y^2}$ and d_{xy} orbitals. The d_{z^2}, d_{xz}, and d_{yz} orbitals are lower than in an octahedral complex because ligands are not present on the z axis in a square planar complex. The d orbital splitting for a square planar complex is

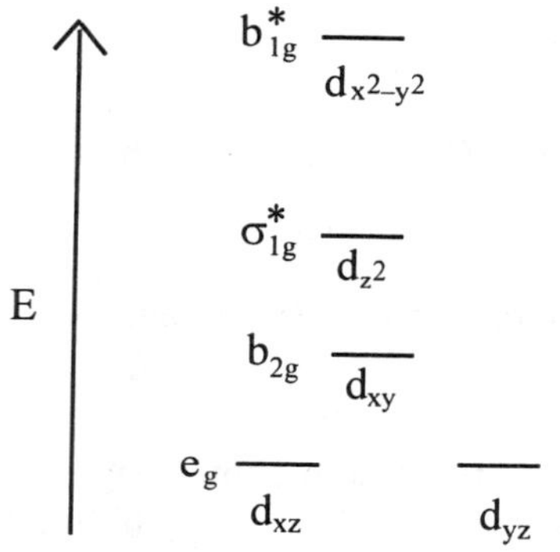

Square planar complexes are favored over octahedral and tetrahedral complexes with strong ligands because it requires less energy to pair electrons in square planar orbitals than to have them in the high-energy antibonding orbitals of an octahedral or tetrahedral complex.

SELF-TEST

Answer questions 1-3 as true or false. If a statement is false change the underlined word(s) so it is true.

1. The melting points of representative metals tend to decrease from <u>top to bottom</u> in a group.

2. One measure of the strength of a metallic bond is the enthalpy of <u>ionization</u>.

3. Most metals form <u>basic</u> oxides.

For questions 4-8 underline the correct word or words.

4. In general, as you move from left to right across any one period the metal-oxygen bond of the metal oxide becomes (more, less) ionic. As the oxidation state of any one metal increases the ionic character of the metal-oxygen bond (increases, decreases).

5. Large atoms with few electrons tend to have (small, large) enthalpies of sublimation and ionization.

6. Transition metal complexes tend to be colored due to the excitation of (p, d) electrons.

7. The energy gap (difference in energy between the conduction and valence bands) in a metallic conductor or semiconductor is (greater, smaller) than in an insulator.

8. The number of electrons in the conduction band of a metallic conductor is (greater, less) than the number of electrons in the conduction band of an insulator.

9. Conductivity data indicate that 1 mol of $CoCl(NO_2)_2{\bullet}4H_2O$ produces 2 mol of ions in solution. Solutions of the compound form a precipitate on addition of $AgNO_3$.
   (a) Write the formula of the coordination compound.
   (b) Give the coordination number and oxidation state of the cobalt ion.

10. Draw all the isomers of $FeBrCl(NO_2)(NH_3)_3$. Identify enantiomer pairs.

11. Cobalt forms a complex with the hexadentate ligand ethylenediaminetetraacetate (EDTA), $[Co(EDTA)]^-$. Draw its optical isomers.

12. The color of $[Cr(H_2O)_6]Cl_3$ is violet while $[Cr(NH_3)_6]Cl_3$ is yellow. Which ligand is higher in the spectrochemical series?

13. The Fe^{2+} ion tends to be oxidized in natural water systems to Fe^{3+}. However, in a waste water stream containing KCN the Fe^{2+} ion is stabilized. Suggest a reason. Hint: Recall there is a stability associated with filled or half-filled sets of orbitals with the same energy.

14. $[Ni(CN)_4]^{2-}$ is square planar while $[NiCl_4]^{2-}$ is tetrahedral. Use MO theory to explain the different shapes. (Hint: Compare the d orbital configuration of a square planar and a tetrahedral complex.)

ANSWERS TO SELF-TEST QUESTIONS

1. T
2. F; sublimation
3. T
4. less, decreases
5. small
6. d
7. smaller
8. greater
9. (a) $[Co(NO_2)_2(H_2O)_4]Cl$
   (b) 6, 3+

10.

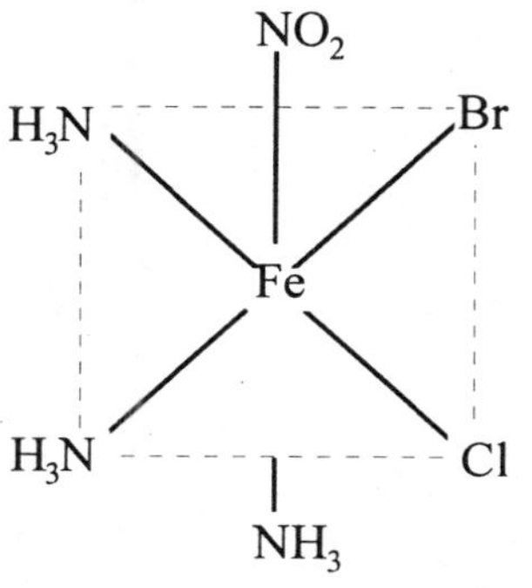

enantiomers

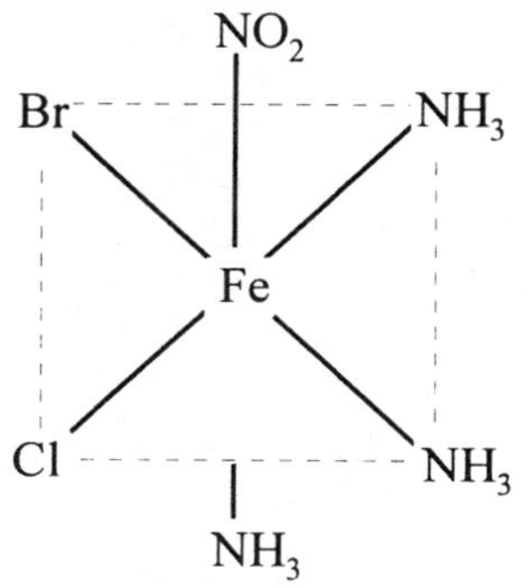

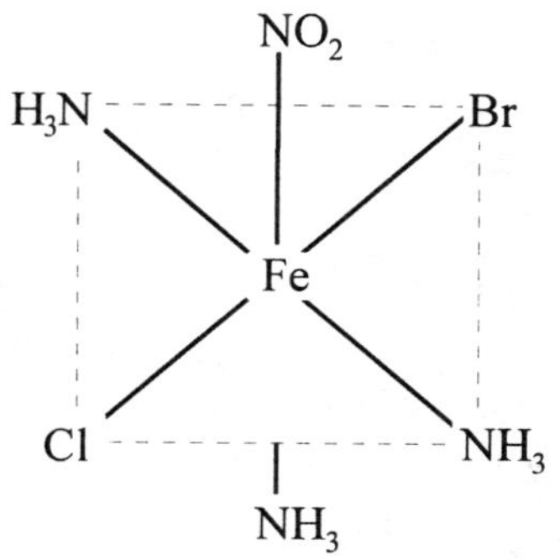

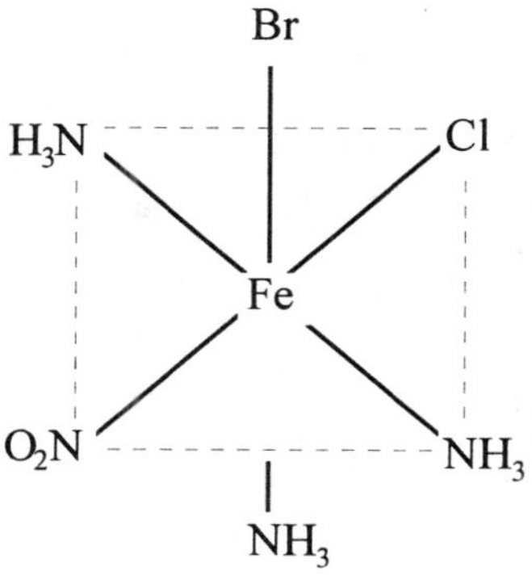

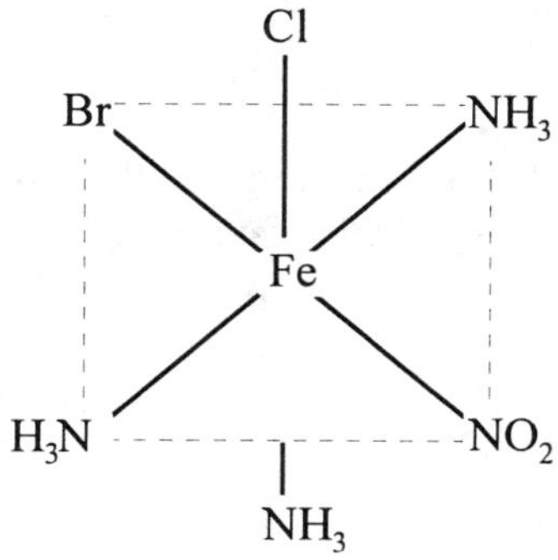

11.

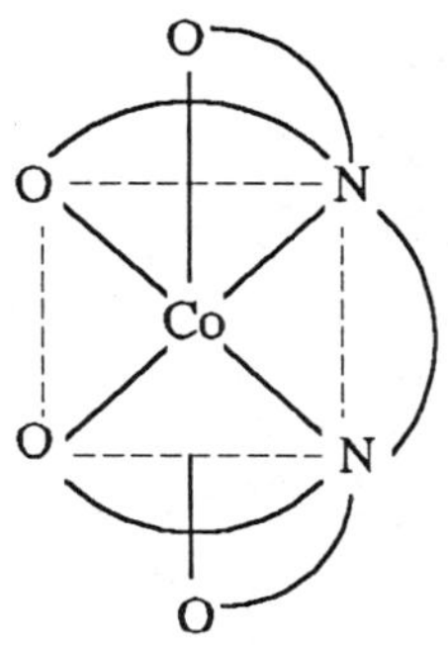

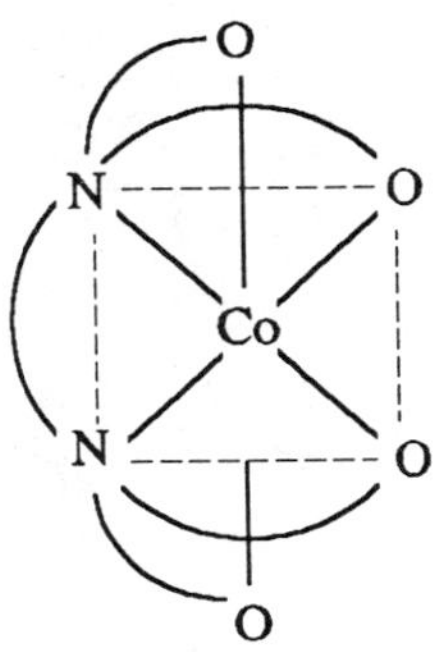

12. Yellow is a longer wavelength light than violet so that $[Cr(NH_3)_6]^{3+}$ absorbs shorter wavelength light (greater energy) than $[Cr(H_2O)_6]^{3+}$. Therefore, $[Cr(NH_3)_6]^{3+}$ has a larger orbital splitting energy and <u>NH_3</u> is higher (stronger-field ligand) in the spectrochemical series.

13. CN^- is a strong-field ligand so that the six d electrons of Fe^{2+} will <u>completely fill</u> the three lowest energy orbitals.

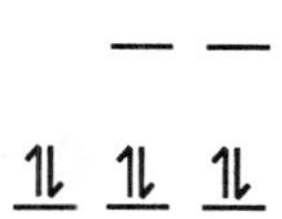

14.

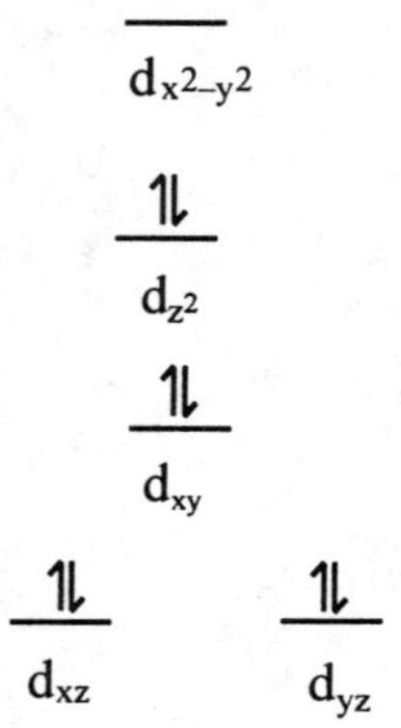

square planar
$[Ni(CN)_4]^{2-}$

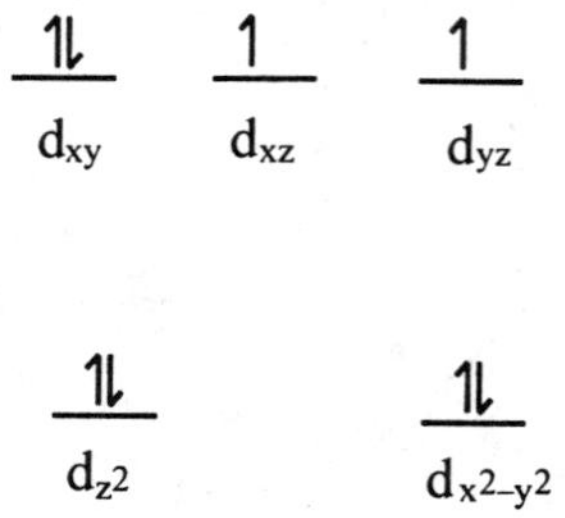

tetrahedral
$[NiCl_4]^{2-}$

A strong-field ligand such as CN^- will cause a large orbital splitting energy in the tetrahedral complex so that the pairing of electrons in the square planar electron configuration is energetically favorable.

SURVEY OF THE ELEMENTS 4
METALS AND METALLURGY

SELF-ASSESSMENT

Metallurgy (Section S4.1)

1. What are the main steps in obtaining a pure metal from its ore?

2. Why is the Bayer process used to produce pure aluminum oxide from bauxite and not iron oxide from its ore?

3. Write equations for the cyanide leaching process for preparing gold.

Metals of the p block (Section S4.2) **and Group 2B: Zinc, Cadmium, and Mercury** (Section S4.3)

For questions 4–9 indicate the metal described by the given statement. Your choices are Al, Sn, Pb, Bi, Zn, Cd, or Hg.

4. An oxide coats this metal, protecting it from water and air. The oxide, however, dissolves in strong acid or base solution. _____________

5. This metal is very toxic. It is present in storage batteries, paints, and solder. Of its two oxidation states the lower is more stable. _____________

6. This metal is a liquid at room temperature and pressure and is relatively inert. As a result it is used in barometers and thermometers. _____________

7. The oxide and hydroxide of this metal are amphoteric. Its fluoride compound is used in toothpaste.

8. It is a very toxic heavy metal whose concentration is increasing in the environment. Its chemical properties are similar to those of zinc.

9. It is the poorest conductor of all metals. It reacts with HCl to form an insoluble compound that contains the same ion found in Pepto-Bismol.

Group 1B: The Coinage Metals (Section S4.4)

Fill in the blanks with Cu, Ag, or Au.

10. The most active Group 1B metal is _____________.

11. When solutions of its salts are treated with base ____________ precipitates an oxide rather than a hydroxide.

12. ____________ does not rust, corrode, or tarnish.

The Iron Triad (Section S4.5)

13. Complete and balance the following equations:
   (a) $Co(s) + O_2(g) \rightarrow$
   (b) $Ni^{2+}(aq) + OH^-(aq) \rightarrow$
   (c) $Co(s) + HCl(1M) \rightarrow$
   (d) $Ni(OH)_2(s) + NH_3(aq) \rightarrow$
   (e) $Fe^{2+}(aq) + O_2(g) + H^+(aq) \rightarrow$

14. The following statements are true of Fe, Co, and Ni except
   (a) All three metals displace hydrogen from acids.
   (b) All three metals burn in oxygen.
   (c) They all exhibit a +2 oxidation state.
   (d) All of these metals are obtained by roasting their sulfide ores and reducing the resulting oxides.

Chromium and Manganese (Section S4.6)

15. Indicate whether the statements given below refer to Cr, Mn, or both.
   (a) It forms an oxide coating that protects it from corrosion.
   (b) Its compounds that contain the metal in the +7 oxidation state are oxidizing agents.
   (c) Its compounds that contain the metal in the +4 oxidation state are oxidizing agents.
   (d) Some of its insoluble compounds are used as pigments.

16. Write balanced equations for the following reactions:
   (a) dichromate ion + concentrated sulfuric acid
   (b) manganese + water
   (c) chromium(II) ions + oxygen gas in acid solution
   (d) manganese in acid solution.

ANSWERS TO SELF-ASSESSMENT PROBLEMS

If you missed an answer, be sure to study the relevant section in the textbook and study guide.

1. (1) concentrating the ore (2) extracting the metal (3) refining (purifying) the crude metal

2. Al_2O_3 is amphoteric and therefore dissolves in NaOH solution.

3. $4Au(s) + 8CN^-(aq) + O_2(g) + 2H_2O(l) \rightarrow 4Au(CN)_2^-(aq) + 4OH^-(aq)$
   $2Au(CN)_2^-(aq) + Zn(s) \rightarrow 2Au(s) + Zn(CN)_4^{2-}(aq)$

4. Al

5. Pb

6. Hg

7. Sn

8. Cd

9. Bi

10. Cu

11. Ag

12. Au

13. (a) $3Co(s) + 2O_2(g) \rightarrow Co_3O_4(s)$
(b) $Ni^{2+}(aq) + 2OH^-(aq) \rightarrow Ni(OH)_2(s)$
(c) $Co(s) + 2HCl(aq) \rightarrow CoCl_2(aq) + H_2(g)$
(d) $Ni(OH)_2(s) + 6NH_3(aq) \rightarrow Ni(NH_3)_6^{2+}(aq) + 2OH^-(aq)$
(e) $4Fe^{2+}(aq) + O_2(g) + 4H^+(aq) \rightarrow 4Fe^{3+}(aq) + 2H_2O(l)$

14. d

15. (a) Cr (b) Mn (c) Cr (d) Cr and Mn

16. (a) $Cr_2O_7^{2-}(aq) + 2H^+(aq) \rightarrow 2CrO_3(s) + H_2O(l)$
(b) $Mn(s) + 2H_2O(l) \rightarrow Mn(OH)_2(aq) + H_2(g)$
(c) $4Cr^{2+}(aq) + O_2(g) + 4H^+(aq) \rightarrow 4Cr^{3+}(aq) + 2H_2O(l)$
(d) $Mn(s) + 2H^+(aq) \rightarrow Mn^{2+}(aq) + H_2(g)$

REVIEW

Metallurgy (Section S4.1)

Learning Objectives

1. List the principal steps in obtaining a pure metal from its ore.
2. Describe at least three physical methods for concentrating an ore.
3. Describe and write equations for the Bayer process for obtaining pure aluminum oxide from bauxite.
4. Write equations for the roasting of sulfide ores.
5. Describe and write equations for the Dow process.
6. Describe and write equations for the smelting of iron ores in a blast furnace.
7. Write equations for the nonelectrolytic reduction of KCl, PbO, WO_3, and Cr_2O_3.
8. Describe and write equations for the cyanide leaching process for preparing gold and silver.
9. Describe and write equations for the electrorefining of copper.

Review

Metallurgy is the science of removing metals from their ores (rock or mineral containing the metal), refining them, and making alloys (mixtures of metals).

Concentrating the Ore

Methods for concentrating an ore are
- gravity separation (as in panning for gold)
- magnetic separation (used for ferromagnetic minerals)

- flotation (used for sulfide ores)
- chemical separation

An example of chemical separation is the Bayer process for obtaining aluminum oxide from bauxite:

$$Al_2O_3(s) + 2OH^-(aq\ 30\%) + 3H_2O(l) \rightarrow 2Al(OH)_4^-(aq)$$

The solution is then filtered, cooled, and diluted to reduce $[OH^-]$ so that $Al(OH)_3$ precipitates. The anhydrous oxide is obtained from heating:

$$2Al(OH)_3(s) \xrightarrow{\text{heat}} Al_2O_3(s) + 3H_2O(g)$$

Obtaining the Metal

Pretreatment changes a mineral to a form that can be reduced.
Roasting of sulfides is an example in which reducible oxides are formed:

$$2PbS(s) + 3O_2(g) \xrightarrow{\text{heat}} 2PbO(s) + 2SO_2(g)$$

The Dow process is a pretreatment method for obtaining magnesium metal from seawater. The equations involved are

$$\underset{\text{shells}}{CaCO_3(s)} \xrightarrow{\text{heat}} \underset{\text{lime}}{CaO} + CO_2(g)$$

$$CaO(s) + H_2O(l) \rightarrow \underset{\text{saturated solution}}{Ca(OH)_2(aq)}$$

$$\underset{\text{seawater}}{Mg^{2+}(aq)} + Ca(OH)_2(aq) \rightarrow Mg(OH)_2(s) + Ca^{2+}(aq)$$

The $Mg(OH)_2$ is filtered and converted into the chloride, which is then used for electrolysis.

$$Mg(OH)_2(s) + 2HCl(aq) \rightarrow MgCl_2(aq) + 2H_2O(l)$$

Reduction

Electrolytic reduction is used for obtaining active metals (Li, Na, Mg, Ca, and Al).

Chemical reduction is used for some other metals such as iron. Smelting is an example of chemical reduction in which a molten metal is obtained from a high-temperature chemical reduction. Carbon in the form of coke is the common reducing agent for obtaining Cu, Sn, Pb, Zn, Mn, Co, Ni, and Fe. The equation is

$$ZnO(s) + C(s) \xrightarrow{\text{heat}} Zn(l) + CO(g)$$

A blast furnace is used for smelting iron ore: blasts of heated air or oxygen are blown through a mixture of crushed ore, coke, and limestone. Carbon is oxidized to carbon monoxide, which reduces the ore to metal. The net equation is

$$Fe_2O_3(s) + 3CO(g) \rightarrow 2Fe(l) + 3CO_2(g)$$

Also, $Fe_2O_3(s) + 3C(s) \rightarrow 2Fe(l) + 3CO(g)$

Limestone removes silica from the ore according to

$$CaCO_3(s) \xrightarrow{\text{heat}} CaO(s) + CO_2(g)$$

$$CaO(s) + SiO_2(s) \rightarrow \underset{\text{slag}}{CaSiO_3(l)}$$

Another example of chemical treatment of ores is the cyanide leaching process for gold and silver.

The cyanide ion complexes and stabilizes oxidized gold:

$$4Au(s) + 8CN^-(aq) + O_2(g) + 2H_2O(l) \rightarrow 4Au(CN)_2^-(aq) + 4OH^-(aq))$$

Zinc is then used to displace gold from the complex:

$2Au(CN)_2^-(aq) + Zn(s) \rightarrow 2Au(s) + Zn(CN)_4^{2-}(aq)$

Refining the Crude Metal

Various techniques are available for separating impurities from a crude metal. Electrorefining is used to obtain a very pure product. Figure S4.3 in your text shows the electrolytic cell used to produce copper. The half-equations are

anode: $\underset{\text{(crude copper)}}{Cu(s)} \rightarrow Cu^{2+}(aq) + 2e^-$

The copper ions then travel to the cathode where they form copper metal:

cathode: $Cu^{2+}(aq) + 2e^- \rightarrow Cu(s)$

Metals of the p Block (Section S4.2)

Learning Objectives

1. Describe how aluminum is ordinarily protected from attack by air and water.
2. Write equations for the reactions of Al and Al_2O_3 with strong acid and strong base.
3. Write Lewis structures for Al_2Cl_6 and other aluminum halides.
4. Describe the periodic trends in Groups 3A and 4A; include physical properties, oxidation state, chemical activity, and nature of the oxides.
5. Describe the occurrence and preparation of tin, lead, and bismuth.
6. Write equations for the reactions of Sn, Pb, and Bi with oxygen, sulfur, and the halogens.
7. Write equations for the reactions of SnO, PbO, SnO_2, and PbO_2 with strong acid and strong base.
8. Write an equation for the preparation of PbO_2.
9. Write an equation for the reaction of Bi_2O_3 with hydrochloric acid.
10. Give the name, formula, and commercial use of at least one compound of aluminum, tin, lead, and bismuth.

Review

General Information

The most important p-block metals are aluminum (Al) in Group 3A, tin (Sn), lead (Pb) in Group 4A, and bismuth (Bi) in Group 5A. Aluminum is an active metal that is protected by an adherent oxide film. Its aqueous ion is acidic and its oxide is amphoteric. The anhydrous chloride, bromide, and iodide are covalent. Gallium (Ga), indium (In), and thallium (Tl) show a trend toward less covalence and increasing stability for the +1 oxidation state. There are two allotropes of tin: white tin, which is metallic and stable at room temperature, and gray tin, which is nonmetallic and stable below 13.2°C. Tin forms both Sn(II) and Sn(IV) compounds, but lead and bismuth favor their lower oxidation states. SnO, PbO, and SnO_2 are amphoteric; Bi_2O_3 is basic. The properties and uses of these metals and their important compounds are summarized for you in Table S4.1 given below.

Note: Table S4.1 is only a brief outline and you should make your own additions as you study Chapter S4 in your text.

Table S4.1: Metals of the p Block

Occurrence and Physical Properties	Chemical Properties
Aluminum (Al)	
• third most abundant element in earth's crust (present as aluminosilicates in feldspars, micas, and clays) • conductive • lustrous	• no d electrons • Al_2O_3 forms on surface of metal • dissolves in acid: $2Al(s) + 6H^+(aq) \rightarrow 2Al^{3+}(aq) + 3H_2(g)$ • in solution Al^{3+} is hydrated and is acidic: $2Al(H_2O)_6^{3+}(aq) + 3CO_3^{2-}(aq) \rightarrow 2Al(OH)_3(H_2O)_3(s) + 3H_2O(l) + 3CO_2(g)$
Tin (Sn)	
• in ores (as SnO_2 (cassiterite) and as PbS (galena)) • See Table S4.4 in text on physical properties • hard, shiny, springy • nontoxic • exists mostly as white tin • allotrope (gray tin) is nonmetallic and stable at low temperatures	• reacts with nonmetals • forms Sn(IV) compounds (SnO_2, $SnCl_4$) • reacts with HNO_3 to form SnO_2
Lead (Pb)	
• dark, soft, malleable • very toxic	• favors lower oxidation state (Pb(II)), Pb(IV) is rare • reacts with nonmetals to form Pb(II) compounds (PbO, $PbCl_2$) • reacts with HNO_3 to form $Pb(NO_3)_2$
Bismuth (Bi)	
• in sulfide ores • see Table S4.4 in text on physical properties • reddish, brittle • poorest conductor of all metals	• favors lower oxidation states • reacts with nonmetals to form Bi(III) compounds (Bi_2O_3, Bi_2S_3) • oxidized by HNO_3 to form $Bi(NO_3)_3$

Uses	Important Compounds and Their Reactions
Aluminum (Al) • commercial drain cleaners • in alloys for lightweight construction	Al_2O_3 and $Al(OH)_3$ are amphoteric: $Al_2O_3(s) + 2OH^-(aq, conc.) + 3H_2O(l) \rightarrow 2Al(OH)_4^-(aq)$
Aluminum Salts • baking powder ($NaAl(SO_4)_2$) • antiperspirants ($Al_2(OH)_5Cl{\cdot}2H_2O$, $Al_2(SO_4)_3{\cdot}18H_2O$ • water treatment (alum: $KAl(SO_4)_2{\cdot}12H_2O$) • antacids, analgesics	$Al_2O_3(s) + 6H^+(aq) \rightarrow 2Al^{3+}(aq) + 3H_2O(l)$

Uses

Tin (Sn)
- used in organ pipes
- food containers (tin cans)

Important Compounds and Their Reactions

$\underline{SnO}$ and $\underline{Sn(OH)_2}$ are amphoteric:
$SnO(s) + 2H^+(aq) \rightarrow Sn^{2+}(aq) + H_2O(l)$
$SnO(s) + OH^-(aq) + H_2O(l) \rightarrow Sn(OH)_3^-(aq)$

$\underline{SnO_2}$ dissolves in base:
$SnO_2(s) + 2OH^-(aq) + 2H_2O(l) \rightarrow Sn(OH)_6^{2-}(aq)$

SnO_2 forms anionic complexes in acid:
$SnO_2(s) + 6HCl(aq) \rightarrow 2H^+(aq) + SnCl_6^{2-}(aq) + 2H_2O(l)$

Lead (Pb)
- pipes, cable sheathing
- ammunition
- alloys of Sn and Pb in metal and solder

- $\underline{PbO}$ and $\underline{Pb(OH)_2}$ are amphoteric (equations analogous to SnO and $Sn(OH)_2$ in acid and base)
- PbO_2 dissolves in base (equation analogous to SnO_2 in base)

Bismuth (Bi)
- in alloys used for fuses
- bismuthyl ion (BiO^+) in Pepto-Bismol

$\underline{BiO_3}$ (basic oxide)
- reacts with acid to give insoluble compounds (BiOCl)
- At very low pH, $BiOCl(s) + 2HCl(aq) \rightarrow Bi^{3+}(aq) + 3Cl^-(aq) + H_2O(l)$

Compounds
- $\underline{SnF_2}$ (stannous fluoride) used in toothpaste
- $\underline{PbO}$ is a yellow pigment used in pottery
- $\underline{PbO_2}$ used in lead storage battery (see Section 20.5)
- $\underline{Pb_3O_4}$ is a red pigment in paints

Group 2B: Zinc, Cadmium, and Mercury (Section S4.3)

Learning Objectives

1. Explain why zinc, cadmium, and mercury resemble the Group 2A elements that precede them in their periods.
2. Describe the periodic trends in Group 2B; include physical properties, oxidation states, chemical activity, and nature of the oxides.
3. Write equations for the preparation of zinc, cadmium, and mercury from their sulfide ores.
4. Write equations for the reactions of zinc, cadmium, and mercury with both oxidizing and nonoxidizing acids.
5. Write equations for the reactions of ZnO, CdO, and HgO with strong acid, strong base, and aqueous ammonia.
6. Give the name, formula, and commercial use of at least one compound of zinc, cadmium, and mercury.

Review

General Information

Zinc, cadmium, and mercury resemble Group 2A elements in that only their s^2 valence electrons are involved in bonding. Like other transition elements, they form numerous complexes, and their activity decreases toward the bottom of the group. Mercury is the only metal that is liquid at room temperature. The usual oxidation state of these elements is +2, but mercury also has a +1 oxidation state in compounds containing Hg_2^{2+}. Zinc and cadmium displace hydrogen from nonoxidizing acids; all three metals are dissolved by oxidizing acids such as HNO_3 and H_2SO_4. Zinc oxide is amphoteric, while the oxides of cadmium and mercury are basic. Zinc and cadmium oxides form ammine complexes. The compounds of zinc are essentially ionic. Cadmium compounds exhibit some covalence, and most mercury compounds have a high degree of covalent character.

Table S4.2 below lists <u>some</u> of the properties and uses of these metals and their important compounds.

Table S4.2: Properties and Uses of Group 2B; Zn, Cd, and Hg

<u>Occurrence and Physical Properties</u>

Zinc (Zn)
- in sulfide ores (ores usually contain Cd as well)
- blue-white metal
- brittle at room temperature (see Table S4.5 of text)

Cadmium (Cd)
- in sulfide ores
- blue-white metal
- soft, can be cut with a knife (See Table S4.5 of text)

Mercury (Hg)
- in sulfide ores
- liquid at room temperature and pressure
- dissolves most metals (See Table S4.5 of text)

<u>Chemical Properties</u>

Zinc (Zn)
- exhibits only a +2 oxidation state in compounds
- compounds are essentially ionic
- forms halides, oxide, and sulfide
- displaces hydrogen from 1M $H^+(aq)$:
  $Zn(s) + 2HCl \rightarrow ZnCl_2(aq) + H_2(g)$
- reacts with oxidizing acids:
  $3Zn(s) + 8HNO_3(aq) \rightarrow 3Zn(NO_3)_2(aq) + 2NO(g) + 4H_2O(l)$

Cadmium (Cd)
- similar to zinc
- exhibits +2 oxidation state in compounds
- compounds show more covalent character than zinc compounds
- forms oxide, halides, and sulfide
- displaces hydrogen from $1M$ $H^+(aq)$
- reacts with oxidizing acids

Mercury (Hg)
- exhibits +2 and +1 oxidation states in compounds
- most compounds show high degree of covalent character
- forms oxides, halides, and sulfides
- reacts with oxidizing acids:
  $3Hg(l) + 8HNO_3(aq) \xrightarrow{\text{excess acid}} 3Hg(NO_3)_2(aq) + 2NO(g) + 4H_2O(l)$
  $6Hg(l) + 8HNO_3(aq) \xrightarrow{\text{excess Hg}} 3Hg_2(NO_3)_2(aq) + 2NO(g) + 4H_2O(l)$

<u>Uses</u>

Zinc (Zn)
- in alloys
- coating in galvanized iron
- anode in flashlight batteries (Section 20.6)
- trace nutrient

<u>Important Compounds and Their Reactions</u>

- <u>ZnO and $Zn(OH)_2$</u> are amphoteric (See Section 18.3)
- ZnO dissolves in $NH_3(aq)$:
  $ZnO(s) + 4NH_3(aq) + H_2O(l) \rightarrow Zn(NH_3)_4^{2+}(aq) + 2OH^-(aq)$

Uses

Cadmium, Cd
- similar to Zn
- in alloys
- protective coating for iron and steel
- control rods of nuclear reactors

Mercury (Hg)
- in thermometers, barometers
- form amalgamates (dental amalgam)

Compounds

ZnO
- white pigment
- astringent, deodorant, ointment

ZnS
- phosphor in TV screens

Hg_2Cl_2 (calomel)
- used in the standard calomel electrode (Section 20.4)

Important Compounds and Their Reactions

CdO is basic
- dissolves in $NH_3(aq)$:
  $CdO(s) + 4NH_3(aq) + H_2O(l) \rightarrow Cd(NH_3)_4^{2+}(aq) + 2OH^-(aq)$

HgO is basic and does not form an ammine complex

Group 1B: The Coinage Metals (Section S4.4)

Learning Objectives

1. Describe periodic trends within the copper group; include physical properties, oxidation states, chemical activity, and nature of the oxides.
2. Write equations for the preparation of copper metal from its sulfide ore.
3. Write equations for the reactions of copper, silver, and gold with chlorine, airborne H_2S, and oxidizing acids.
4. Write equations for the reactions of silver halides with aqueous ammonia, aqueous sodium thiosulfate, and light.
5. Write equations for the reactions of aqueous Cu^{2+} and Ag^+ ions with ammonia and sodium hydroxide.
6. Give the name, formula, and commercial use of at least one compound of copper, silver, and gold.

Review

General Information

The coinage metals, copper (Cu), silver (Ag), and gold (Au), are the most malleable, ductile, and conductive of all metals. Copper is obtained from sulfide ores. Gold and silver are obtained from the anode sludge formed during the electrorefining of copper. The most common oxidation state of copper is +2; the +1 state is stable only in certain complexes and insoluble compounds. The usual oxidation state of silver is +1; gold exists in the +1 and +3 states. Copper and silver react with airborne sulfides; copper reacts with moist air to form a green basic copper carbonate. All three metals are attacked by chlorine. The metals do not displace hydrogen from acids. Copper and silver are dissolved by oxidizing acids; gold is dissolved by a mixture of concentrated HCl and HNO_3 called aqua regia.

Table S4.3 lists some of the properties and uses of the coinage metals and their compounds.

Table S4.3: The Coinage Metals

Occurrence and Physical Properties

Copper (Cu)
- in sulfide ores
- malleable, ductile
- conductive

Silver (Ag)
- malleable, ductile
- conductive

Gold (Au)
- most malleable, and ductile of all metals
- conductive

Chemical Properties

Copper (Cu)
- usual oxidation state is +2
- Cu(I) is stable in sparingly soluble compounds (Cu_2S, CuI) and in complexes ($Cu(CN)_2^-$)
- most active of coinage metals:
  $2Cu(s) + 4H^+(aq) + O_2(g) \rightarrow 2Cu^{2+}(aq) + 2H_2O(l)$
- oxidized by HNO_3 and H_2SO_4:
  $Cu(s) + 4H^+(aq) + SO_4^{2-}(aq) \rightarrow Cu^{2+}(aq) + SO_2(g) + 2H_2O(l)$
- reacts with Cl_2 to form $CuCl_2$
- tarnished by natural sulfides

Silver (Ag)
- usual oxidation state is +1
- reacts with chlorine to form AgCl
- tarnished by natural sulfides and sulfur compounds in air:
  $4Ag(s) + 2H_2S(g) + O_2(g) \rightarrow \underset{\text{tarnish}}{2Ag_2S(s)} + 2H_2O(g)$

Gold (Au)
- does not rust, corrode, or tarnish
- dissolves in aqua regia:
  $Au(s) + 4H^+(aq) + NO_3^-(aq) + 4Cl^-(aq) \rightarrow AuCl_4^-(aq) + NO + 2H_2O(l)$
- reacts with chlorine to form $AuCl_3$ (See Table S4.6 in text for a list of properties)

Uses

Copper (Cu)
- blister copper (98–99% pure) used for plumbing, decorative purposes
- electrical wiring (electrorefined copper)

Silver (Ag)
- dental and other alloys (Table S4.1of text)
- printed circuits
- electrical contacts
- batteries
- coins and jewelry

Important Ions and Their Reactions

Copper (Cu)
- Low concentration of Cu^{2+} can be detected by the formation of $Cu(NH_3)_4^{2+}$ (blue)
  $Cu^{2+}(aq) + 4NH_3(aq) \rightarrow Cu(NH_3)_4^{2+}(aq)$
- Cu^+ ion disproportionates in solution:
  $2Cu^+(aq) \rightarrow Cu^{2+}(aq) + Cu(s)$

Silver (Ag)
- precipitates an oxide $2Ag^+(aq) + 2OH^-(aq) \rightarrow \underset{\text{black}}{Ag_2O(s)} + H_2O(l)$
- Ag^+ identified in solution by forming white solid that dissolves in $NH_3(aq)$: $Ag^+(aq) + Cl^-(aq) \rightarrow AgCl(s)$
  $AgCl(s) + 2NH_3(aq) \rightarrow Ag(NH_3)_2^+(aq) + Cl^-(aq)$
- Silver salts darken in light: $\underset{\text{white}}{2AgCl(s)} \xrightarrow{hv} \underset{\text{black}}{2Ag(s)} + Cl_2(g)$

Uses

Gold (Au)
- electric devices
- coins
- jewelry
- international currency standard

Compounds

$CuSO_4 \cdot 5H_2O$
- copper plating
- fungicide, algicide

$AgNO_3$
- silver platings
- silver mirrors
- disinfectant, astringent

AgBr
- used in film

chlorauric acid ($HAuCl_4$)
- gold plating
- photography
- in manufacture of ruby glass

The Iron Triad (Section S4.5)

Learning Objectives

1. Describe, including equations where pertinent, the production of pig iron and steel.
2. Explain how quenching and tempering affect the properties of a steel.
3. Write equations for the reactions of iron, cobalt, and nickel with oxygen and nonoxidizing acids.
4. Write equations for the reactions of aqueous Fe^{2+}, Co^{2+}, and Ni^{2+} with sodium hydroxide, ammonia, and sulfide ion.
5. Describe analytical tests for identifying Fe^{3+}, Co^{2+}, and Ni^{2+}.
6. Give the name, formula, and commercial use of at least one compound of iron, cobalt, and nickel.

Review

General Information

The iron triad consists of iron (Fe), cobalt (Co), and nickel (Ni). "Pig iron" is prepared by the reduction of oxide ores in a blast furnace. Steels are ferrous alloys containing controlled amounts of carbon (up to 1.7%). Other metals are usually added to impart some desired mix of properties. In the United States most steel is made by the basic oxygen process. The properties of steel can be modified by quenching and tempering. Cobalt and nickel are obtained from sulfide ores. All members of the iron triad display some degree of ferromagnetism; all burn in oxygen, displace hydrogen from acids, and form metal carbonyls when heated with carbon monoxide. The most common oxidation states of iron and cobalt are +2 and +3; the most common state of nickel is +2. The +3 state of iron is more stable than the +2 state in simple iron salts. The +2 state of cobalt is more stable than the +3 except when the ion is complexed by ligands that are stronger than H_2O.

Table S4.4 lists some of the properties and uses of the iron triad and their compounds.

Table S4.4: The Iron Triad

Occurrence and Physical Properties

Iron (Fe)
- fourth most abundant element in the earth's crust
- primary iron ores are hematite (Fe_2O_3), magnetite (Fe_3O_4), and siderite ($FeCO_3$)
- ferromagnetic

Cobalt (Co)
- occurs in sulfide ores
- ferromagnetic

Nickel (Ni)
- may be a major component of earth's core
- in sulfide ore
- ferromagnetic

Chemical Properties

Iron (Fe)
- burns in oxygen to form Fe_2O_3 and/or Fe_3O_4
- displaces hydrogen from acids:
  $Fe(s) + 2H^+(aq) \rightarrow Fe^{2+}(aq) + H_2(g)$
  (See Table S4.8 in your text)

Cobalt (Co)
- common oxidation states are +2 and +3 (both in Co_3O_4).
- burns in oxygen to form Co_3O_4
- +2 state when not complexed by a ligand stronger than H_2O ($Co(OH)_2$ and CoS)
- +3 state when ligand is stronger than H_2O ($Co(NH_3)_6^{3+}$)
- displaces hydrogen from acids (See Table S4.8 in your text)

Nickel (Ni)
- common oxidation state is +2
- +4 oxidation state is rare (strong oxidizing agent)
- burns in oxygen to form NiO
- reacts with CO: $Ni(s) + 4CO(g) \underset{}{\overset{\text{heat}}{\rightleftharpoons}} Ni(CO)_4(g)$
- displaces hydrogen from acids (See Table S4.8 in text)

Uses

Iron (Fe)

Cast iron
- iron pipes
- hot water heaters
- car engine block

Steels (0.1% – 1.5% carbon)

Cobalt (Co)
- in steels and other alloys
- as a catalyst
- trace nutrient

Important Ions and Their Reactions

Iron (Fe)
- $Fe^{2+}(aq)$ oxidized by dissolved oxygen:
  $4Fe^{2+}(aq) + O_2(g) + 4H^+(aq) \rightarrow 4Fe^{3+}(aq) + 2H_2O(l)$
- $Fe^{3+}(aq)$ reduced by sulfide ion:
  $2Fe^{3+}(aq) + 3S^{2-}(aq) \rightarrow \underset{\text{black}}{2FeS(s)} + S(s)$
- $Fe^{3+}(aq)$ can be identified by adding KSCN:
  $Fe^{3+}(aq) + \underset{\text{colorless}}{SCN^-(aq)} \rightleftharpoons \underset{\text{red}}{Fe(SCN)^{2+}}$
- displaces hydrogen from acids

Cobalt (Co)
- oxidation of $Co^{2+}(aq)$
  $4Co^{2+}(aq) + 24NH_3(aq) + O_2(g) + 4H^+(aq) \rightarrow$
  $\underset{\text{yellow}}{4Co(NH_3)_6^{3+}(aq)} + 2H_2O(l)$
- Co^{2+} can be identified by the reaction
  $Co^{2+}(aq) + 7KNO_2(aq) + 2H^+(aq) \rightarrow$
  $\underset{\text{yellow}}{K_3[Co(NO_2)_6](s)} + 4K^+(aq) + NO(g) + H_2O(l)$

Uses

Nickel (Ni)

- in steel and other alloys
- as a catalyst
- in Ni-Cd cell (Section 20.5)

Important Ions and Their Reactions

Ni^{2+}

- reacts with base to form $Ni(OH)_2$ (green)
- reacts with sulfide ion to form NiS (black)
- forms an insoluble red complex with dimethylglyoxime (used for identification of $Ni^{2+}(aq)$

$Ni(OH)_2$

- dissolves in NH_3:

$$\underset{\text{green}}{Ni(OH)_2(s)} + 6NH_3(aq) \rightarrow \underset{\text{blue}}{Ni(NH_3)_6{}^{2+}(aq)} + 2OH^-(aq)$$

Uses

Compounds

$FeSO_4$

- electroplating
- iron nutritional supplement

Fe_2O_3

- red pigment in paints, rubber and ceramics
- in magnetic tapes

Chromium and Manganese (Section S4.6)

Learning Objectives

1. Write equations for the reduction of chromite ore, pyrolusite ore, and Cr_2O_3.
2. Write equations for the reactions of chromium and manganese with nonoxidizing acids.
3. Write equations for the reactions of aqueous Cr^{3+} and Mn^{2+} with sodium hydroxide.
4. Write an equation describing the chromate-dichromate equilibrium.
5. Give the name, formula, and commercial use of at least one compound of chromium and manganese.

Review

General Information

Chromium and magnesium can be prepared by the reduction of Cr_2O_3 and MnO_2, respectively. The uses of chromium are based on its resistance to corrosion. Manganese is used to give hardness to steel. Chromium and manganese are active metals. Manganese displaces hydrogen from water, but chromium is protected by an oxide film. Both metals exhibit a number of oxidation states and a variety of colorful ions and compounds. Important chromium ions include Cr^{2+}, Cr^{3+}, $CrO_4{}^{2-}$, and $Cr_2O_7{}^{2-}$. Chromium(VI) compounds such as K_2CrO_4 and $K_2Cr_2O_7$ are oxidizing agents. Important manganese ions include Mn^{2+}, $MnO_4{}^{2-}$, and $MnO_4{}^-$. Manganese(VII) compounds such as $KMnO_4$ are strong oxidizing agents.

Table S4.5 lists some of the properties and uses of Cr and Mn.

Table S4.5: Chromium and Manganese

Occurrence and Physical Properties

Chromium (Cr)
- principal ore is chromite ($FeCr_2O_4$)
- shiny, hard, brittle

Manganese (Mn)
- principal ore is pyrolusite (contains MnO_2)
- looks like iron but more brittle

Chemical Properties

Chromium (Cr)
- oxidation states range from +2 to +6 (See Table S4.9 in your text.)
- active metal:

$$Cr(s) + 2H^+(aq) \rightarrow \underset{\text{blue}}{Cr^{2+}}(aq) + H_2(g)$$

- like Al protected from corrosion by an oxide film:

$$4Cr(s) + 3O_2(g) \rightarrow 2Cr_2O_3(s)$$

Manganese (Mn)
- oxidation states range from +2 to +7 (See Table S4.9 in your text)
- dissolves in HCl and H_2SO_4

$$Mn(s) + 2H^+(aq) \rightarrow Mn^{2+}(aq) + H_2(g)$$

Uses

Chromium (Cr)
- protective plating
- in ferrochrome (Fe and Cr) component of stainless steel

Important Compounds and Ions and Their Reactions

- chromite ion ($Cr(OH)_4^-$) oxidized in basic solution by H_2O_2:

$$2Cr(OH)_4^-(aq) + 3H_2O_2(aq) + 2OH^-(aq) \rightarrow 2CrO_4^{2-}(aq) + 8H_2O(l)$$

- chromate ion (CrO_4^{2-}, yellow) exists in neutral and basic solution
- dichromate ion ($Cr_2O_7^{2-}$, orange) forms a bright red precipitate in concentrated $H_2SO_4(aq)$:

$$Cr_2O_7^{2-}(aq) + 2H^+(aq) \rightarrow \underset{\text{red}}{2CrO_3}(s) + H_2O(l)$$

(above solution is a powerful oxidizing agent)

- All Cr(VI) compounds are oxidizing agents: chromate ion predominates in neutral and basic solutions, dichromate ion predominates in acidic solutions:

$$\underset{\text{yellow}}{2CrO_4^{2-}}(aq) + 2H^+(aq) \underset{\text{base}}{\overset{\text{acid}}{\rightleftharpoons}} \underset{\text{orange}}{Cr_2O_7^{2-}}(aq) + H_2O(l)$$

$\underline{Cr(OH)_3}$ is amphoteric:

$$Cr(OH)_3(s) + OH^-(aq) \rightarrow \underset{\text{green}}{Cr(OH)_4^-}(aq)$$

$\underline{Cr^{2+}\text{ ion}}$ is readily oxidized to Cr^{+3}:

$$4Cr^{2+}(aq) + O_2(g) + 4H^+(aq) \rightarrow 4Cr^{3+}(aq) + 2H_2O(l)$$

hydrated Cr^{3+} ion is acidic:

$$Cr(H_2O)_6^{3+}(aq) + 3OH^-(aq) \rightarrow \underset{\text{gray - green}}{Cr(H_2O)_3(OH)_3}(s) + 3H_2O(l)$$

Uses

Manganese (Mn)
- gives strength and hardness to steel and other alloys

Important Compounds and Ions and Their Reactions

Mn^{2+} ion
- $Mn(H_2O)_6^{2+}(aq)$ is pink
- MnS is pink
- $Mn(OH)_2$ is a sparingly soluble base and oxidizes in air to form $Mn(OH)_3$

MnO_2 (covalent oxide)
- does not react with ordinary acids and bases

MnO_4^- ion (permanganate ion)
- oxidizing agent
- reduced to Mn^{2+} in acid solution
- reduced to $MnO_2(s)$ in base

Uses

MnO_2
- component of dry cells
- as a pigment (black-brown)
- coloring agent in amethyst glass

$KMnO_4$
- antiseptic, antifungal properties

$KCrO_4$, $K_2Cr_2O_7$
- corrosion inhibitors
- $Cr_2O_7^{2-}$ in acid forms a powerful oxidizing solution used as a cleaning solution for laboratory glassware

Cr_2O_3
- green pigment for glass and printing banknotes

SELF-TEST

Answer Questions 1–7 as true or false. If a statement is false change the underlined words so it becomes true.

1. The Dow process is a pretreatment method for obtaining aluminum.

2. Smelting is the heating of a mineral in dry air.

3. Al is the only p-block metal that has no d electrons.

4. Lower oxidation states become more common toward the bottom of a group.

5. Lead forms a nonmetallic allotrope below 13.2°C.

6. Bismuth is the only metal of Group 3A.

7. The oxide of <u>mercury</u> does not form an ammine complex.

8. Complete and balance the following equations
(a) $Ag^{+}(aq) + OH^{-}(aq) \rightarrow$? + ?
(b) $AgCl(s) + NH_3(aq) \rightarrow$? + ?
(c) $Cu(s) + H^{+}(aq) + O_2(g) \rightarrow$? + ?
(d) $Au(s) + H^{+}(aq) + NO_3^{-}(aq) + Cl^{-}(aq) \rightarrow AuCl_4^{-}(aq) + NO(g) + 2H_2O(l)$

9. All of the following statements are true about both manganese and chromium except
(a) They are active metals; they both dissolve in acid solution.
(b) They both form colored compounds.
(c) They both form a +7 oxidation state.
(d) They are both used in alloys with iron.

10. Give the formula for the chromium compound or ion formed after each reaction.
$$Cr(H_2O)_6^{3+}(aq) \xrightarrow{OH^-} ? \xrightarrow{\text{excess } OH^-} ? \xrightarrow[\text{basic solution}]{H_2O_2} ?$$

ANSWERS TO SELF-TEST QUESTIONS

1. F; Mg
2. F; roasting
3. T
4. T
5. F; tin
6. F; 5A
7. T
8. (a) $2Ag^{+}(aq) + 2OH^{-}(aq) \rightarrow Ag_2O(s) + H_2O(l)$
(b) $AgCl(s) + 2NH_3(aq) \rightarrow Ag(NH_3)_2^{+}(aq) + Cl^{-}(aq)$
(c) $2Cu(s) + 4H^{+}(aq) + O_2(g) \rightarrow 2Cu^{2+}(aq) + 2H_2O(l)$
(d) $Au(s) + 4H^{+}(aq) + 4Cl^{-}(aq) + NO_3^{-}(aq) \rightarrow AuCl_4^{-}(aq) + NO(g) + 2H_2O(l)$
9. (c)
10. $Cr(OH)_3$; $Cr(OH)_4^{-}$; CrO_4^{2-}

CHAPTER 22
NUCLEAR CHEMISTRY

SELF-ASSESSMENT

Natural Radioactivity (Section 22.1)

1. All of the following statements are true about alpha particles except
   (a) They are helium nuclei.
   (b) They have a greater penetrating ability than gamma rays.
   (c) They are deflected in an electric field.
   (d) They are heavier than beta particles.

Choose the phrase that correctly completes the statement.

2. Beta particles
   (a) do not cause ionization.
   (b) are high-energy electrons emanating from unstable nuclei.
   (c) are not deflected by an electric field.
   (d) cannot be detected by a Geiger-Müller counter.

3. Gamma radiation
   (a) emanates from nuclei whose energies are above the ground state.
   (b) is deflected in an electric field.
   (c) consists of high-energy particles.
   (d) cannot be detected by a scintillation counter.

Equations for Nuclear Reactions (Section 22.2)

4. Write balanced nuclear equations for
   (a) the beta decay of potassium-40.
   (b) positron emission by potassium-40.
   (c) the alpha decay of uranium-238.
   (d) the emission of gamma radiation by cobalt-60.

5. Indicate the effect of each of the emissions above on the number of protons and the number of neutrons in an unstable nucleus.

Nuclear Bombardment Reactions (Section 22.3)

6. Write balanced nuclear equations for

(a) ^{14}N (n, P) ^{14}C

(b) ^{96}Mo (D, n) ^{97}Tc

(c) ^{9}Be (α, n) ^{12}C

7. Give the complete symbol for the element formed in each nuclear bombardment reaction.

(a) $^{12}_{6}C + ^{4}_{2}He \rightarrow ? + ^{1}_{1}H$

(b) $^{235}_{92}U + ^{1}_{0}n \rightarrow ^{140}_{54}Xe + ? + 2^{1}_{0}n$

(c) $^{7}_{3}Li + ^{1}_{1}H \rightarrow ^{4}_{2}He + ?$

Activity and Half-Life (Section 22.4)

8. Radioactive gold-198 is used in radiation therapy. If 100 mCi of gold-198 is given to a patient what would be its activity one week later? The half-life of gold-198 is 2.7 days. Assume that the only loss of gold-198 is from nuclear decay.

9. The $^{206}Pb/^{238}U$ ratio in a rock removed from the Grand Canyon in Arizona was measured to be 0.170. Estimate the age of the rock. The half-life for ^{238}U is 4.46×10^{9} years.

10. The carbon-14 activity of a piece of wood removed from an ancient dwelling is only 0.721 times that of wood cut today. Find the approximate age of the wood. The half-life of carbon-14 is 5730 y.

Nuclear Stability (Section 22.5)

Underline the correct word or words in questions 11–14.

11. Gamma ray photons are emitted when nuclei undergo transitions from (higher, lower) energy levels to (higher, lower) energy levels.

12. Most stable nuclei have an (even, odd) number of protons and/or neutrons.

13. Potassium-42 lies above the stability band. It may undergo (beta, alpha) decay.

14. Lead-206 is stable while lead-210 is not. Write a nuclear equation for a possible mode of decay. (Hint: compare the n/p ratio of the stable nuclide with the radioactive nuclide.)

Mass-Energy Conversions (Section 22.6)

15. Use Einstein's famous equation to calculate the energy released in (a) joules and in (b) megaelectron volts (1 u = 931.5 MeV) when 1 mol of plutonium-239 decays according to

$$^{239}_{94}Pu \rightarrow ^{235}_{92}U + ^{4}_{2}He$$

The masses are

plutonium-239	239.0006 u
uranium-235	234.9934 u
alpha particle	4.00150 u

16. Calculate the binding energy per nucleon in MeV (1 u = 931.5 MeV) for an alpha particle. The masses are

neutron	=	1.00866 u
proton	=	1.00728 u
alpha particle	=	4.00150 u

17. Indicate whether the following transformations would release or absorb energy.

(a) $^{236}_{92}U \rightarrow ^{144}_{54}Xe + ^{90}_{38}Sr + 2^{1}_{0}n$

(b) $^{12}_{6}C + ^{4}_{2}He \rightarrow ^{16}_{8}O$

(c) $^{226}_{88}Ra \rightarrow ^{4}_{2}He + ^{222}_{86}Rn$

(d) $^{14}_{7}N \rightarrow 7^{1}_{0}n + 7^{1}_{1}H$

Fission (Section 22.7)

18. List the advantages and disadvantages of nuclear reactors as a source of energy.

19. What is a breeder reactor? What are the major disadvantages of a breeder reactor?

Fusion (Section 22.8)

20. List the advantages of fusion reactors versus fission reactors as a source of power.

21. Describe thermonuclear bombs.

ANSWERS TO SELF-ASSESSMENT PROBLEMS

If you missed an answer, be sure to study the relevant section in your text and study guide.

1. b

2. b

3. a

4. (a) $^{40}_{19}K \rightarrow ^{40}_{20}Ca + ^{0}_{-1}e$

   (b) $^{40}_{19}K \rightarrow ^{40}_{18}Ar + ^{0}_{1}e$

   (c) $^{238}_{92}U \rightarrow ^{4}_{2}He + ^{234}_{90}Th$

   (d) $^{60}_{27}Co \rightarrow ^{60}_{27}Co + ^{0}_{0}\gamma$

5. Beta emission: protons up 1, neutrons down 1
   Positron emission: protons down 1, neutrons up 1
   Alpha emission: protons down 2, neutrons down 2
   Gamma rays: no effect

6. (a) $^{14}_{7}N + ^{1}_{0}n \rightarrow ^{14}_{6}C + ^{1}_{1}H$

   (b) $^{96}_{42}Mo + ^{2}_{1}H \rightarrow ^{97}_{43}Tc + ^{1}_{0}n$

   (c) $^{9}_{4}Be + ^{4}_{2}He \rightarrow ^{12}_{6}C + ^{1}_{0}n$

7. (a) $^{15}_{7}N$ (b) $^{94}_{38}Sr$ (c) $^{4}_{2}He$

8. $k = 0.693/2.7 \text{ days} = 0.257 \text{ d}^{-1}$

$$\ln \frac{N}{N_o} = -0.257 \text{ d}^{-1} \times 7 \text{ d} = -1.80$$

$$\frac{N}{N_o} = 0.17;\ N = 0.17 \times 100 \text{ mCi} = \underline{17 \text{ mCi}}$$

9. $$N/N_o = \frac{U}{U + Pb} = \frac{1.00}{1.00 + 0.17} = 1.00/1.17 = 0.855$$

$$t = \frac{-\ln(0.855)}{0.693} \times 4.46 \times 10^9 \text{ y} = \underline{1.01 \times 10^9 \text{ y}}$$

10. $$N/N_o = \frac{0.721\ N_o}{N_o}$$

$$t = \frac{-\ln(0.721)}{0.693} \times 5730 \text{ y} = \underline{2.70 \times 10^3 \text{ y}}$$

11. higher, lower

12. even

13. beta

14. $^{210}_{82}Pb \rightarrow {}^{0}_{-1}e + {}^{210}_{83}Bi$ (n/p ratio of ^{210}Pb too high)

15. (a) mass defect = 239.0006 u – (234.9934 + 4.00150) u $= 5.7 \times 10^{-3} \text{ u} \times \dfrac{1 \text{ g}}{6.022 \times 10^{23} \text{ u}} = 9.46 \times 10^{-27} \text{ g}$

$E = (9.46 \times 10^{-30} \text{ kg})(3.0 \times 10^8 \text{ m/s})^2 = 8.5 \times 10^{-13} \text{ J}$

8.5×10^{-13} J/nucleus $\times$ 6.02×10^{23} nuclei/1 mol = $\underline{5.1 \times 10^{11} \text{ J/mol}}$

(b) 5.7×10^{-3} u $\times$ 931.5 MeV/1 u = 5.3 MeV

5.3 MeV/nucleus $\times$ 6.02×10^{23} nuclei/1 mol = $\underline{3.2 \times 10^{24} \text{ MeV/mol}}$

16. 2 n $\times$ 1.00866 u/1 n + 2 p $\times$ 1.00728 u/1 p – 4.00150 u = 0.03038 u

$$\text{binding energy/nucleon} = \frac{0.03038 \text{ u} \times 931.5 \text{ meV} / 1 \text{ u}}{4 \text{ nucleons}} = \underline{7.075 \text{ MeV/nucleon}}$$

17. (a), (b), and (c) release energy; (d) absorbs energy

REVIEW

Natural Radioactivity (Section 22.1)

Learning Objectives

1. State the nature of alpha, beta, and gamma radiation, and describe how each form of radiation behaves in an electric field.
2. Compare alpha, beta, and gamma radiation in terms of average energy and penetrating ability.
3. Distinguish between ionization counters and scintillation counters.

Review

Unstable nuclei emit photons and energetic particles and are said to be radioactive. When the emissions of naturally occurring unstable nuclei are passed through an electric field they separate into three distinct streams: alpha particles are deflected toward the negative plate, beta particles are deflected toward the positive plate, and gamma rays pass through unaffected. These types of radiation are described below:

Emission	Kind and Symbol	Charge	Penetrating Ability
alpha particle	helium nucleus $^{4}_{2}He$ or α	+2	Stopped by clothing, and by skin
beta particle	high-energy electron $^{0}_{-1}e$ or β	-1	Penetrates several millimeters into flesh
gamma rays	high-energy photons $^{0}_{0}\gamma$	0	Passes through human body and through several feet of concrete

Measuring Devices

Radiation causes ionization of the medium it is penetrating. Hence, an ionization counter detects the presence of ionizing radiation. Each emanation causes a surge in current which triggers a meter or causes audible sounds (Geiger-Müller counter).

A scintillation counter records the presence of a particle or photon by a flash from a phosphor. Because the intensity of the flash is in proportion to the energy of the absorbed emanation, a scintillation counter determines the energies of the emanations, which identify the radioisotope.

Equations for Nuclear Reactions (Section 22.2)

Learning Objectives

1. Describe the atomic and mass number changes that accompany the emission of alpha particles, beta particles, and positrons.
2. Write balanced equations for nuclear reactions involving alpha particles, beta particles, and positrons.

Review

In a <u>nuclear reaction</u> the nucleus changes. The nuclear equation shows all of the nuclei and particles involved in the reaction. Two important rules must be obeyed in writing balanced nuclear equations:

1. The sum of the mass numbers is constant.
2. The sum of the atomic numbers is constant.

Some examples of nuclear equations are given below.

(a) <u>Alpha emission</u>

Radium-226 decays by releasing alpha particles to form radon-222. The equation is

$${}^{226}_{88}\text{Ra} \rightarrow {}^{4}_{2}\text{He} + {}^{222}_{86}\text{Rn}$$

On each side of the equation the sum of the atomic numbers is 88 and the sum of the mass numbers is 226. Emission of an alpha particle reduces the atomic number of the radioactive nucleus by two units and the mass number by four units.

(b) <u>Beta emission</u>

Krypton-87 releases beta particles to form rubidium-86. The equation is

$${}^{87}_{36}\text{Kr} \rightarrow {}^{0}_{-1}\text{e} + {}^{87}_{37}\text{Rb}$$

On each side of the equation the sum of the atomic numbers is 87 and the sum of the mass numbers is 36. Emission of a beta particle has the effect of increasing the atomic number of the nucleus by one unit while leaving the mass number unchanged.

(c) <u>Positron emission</u>

A positron has the same mass of an electron but has the charge of a proton. Its symbol is

$${}^{0}_{1}\text{e}$$

Phosphorus-30 emits a positron to form silicon-30. The equation is

$${}^{30}_{15}\text{P} \rightarrow {}^{0}_{1}\text{e} + {}^{30}_{14}\text{Si}$$

The sum of the atomic numbers is 15 and the sum of the mass numbers is 30. Positron emission reduces the atomic number of the nucleus by one unit while leaving the mass number unchanged.

Practice Drill on Balancing Nuclear Equations: Fill in the blanks for nuclear equations (a)–(c) given below. Use a periodic table when necessary.

(a) ${}^{\_}_{\_}\text{Th} \rightarrow {}^{0}_{-1}\text{e} + {}^{234}_{\_}\text{Pa}$

(b) ${}^{230}_{\_}\text{Th} \rightarrow {}^{\_}_{\_}\text{He} + {}^{\_}_{\_}\_$

(c) ${}^{\_}_{11}\_ \rightarrow {}^{22}_{\_}\_ + {}^{\_}_{1}\text{e}$

<u>Answers</u>:

(a) ${}^{234}_{90}\text{Th} \rightarrow {}^{0}_{-1}\text{e} + {}^{234}_{91}\text{Pa}$ (b) ${}^{230}_{90}\text{Th} \rightarrow {}^{4}_{2}\text{He} + {}^{226}_{88}\text{Ra}$ (c) ${}^{22}_{11}\text{Na} \rightarrow {}^{22}_{10}\text{Ne} + {}^{0}_{1}\text{e}$

Nuclear Bombardment Reaction (Section 22.3)

Learning Objectives

1. Write equations for nuclear bombardment reactions.
2.* Describe and diagram the operation of a cyclotron and linear accelerator.

Review

<u>Transmutation</u> is the changing of one element into another. It can occur by bombarding certain nuclei with streams of energetic particles. The transuranium elements, the elements that follow uranium in the periodic table, can only be artificially produced by nuclear bombardment. Some lighter elements are also prepared from nuclear bombardment. An example of transmutation is the bombardment of bismuth-209 with alpha particles to produce astatine-211:

$$^{209}_{83}\text{Bi} \rightarrow {}^{4}_{2}\text{He} + {}^{211}_{85}\text{At} + 2{}^{1}_{0}\text{n}$$

The above reaction can be abbreviated as

$$^{209}\text{Bi} \ (\alpha, 2\text{n}) \ ^{211}\text{At}$$

original bombarding particle nucleus
nucleus particle emitted formed

 Practice: Write nuclear reactions for
 (a) $^{6}\text{Li}(\text{n}, \alpha)\,^{3}\text{H}$ (c) $^{107}\text{Ag}(\text{n}, 2\text{n})\,^{106}\text{Ag}$

 (b) $^{12}\text{C}(\alpha)\,^{16}\text{O}$

 <u>Answers</u>:
 (a) $^{6}_{3}\text{Li} \rightarrow {}^{1}_{0}\text{n} + {}^{3}_{1}\text{H} + {}^{4}_{2}\text{He}$ (c) $^{107}_{47}\text{Ag} + {}^{1}_{0}\text{n} \rightarrow {}^{106}_{47}\text{Ag} + 2{}^{1}_{0}\text{n}$

 (b) $^{12}_{6}\text{C} + {}^{4}_{2}\text{He} \rightarrow {}^{16}_{8}\text{O}$

Activity and Half-Life (Section 22.4)

Learning Activities

1. Use first-order kinetics in calculations relating half-life, elapsed time, and the remaining fraction of radioactive atoms.
2. Calculate the age of a carbon-containing artifact from its carbon-14 beta activity.
3. Calculate the age of an object from the isotope ratio of a radionuclide and its decay product.
4.* Use RBE factors to calculate the biologically effective radiation dose in rems from the actual dose in rads, and vice-versa.

Review

The rate of nuclear decay is the number of nuclei disintegrating per unit time and is called the <u>activity</u> of a radioactive sample. The <u>becquerel</u> is the SI unit for activity and is equal to one disintegration per second (dps). A commonly used non-SI unit is the curie (Ci):

1 Bq = 1 dps
1 Ci = 3.7×10^7 Bq

The rate of radioactive decay is first order, so the activity of a sample is proportional to the number of radionuclides present:

activity = kN   N = number of radionuclides
k = rate constant

The fraction of radionuclides remaining after passage of time t is given by

$$\ln \frac{N}{N_o} = -kt$$ N_o = number of radionuclides at t = 0

The half-life of a radionuclide (the time it takes for half of the sample to decay) is related to the rate constant by

$k \times t_{1/2} = 0.693$

Note: The above equations come from first-order kinetics. (See 14.4 for a review)

Example 22.1: Cesium-137 is a waste product of nuclear reactors. Calculate the fraction of the radioisotope remaining after the passage of 100 y. The half-life of cesium-137 is 30.0 y.

Given: $t_{1/2} = 30.0$ y
t = 100 y

Unknown: N/N_o

Plan:
1. Use the given half-life to find the rate constant, k.
2. Substitute the values for k and t into the first-order rate equationand solve for N/N_o.

Calculations:
1. $$k = \frac{0.693}{t_{1/2}} = \frac{0.693}{30.0\ y} = 0.0231\ y^{-1}$$

2. $$\ln \frac{N}{N_o} = -0.0231\ y^{-1} \times 100\ y = -2.31$$

$$\frac{N}{N_o} = e^{-2.31} = 0.099$$

The fraction remaining is 0.099 or 9.9%

Practice: Use the half-life of cesium-137 to calculate the amount of time it would take for only 1.00% of the radioisotope to remain. Answer: 199 y

Since the mass of a given substance is proportional to the number of atoms present, the ratio by atom count, N/N_o, can be replaced by the ratio by mass of the radioactive sample at t and t_o.

Example 22.2: Indium-113 is used in nuclear medicine. It has a half-life of 102 min. If 20.0 mg of In-113 are used how much will remain after 1.00 h?

Answer: Follow the steps given in Example 22.1 above to find N/N_o.

$$k = \frac{0.693}{102 \text{ min}} = 6.79 \times 10^{-3}\text{min}^{-1}$$

$$\ln \frac{N}{N_o} = -6.79 \times 10^{-3}\text{min}^{-1} \times 60.0 \text{ min} = -0.407$$

$$\frac{N}{N_o} = 0.666$$

Substitute $N_o = 20.0$ mg and calculate N; $N = 20.0 \text{ mg} \times 0.666 = \underline{13.3 \text{ mg}}$

The activity of a sample is proportional to the number of radioactive nuclei present. Hence, the ratio of activities at two different times equals N/N_o. See Example 22.3 and 22.4 given below.

Radioisotope Dating

By measuring the activity of naturally occurring radioisotopes the age of radioactive samples can often be determined. Solving the nuclear decay equation for t gives

$$\ln \frac{N}{N_o} = -kt$$

$$t = \frac{-\ln(N/N_o)}{k}$$

Substituting $k = 0.693/t_{1/2}$ and simplifying gives

$$t = \frac{-\ln(N/N_o)}{0.693} \times t_{1/2}$$

The above equation can be used to calculate the unknown age of a radioactive sample. For example, the concentration of carbon-14, a beta emitter remains fairly constant in living tissue. Its activity is about 15.3 dps. After the tissue dies the carbon-14 will decay according to the first-order equation given above. Hence, by measuring the current activity of the dead tissue its approximate age can be calculated.

Example 22.3: Radiocarbon dating is used to determine the age of one gram of an unknown bone sample. In living tissue each gram has a constant beta ray activity of 15.3 disintegrations per minute. The bone sample has a decay rate of 1.75 dps per gram. How old is the bone? Use 5730 y for the half-life of carbon-14.

Given: $N = 1.75$ dps
$N_o = 15.3$ dps
$t_{1/2} = 5730$ y

Unknown: t

Plan: Substitute the given information into the decay equation.

Calculation: $$t = \frac{-\ln(N/N_o)}{0.693} \times t_{1/2} = \frac{-\ln(1.75/15.3)}{0.693} \times 5730 \text{ y} = \underline{1.79 \times 10^4 \text{ y}}$$

Other radioisotopes are used for dating.

Uranium-238, for example, occurs naturally in some rocks and decays in a series of steps to nonradioactive lead-206. Hence the $^{206}Pb/^{238}U$ ratio can be used to estimate the age of some rocks.

Example 22.4: A geologist measures the $^{206}Pb/^{238}U$ ratio in a rock found in the Grand Canyon to be 1/1.50. Find the age of the rock. The half-life of ^{238}U is 4.46×10^9 y.

Answer: The 1 to 1.50 ratio means that 100 atoms of ^{206}Pb are present for every 150 atoms of ^{238}U. Hence, there were originally 250 atoms of $^{238}U(N_o)$ and currently 150 atoms of $^{238}U(N)$. The N/N_o ratio is therefore 150/250 or 0.600.

Substituting this value and the given half-life into the equation for calculating age gives

$$t = \frac{-\ln(N/N_o)}{0.693} \times t_{1/2} = \frac{-\ln(0.600)}{0.693} \times 4.46 \times 10^9 \text{ y} = \underline{3.29 \times 10^9 \text{ y}}$$

Nuclear Stability (Section 22.5)

Learning Objectives

1. Describe the role of magic numbers in accounting for the stability of a nucleus.
2. Describe the source of gamma rays in nuclear decay reactions.
3. Use the stability band and neutron/proton ratios to predict possible decay modes for a radionuclide.

Review

The nuclear force holds nucleons (protons and neutrons) together. Although this force is thought to be the most powerful in nature, the number of protons in the known stable nuclei does not exceed 83.

Nuclear Energy Levels

Nucleons, like electrons, pair in discrete energy levels. Hence, stable nuclei tend to have an even number of neutrons, or protons, or both. These magic numbers are 2, 8, 20, 28, 50, or 82 for either protons or neutrons (126 is "magic" for neutrons only).

Gamma photons are emitted when a nucleus formed from a nuclear reaction is in an excited state and then makes its way down to its ground state. The transitions are analogous to electronic transitions in that only photons with discrete energies and frequencies are released.

The Stability Band

When the number of neutrons in stable nuclei is plotted against the number of protons, a narrow band called the stability band is formed. (See Figure 22.15 in your text) It shows that the neutron/proton ratio is about 1 for stable light nuclei and about 1.5 for stable heavy nuclei.

The n/p ratio for radionuclides falls outside of the stability band. These nuclides therefore decay until they form stable nuclides. The mode of decay that a radionuclide favors depends on whether it lies above or below the stability band.

<u>If the radionuclide lies below the stability band (n/p ratio is too low), the mode of decay will raise the n/p ratio.</u>

1. Alpha emission raises the n/p ratio:

$$^{210}_{84}\text{Po} \rightarrow \,^{4}_{2}\text{He} + \,^{206}_{82}\text{Pb}$$

$$\text{n/p} = \frac{210-84}{84} = \underline{1.50} \qquad \text{n/p} = \frac{206-82}{82} = \underline{1.51}$$

2. Positron emission diminishes the number of protons by one while increasing the number of neutrons thus raising the n/p ratio:

$$^{78}_{35}\text{Br} \rightarrow \,^{78}_{34}\text{Se} + \,^{0}_{1}\text{e}$$

$$\text{n/p} = \frac{78-35}{35} = \underline{1.23} \qquad \text{n/p} = \frac{78-34}{34} = \underline{1.29}$$

3. Electron capture is the combining of a 1s extranuclear electron with a proton to form a neutron. Hence, the number of protons is decreased by one while the number of neutrons is increased by one raising the n/p ratio:

$$^{7}_{4}\text{Be} + \,^{0}_{-1}\text{e} \rightarrow \,^{7}_{3}\text{Li}$$

$$\text{n/p} = \frac{7-4}{4} = \underline{0.75} \qquad \text{n/p} = \frac{7-3}{3} = \underline{1.33}$$

4. Proton emission raises the n/p ratio by lowering the number of protons by one:

$$^{43}_{21}\text{Sc} \rightarrow \,^{42}_{20}\text{Ca} + \,^{1}_{1}\text{H}$$

$$\text{n/p} = \frac{43-21}{21} = \underline{1.05} \qquad \text{n/p} = \frac{42-20}{20} = \underline{1.10}$$

If the radionuclide lies above the stability band (n/p ratio is too high) the mode of decay will lower the n/p ratio.

1. Beta emission has the effect of converting a neutron into a proton and an electron. Thus the number of protons is increased by one while the number of neutrons is decreased by one and the n/p ratio is lowered. Beta emission is the most favored mode of decay for radionuclides with high n/p ratios.

$$^{210}_{83}\text{Bi} \rightarrow \,^{210}_{84}\text{Po} + \,^{0}_{-1}\text{e}$$

$$\text{n/p} = \frac{210-83}{83} = \underline{1.53} \qquad \text{n/p} = \frac{210-84}{84} = \underline{1.50}$$

2. Neutron emission is rare and tends to be observed in heavy artificially produced radionuclides.

$$^{87}_{36}\text{Kr} \rightarrow \,^{86}_{36}\text{Kr} + \,^{1}_{0}\text{n}$$

$$\text{n/p} = \frac{87-36}{36} = \underline{1.42} \qquad \text{n/p} = \frac{86-36}{36} = \underline{1.39}$$

Practice Drill–Decay Modes of Radionuclides: Write a nuclear equation for a possible decay mode for each nuclide given below.

(a) Thorium-234 lies above the stability band.
(b) Radium-226 lies below the stability band.
(c) Carbon-14 lies above the stability band.

Answers:
(a) $^{234}_{90}Th \rightarrow ^{234}_{91}Pa + ^{0}_{-1}e$ (b) $^{236}_{88}Ra \rightarrow ^{222}_{86}Rn + ^{4}_{2}He$ (c) $^{14}_{6}C \rightarrow ^{14}_{7}N + ^{0}_{-1}e$

Mass-Energy Conversions (Section 22.6)

Learning Objectives

1. Use nuclear masses to calculate the energy released during a nuclear reaction.
2. Use nuclear masses to calculate the mass defect, the binding energy, and the binding energy per nucleon of any nucleus.
3. Sketch the binding energy curve and describe how the binding energy per nucleon varies with mass number.
4. Explain why both fission and fusion reactions release energy.

Review

Mass is always lost during a nuclear decay reaction, that is, the sum of the masses of products is always less than the sum of the masses of reactants. This is because a small quantity of mass is converted into kinetic or radiant energy. For example, polonium-212 emits an alpha particle according to the equation

$$^{212}_{84}Po \rightarrow ^{4}_{2}He + ^{208}_{82}Pb$$

Atomic masses: 211.9889 u $\underbrace{4.0015\ u \quad 207.9766\ u}_{211.9781\ u}$

Hence, 211.9889 u – 211.9781 u or 0.0108 u is lost for each decay of a polonium-212 nucleus. This mass is converted into energy the magnitude of which can be calculated from the Einstein mass-energy equation

$E = mc^2$

E = energy equivalent of mass difference (J)
m = mass lost or mass difference (kg)
c = speed of light = 2.9979×10^8 m/s

The energy released in the decay of a polonium-212 nucleus is

$$E = \left[0.0108\ u \times \frac{1\ g}{6.0221 \times 10^{23}\ u} \times \frac{1\ kg}{10^3\ g}\right] \times (2.9979 \times 10^8\ m/s)^2 = \underline{1.61 \times 10^{-12}\ J}$$

Because the energy equivalent is small in a nuclear reaction, the electron volt and megaelectron volt are often used:

1 u = 931.47 MeV

This relationship can be used to convert mass lost into megaelectron volts directly.

$$0.0108\ u \times \frac{931.5\ MeV}{1\ u} = \underline{10.1\ MeV}$$

Practice: Use $E = mc^2$ to calculate the energy equivalent of 1 u in joules.
Answer: 1.492×10^{-10} J/u

Binding Energy

The binding energy is the energy required to separate a nucleus into its component protons and neutrons. The binding energy of a nucleus can be calculated by finding the mass defect, the difference in mass between the sum of the masses of nucleons and the nucleus, and substituting the value into $E = mc^2$. (The relationship 1 u = 931.5 MeV can also be used.)

Example 22.5: Calculate the binding energy of the carbon-12 nucleus in megaelectron volts. Masses are: C = 11.9967 u; proton = 1.00728 u; neutron = 1.00866 u.

Answer:

$$\text{mass defect} = \left[6\ \text{protons} \times \frac{1.00728\ \text{u}}{1\ \text{proton}}\right] + \left[6\ \text{neutrons} \times \frac{1.00866\ \text{u}}{1\ \text{neutron}}\right] - 11.9967\ \text{u} = 0.0989\ \text{u}$$

$$\text{binding energy} = 0.0989\ \text{u} \times \frac{931.5\ \text{MeV}}{1\ \text{u}} = \underline{92.1\ \text{MeV}}$$

The binding energy per nucleon is often used to compare the stability of different nuclei. Its value for carbon-12 is

$$\frac{92.1\ \text{MeV}}{12\ \text{nucleons}} = \underline{7.68\ \text{MeV/nucleon}}$$

Generally, the greater the binding energy per nucleon the more stable the nucleus. When the binding energies per nucleon are plotted versus mass number (See Figure 22.16 in your text) for stable nuclei several important trends are observed:

1. For the light elements there is an increase of binding energy per nucleon with increasing mass number peaking around iron-56.
2. For heavy elements there is a decrease of binding energy per nucleon with increasing mass number.
3. The increase of binding energy per nucleon occurs more rapidly for light elements than the decrease of binding energy per nucleon for heavy elements on the curve.

These trends tell us the following:

Energy will be released when very heavy nuclei split into lighter nuclei. This is called a fission reaction.

Energy will be released when very light nuclei combine to form heavier nuclei. This is a fusion reaction.

Fusion reactions will generate more energy than fission reactions. (See item 3 above.)

Fission (Section 22.7)

Learning Objectives

1. List some fissionable nuclei and write typical fission equations.
2. Explain how a fission reaction becomes self-sustaining.
3. Compare the advantages and disadvantages of ordinary fission reactors and breeder reactors.
4. Diagram a reactor core and state the function of the fuel rods, the control rods, and the moderator.

Review

The fission of a heavy nucleus into two or more lighter nuclei releases energy. The fission of uranium-235 and plutonium-239 is used to produce nuclear energy.

Uranium-235 must first absorb a neutron to form uranium-236, which then undergoes fission reactions. A few of these nuclear reactions are

$$^{235}_{92}U + ^{1}_{0}n \rightarrow ^{236}_{92}U \rightarrow \begin{cases} ^{141}_{56}Ba + ^{92}_{36}Kr + 3^{1}_{0}n \\ ^{144}_{54}Xe + ^{90}_{38}Sr + 2^{1}_{0}n \\ ^{144}_{55}Cs + ^{90}_{37}Rb + 2^{1}_{0}n \\ \text{other nuclides} \end{cases}$$

Note that each fission event produces more neutrons than it consumes, which can then trigger further fission reactions. A nuclear chain reaction occurs when the fission reactions are self-sustaining. One requirement is that the critical mass of fissionable material must be present. The critical mass is the minimum mass needed to sustain a chain reaction.

When the chain reaction is not controlled it will produce an enormous amount of energy. Nuclear energy is generally measured in units of megatons, the energy equivalent of 10^6 tons of TNT.

Nuclear Reactors

Nuclear reactors control the amount of energy released from fission reactions.

Advantages of Nuclear Reactors as a Source of Energy
- produces much more energy than coal or oil
- does not emit the pollutants to the atmosphere (SO_2, NO_2, etc.) that cause acid rain and other environmental problems

Disadvantages of Nuclear Reactors as a Source of Energy
- U_3O_8 pellets must be enriched in uranium-235
- spent fuel must be reprocessed
- radioactive wastes must be safely stored
- a malfunction may release harmful radionuclides into the atmosphere

Breeder Reactors

The supply of uranium-235 is limited. A breeder reactor produces nuclear fuel instead of just consuming it. The reactions are

$$^{238}_{92}U + ^{1}_{0}n \rightarrow ^{239}_{92}U \rightarrow ^{239}_{93}Np + ^{0}_{-1}e$$

$$\uparrow \text{ (From } ^{235}U \text{ fission)} \qquad ^{239}_{93}Np \downarrow \; \boxed{^{239}_{94}Pu} + ^{0}_{-1}e$$

Uranium-238 is abundant and is used to produce plutonium-239, a fissionable material. Plutonium-239 however, is a very potent carcinogen and is also used in nuclear weapons so that it must be well guarded. Consequently, breeder reactors are not commercially desirable sources of energy.

How a Nuclear Reactor Works

See Figure 22.21 in your text. The major components are

Fuel rods: made of U_3O_8 enriched in uranium-235

Control rods: made of cadmium or boron, materials that absorb neutrons to control the rate of fission

Moderator: made of graphite, light water (H_2O) or heavy water (D_2O), substances that slow neutrons to energies that promote fission. (Most reactors in the U.S. use light water as a moderator and a coolant, graphite having the disadvantage of being combustible.)

Coolant: light or heavy water, also liquid sodium is used (the water becomes steam and is used to drive turbines that generate electricity).

Fusion (Section 22.8)

Learning Objectives

1. Write the equation for the deuterium-tritium fusion reaction.
2. Describe the potential advantages of fusion power and the difficulties associated with its development.

Review

Fusion or thermonuclear reactions require extremely high temperatures to occur. The nuclei must have large kinetic energies upon collision so that they can get close enough for the nuclear force to take hold.

The deuterium-tritium fusion reaction

$$^{2}_{1}H + ^{3}_{1}H \rightarrow ^{4}_{2}He + ^{1}_{0}n + 17.6\ MeV$$

is presently being investigated for use in a commercial fusion reactor.

Advantages of fusion reactors over fission reactors include
- ten times more energy is released per gram of deuterium compared to energy per gram of uranium-235
- deuterium can be obtained from seawater, providing an essentially inexhaustible supply
- no long-lived radioisotopes are produced
- the only radioisotope that could leak into the environment is the relatively harmless tritium

The major obstacle to the development of a deuterium-tritium reactor is maintaining the extraordinary temperatures required for fusion, temperatures on the order of 10^6 K.

SELF-TEST

1. Write a balanced nuclear equation for
   (a) the alpha decay of radon-220
   (b) fission of uranium-236 to form samarium-160, zinc-72, and neutrons
   (c) $^{239}Pu\ (\alpha, n)\ ^{242}Cm$
   (d) fusion of two protons to form deuterium and a positron

2. Which of the following nuclear reactions are unlikely to occur spontaneously? (Carbon-12, sodium-23, and iodine-127 are the most stable isotopes of the reactant elements.)

   (a) $^{11}_{6}C \rightarrow ^{11}_{7}N + ^{0}_{-1}e$
   (b) $^{22}_{11}Na + ^{0}_{-1}e \rightarrow ^{22}_{10}Ne$
   (c) $^{3}_{1}H \rightarrow ^{3}_{2}He + ^{0}_{-1}e$
   (d) $^{129}_{53}I \rightarrow ^{129}_{52}Te + ^{0}_{1}e$

3. A nuclear reactor in a few weeks time accumulates an amount of strontium-90 equivalent to that produced by the detonation of the atomic bomb over Hiroshima. What fraction of the strontium-90 activity remains after the passage of 25 y? The half-life of the radioisotope is 29 y.

4. Calculate the minimum energy in megaelectron volts needed to separate a tritium nucleus into its component nucleons.
neutron = 1.00866 u
proton = 1.00728 u
hydrogen-3 = 3.01550 u
and 1 u = 931.5 MeV.

5. How many kilograms of coal would have to be burned to produce the same energy as a gram of deuterium undergoing the fusion reaction?
$^{2}_{1}H + ^{3}_{1}H \rightarrow ^{4}_{2}He + ^{1}_{0}n$
Every gram of coal releases about 30 kJ of energy. The relevant masses are

$^{2}_{1}H = 2.0135$ $^{4}_{2}He = 4.00150$ u

$^{3}_{1}H = 3.01550$ $^{1}_{0}n = 1.00866$ u

and 1 u = 1.4924×10^{-10} J

6. Use the belt of stability to explain why neutrons are often emitted when heavy nuclei undergo fission reactions.

7. The carbon-14 activity of heartwood taken from a giant sequoia tree is 10.8 dps per gram while the wood from the outer portion of the tree is 15.3 dps per gram. Estimate the age of the tree. (Hint: Heartwood is laid down at the beginning of the life of a tree and is no longer living.) Use 5730 y for the half-life of carbon-14.

8. (a) Polonium-213 decays through an intermediate atom to form the stable isotope bismuth-209. Write the equations for the decay of polonium-213 to bismuth-209.
(b) The natural decay series of uranium-238 ends with the stable isotope lead-206. How many alpha and beta particles are emitted in total?

Underline the correct word or words for 9–12.

9. Boron is frequently used in nuclear reactors to (absorb neutrons, slow down neutrons).

10. A breeder reactor produces the fissionable isotope (uranium-235, plutonium-239).

11. Long-lived radioisotopes are produced in (fusion, fission) reactors.

12. A (scintillation, Geiger) counter not only detects radioactivity but can be used to identify the radioisotope.

13. What would be the argon-40/potassium-40 ratio in a sedimentary rock layer containing a one-billion-year old skull? The half-life of potassium-40 is 1.25×10^{9} y.

ANSWERS TO SELF-TEST QUESTIONS

1. (a) $^{220}_{86}Rn \rightarrow ^{216}_{84}Po + ^{4}_{2}He$
(b) $^{236}_{92}U \rightarrow ^{160}_{62}Sm + ^{72}_{30}Zn + 4^{1}_{0}n$
(c) $^{239}_{94}Pu + ^{4}_{2}He \rightarrow ^{242}_{96}Cm + ^{1}_{0}n$
(d) $^{1}_{1}H + ^{1}_{1}H \rightarrow ^{2}_{1}H + ^{0}_{1}e$

2. (a) and (d)

3. $\ln(N/N_o) = -(0.693/29\ y) \times 25\ y = -0.60$; $N/N_o = \underline{0.55}$

4. $^{3}_{1}H + \text{energy} \rightarrow {}^{1}_{1}H + 2^{1}_{0}n$

   difference in mass = (2 × 1.00866 u) + 1.00728 u – 3.01550 u = 0.00910 u

   energy = 0.00910 × 931.5 MeV/1 u = <u>8.48 MeV</u>

5. mass defect = (4.00150 u + 1.00866 u) – (2.0135 + 3.01550) = 0.01884 u/deuterium

$$1\ g\ ^{2}_{1}H \times \frac{6.022 \times 10^{23}\ u\ ^{2}_{1}H}{1\ g\ ^{2}_{1}H} \times \frac{1\ ^{2}_{1}H}{2.0135\ u\ ^{2}_{1}H} \times \frac{0.01884\ u}{1\ ^{2}_{1}H\ \text{reacting}}$$
$$\times \frac{1.4924 \times 10^{-7}\ kJ}{1\ u} \times \frac{1\ g\ \text{coal}}{30\ kJ} \times \frac{1\ kg}{10^{3}\ g} = \underline{2.8 \times 10^{10}\ \text{kg of coal}}$$

6. n/p ratio is higher for heavier elements.

7. $$t = \frac{-\ln(10.8/15.3)}{0.693} \times 5730\ y = \underline{2.88 \times 10^{3}\ y}$$

8. (a) $^{213}_{84}Po \rightarrow {}^{209}_{82}Pb + {}^{4}_{2}He$ $\qquad$ $^{209}_{82}Pb \rightarrow {}^{209}_{83}Bi + {}^{0}_{-1}e$

   (b) The net reaction for the uranium-238 decay series is

   $^{238}_{92}U \rightarrow {}^{206}_{82}Pb + 8\ {}^{4}_{2}He + 6\,{}^{0}_{-1}e$

9. absorb neutrons

10. plutonium-239

11. fission

12. scintillation

13. $$1 \times 10^{9}\ y = \frac{-\ln(N/N_o)}{0.693} \times 1.25 \times 10^{9}\ y$$

    $N/N_o = 0.574$

$$\frac{\text{argon-40}}{\text{potassium-40}} = \frac{0.426}{0.574} = \underline{0.742}$$

CHAPTER 23
ORGANIC CHEMISTRY AND THE CHEMICALS OF LIFE

SELF-ASSESSMENT

Saturated Hydrocarbons: Alkanes (Section 23.1)

1. Write the structural formula for
   (a) 3-ethyl-2-methylheptane
   (b) 3-ethyl-3,5,6-trimethyloctane
   (c) 2,3-dimethylpentane
   (d) methylcyclopentane

2. Draw the three open chain structural isomers of C_5H_{12} (see Section 10.3) and give their IUPAC names.

Unsaturated Hydrocarbons: Alkenes and Alkynes (Section 23.2)

3. Write the structural formula for
   (a) cis-1-bromo-1-butene
   (b) trans-3-hexene
   (c) 2-methyl-3-heptyne

4. Draw *cis* and *trans* isomers (see Section 10.3) for
   (a) 2-pentene
   (b) 4-methyl-2-pentene
   (c) 1,2-dibromoethene

5. (a) How many carbon and hydrogen atoms are in

   (i) 

   (ii) 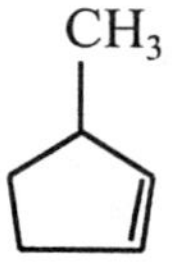

   (b) Which of the following are isomers?

   (i) 

   (ii) $CH_2{=}CH{-}CH_2{-}CH_3$

   (iii) $CH_3-\underset{\displaystyle CH_3}{\underset{|}{CH}}-CH_3$

   (iv) 

6. Write balanced equations for the reaction of 1 mol of 1-butyne with
   (a) 1 mol H_2 in the presence of Ni
   (b) excess Cl_2
   (c) excess HBr

7. Vinyl chloride, $CH_2{=}CHCl$, is used to form the addition polymer PVC (polyvinylchloride). Give the structure of the polymer.

8. The structure of polystyrene is

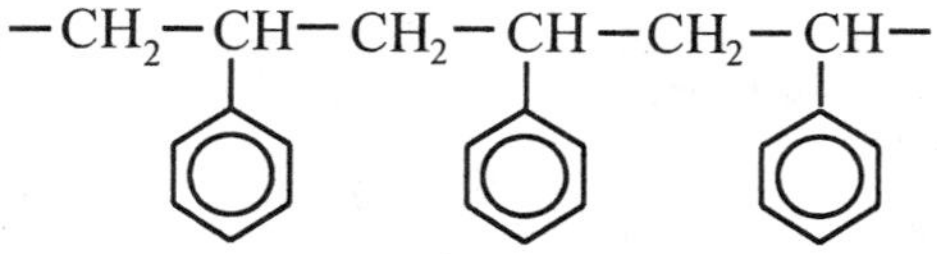

   Give the structure of the monomer.

Aromatic Compounds (Section 23.3)

9. All of the following statements are true about benzene except
   (a) It has delocalized pi electrons.
   (b) It undergoes substitution reactions.
   (c) It undergoes addition reactions like those of alkenes.
   (d) It is a planar molecule with pi electron density above and below the aromatic ring.

10. Name the isomers of bromotoluene and draw their structures.

Functional Groups (Section 23.4)

11. Indicate the functional groups in each of the following compounds:

    (a) Vinethene (an anesthetic): $CH_2{=}CH{-}O{-}CH{=}CH_2$

    (b) Lactic acid (in sour milk): $CH_3{-}\underset{\displaystyle OH}{\underset{|}{CH}}{-}\underset{\displaystyle O}{\underset{\|}{C}}{-}OH$

    (c) estrone (a female hormone):

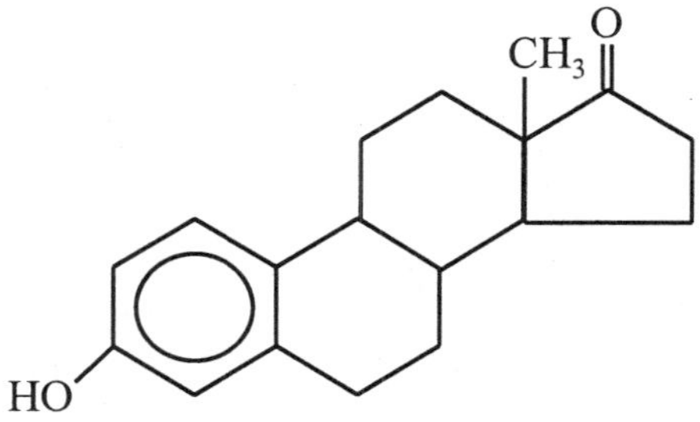

12. Fill in the formulas for the oxidized products for the following reactions:

    (a) $CH_3{-}CH_2{-}CH_2{-}CH_2{-}OH \rightarrow\ ?\ \rightarrow\ ?$

    (b) $CH_3{-}\underset{\displaystyle OH}{\underset{|}{CH}}{-}CH_3 \rightarrow\ ?$

13. True or false? If a statement is false, change the underlined word(s) to make it true.
    (a) Phenol is a weak acid.
    (b) Ethanol can be made from the fermentation of carbohydrates.
    (c) Glycerol contains two hydroxy groups.
    (d) Isopropyl alcohol is reduced to acetone.
    (e) Ketones are reducing agents.

The Ester Linkage (Section 23.5)

14. Write structural formulas for the acid and alcohol that can be used to form the following esters:

(a) 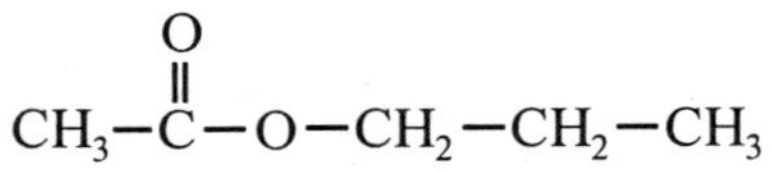 $CH_3-\overset{O}{\overset{\|}{C}}-O-CH_2-CH_2-CH_3$

(b) 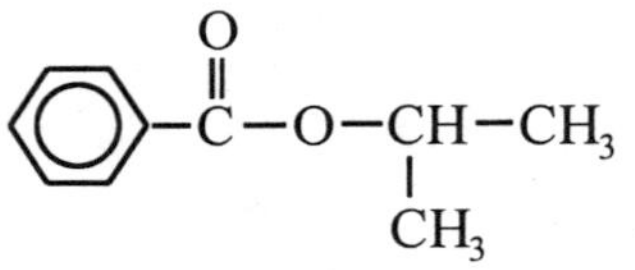

(c) 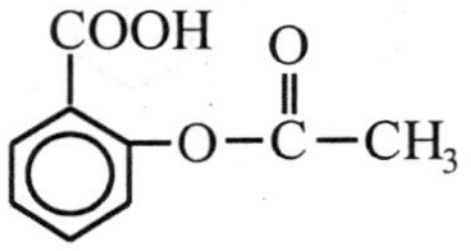

(d) isobutyl acetate

15. Upon treatment with sodium hydroxide a particular triglyceride ester yields sodium salts of stearic ($CH_3(CH_2)_{16}COOH$), myristic ($CH_3(CH_2)_{12}COOH$), and palmitic ($CH_3(CH_2)_{14}COOH$) acids. Write one possible structure for the ester. Do you think the triglyceride is a fat or an oil?

Amines (Section 23.6)

16. Label the following amines as primary, secondary, or tertiary and write their reactions with strong acids:

(a) ethylamine

(b) 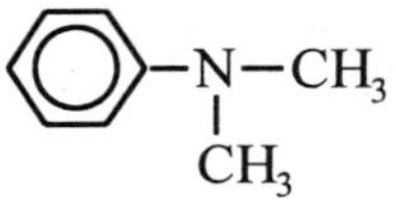

(c) 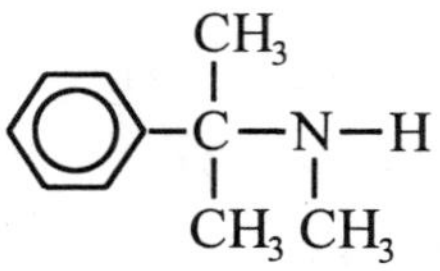

17. Star the asymmetric carbon in valine (val) 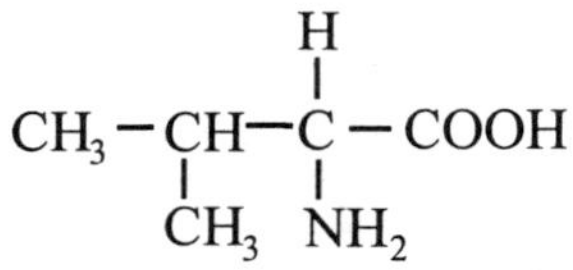 $CH_3-\underset{CH_3}{\underset{|}{CH}}-\overset{H}{\overset{|}{\underset{NH_2}{\underset{|}{C}}}}-COOH$

The Amide Linkage (Section 23.7)

18. Draw the tripeptide gly-ala-val. (R=H for glycine and R=CH_3 for alanine. See question 17 above for the structure of valine.) Indicate the peptide linkages in your structure.

19. How many other tripeptides contain the same three amino acid residues as the one in question 18? List them, using the three letter symbols.

Carbohydrates (Section 23.8)

20. All of the following statements are true except
 (a) All monosaccharides are aldehyde sugars.
 (b) A dissacharide is a condensation product of two monosaccharides.
 (c) Cellulose is a large polymer of glucose.
 (d) The glucose used by the body has the D configuration.

21. What is a glycosidic linkage and how does it form?

ANSWERS TO SELF-ASSESSMENT PROBLEMS

If you missed an answer <u>be sure</u> to study the relevant section.

1. (a) $CH_3-\overset{\overset{\displaystyle CH_3}{|}}{CH}-\overset{\overset{\displaystyle CH_2-CH_3}{|}}{CH}-CH_2-CH_2CH_2CH_3$ (c) $CH_3-\overset{\overset{\displaystyle CH_3}{|}}{CH}-\overset{\overset{\displaystyle CH_3}{|}}{CH}-CH_2-CH_3$

 (b) $CH_3-CH_2-\overset{\overset{\displaystyle CH_2CH_3}{|}}{\underset{\underset{\displaystyle CH_3}{|}}{C}}-CH_2-\overset{\overset{\displaystyle CH_3}{|}}{CH}-\overset{\overset{\displaystyle CH_3}{|}}{CH}-CH_2-CH_3$ (d)

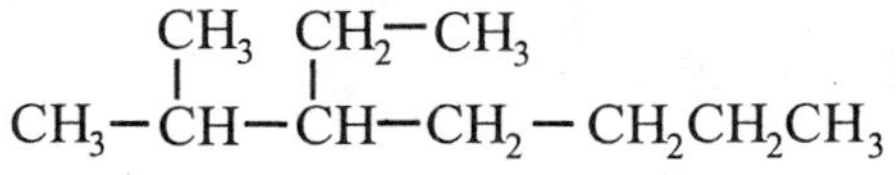

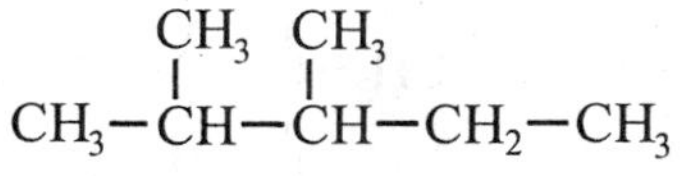

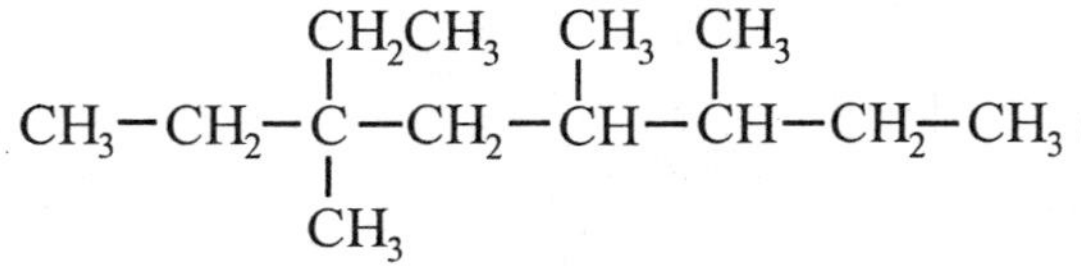

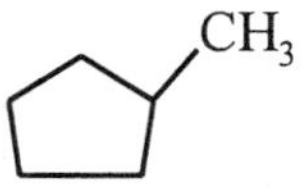

2. $CH_3–CH_2–CH_2–CH_2–CH_3$ (pentane) $CH_3-\overset{\overset{\displaystyle CH_3}{|}}{\underset{\underset{\displaystyle CH_3}{|}}{C}}-CH_3$ (2, 2-dimethylpropane)

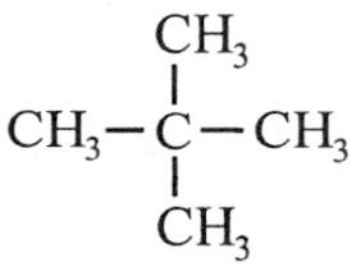

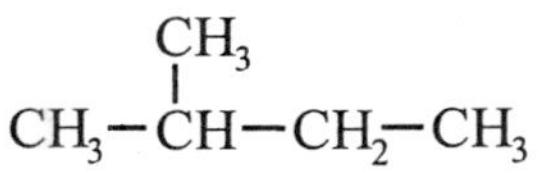

 $CH_3-\overset{\overset{\displaystyle CH_3}{|}}{CH}-CH_2-CH_3$ (2-methylbutane)

3. (a)
 cis

 (c) $CH_3-\overset{\overset{\displaystyle CH_3}{|}}{CH}-C\equiv C-CH_2-CH_2-CH_3$

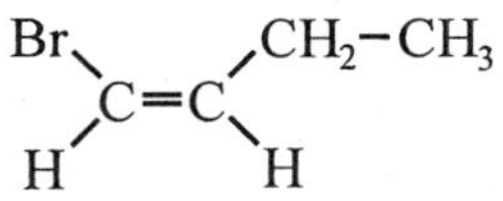

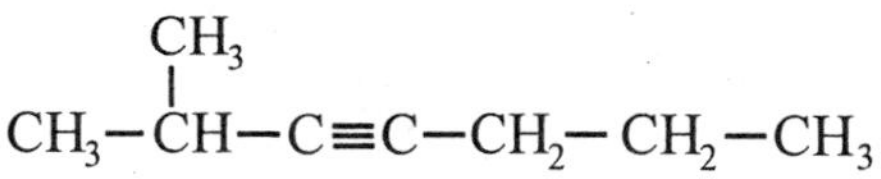

 (b)
 trans

4. (a)
 cis

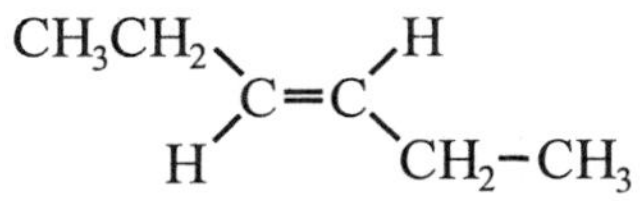

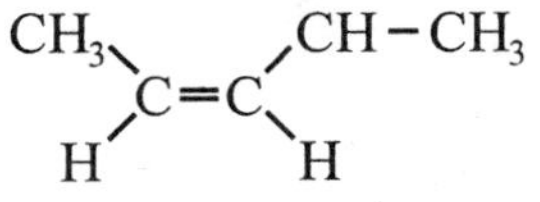

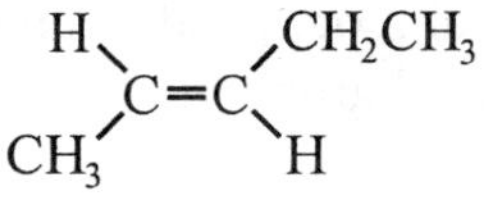

 trans

 (b)
 cis

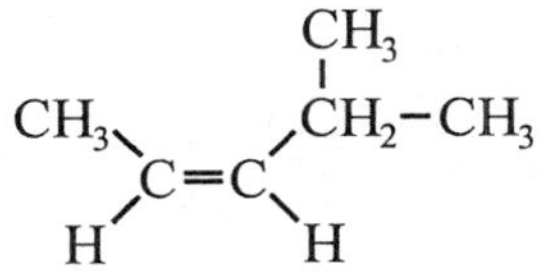

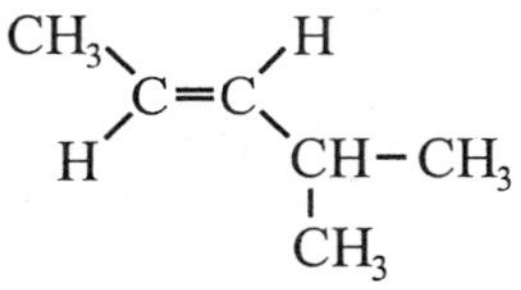

 trans

(c) 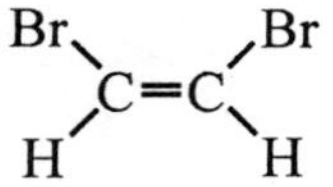

cis

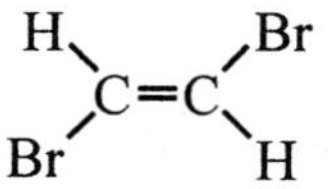

trans

5. (a) (i) 6C, 6H (ii) 6C, 10H
   (b) (i), (ii), (iv) are isomers

6. (a) $CH_3–CH_2–C≡CH + H_2 \xrightarrow{Ni} CH_3–CH_2–CH=CH_2$
   (b) $CH_3–CH_2–C≡CH + 2Cl_2 \rightarrow CH_3-CH_2-CCl_2-CCl_2-H$
   (c) $CH_3–CH_2–C≡CH + 2HBr \rightarrow CH_3-CH_2-CBr_2-CH_2-H$ (Markovnikov's Rule)

7. PVC $-CH(Cl)-CH_2-CH(Cl)-CH_2-CH(Cl)-CH_2-$ or $\left[-CH(Cl)-CH_2-\right]_n$

8. 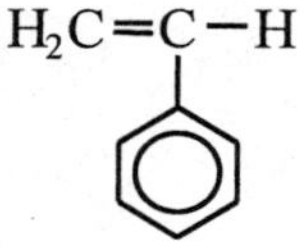

9. (c)

10. 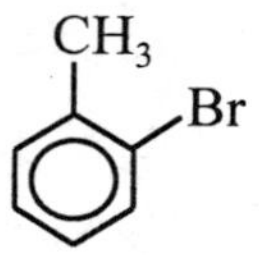

o-bromotoluene or
2-bromotoluene

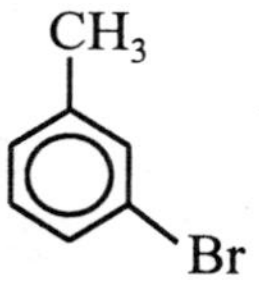

m-bromotoluene or
3-bromotoluene

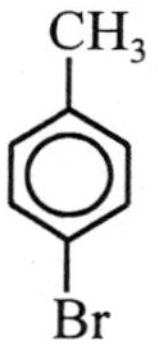

p-bromotoluene or
4-bromotoluene

11. (a) two double bonds and an ether linkage
    (b) hydroxy group (alcoholic functional group) and a carboxyl group
    (c) phenolic group and a ketone carbonyl group

12. (a) $CH_3-CH_2-CH_2-C(=O)-H$, $CH_3-CH_2-CH_2-C(=O)-OH$
    (b) $CH_3-C(=O)-CH_3$

13. (a) T
    (b) T
    (c) F, three
    (d) F; oxidized
    (e) F, Aldehydes

14. (a) acid; CH_3-COOH: alcohol; $CH_3-CH_2-CH_2-OH$

(b) acid;  : alcohol; 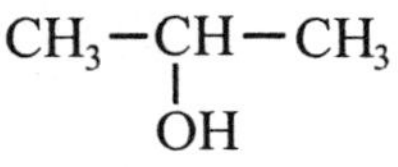

(c) acid; CH_3-COOH: a phenol; 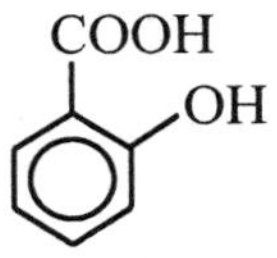

(d) acid; CH_3-COOH: alcohol; 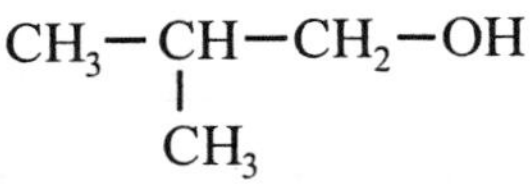

15. 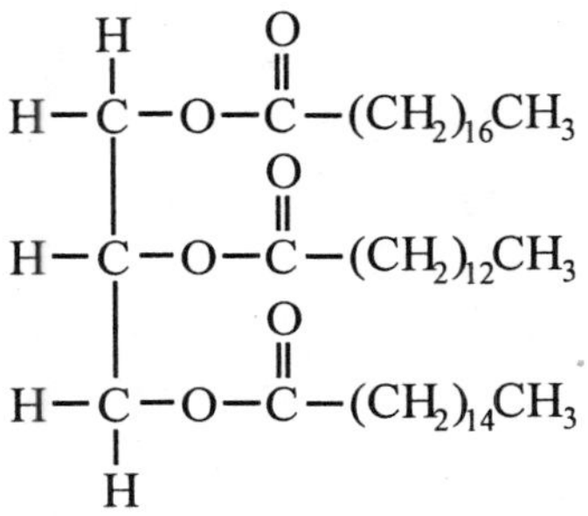 fat (saturated fatty acids)

16. (a) primary (b) tertiary (c) secondary

(a) $CH_3CH_2NH_2 + H^+ \rightarrow CH_3CH_2NH_3^+$

(b) 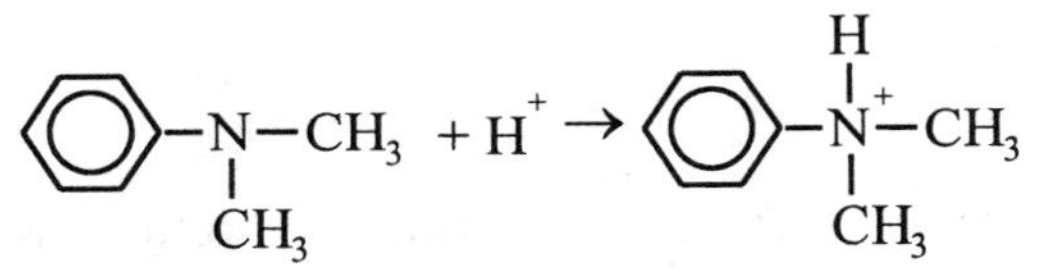

(c) 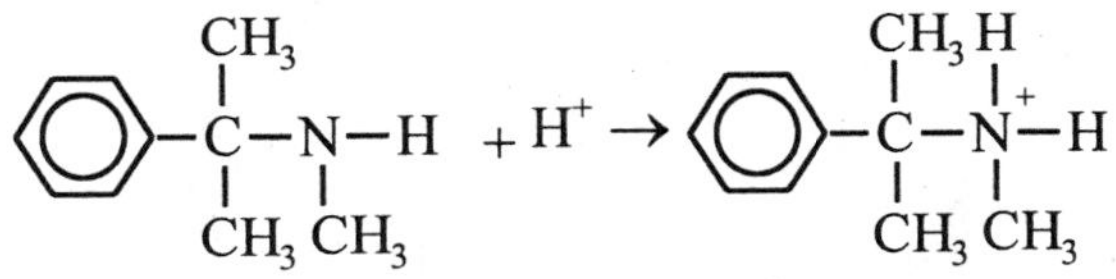

17. 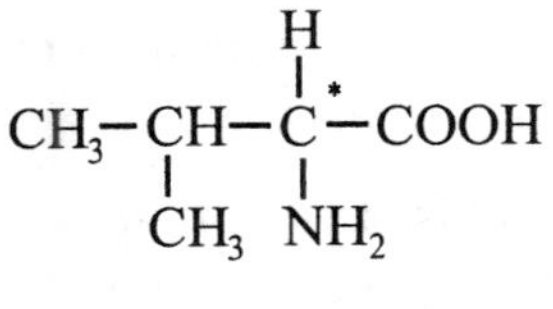

18. 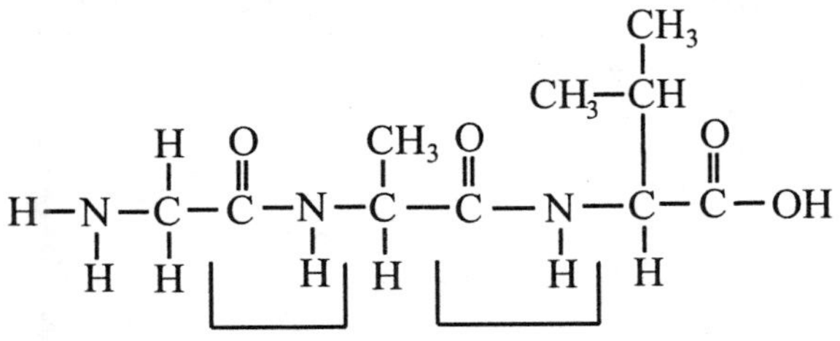

peptide linkages

19. five: ala-gly-val ala-val-gly
val-gly-ala gly-ala-val
gly-val-ala val-ala-gly

20. (a)

21. A glycosidic linkage is the oxygen bridge formed by the condensation reaction of two monosaccharides. (See Figure 23.33 in your text)

REVIEW

The large number and variety of carbon compounds is due to three factors:
(1) the stability of the carbon-carbon bond so that long chains can form,
(2) the four valence electrons of carbon so that branched chains and bonds to other elements can form,
(3) the small size of the carbon atom so that multiple bonds are possible.

Saturated Hydrocarbons: Alkanes (Section 23.1)

Learning Objectives

1. Give the names and structural formulas for straight-chain and branched-chain alkanes with one to ten carbon atoms.
2. Name alkanes, given their structural formulas; write structural formulas for alkanes, given their names.
3. Give names, structural formulas, and polygon notation for cycloalkanes.
4. Draw the boat and chair configurations of cyclohexane.
5. Draw the basic shape of a steroid molecule.
6. Write equations with structural formulas to illustrate the light-catalyzed reactions of chlorine and bromine with alkanes.

Review

The saturated hydrocarbons, or alkanes, contain only single bonds. The general formula for alkanes is C_nH_{2n+2} where n is the number of carbon atoms in the molecule. (Table 23.1 in this study guide lists the first ten members of the straight-chain alkane series and their states at room temperature.) Note that the straight-chain alkanes are called "normal" so the alkane name is preceded by n-. You should know these names.

An alkyl group is an alkane molecule missing one hydrogen atom: for example, methane CH_4 minus one hydrogen becomes the methyl group $-CH_3$. Note that an alkyl group is not a molecule by itself, but is attached to a molecule.

Table 23.1 Straight-Chain Alkanes

Formula	Name	State at Room Temperature
CH_4	methane	Gas
C_2H_6	ethane	
C_3H_8	propane	
C_4H_{10}	butane	
C_5H_{12}	pentane	Liquid
C_6H_{14}	hexane	
C_7H_{16}	heptane	
C_8H_{18}	octane	
C_9H_{20}	nonane	
$C_{10}H_{22}$	decane	

Alkane Nomenclature

Naming alkanes according to the IUPAC system involves only a few rules. Let us apply these rules to name

$$CH_3-CH-CH-CH_2-CH_3$$
$$\quad\ \ | \quad\ |$$
$$\quad\ CH_3\ \ CH_3$$

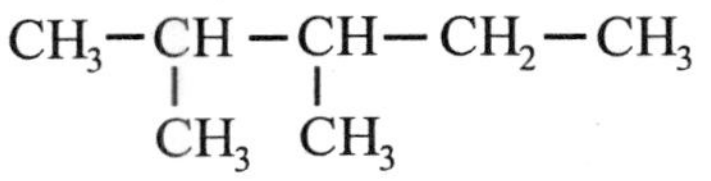

(1) Identify the longest continuous chain of carbon atoms in the molecule. Use the alkane name that pertains to this number of carbon atoms.

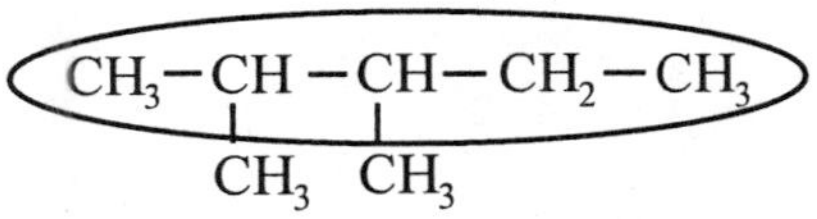

$$CH_3-CH-CH-CH_2-CH_3 \qquad pentane$$
$$\quad\ \ | \quad\ |$$
$$\quad\ CH_3\ \ CH_3$$

(2) Identify substituent groups (groups or atoms other than hydrogen) that are joined to the longest chain. Add the names of these substituents as a prefix along with di- tri- and so forth to indicate the number of each substituent.

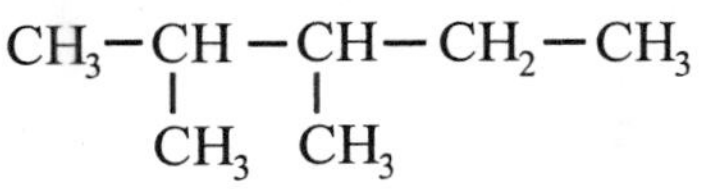

$$CH_3-CH-CH-CH_2-CH_3 \qquad dimethylpentane$$
$$\quad\ \ | \quad\ |$$
$$\quad\ CH_3\ \ CH_3$$

(3) Number the carbon atoms in the parent chain so as to give the lowest possible numbers to the carbon atoms bonded to the substituents.

$$\overset{1}{CH_3}-\overset{2}{CH}-\overset{3}{CH}-\overset{4}{CH_2}-\overset{5}{CH_3} \qquad 2,3\text{-dimethylpentane}$$
$$\quad\ \ | \quad\ |$$
$$\quad\ CH_3\ \ CH_3$$

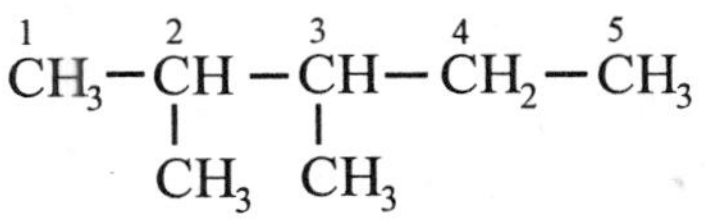

Practice Drill on Naming Alkanes: Give the names of the following alkanes following the three steps given above. Note that different substituents on the carbon chain are written in alphabetical order.

Structure	Longest Chain	Number and Name of Each Substituent	Position of Each Substituent

(a)
$$CH_3-CH_2-\underset{\underset{CH_2CH_3}{|}}{CH}-\underset{\underset{}{\overset{\overset{CH_3}{|}}{CH}}}-CH_3$$

(b)
$$CH_3-\underset{\underset{CH_2-CH_2-CH_3}{|}}{CH}-CH_2-\underset{\overset{\overset{CH_3}{|}}{CH}}{}-CH_3$$

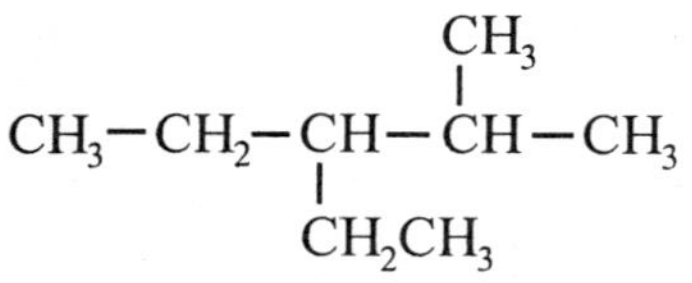

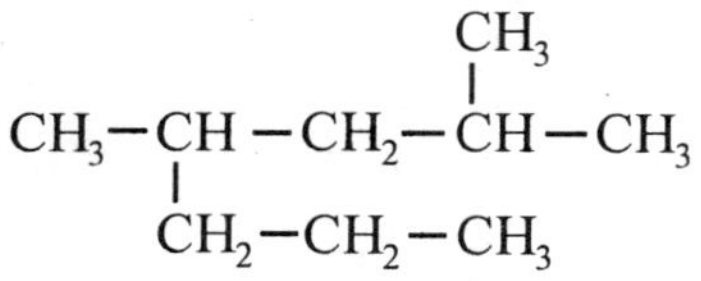

Answers: (a) 3-ethyl-2-methylpentane (b) 2,4-dimethylheptane

Cyclic Alkanes

A cycloalkane is a ring of carbon atoms. The general formula of cycloalkanes is C_nH_{2n}. They are named by preceding their alkane name with cyclo. For example

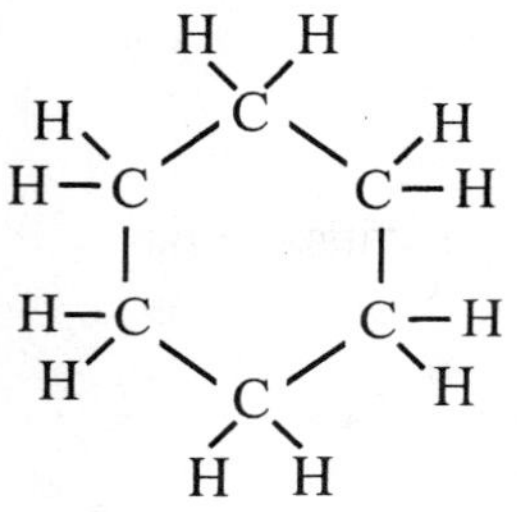

or

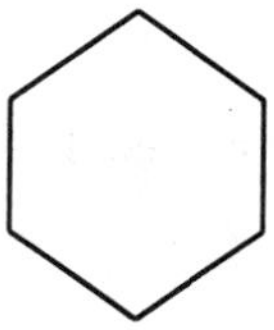

cyclohexane

The cyclohexane ring is not planar; it has boat and chair conformations. (See Figure 23.1 in your text.) In these conformations bond angles are tetrahedral. Smaller cycloalkanes such as cyclopropane are much less stable because they cannot adopt these puckered conformations with tetrahedral bond angles.

Reactions of Alkanes

Alkanes react with few substances. See Table 23.2 in this study guide for a list of some reactions of hydrocarbons.

Table 23.2 Some Important Reactions of Hydrocarbons

Class of Compounds	General Formula (Functional Group)	Some Reactions
Alkanes	C_nH_{2n+2}	Combustion

$$2C_8H_{18} + 25O_2 \xrightarrow{\text{heat}} 16CO_2 + 18H_2O$$

Halogen Substitution

$$CH_3CH_2CH_3 + Br_2 \xrightarrow{h\nu} CH_3CH_2CH_2Br + HBr$$

(Other brominated compounds are formed as well)

Class of Compounds	General Formula (Functional Group)	Some Reactions
Alkenes and Alkynes	C_nH_{2n} (C=C) C_nH_{2n-2} (C≡C)	Addition

Hydrogen (hydrogenation)

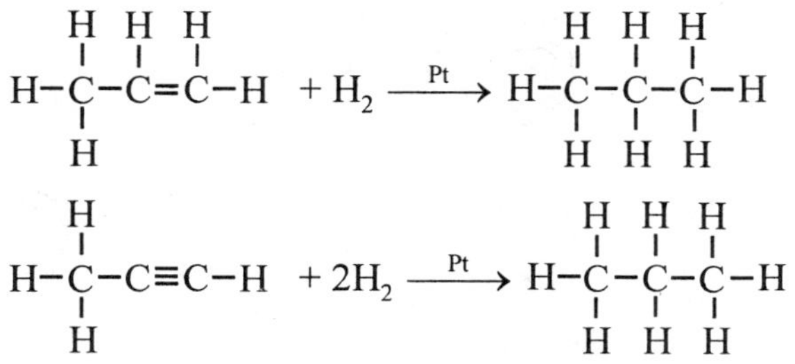

Halogen (halogenation)

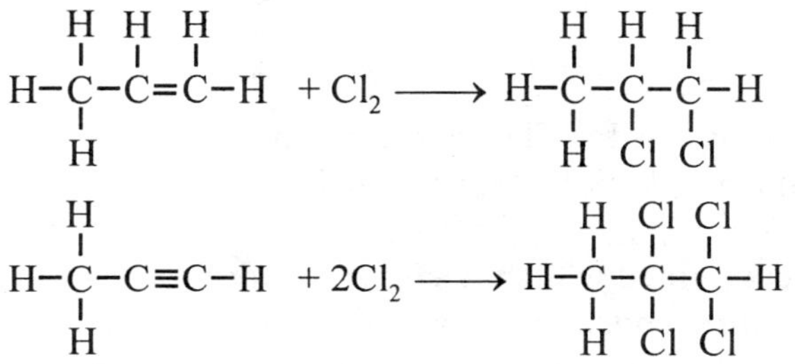

Hydrogen Halide

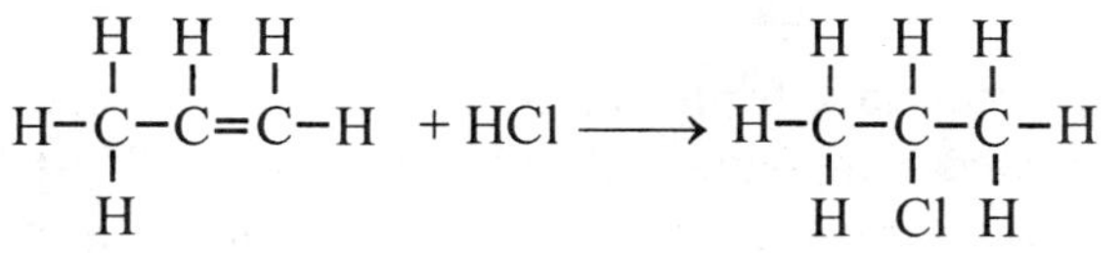

Note 2-chloropropane is formed and not 1-chloropropane. Markovnikov's rule tells us that the hydrogen will bond to the carbon that is already bonded to the most hydrogen atoms.

Class of Compounds	General Formula (Functional Group)	Some Reactions
		Polymerization
		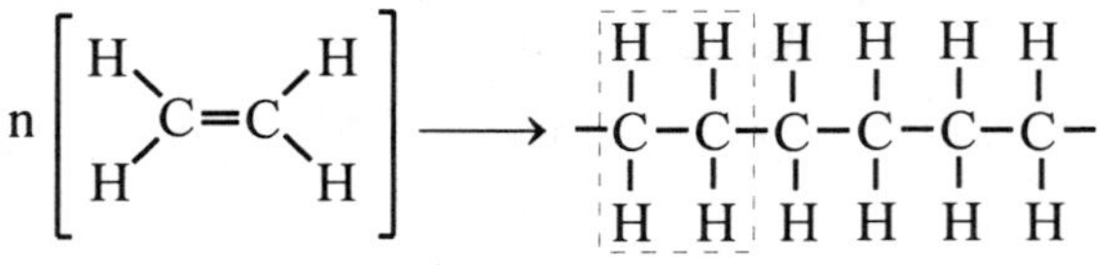 unit is repeated (polymer)
Aromatic Hydrocarbons		Substitution 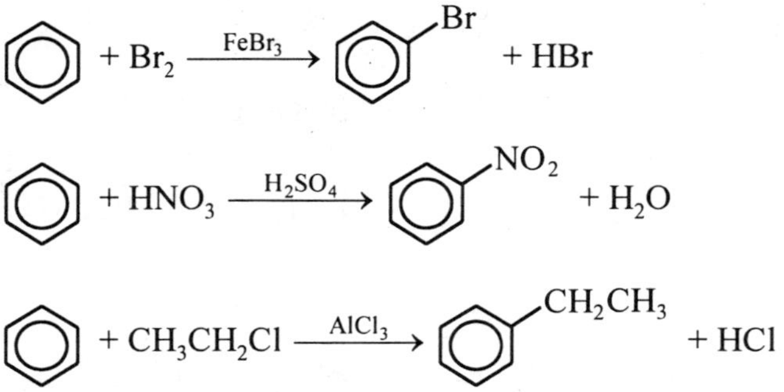

Unsaturated Hydrocarbons: Alkenes and Alkynes (Section 23.2)

Learning Objectives

1. Give names and structural formulas for alkenes and alkynes.
2. Write equations to illustrate the addition of hydrogen, halogens, and hydrogen halides to hydrocarbons with double and triple bonds.
3. Write formulas for polymers formed with ethylene and its derivatives.
4. Given the formula for an addition polymer, identify the monomer.
5. Use equations with structural formulas to show the chain mechanism of addition polymerization.

Review

Alkenes contain double bonds; alkynes have triple bonds. Any hydrocarbon that contains one or more multiple bonds is unsaturated. The simplest unsaturated compound is the alkene, ethene, $CH_2{=}CH_2$ (also called ethylene). Adding CH_2 units to ethylene generates the alkene series (C_nH_{2n}). The simplest alkyne is ethyne, HC≡CH, also called acetylene. Adding CH units generates the alkyne series C_nH_{2n-2}.

Alkenes and alkynes are named by replacing the *ane* in the alkane series with *ene* for alkenes and *yne* for alkynes. In addition you must indicate the position of the multiple bond by assigning it the lowest possible number. Some examples are given below.

$CH_2{=}CH{-}CH_2{-}CH_3$	$CH_3{-}C{\equiv}C{-}CH_2{-}CH_3$	$CH_2{=}C(CH_3){-}CH_2{-}CH_2{-}CH_3$
1-butene	2-pentyne	2-methyl-1-pentene

Cyclic alkenes and dienes also exist. Cyclohexene is an example and can be written as

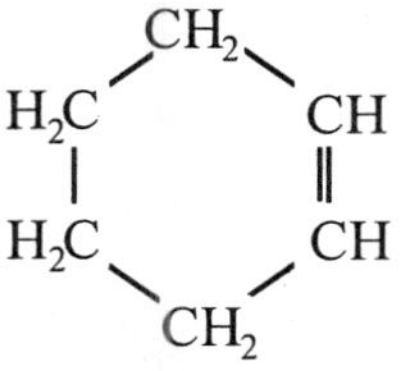

or

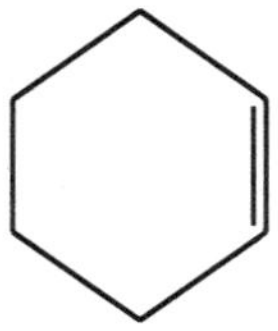

Alkenes and alkynes because they are unsaturated can undergo various reactions. See Table 23.2 in this study guide for a list of some of these reactions.

Aromatic Hydrocarbons (Section 23.3)

1. Give the structural formulas of benzene, napthalene, and various substituted benzenes.
2. Identify substituents on a benzene ring as *ortho*, *meta*, or *para*.

Review

Aromatic hydrocarbons contain benzene rings. Benzene is usually drawn as

Recall that the six pi electrons are delocalized.

A benzene derivative is formed by replacing one or more hydrogen atoms in benzene by other groups or atoms. For example

toluene

bromobenzene

phenol

Isomers are possible when two groups other than hydrogen are bonded to the benzene ring. These isomers are distinguished by the prefixes o- (ortho, adjacent), m- (meta, separated by one carbon) and p- (para, opposite). For example the three isomers of dichlorobenzene are

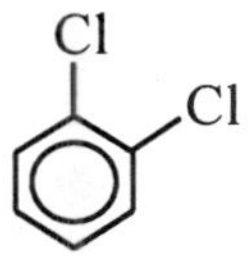

o-dichlorobenzene

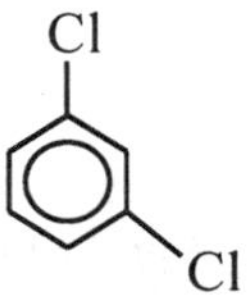

m-dichlorobenzene

p-dichlorobenzene

Practice: Draw the three isomers for bromotoluene.

Reactions of Aromatic Compounds

Aromatic compounds generally undergo reactions that preserve the delocalized pi electron structure. These reactions are generally substitution reactions in which the hydrogens on the benzene ring are replaced by other groups or atoms. See Table 23.2 in the study guide for examples.

Functional Groups (Section 23.4)

Learning Objectives

1. Identify the functional groups in a structural formula.
2. Use the structural formula of a compound to identify it as an ether; primary, secondary, or tertiary alcohol; phenol; aldehyde; ketone; or carboxylic acid.
3. Write the structural formulas for all the possible alcohols, aldehydes, ketones, and carboxylic acids with a given carbon skeleton.
4. Give the formulas, properties, and uses of methanol, ethanol, isopropyl alcohol, ethylene glycol, glycerol, and phenol.
5. List some of the sources of methanol and ethanol.
6. Describe the molecular structure of saturated and unsaturated fatty acids.
7. List the oxygen derivatives of a given hydrocarbon in order of increasing oxidation number of carbon.
8. Explain why glucose is a reducing agent.

Review

The chemical and physical properties of organic compounds are determined largely by functional groups, combinations of atoms that serve as reactive sites. See Table 23.4 in your text for a list of functional groups. Many of these functional groups contain oxygen.

The major classes of organic compounds that contain oxygen are ethers, alcohols and phenols, aldehydes and ketones, and carboxylic acids and their derivatives.

An ether contains an oxygen atom bonded to two hydrocarbon groups:

$$R\overset{O}{\diagup}\diagdown R'$$

$$CH_3-CH_2\overset{O}{\diagup}\diagdown CH_2-CH_3$$

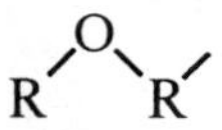

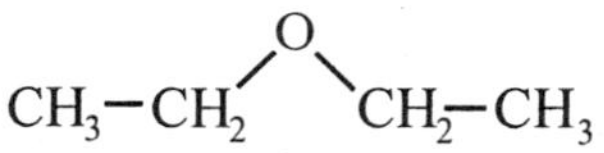

ethyl ether

An alcohol contains an –OH (hydroxy) group covalently bonded to a hydrocarbon group:

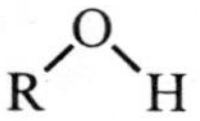

$$R\overset{O}{\diagup}\diagdown H$$

$CH_3–OH$
methanol
(methyl alcohol)

$CH_3–CH_2–OH$
ethanol
(ethyl alcohol)

Note: The IUPAC name adds -ol as a suffix to the hydrocarbon root.

Primary alcohols have –OH bonded to an end carbon atom. Ethanol and 1-butanol are examples:

$CH_3–CH_2–CH_2–CH_2–OH$

1-butanol
(n-butyl alcohol)

Secondary and tertiary alcohols contain –OH bonded to a carbon atom which is bonded to two and three other carbon atoms respectively. Examples are

OH
|
CH₃—CH—CH₂—CH₃

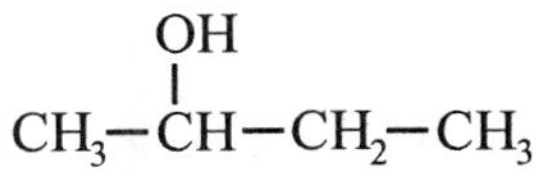

2-butanol
(sec-butyl alcohol)

OH
|
CH₃—C—CH₃
|
CH₃

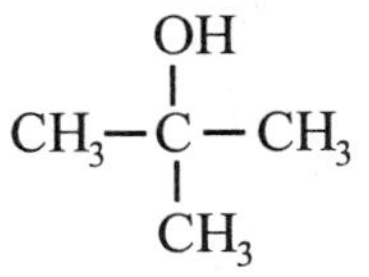

2-methyl-2-propanol
(t-butyl alcohol)

Practice: Write the structure of secondary and tertiary alcohols that have five carbon atoms.

Hydrogen bonding occurs in alcohols and accounts for the relatively high boiling points of light alcohols and their miscibility with water.

Polyhydroxy alcohols have two or more –OH groups. Ethylene glycol, used as an antifreeze in automobile engines, is an example:

H
|
H—C—OH
|
H—C—OH
|
H

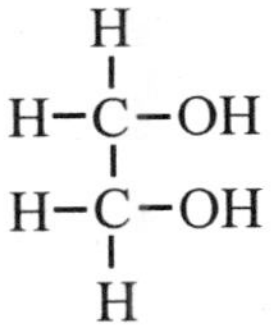

A phenol contains an –OH group bonded to an aromatic ring. Phenol has the structure

Phenols, unlike alcohols, tend to be weak acids due to the electron withdrawing ability of the aromatic ring.

Aldehydes and Ketones

The functional group in aldehydes and ketones is the carbonyl group \diagdownC=O

O
‖
R—C—H

aldehyde

O
‖
R—C—R′

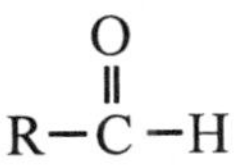

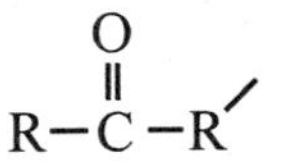

ketone

Examples are

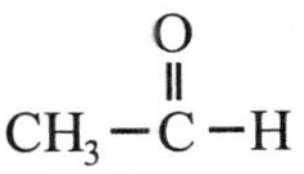

O
‖
CH₃—C—H

acetaldehyde

O
‖
CH₃—C—CH₃

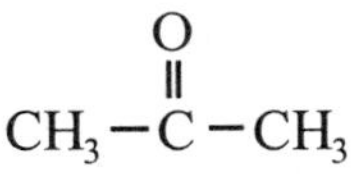

acetone

Carboxylic Acids

The functional group of carboxylic acids is the carboxyl group

O
‖
—C—OH

O
‖
R—C—OH

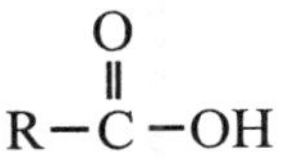

carboxylic acid

A familiar example is acetic acid:

$$CH_3-\overset{\overset{O}{\|}}{C}-OH \quad \text{or } CH_3COOH$$

Some carboxylic acids are listed in Table 23.5 of your text.

Oxidation and Reduction

In organic reactions oxidation usually means oxygen (O) atoms are acquired or hydrogen (H) atoms are lost, while reduction means the reverse. For example,

$$CH_3{-}CH_2{-}CH_3 \xrightarrow{[O]} CH_3{-}CH_2{-}CH_2{-}OH \xrightarrow{[O]} CH_3-CH_2-\overset{\overset{O}{\|}}{C}-H \xrightarrow{[O]} CH_3-CH_2-\overset{\overset{O}{\|}}{C}-OH$$

Each molecule is an oxidation product of the molecule to its left and a reduction product of the molecule to its right.

The Ester Linkage (Section 23.5)

Learning Objectives

1. Identify an ester from its structural formula.
2. Write the equation for the formation of an ester from a given alcohol and acid.
3. Given the formula of an ester, write the equation for its hydrolysis.
4. Write the formula of the polyester formed from a given dialcohol and diacid.
5. Given the formula of a polyester, write the formulas of its monomers.
6. Write the structural formula of a triglyceride formed from given fatty acids, and predict whether it is a fat or an oil at room temperature.
7. Write an equation to show how soap is made from fat.

Review

Esters are made from condensation reactions between carboxylic acids and alcohols:

$$R-\overset{\overset{O}{\|}}{C}-OH + R'OH \rightarrow R-\overset{\overset{O}{\|}}{C}-O-R'$$

For example,

$$CH_3-\overset{\overset{O}{\|}}{C}-OH \quad + \; CH_3OH \;\rightarrow\; CH_3-\overset{\overset{O}{\|}}{C}-O-CH_3$$

acetic acid methyl alcohol methyl acetate

Many esters have pleasant odors and tastes and are used in perfumes and artificial flavors.

The ester linkage is the group $-\overset{\overset{O}{\|}}{C}-O-$

An ester linkage is present in aspirin

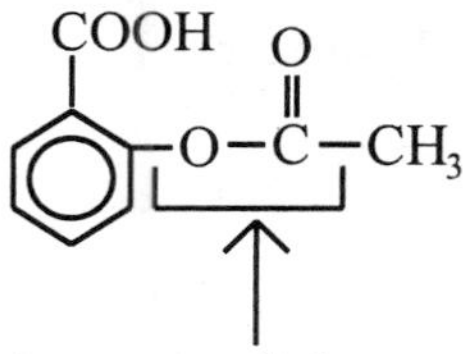

ester linkage

Polyesters contain many ester linkages; they are polymers that are formed by condensation of a dicarboxylic acid (two COOH groups) and a dihydroxy alcohol (two OH groups).

Fats and Oils

Triglycerides are esters of fatty acids and glycerol. The general formula is

$$
\begin{array}{c}
H \\
| \\
H-C-O-\overset{\overset{\displaystyle O}{\|}}{C}-R_1 \\
| \\
H-C-O-\overset{\overset{\displaystyle O}{\|}}{C}-R_2 \\
| \\
H-C-O-\overset{\overset{\displaystyle O}{\|}}{C}-R_3 \\
| \\
H
\end{array}
$$

ester linkages fatty acid chains

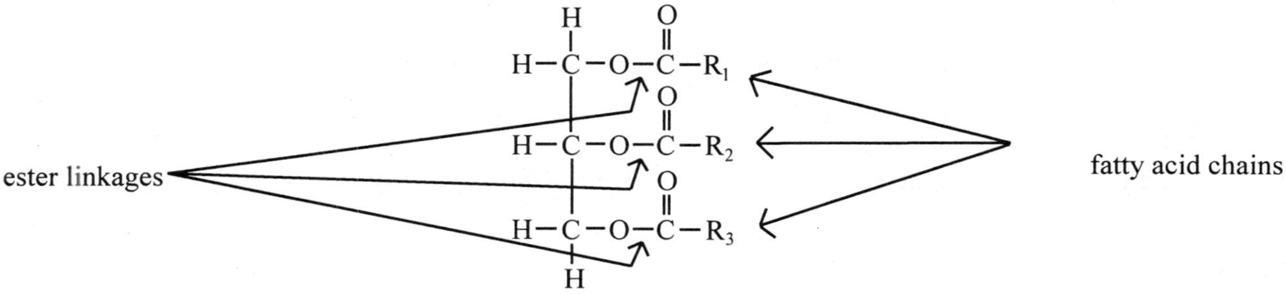

Fats are triglycerides formed from saturated fatty acids and glycerol and are solids at room temperature. Fats are a source of concentrated energy in the body. They provide 9 kcal/g compared to 4 kcal/g for carbohydrates and proteins.

Oils are also triglycerides but are formed from unsaturated fatty acids and glycerol and are generally liquids at room temperature.

Esters undergo hydrolysis in acidic or basic solution to form the parent acid and alcohol. Basic hydrolysis of a fat is called saponification. One product is soap, which is the salt of the fatty acid. An example of saponification is shown below:

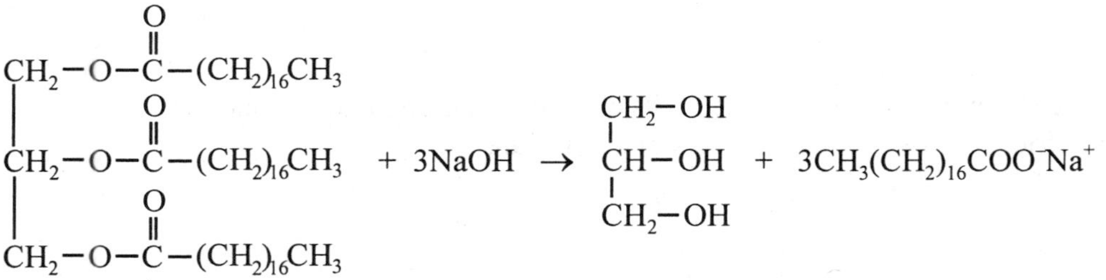

$$
\begin{array}{l}
CH_2-O-\overset{\overset{\displaystyle O}{\|}}{C}-(CH_2)_{16}CH_3 \\
| \\
CH_2-O-\overset{\overset{\displaystyle O}{\|}}{C}-(CH_2)_{16}CH_3 \quad + \; 3NaOH \;\rightarrow \\
| \\
CH_2-O-\overset{\overset{\displaystyle O}{\|}}{C}-(CH_2)_{16}CH_3 \\
\qquad\qquad fat
\end{array}
\qquad
\begin{array}{l}
CH_2-OH \\
| \\
CH-OH \quad + \; 3CH_3(CH_2)_{16}COO^-Na^+ \\
| \\
CH_2-OH \\
glycerol \qquad\quad sodium\ stearate,\ a\ soap
\end{array}
$$

Amines (Section 23.6)

Learning Objectives

1. Write general formulas for primary, secondary, and tertiary amines.
2. Write equations for the reactions of amines with strong acids.
3. Write the general formula of an α-amino acid.
4. Identify a chiral organic molecule from its fomula, and identify the asymmetric carbon atom.

Review

Amines are derivatives of ammonia (NH_3) in which one or more hydrogen atoms have been replaced by hydrocarbon groups. Examples are

$$CH_3-CH_2-\overset{\overset{\displaystyle H}{|}}{N}-H \qquad H-\overset{\overset{\displaystyle CH_3}{|}}{N}-CH_3 \qquad CH_3-\overset{\overset{\displaystyle CH_3}{|}}{N}-CH_3$$

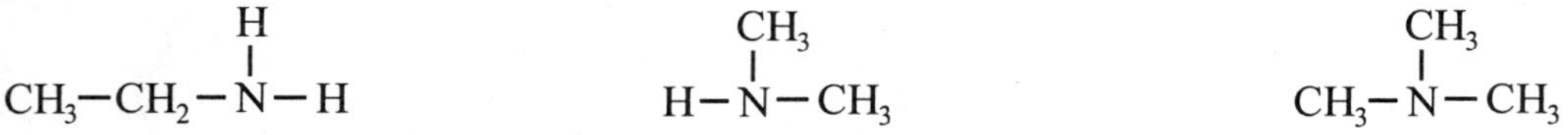

ethyl amine
(primary amine)

dimethyl amine
(secondary amine)

trimethyl amine
(tertiary amine)

Like ammonia amines are weak bases and will neutralize strong acids to form salts:

$$H-\overset{\overset{\displaystyle H}{|}}{N}-CH_3 \;+\; H^+ \;\rightarrow\; H-\overset{\overset{\displaystyle H}{|}}{\underset{\underset{\displaystyle H}{|}}{N}}\overset{+}{-}CH_3$$

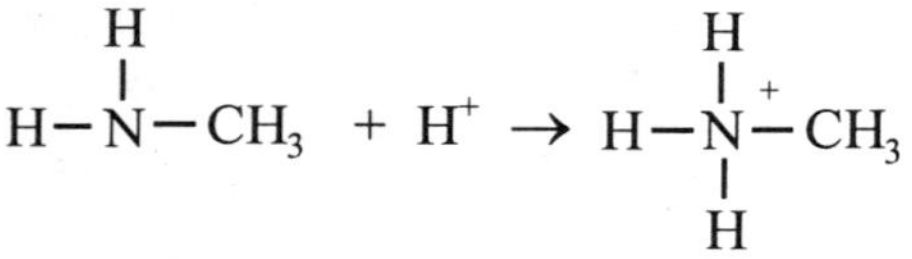

The functional group in a primary amine is $-NH_2$, the amino group.

Amino Acids: Chirality

Proteins are made of α-amino acids. The general structure of these amino acids is given by

$$H-\overset{\overset{\displaystyle R}{|}}{\underset{\underset{\displaystyle NH_2}{|}}{N}}-COOH$$

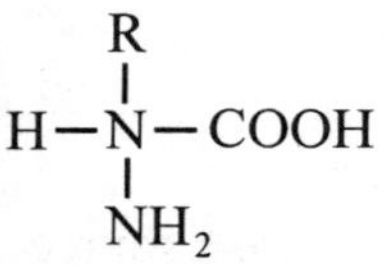

Note that at least two functional groups are present in amino acids, the carboxyl group and the amino group. Table 23.8 in the text lists the structure of 20 amino acids that occur in natural proteins.

Most amino acid molecules cannot be superimposed on their mirror images and are said to be chiral. Each of these amino acids is therefore an optical isomer or an enantiomer (see Section 21.5). Alanine is a chiral amino acid; its structure is

$$H-\overset{\overset{\displaystyle CH_3}{|}}{\underset{\underset{\displaystyle NH_2}{|}}{C}}\overset{*}{-}COOH$$

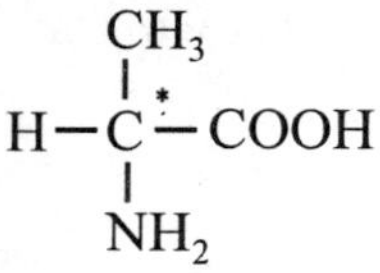

The carbon next to the asterisk in the above structure gives the molecule its chirality. A carbon atom bonded to four different groups is called chiral or asymmetric. The two optical isomers of alanine are

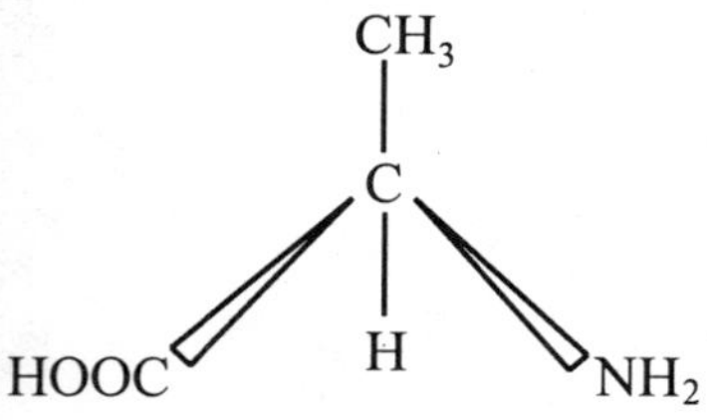

D-alanine L-alanine

Only the L-amino acids are used to build proteins. Generally, when biomolecules are chiral only one of the enantiomers has biochemical activity.

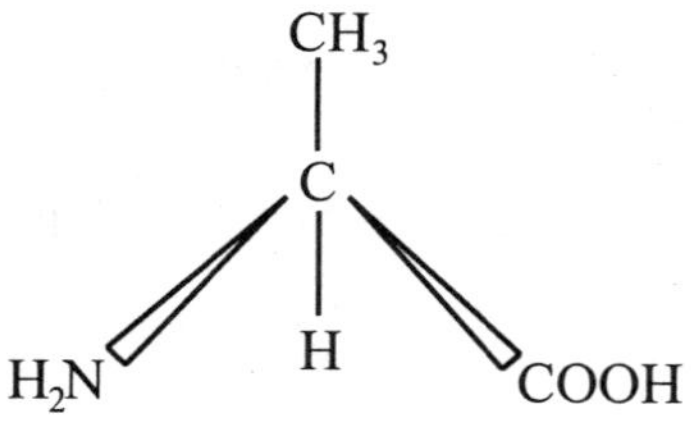

The Amide Linkage (Section 23.7)

Learning Objectives

1. Identify an amide linkage in a structural formula.
2. Write the equation for the formation of an amide from a given acid and amine.
3. Write the formula of the polyamide produced by a given diacid and diamine.
4. Given a polyamide formula, write the formulas of its monomers.
5. Write the structural formula for a polypeptide formed from given amino acids in a specified sequence.
6.* Explain what is meant by primary, secondary, tertiary, and quaternary components of protein structure.
7.* Describe two common types of secondary protein structure.
8.* List and describe the interactions that maintain secondary, tertiary, and quaternary protein structures.

Review

An amide is formed from the condensation reaction between a carboxyl group

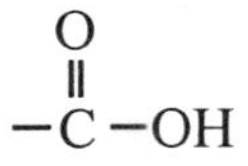

$$\overset{\displaystyle O}{\overset{\displaystyle \|}{-C}}-OH \quad \text{and an amino group } -NH_2. \text{ For example,}$$

$$CH_3-\overset{O}{\overset{\|}{C}}-OH \quad + \quad H-\overset{}{\underset{H}{N}}-CH_3 \quad \rightarrow \quad CH_3-\overset{O}{\overset{\|}{C}}-\overset{}{\underset{H}{N}}-CH_3$$

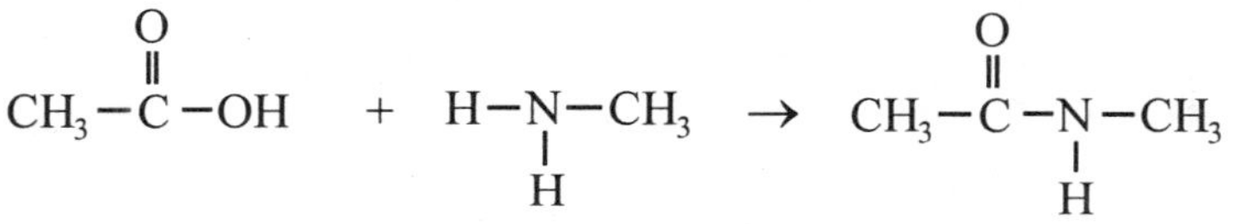

amide

The amide linkage is $-\overset{O}{\overset{\|}{C}}-\overset{}{\underset{|}{N}}-$

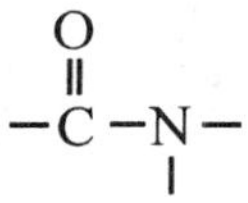

Tylenol contains an amide linkage. Nylon is a polyamide; the nylons are polymers that contain amide linkages joining diacid molecules (two COOH groups per molecule) to diamine molecules.

Polypeptides

A peptide contains two or more acids joined by amide linkages. An amide linkage between amino acids is called a peptide linkage. For example,

peptide linkage

$$H-\overset{H}{\underset{H}{\overset{|}{N}}}-\overset{H}{\underset{H}{\overset{|}{C}}}-\overset{O}{\overset{\|}{C}}-OH \quad + \quad H-\overset{H}{\underset{CH_3}{\overset{|}{N}}}-\overset{H}{\underset{}{\overset{|}{C}}}-\overset{O}{\overset{\|}{C}}-OH \quad \rightarrow \quad H-\overset{H}{\underset{H}{\overset{|}{N}}}-\overset{H}{\underset{H}{\overset{|}{C}}}-\overset{O}{\overset{\|}{C}}-\overset{H}{\underset{H}{\overset{|}{N}}}-\overset{CH_3}{\underset{H}{\overset{|}{C}}}-\overset{O}{\overset{\|}{C}}-OH$$

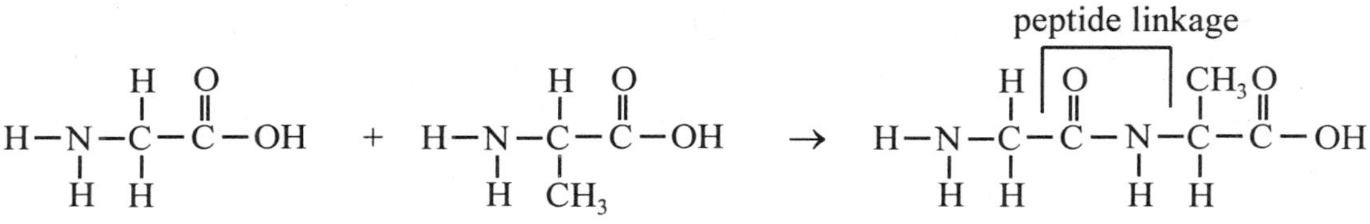

glycine alanine glycyl alanine (a dipeptide)

Each amino acid in the peptide is called a residue. A peptide with two residues is a dipeptide, three residues a tripeptide and so forth. A polypeptide contains many residues. Proteins are polypeptides that generally contain 50 or more amino acid residues.

Three letters are used to indicate the presence and position of an amino acid residue in a peptide. For example glycyl alanine, which is drawn for you above, is called gly-ala. Table 23.9 in your text lists the three letter symbols used for the important amino acids.

Caution: gly-ala is not the same dipeptide as ala-gly. In gly-ala the amino group of the dipeptide is attached to the glycine residue and the carboxyl group to the alanine residue. The reverse is true for ala-gly.

> **Practice:** How many different tripeptides contain two glycine residues and one alanine residue? Draw their structures. Answer: Three

Carbohydrates (Section 23.8)

Learning Objectives

1. Write the general formula for a monosaccharide.
2. Write molecular formulas and give examples of a hexose and a pentose.
3. Identify an aldose or ketose from its open ring structural formula.
4. List four naturally occurring hexoses.
5. Use structural formulas to show the equilibrium between closed and open ring forms of glucose.
6. List three natural disaccharides and their monosaccharide components.
7. Describe the glycosidic linkage, and show how two hexoses condense to form a disaccharide.
8. Describe the occurrence and biochemical function of starch, glycogen, and cellulose.
9. Describe the difference between amylose and amylopectin, the components of starch.
10. Explain the differences in the properties of starch and cellulose on the basis of structure.

Review

Carbohydrates received their general name because many have the formula $C_x(H_2O)_y$ so that they appear to be hydrates of carbon. However, we now know that they are either polyhydroxy aldehydes or ketones or they are substances that hydrolyze into polyhydroxy aldehydes or ketones.

The simplest carbohydrates are called monosaccharides; they are the simple sugars and they cannot be hydrolyzed. The monosaccharides found in food are mostly hexoses and have the general formula $C_6H_{12}O_6$. Examples include glucose, which is an aldose (aldehyde sugar), and fructose, a ketose (ketone sugar).

There are four asymmetric (chiral) carbon atoms in a $C_6H_{12}O_6$ aldose; therefore, eight pairs of enantiomers are possible. Each member of a pair is classified as D or L depending on the configuration around the assymetric carbon atom farthest from the

>C=O group. D-glucose is one of these isomers and its configuration is given below:

```
     CHO
      |
  H—C—OH
      |
 HO—C—H
      |
  H—C—OH        D-glucose
      |
  H—C—OH
      |
     CH2OH
```

D-glucose is the hexose produced by green plants and used in the body. Note that in solution these aldose sugars exist as six membered rings. (See Figure 23.31 in your text for the cyclic structure of glucose.)

A disaccharide is the condensation product of two monosaccharides. The two monosaccharides are joined by a glycosidic linkage (see Figure 23.33 in your text). Disaccharides undergo hydrolysis to yield the two monosaccharides back again.

Sucrose in your digestive tract is hydrolyzed to form glucose and fructose. Lactose and maltose are other examples of disaccharides.

Starch, glycogen, and cellulose are examples of polysaccharides containing only glucose. Cellulose, the structural polysaccharide in plants is the most abundant glucose polymer.

SELF-TEST

1. Give the structural formula for each of the following
   (a) 3,4-dimethyl-2-hexene
   (b) o-nitrotoluene
   (c) acetone
   (d) butyric acid
   (e) ethyl acetate
   (f) isobutyl amine

2. Match the compounds in column A with the statements in column B

A		B	
1.	alanyl glycine	(a)	One mole reacts with 1 mol of H_2 to form an alkane.
2.	2-butene	(b)	undergoes substitution reactions
3.	propyne	(c)	hydrolyzes to form two different amino acids
4.	methyl acetate	(d)	reacts with excess HBr to form 2,2-dibromopropane
5.	phenol	(e)	hydrolyzes to form an acid and an alcohol

3. Write the structure of the organic compound(s) formed in each reaction

(a) 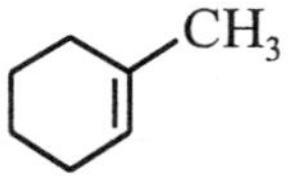 + HBr →

(b) $CH_3{-}CH_2{-}C{\equiv}CH + 2Br_2 \rightarrow$

(c) $H{-}\overset{\overset{\large Cl}{|}}{C}{=}CH_2 \rightarrow$ (addition polymerization)

(d)
$$\begin{array}{l} CH_2{-}O{-}\overset{O}{\overset{\|}{C}}{-}(CH_2)_{12}{-}CH_3 \\ | \\ CH_2{-}O{-}\overset{O}{\overset{\|}{C}}{-}(CH_2)_{12}{-}CH_3 \quad + \; 3NaOH \; \rightarrow \\ | \\ CH_2{-}O{-}\overset{O}{\overset{\|}{C}}{-}(CH_2)_{12}{-}CH_3 \end{array}$$

(e)
$$CH_3{-}\underset{\underset{CH_3}{|}}{CH}{-}\overset{O}{\overset{\|}{C}}{-}H \xrightarrow{[O]}$$

(f) 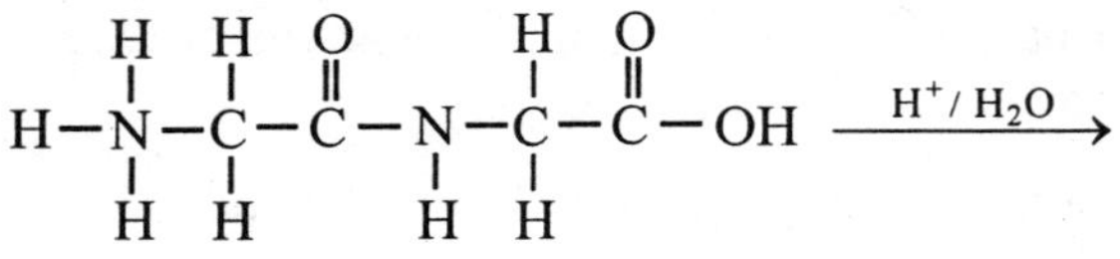

4. Draw
   (a) the two optical isomers of valine
   (b) the *cis-trans* isomers of 2-butene
   (c) the three structural isomers of chlorotoluene

5. Give the structures of the monomers used to synthesize the condensation polymers.

(a) 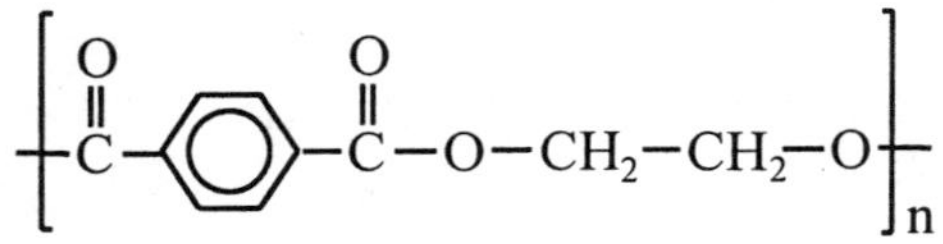

Dacron

(b) 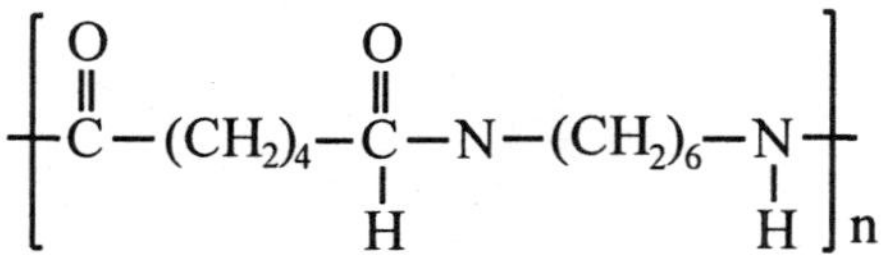

Nylon

6. Rank the following in order of increasing acidity: HCl, CH_3NH_2, CH_3COOH, CH_3OH

Answer 7-10 as True or False. If an answer is false, change the underlined word(s) to make it true.

7. <u>Ethanol</u> neutralizes strong bases.
8. <u>Starch</u> hydrolyzes to form glucose molecules.
9. <u>Glycine</u>, the amino acid in which <u>R=H</u>, has D and L isomers.
10. Proteins are <u>condensation</u> polymers.

ANSWERS TO SELF-TEST QUESTIONS

1. (a) 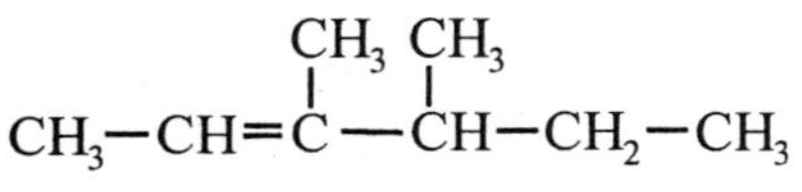

   (d) $CH_3–CH_2–CH_2–COOH$

   (b) 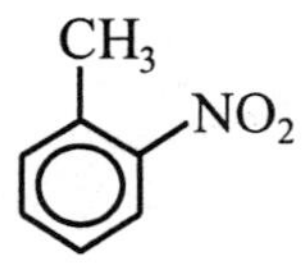

   (e) 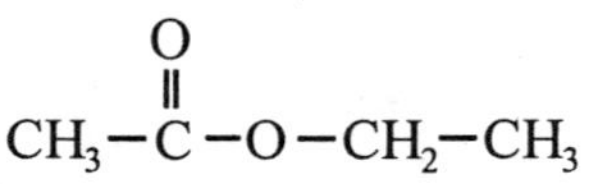

   (c) 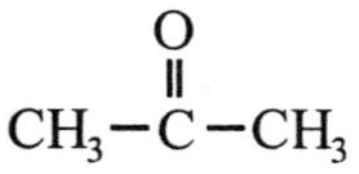

   (f) 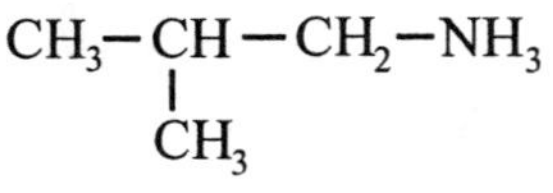

2. 1. (c) 2. (a) 3. (d) 4. (e) 5. (b)

3. (a) 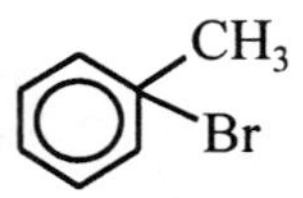

   (b) 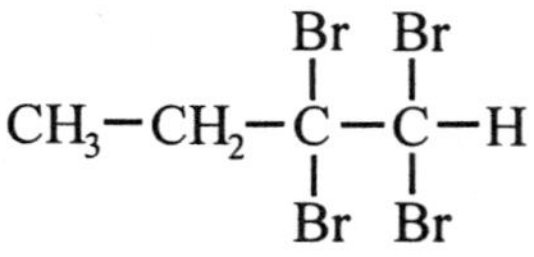

(c)

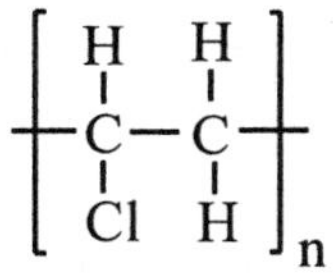

$$\left[\begin{array}{cc} H & H \\ | & | \\ C - C \\ | & | \\ Cl & H \end{array}\right]_n$$

PVC

(d)

$$\begin{array}{l} CH_2-OH \\ | \\ CH-OH \\ | \\ CH_2-OH \end{array} + 3CH_3(CH_2)_{12}COO^-Na^+$$

(e)

$$CH_3-\underset{\underset{CH_3}{|}}{CH}-\overset{\overset{O}{\|}}{C}-OH$$

(f) $2H_3N^+-CH_2-COOH$

4. (a)

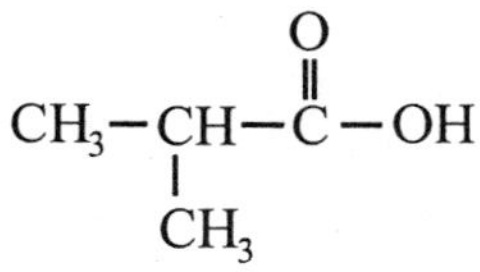

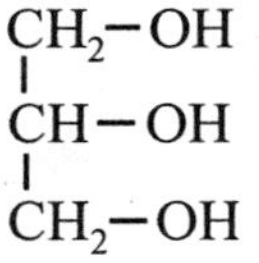

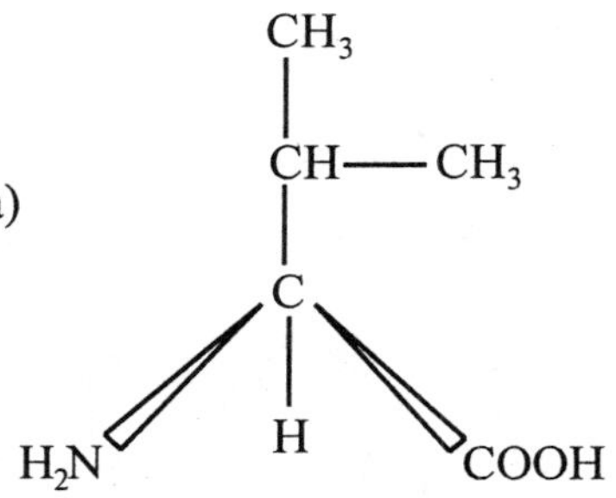

L-valine

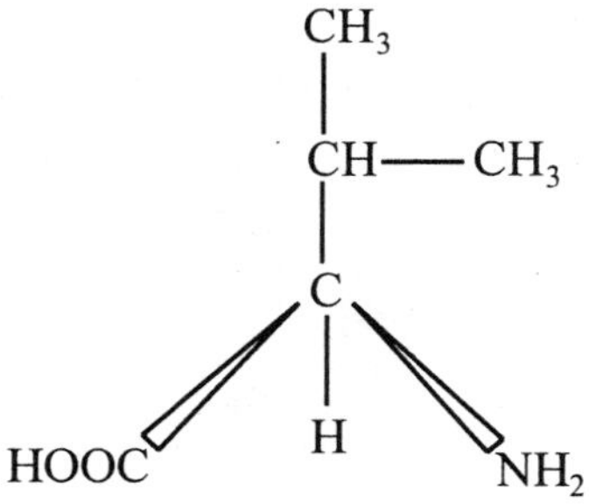

D-valine

(b)

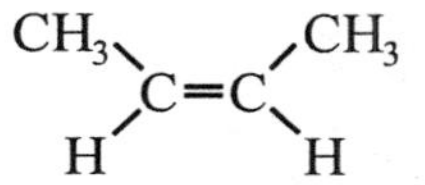

cis

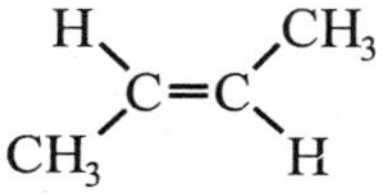

trans

(c)

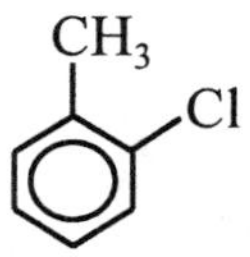

ortho

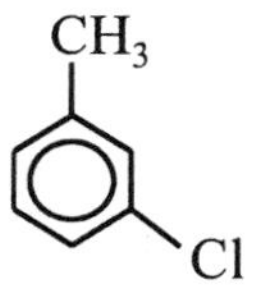

meta

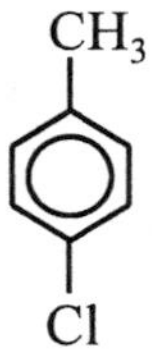

para

5. (a)

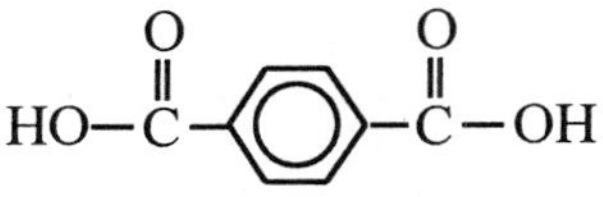

$$HO-\overset{\overset{O}{\|}}{C}-\langle \rangle-\overset{\overset{O}{\|}}{C}-OH$$
$$HO-CH_2-CH_2-OH$$

(b) $HO-\overset{\overset{O}{\|}}{C}-(CH_2)_4-\overset{\overset{O}{\|}}{C}-OH$

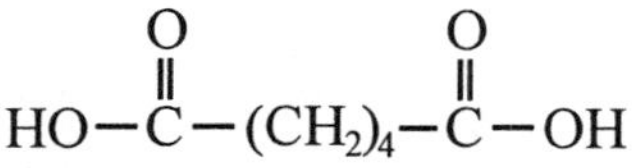

$H_2N-(CH)_6-NH_2$

6. $CH_3NH_2, CH_3OH, CH_3COOH, HCl$

7. phenol

8. T

9. Alanine, $R=CH_3$ or any other amino acid in Table 23.8 of your text.

10. T